T0252331

UNCERTAINTY AND INFORMATION

UNCERTAINTY AND INFORMATION
Foundations of Generalized Information Theory

George J. Klir

Binghamton University—SUNY

WILEY-INTERSCIENCE

A JOHN WILEY & SONS, INC., PUBLICATION

Copyright © 2006 by John Wiley & Sons, Inc. All rights reserved

Published by John Wiley & Sons, Inc., Hoboken, New Jersey
Published simultaneously in Canada

No part of this publication may be reproduced, stored in a retrieval system, or transmitted in any form or by any means, electronic, mechanical, photocopying, recording, scanning, or otherwise, except as permitted under Section 107 or 108 of the 1976 United States Copyright Act, without either the prior written permission of the Publisher, or authorization through payment of the appropriate per-copy fee to the Copyright Clearance Center, Inc., 222 Rosewood Drive, Danvers, MA 01923, (978) 750-8400, fax (978) 750-4470, or on the web at www.copyright.com. Requests to the Publisher for permission should be addressed to the Permissions Department, John Wiley & Sons, Inc., 111 River Street, Hoboken, NJ 07030, (201) 748-6011, fax (201) 748-6008, or online at http://www.wiley.com/go/permission.

Limit of Liability/Disclaimer of Warranty: While the publisher and author have used their best efforts in preparing this book, they make no representations or warranties with respect to the accuracy or completeness of the contents of this book and specifically disclaim any implied warranties of merchantability or fitness for a particular purpose. No warranty may be created or extended by sales representatives or written sales materials. The advice and strategies contained herein may not be suitable for your situation. You should consult with a professional where appropriate. Neither the publisher nor author shall be liable for any loss of profit or any other commercial damages, including but not limited to special, incidental, consequential, or other damages.

For general information on our other products and services or for technical support, please contact our Customer Care Department within the United States at (800) 762-2974, outside the United States at (317) 572-3993 or fax (317) 572-4002.

Wiley also publishes its books in a variety of electronic formats. Some content that appears in print may not be available in electronic formats. For more information about Wiley products, visit our web site at www.wiley.com.

Library of Congress Cataloging-in-Publication Data:

Klir, George J., 1932–
    Uncertainty and information : foundations of generalized information theory / George J. Klir.
        p.   cm.
    Includes bibliographical references and indexes.
    ISBN-13: 978-0-471-74867-0
    ISBN-10: 0-471-74867-6
    1. Uncertainty (Information theory)   2. Fuzzy systems.   I. Title.
Q375.K55 2005
033′.54—dc22

2005047792

A book is never finished.
It is only abandoned.

—Honoré De Balzac

CONTENTS

PREFACE

The concepts of *uncertainty* and *information* studied in this book are tightly interconnected. Uncertainty is viewed as a manifestation of some information deficiency, while information is viewed as the capacity to reduce uncertainty. Whenever these restricted notions of uncertainty and information may be confused with their other connotations, it is useful to refer to them as *information-based uncertainty* and *uncertainty-based information*, respectively.

The restricted notion of uncertainty-based information does not cover the full scope of the concept of information. For example, it does not fully capture our common-sense conception of information in human communication and cognition or the algorithmic conception of information. However, it does play an important role in dealing with the various problems associated with systems, as I already recognized in the late 1970s. It is this role of uncertainty-based information that motivated me to study it.

One of the insights emerging from systems science is the recognition that scientific knowledge is organized, by and large, in terms of systems of various types. In general, systems are viewed as relations among states of some variables. In each system, the relation is utilized, in a given purposeful way, for determining unknown states of some variables on the basis of known states of other variables. Systems may be constructed for various purposes, such as prediction, retrodiction, diagnosis, prescription, planning, and control. Unless the predictions, retrodictions, diagnoses, and so forth made by the system are unique, which is a rather rare case, we need to deal with predictive uncertainty, retrodictive uncertainty, diagnostic uncertainty, and the like. This respective uncertainty must be properly incorporated into the mathematical formalization of the system.

In the early 1990s, I introduced a research program under the name "generalized information theory" (GIT), whose objective is to study information-based uncertainty and uncertainty-based information in all their manifestations. This research program, motivated primarily by some fundamental issues emerging from the study of complex systems, was intended to expand classical information theory based on probability. As is well known, the latter emerged in 1948, when Claude Shannon established his measure of probabilistic uncertainty and information.

GIT expands classical information theory in two dimensions. In one dimension, additive probability measures, which are inherent in classical information theory, are expanded to various types of nonadditive measures. In the other dimension, the formalized language of classical set theory, within which probability measures are formalized, is expanded to more expressive formalized languages that are based on fuzzy sets of various types. As in classical information theory, uncertainty is the primary concept in GIT, and information is defined in terms of uncertainty reduction.

Each uncertainty theory that is recognizable within the expanded framework is characterized by: (a) a particular *formalized language* (classical or fuzzy); and (b) a *generalized measure* of some particular type (additive or nonadditive). The number of possible uncertainty theories that are subsumed under the research program of GIT is thus equal to the product of the number of recognized formalized languages and the number of recognized types of generalized measures. This number has been growing quite rapidly. The full development of any of these uncertainty theories requires that issues at each of the following four levels be adequately addressed: (1) the theory must be formalized in terms of appropriate *axioms*; (2) a *calculus* of the theory must be developed by which this type of uncertainty can be properly manipulated; (3) a justifiable way of *measuring* the amount of uncertainty (predictive, diagnostic, etc.) in any situation formalizable in the theory must be found; and (4) various *methodological aspects* of the theory must be developed.

GIT, as an ongoing research program, offers us a steadily growing inventory of distinct uncertainty theories, some of which are covered in this book. Two complementary features of these theories are significant. One is their great and steadily growing *diversity*. The other is their *unity*, which is manifested by properties that are invariant across the whole spectrum of uncertainty theories or, at least, within some broad classes of these theories. The growing diversity of uncertainty theories makes it increasingly more realistic to find a theory whose assumptions are in harmony with each given application. Their unity allows us to work with all available theories as a whole, and to move from one theory to another as needed.

The principal aim of this book is to provide the reader with a comprehensive and in-depth overview of the two-dimensional framework by which the research in GIT has been guided, and to present the main results that have been obtained by this research. Also covered are the main features of two *classical information theories*. One of them, covered in Chapter 3, is based on the concept of *probability*. This classical theory is well known and is extensively covered in the literature. The other one, covered in Chapter 2, is based on the dual concepts of *possibility* and *necessity*. This classical theory is older and more fundamental, but it is considerably less visible and has often been incorrectly dismissed in the literature as a special case of the probability-based information theory. These two classical information theories, which are formally incomparable, are the roots from which distinct generalizations are obtained.

Principal results regarding generalized uncertainty theories that are based on classical set theory are covered in Chapters 4–6. While the focus in Chapter 4 is on the common properties of uncertainty representation in all these theories, Chapter 5 is concerned with special properties of individual uncertainty theories. The issue of how to measure the amount of uncertainty (and the associated information) in situations formalized in the various uncertainty theories is thoroughly investigated in Chapter 6. Chapter 7 presents a concise introduction to the fundamentals of fuzzy set theory, and the fuzzification of uncertainty theories is discussed in Chapter 8, in both general and specific terms. Methodological issues associated with GIT are discussed in Chapter 9. Finally, results and open problems emerging from GIT are summarized and assessed in Chapter 10.

The book can be used in several ways and, due to the universal applicability of GIT, it is relevant to professionals in virtually any area of human affairs. While it is written primarily as a textbook for a one-semester graduate course, its utility extends beyond the classroom environment. Due to the comprehensive and coherent presentation of the subject and coverage of some previously unpublished results, the book is also a useful resource for researchers. Although the treatment of uncertainty and information in the book is mathematical, the required mathematical background is rather modest: the reader is only required to be familiar with the fundamentals of classical set theory, probability theory and the calculus. Otherwise, the book is completely self-contained, and it is thus suitable for self-study.

While working on the book, clarity of presentation was always on my mind. To achieve it, I use examples and visual illustrations copiously. Each chapter is also accompanied by an adequate number of exercises, which allow readers to test their understanding of the studied material. The main text is only rarely interrupted by bibliographical, historical, or any other references. Almost all references are covered in specific Notes, organized by individual topics and located at the end of each chapter. These notes contain ample information for further study.

For many years, I have been pursuing research on GIT while, at the same time, teaching an advanced graduate course in this area to systems science students at Binghamton University in New York State (SUNY-Binghamton). Due to rapid developments in GIT, I have had to change the content of the course each year to cover the emerging new results. This book is based, at least to some degree, on the class notes that have evolved for this course over the years. Some parts of the book, especially in Chapters 6 and 9, are based on my own research.

It is my hope that this book will establish a better understanding of the very complex concepts of information-based uncertainty and uncertainty-based information, and that it will stimulate further research and education in the important and rapidly growing area of generalized information theory.

Binghamton, New York   Georgе J. Klir
December 2004

ACKNOWLEDGMENTS

Over more than three decades of my association with Binghamton University, I have had the good fortune to advise and work with many outstanding doctoral students. Some of them contributed in a significant way to generalized information theory, especially to the various issues regarding uncertainty measures. These students, whose individual contributions to generalized information theory are mentioned in the various notes in this book, are (in alphabetical order): David Harmanec, Masahiko Higashi, Cliff Joslyn, Matthew Mariano, Yin Pan, Michael Pittarelli, Arthur Ramer, Luis Rocha, Richard Smith, Mark Wierman, and Bo Yuan. A more recent doctoral student, Ronald Pryor, read carefully the initial version of the manuscript of this book and suggested many improvements. In addition, he developed several computer programs that helped me work through some intricate examples in the book. I gratefully acknowledge all this help.

As far as the manuscript preparation is concerned, I am grateful to two persons for their invaluable help. First, and foremost, I am grateful to Monika Fridrich, my Editorial Assistant and a close friend, for her excellent typing of a very complex, mathematically oriented manuscript, as well as for drawing many figures that appear in the book. Second, I am grateful to Stanley Kauffman, a graphic artist at Binghamton University, for drawing figures that required special skills.

Last, but not least, I am grateful to my wife, Milena, for her contribution to the appearance of this book: it is one of her photographs that the publisher chose to facilitate the design for the front cover. In addition, I am also grateful for her understanding, patience, and encouragement during my concentrated, disciplined and, at times, frustrating work on this challenging book.

1

INTRODUCTION

> The mind, once expanded to the dimensions of larger ideas, never returns to its original size.
>
> —Oliver Wendel Holmes

1.1. UNCERTAINTY AND ITS SIGNIFICANCE

It is easy to recognize that uncertainty plays an important role in human affairs. For example, making everyday decisions in ordinary life is inseparable from uncertainty, as expressed with great clarity by George Shackle [1961]:

> In a predestinate world, decision would be *illusory*; in a world of a perfect foreknowledge, *empty*, in a world without natural order, *powerless*. Our intuitive attitude to life implies non-illusory, non-empty, non-powerless decision. . . . Since decision in this sense excludes both perfect foresight and anarchy in nature, it must be defined as choice in face of bounded uncertainty.

Conscious decision making, in all its varieties, is perhaps the most fundamental capability of human beings. It is essential for our survival and well-being. In order to understand this capability, we need to understand the notion of uncertainty first.

In decision making, we are uncertain about the future. We choose a particular action, from among a set of conceived actions, on the basis of our *antici-*

Uncertainty and Information: Foundations of Generalized Information Theory, by George J. Klir
© 2006 by John Wiley & Sons, Inc.

pation of the consequences of the individual actions. Our anticipation of future events is, of course, inevitably subject to uncertainty. However, uncertainty in ordinary life is not confined to the future alone, but may pertain to the past and present as well. We are uncertain about past events, because we usually do not have complete and consistent records of the past. We are uncertain about many historical events, crime-related events, geological events, events that caused various disasters, and a myriad of other kinds of events, including many in our personal lives. We are uncertain about present affairs because we lack relevant information. A typical example is *diagnostic uncertainty* in medicine or engineering. As is well known, a physician (or an engineer) is often not able to make a definite diagnosis of a patient (or a machine) in spite of knowing outcomes of all presumably relevant medical (or engineering) tests and other pertinent information.

While ordinary life without uncertainty is unimaginable, science without uncertainty was traditionally viewed as an ideal for which science should strive. According to this view, which had been predominant in science prior to the 20th century, uncertainty is incompatible with science, and the ideal is to completely eliminate it. In other words, uncertainty is unscientific and its elimination is one manifestation of progress in science. This traditional attitude toward uncertainty in science is well expressed by the Scottish physicist and mathematician William Thomson (1824–1907), better known as Lord Kelvin, in the following statement made in the late 19th century (*Popular Lectures and Addresses*, London, 1891):

> In physical science a first essential step in the direction of learning any subject is to find principles of numerical reckoning and practicable methods for measuring some quality connected with it. I often say that when you can measure what you are speaking about and express it in numbers, you know something about it; but when you cannot measure it, when you cannot express it in numbers, your knowledge is of meager and unsatisfactory kind; it may be the beginning of knowledge but you have scarcely, in your thought, advanced to the state of science, whatever the matter may be.

This statement captures concisely the spirit of science in the 19th century: scientific knowledge should be expressed in precise numerical terms; imprecision and other types of uncertainty do not belong to science. This preoccupation with precision and certainty was responsible for neglecting any serious study of the concept of uncertainty within science.

The traditional attitude toward uncertainty in science began to change in the late 19th century, when some physicists became interested in studying processes at the molecular level. Although the precise laws of *Newtonian mechanics* were relevant to these studies in principle, they were of no use in practice due to the enormous complexities of the systems involved. A fundamentally different approach to deal with these systems was needed. It was eventually found in statistical methods. In these methods, specific manifestations of microscopic entities (positions and moments of individual molecules) were replaced with their statistical averages. These averages, calculated under certain reasonable assumptions, were shown to represent relevant macroscopic entities such as temperature and pressure. A new field of physics, *statistical mechanics*, was an outcome of this research.

Statistical methods, developed originally for studying motions of gas molecules in a closed space, have found utility in other areas as well. In engineering, they have played a major role in the design of large-scale telephone networks, in dealing with problems of engineering reliability, and in numerous other problems. In business, they have been essential for dealing with problems of marketing, insurance, investment, and the like. In general, they have been found applicable to problems that involve large-scale systems whose components behave in a highly random way. The larger the system and the higher the randomness, the better these methods perform.

When statistical mechanics was accepted, by and large, by the scientific community as a legitimate area of science at the beginning of the 20th century, the negative attitude toward uncertainty was for the first time revised. Uncertainty became recognized as useful, or even essential, in certain scientific inquiries. However, it was taken for granted that uncertainty, whenever unavoidable in science, can adequately be dealt with by probability theory. It took more than half a century to recognize that the concept of uncertainty is too broad to be captured by probability theory alone, and to begin to study its various other (nonprobabilistic) manifestations.

Analytic methods based upon the calculus, which had dominated science prior to the emergence of statistical mechanics, are applicable only to problems that involve systems with a very small number of components that are related to each other in a predictable way. The applicability of statistical methods based upon probability theory is exactly opposite: they require systems with a very large number of components and a very high degree of randomness. These two classes of methods are thus complementary. When methods in one class excel, methods in the other class totally fail. Despite their complementarity, these classes of methods can deal only with problems that are clustered around the two extremes of complexity and randomness scales. In his classic paper "Science and Complexity" [1948], Warren Weaver refers to them as problems of *organized simplicity* and *disorganized complexity*, respectively. He argues that these classes of problems cover only a tiny fraction of all conceivable problems. Most problems are located somewhere between the two extremes of complexity and randomness, as illustrated by the shaded area in Figure 1.1. Weaver calls them problems of *organized complexity* for reasons that are well described in the following quote from his paper:

The new method of dealing with disorganized complexity, so powerful an advance over the earlier two-variable methods, leaves a great field untouched. One is tempted to oversimplify, and say that scientific methodology went from one extreme to the other—from two variables to an astronomical number—and

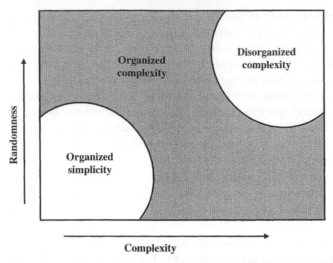

Figure 1.1. Three classes of systems and associated problems that require distinct mathematical treatments [Weaver, 1948].

left untouched a great middle region. The importance of this middle region, moreover, does not depend primarily on the fact that the number of variables is moderate—large compared to two, but small compared to the number of atoms in a pinch of salt. The problems in this middle region, in fact, will often involve a considerable number of variables. The really important characteristic of the problems in this middle region, which science has as yet little explored and conquered, lies in the fact that these problems, as contrasted with the disorganized situations with which statistics can cope, show the essential feature of *organization*. In fact, one can refer to this group of problems as those of *organized complexity*. . . . These new problems, and the future of the world depends on many of them, require science to make a third great advance, an advance that must be even greater than the nineteenth-century conquest of problems of organized simplicity or the twentieth-century victory over problems of disorganized complexity. Science must, over the next 50 years, learn to deal with these problems of organized complexity.

The emergence of computer technology in World War II and its rapidly growing power in the second half of the 20th century made it possible to deal with increasingly complex problems, some of which began to resemble the notion of organized complexity. However, this gradual penetration into the domain of organized complexity revealed that high computing power, while important, is not sufficient for making substantial progress in this problem domain. It was again felt that radically new methods were needed, methods based on fundamentally new concepts and the associated mathematical theories. An important new concept (and mathematical theories formalizing its various facets) that emerged from this cognitive tension was a *broad concept of uncertainty*, liberated from its narrow confines of probability theory. To introduce this broad concept of uncertainty and the associated mathematical theories is the very purpose of this book.

A view taken in this book is that scientific knowledge is organized, by and large, in terms of systems of various types (or categories in the sense of mathematical theory of categories). In general, *systems* are viewed as relations among states of given variables. They are constructed from our experiential domain for various purposes, such as prediction, retrodiction, extrapolation in space or within a population, prescription, control, planning, decision making, scheduling, and diagnosis. In each system, its relation is utilized in a given purposeful way for determining unknown states of some variables on the basis of known states of some other variables. Systems in which the unknown states are always determined uniquely are called *deterministic systems*; all other systems are called *nondeterministic systems*. Each nondeterministic system involves uncertainty of some type. This uncertainty pertains to the purpose for which the system was constructed. It is thus natural to distinguish predictive uncertainty, retrodictive uncertainty, prescriptive uncertainty, extrapolative uncertainty, diagnostic uncertainty, and so on. In each nondeterministic system, the relevant uncertainty (predictive, diagnostic, etc.) must be properly incorporated into the description of the system in some formalized language.

Deterministic systems, which were once regarded as ideals of scientific knowledge, are now recognized as too restrictive. Nondeterministic systems are far more prevalent in contemporary science. This important change in science is well characterized by Richard Bellman [1961]:

> It must, in all justice, be admitted that never again will scientific life be as satisfying and serene as in days when determinism reigned supreme. In partial recompense for the tears we must shed and the toil we must endure is the satisfaction of knowing that we are treating significant problems in a more realistic and productive fashion.

Although nondeterministic systems have been accepted in science since their utility was demonstrated in statistical mechanics, it was tacitly assumed for a long time that probability theory is the only framework within which uncertainty in nondeterministic systems can be properly formalized and dealt with. This presumed equality between uncertainty and probability was challenged in the second half of the 20th century, when interest in problems of organized complexity became predominant. These problems invariably involve uncertainty of various types, but rarely uncertainty resulting from randomness, which can yield meaningful statistical averages.

Uncertainty liberated from its probabilistic confines is a phenomenon of the second half of the 20th century. It is closely connected with two important generalizations in mathematics: a generalization of the classical measure theory and a generalization of the classical set theory. These generalizations, which are introduced later in this book, enlarged substantially the framework for formalizing uncertainty. As a consequence, they made it possible to conceive of new uncertainty theories distinct from the classical probability theory.

To develop a fully operational theory for dealing with uncertainty of some conceived type requires that a host of issues be addressed at each of the following four levels:

- *Level 1*—We need to find an appropriate *mathematical formalization* of the conceived type of uncertainty.
- *Level 2*—We need to develop a *calculus* by which this type of uncertainty can be properly manipulated.
- *Level 3*—We need to find a meaningful way of *measuring* the amount of relevant uncertainty in any situation that is formalizable in the theory.
- *Level 4*—We need to develop *methodological aspects* of the theory, including procedures of making the various *uncertainty principles* operational within the theory.

Although each of the uncertainty theories covered in this book is examined at all these levels, the focus is on the various issues at levels 3 and 4. These issues are presented in greater detail.

1.2. UNCERTAINTY-BASED INFORMATION

As a subject of this book, the broad concept of uncertainty is closely connected with the concept of information. The most fundamental aspect of this connection is that uncertainty involved in any problem-solving situation is a result of some information deficiency pertaining to the system within which the situation is conceptualized. There are various manifestations of information deficiency. The information may be, for example, incomplete, imprecise, fragmentary, unreliable, vague, or contradictory. In general, these various information deficiencies determine the type of the associated uncertainty.

Assume that we can measure the amount of uncertainty involved in a problem-solving situation conceptualized in a particular mathematical theory. Assume further that this amount of uncertainty is reduced by obtaining relevant information as a result of some action (performing a relevant experiment and observing the experimental outcome, searching for and discovering a relevant historical record, requesting and receiving a relevant document from an archive, etc.). Then, the amount of information obtained by the action can be measured by the amount of reduced uncertainty. That is, the amount of information pertaining to a given problem-solving situation that is obtained by taking some action is measured by the difference between *a priori* uncertainty and *a posteriori* uncertainty, as illustrated in Figure 1.2.

Information measured solely by the reduction of relevant uncertainty within a given mathematical framework is an important, even though restricted, notion of information. It does not capture, for example, the

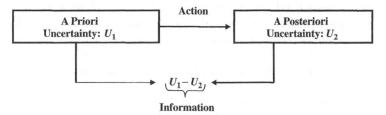

Figure 1.2. The meaning of uncertainty-based information.

common-sense conception of information in human communication and cognition, or the algorithmic conception of information, in which the amount of information needed to describe an object is measured by the shortest possible description of the object in some standard language. To distinguish information conceived in terms of uncertainty reduction from the various other conceptions of information, it is common to refer to it as *uncertainty-based information*.

Notwithstanding its restricted nature, uncertainty-based information is very important for dealing with nondeterministic systems. The capability of measuring uncertainty-based information in various situations has the same utility as any other measuring instrument. It allows us, in general, to analyze and compare systems from the standpoint of their informativeness. By asking a given system any question relevant to the purpose for which the system has been constructed (prediction, retrodiction, diagnosis, etc.), we can measure the amount of information in the obtained answer. How well we utilize this capability to measure information depends of course on the questions we ask.

Since this book is concerned only with uncertainty-based information, the adjective "uncertainty-based" is usually omitted. It is used only from time to time as a reminder or to emphasize the connection with uncertainty.

1.3. GENERALIZED INFORMATION THEORY

A formal treatment of uncertainty-based information has two classical roots, one based on the notion of *possibility*, and one based on the notion of *probability*. Overviews of these two classical theories of information are presented in Chapters 2 and 3, respectively. The rest of the book is devoted to various generalizations of the two classical theories. These generalizations have been developing and have commonly been discussed under the name "Generalized Information Theory" (GIT). In GIT, as in the two classical theories, the primary concept is uncertainty, and information is defined in terms of uncertainty reduction.

The ultimate goal of GIT is to develop the capability to deal formally with any type of uncertainty and the associated uncertainty-based information that we can recognize on intuitive grounds. To be able to deal with each recognized type of uncertainty (and uncertainty-based information), we need to address scores of issues. It is useful to associate these issues with four typical levels of development of each particular uncertainty theory, as suggested in Section 1.1. We say that a particular theory of uncertainty, T, is fully operational when the following issues have been resolved adequately at the four levels:

- *Level 1*—Relevant uncertainty functions, u, of theory T have been characterized by appropriate axioms (examples of these functions are probability measures).
- *Level 2*—A calculus has been developed for dealing with functions u (an example is the calculus of probability theory).
- *Level 3*—A justified functional U in theory T has been found, which for each function u in the theory measures the amount of uncertainty associated with u (an example of functional U is the well-known Shannon entropy in probability theory).
- *Level 4*—A methodology has been developed for dealing with the various problems in which theory T is involved (an example is the Bayesian methodology, combined with the maximum and minimum entropy principles, in probability theory).

Clearly, the functional U for measuring the amount of uncertainty expressed by the uncertainty function u can be investigated only after this function is properly formalized and a calculus is developed for dealing with it. The functional assigns to each function u in the given theory a nonnegative real number. This number is supposed to measure, in an intuitively meaningful way, the amount of uncertainty of the type considered that is embedded in the uncertainty function. To be acceptable as a measure of the amount of uncertainty of a given type in a particular uncertainty theory, the functional must satisfy several intuitively essential axiomatic requirements. Specific mathematical formulation of each of the requirements depends on the uncertainty theory involved. For the classical uncertainty theories, specific formulations of the requirements are introduced and discussed in Chapters 2 and 3. For the various generalized uncertainty theories, these formulations are introduced and examined in both generic and specific terms in Chapter 6.

The strongest justification of a functional as a meaningful measure of the amount of uncertainty of a considered type in a given uncertainty theory is obtained when we can prove that it is the only functional that satisfies the relevant axiomatic requirements and measures the amount of uncertainty in some specific measurement units. A suitable measurement unit is uniquely defined by specifying what the amount of uncertainty should be for a particular (and usually very simple) uncertainty function.

GIT is essentially a research program whose objective is to develop a broader treatment of uncertainty-based information, not restricted to its classical notions. Making a blueprint for this research program requires that a sufficiently broad framework be employed. This framework should encompass a broad spectrum of special mathematical areas that are fitting to formalize the various types of uncertainty conceived.

The framework employed in GIT is based on two important generalizations in mathematics that emerged in the second half of the 20th century. One of them is the generalization of classical measure theory to the theory of *monotone measures*. The second one is the generalization of classical set theory to the theory of *fuzzy sets*. These two generalizations expand substantially the classical, probabilistic framework for formalizing uncertainty, which is based on classical set theory and classical measure theory. This expansion is 2-dimensional. In one dimension, the *additivity* requirement of classical measures is replaced with the less restrictive requirement of *monotonicity* with respect to the subsethood relationship. The result is a considerably broader theory of monotone measures, within which numerous branches are distinguished that deal with monotone measures with various special properties. In the other dimension, the formalized language of classical set theory is expanded to the more expressive language of fuzzy set theory, where further distinctions are based on various special types of fuzzy sets.

The 2-dimensional expansion of the classical framework for formalizing uncertainty theories is illustrated in Figure 1.3. The rows in this figure represent various branches of the theory of monotone measures, while the columns represent various types of formalized languages. An uncertainty theory of a particular type is formed by choosing a particular formalized language and expressing the relevant uncertainty (predictive, prescriptive, etc.) involved in situations described in this language in terms of a monotone measure of a chosen type. This means that each entry in the matrix in Figure 1.3 represents an uncertainty theory of a particular type. The shaded entries indicate uncertainty theories that are currently fairly well developed and are covered in this book.

As a research program, GIT has been motivated by the following attitude toward dealing with uncertainty. One aspect of this attitude is the recognition of multiple types of uncertainty and the associated uncertainty theories. Another aspect is that we should not a priori commit to any particular theory. Our choice of uncertainty theory for dealing with each given problem should be determined solely by the nature of the problem. The chosen theory should allow us to express fully our ignorance and, at the same time, it should not allow us to ignore any available information. It is remarkable that these principles were expressed with great simplicity and beauty more than two millennia ago by the ancient Chinese philosopher Lao Tsu (ca. 600 B.C.) in his famous book *Tao Te Ching* (Vintage Books, New York, 1972):

> *Knowing ignorance is strength.*
>
> *Ignoring knowledge is sickness.*

The *primacy of problems* in GIT is in sharp contrast with the *primacy of methods* that is a natural consequence of choosing to use one particular theory

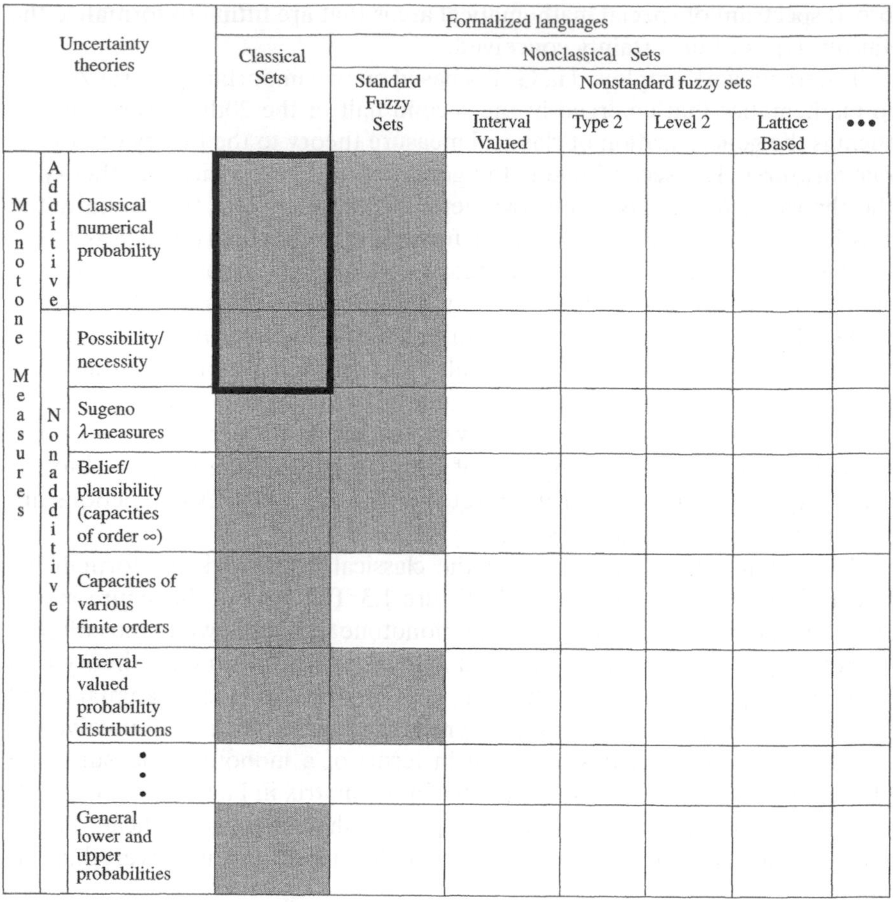

Uncertainty theories			Formalized languages						
			Classical Sets	Nonclassical Sets					
				Standard Fuzzy Sets	Nonstandard fuzzy sets				
					Interval Valued	Type 2	Level 2	Lattice Based	•••
Monotone Measures	Additive	Classical numerical probability							
		Possibility/ necessity							
	Nonadditive	Sugeno λ-measures							
		Belief/ plausibility (capacities of order ∞)							
		Capacities of various finite orders							
		Interval-valued probability distributions							
		• • •							
		General lower and upper probabilities							

Figure 1.3. A framework for conceptualizing uncertainty theories, which is used as a blueprint for research within generalized information theory (GIT).

for all problems involving uncertainty. The primary aim of GIT is to pursue the development of new uncertainty theories, through which we gradually extend our capability to deal with uncertainty honestly: to be able to fully recognize our ignorance without ignoring available information.

1.4. RELEVANT TERMINOLOGY AND NOTATION

The purpose of this section is to introduce names and symbols for some general mathematical concepts, primarily from the area of classical set theory, which are frequently used throughout this book. Names and symbols of many other concepts that are used in the subsequent chapters are introduced locally in each individual chapter.

A *set* is any collection of some objects that are considered for some purpose as a whole. Objects that are included in a set are called its *members* (or *elements*). Conventionally, sets are denoted by capital letters and elements of sets are denoted by lowercase letters. Symbolically, the statement "*a* is a member of set *A*" is written as $a \in A$.

A set is defined by one of three methods. In the first method, members (or elements) of the set are explicitly listed, usually within curly brackets, as in $A = \{1, 3, 5, 7, 9\}$. This method is, of course, applicable only to a set that contains a finite number of elements. The second method for defining a set is to specify a property that an object must possess to qualify as a member of the set. An example is the following definition of set *A*:

$$A = \{x \mid x \text{ is a real number that is greater than 0 and smaller than 1}\}.$$

The symbol | in this definition (and in other definitions in this book) stands for "such that." As can be seen from this example, this method allows us to define sets that include an infinite number of elements.

Both of the introduced methods for defining sets tacitly assume that members of the sets of concern in each particular application are drawn from some underlying *universal set*. This is a collection of all objects that are of interest in the given application. Some common universal sets in mathematics have standard symbols to represent them, such as $\mathbb{N}$ for the set of all natural numbers, $\mathbb{N}_n$ for the set $\{1, 2, 3, \ldots, n\}$, $\mathbb{Z}$ for the set of all integers, $\mathbb{R}$ for the set of all real numbers, and $\mathbb{R}^+$ for the set of all nonnegative real numbers. Except for these standard symbols, letter *X* is reserved in this book to denote a universal set.

The third method to define a set is through a *characteristic function*. If χ_A is the characteristic function of a set *A*, then χ_A is a function from the universal set *X* to the set $\{0, 1\}$, where

$$\chi_A(x) = \begin{cases} 1 & \text{if } x \text{ is a member of } A \\ 0 & \text{if } x \text{ is not an member of } A \end{cases}$$

for each $x \in X$. For the set *A* of odd natural numbers less then 10, the characteristic function is defined for each $x \in \mathbb{N}$ by the formula

$$\chi_A(x) = \begin{cases} 1 & \text{if when } x = 1,3,5,7,9 \\ 0 & \text{otherwise.} \end{cases}$$

Set *A* *is* contained in or is equal to another set *B*, written $A \subseteq B$, if every element of *A* is an element of *B*, that is, if $x \in A$ implies $x \in B$. If *A* is contained in *B*, then *A* is said to be a *subset* of *B*, and *B* is said to be a *superset* of *A*. Two sets are equal, symbolically $A = B$, if they contain exactly the same elements; therefore, if $A \subseteq B$ and $B \subseteq A$ then $A = B$. If $A \subseteq B$ and *A* is not equal to *B*, then *A* is called a *proper subset* of *B*, written $A \subset B$. The negation of each

Table 1.1. Definition of All Subsets, A, of Set $X = \{x_1, x_2, x_3\}$ by Their Characteristic Functions

	x_1	x_2	x_3
A:	0	0	0
	1	0	0
	0	1	0
	0	0	1
	1	1	0
	1	0	1
	0	1	1
	1	1	1

of these propositions is expressed symbolically by a slash crossing the operator. That is $x \notin A$, $A \not\subset B$, and $A \neq B$ represent, respectively, x is not an element of A, A is not a proper subset of B, and A is not equal to B.

The family of *all subsets* of a given set A is called the *power set* of A, and it is usually denoted by $\mathcal{P}(A)$. The family of all subsets of $\mathcal{P}(A)$ is called a *second-order power set of A*; it is denoted by $\mathcal{P}^2(A)$, which stands for $\mathcal{P}(\mathcal{P}(A))$. Similarly, *higher-order power sets* $\mathcal{P}^3(A)$, $\mathcal{P}^4(A)$, ... can be defined.

For any finite universal set, it is convenient to define its various subsets by their characteristic functions arranged in a tabular form, as shown in Table 1.1 for $X = \{x_1, x_2, x_3\}$. In this case, each set, A, of X is defined by a triple $\langle \chi_A(x_1), \chi_A(x_2), \chi_A(x_3) \rangle$. The order of these triples in the table is not significant, but it is useful for discussing typical examples in this book to list subsets containing one element first, followed by subsets containing two elements and so on.

The *intersection* of sets A and B is a new set, $A \cap B$, that contains every object that is simultaneously an element of both the set A and the set B. If $A = \{1, 3, 5, 7, 9\}$ and $B = \{1, 2, 3, 4, 5\}$, then $A \cap B = \{1, 3, 5\}$. The *union* of sets A and B is a new set, $A \cup B$, which contains all the elements that are in set A or in set B. With the sets A and B defined previously, $A \cup B = \{1, 2, 3, 4, 5, 7, 9\}$.

The *complement* of a set A, denoted $\bar{A}$, is the set of all elements of the universal set that are not elements of A. With $A = \{1, 3, 5, 7, 9\}$ and the universal set $X = \{1, 2, 3, 4, 5, 6, 7, 8, 9\}$, the complement of A is $\bar{A} = \{2, 4, 6, 8\}$. A related set operation is the *set difference*, $A - B$, which is defined as the set of all elements of A that are not elements of B. With A and B as defined previously, $A - B = \{7, 9\}$ and $B - A = \{2, 4\}$. The complement of A is equivalent to $X - A$.

All the concepts of set theory can be recast in terms of the *characteristic functions* of the sets involved. For example we have that $A \subseteq B$ if and only if $\chi_A(x) \leq \chi_B(x)$ for all $x \in X$. Similarly,

$$\chi_{A \cap B}(x) = \min\{\chi_A(x), \chi_B(x)\},$$

$$\chi_{A \cup B}(x) = \max\{\chi_A(x), \chi_B(x)\}.$$

The phrase "for all" occurs so often in set theory that a special symbol, $\forall$, is used as an abbreviation. Similarly, the phrase "there exists" is abbreviated as $\exists$. For example, the definition of set equality can be restated as $A = B$ if and only if $\chi_A(x) = \chi_B(x)$, $\forall x \in X$.

The size of a finite set, called its *cardinality*, is the number of elements it contains. If $A = \{1, 3, 5, 7, 9\}$, then the cardinality of A, denoted by $|A|$, is 5. A set may be empty, that is, it may contain no elements. The *empty set* is given a special symbol $\emptyset$; thus $\emptyset = \{\}$ and $|\emptyset| = 0$. When A is finite, then

$$|\mathcal{P}(A)| = 2^{|A|}, \; |\mathcal{P}^2(A)| = 2^{2^{|A|}}, \text{etc.}$$

The most fundamental properties of the set operations of absolute complement, union, and intersection are summarized in Table 1.2, where sets A, B, and C are assumed to be elements of the power set $\mathcal{P}(X)$ of a universal set X. Note that all the equations in this table that involve the set union and intersection are arranged in pairs. The second equation in each pair can be obtained from the first by replacing $\emptyset$, $\cup$, and $\cap$ with X, $\cap$, and $\cup$, respectively, and vice versa. These pairs of equations exemplify a *general principle of duality*: for each valid equation in set theory that is based on the union and intersection operations, there is a corresponding dual equation, also valid, that is obtained by the replacement just specified.

Any two sets that have no common members are called *disjoint*. That is, every pair of disjoint sets, A and B, satisfies the equation

$$A \cap B = \emptyset.$$

Table 1.2. Fundamental Properties of Set Operations

Involution	$\overline{\overline{A}} = A$
Commutativity	$A \cup B = B \cup A$
	$A \cap B = B \cap A$
Associativity	$(A \cup B) \cup C = A \cup (B \cup C)$
	$(A \cap B) \cap C = A \cap (B \cap C)$
Distributivity	$A \cap (B \cup C) = (A \cap B) \cup (A \cap C)$
	$A \cup (B \cap C) = (A \cup B) \cap (A \cup C)$
Idempotence	$A \cup A = A$
	$A \cap A = A$
Absorption	$A \cup (A \cap B) = A$
	$A \cap (A \cup B) = A$
Absorption by X and $\emptyset$	$A \cup X = X$
	$A \cap \emptyset = \emptyset$
Identity	$A \cup \emptyset = A$
	$A \cap X = A$
Law of contradiction	$A \cap \overline{A} = \emptyset$
Law of excluded middle	$A \cup \overline{A} = X$
De Morgan's laws	$\overline{A \cap B} = \overline{A} \cup \overline{B}$
	$\overline{A \cup B} = \overline{A} \cap \overline{B}$

A family of pairwise disjoint nonempty subsets of a set A is called a *partition* on A if the union of these subsets yields the original set A. A partition on A is usually denoted by the symbol $\pi(A)$. Formally,

$$\pi(A) = \{A_i \mid i \in I, \ A_i \subseteq A, A_i \neq \emptyset\}$$

is a partition on A iff (i.e., if and only if)

$$A_i \cap A_j \neq \emptyset$$

for each pair $i, j \in I, i \neq j$, and

$$\bigcup_{i \in I} A_i = A.$$

Members of $\pi(A)$, which are subsets of A, are usually referred to as *blocks* of the partition. Each member of A belongs to one and only one block of $\pi(A)$.

Given two partitions $\pi_1(A)$ and $\pi_2(A)$, we say that $\pi_1(A)$ is a *refinement* of $\pi_2(A)$ iff each block of $\pi_1(A)$ is included in some block of $\pi_2(A)$. The refinement relation on the set of all partitions of A, $\Pi(A)$, which is denoted by $\leq$ (i.e., $\pi_1(A) \leq \pi_2(A)$ in our case), is a partial ordering. The pair $\langle \Pi(A), \leq \rangle$ is a lattice, referred to as the *partition lattice* of A.

Let $\mathcal{A} = \{A_1, A_2, \ldots, A_n\}$ be a family of sets such that

$$A_i \subseteq A_{i+1} \quad \text{for all } i = 1, 2, \ldots, n-1.$$

Then, $\mathcal{A}$ is called a *nested family*, and the sets A_1 and A_n are called the *innermost set* and the *outermost set*, respectively. This definition can easily be extended to infinite families.

The ordered pair formed by two objects x and y, where $x \in X$ and $y \in Y$, is denoted by $\langle x, y \rangle$. The set of all ordered pairs, where the first element is contained in a set X and the second element is contained in a set Y, is called a *Cartesian product* of X and Y and is denoted as $X \times Y$. If, for example, $X = \{1, 2\}$ and $Y = \{a, b\}$, then $X \times Y = \{\langle 1, a \rangle, \langle 1, b \rangle, \langle 2, a \rangle, \langle 2, b \rangle\}$. Note that the size of $X \times Y$ is the product of the size of X and the size of Y when X and Y are finite: $|X \times Y| = |X| \cdot |Y|$. It is not required that the Cartesian product be defined on distinct sets. A Cartesian product $X \times X$ is perfectly meaningful. The symbol X^2 is often used instead of $X \times X$. If, for example, $X = \{0, 1\}$, then $X^2 = \{\langle 0, 0 \rangle, \langle 0, 1 \rangle, \langle 1, 0 \rangle, \langle 1, 1 \rangle\}$. Any subset of $X \times Y$ is called a *binary relation*.

Several important properties are defined for binary relations $R \subseteq X^2$. They are: R is *reflexive* iff $\langle x, x \rangle \in R$ for all $x \in X$; R is *symmetric* iff for every $\langle x, y \rangle \in R$ it is also $\langle y, x \rangle \in R$; R is *antisymmetric* iff $\langle x, y \rangle \in R$ and $\langle y, x \rangle \in R$ implies $x = y$; R is *transitive* iff $\langle x, y \rangle \in R$ and $\langle y, z \rangle \in R$ implies $\langle x, z \rangle \in R$. Relations that are reflexive, symmetric, and transitive are called *equivalence relations*. Relations that are reflexive and symmetric are called *compatibility relations*. Relations that are reflexive, antisymmetric, and transitive are called *partial order-ings*. When R is a partial ordering and $\langle x, y \rangle \in R$, it is common to write $x \leq y$ and say that x precedes y or, alternatively, that x is smaller than or equal to y.

A partial ordering $\leq$ on X does not guarantee that all pairs of elements x, y in X are comparable in the sense that either $x \leq y$ or $y \leq x$. If all pairs of ele-ments are comparable, the partial ordering becomes *total ordering* (or *linear ordering*). Such an ordering is characterized by—in addition to reflexivity, tran-sitivity, and antisymmetry—a property of *connectivity*: for all $x, y \in X, x \neq y$ implies either $x \leq y$ or $y \leq x$.

Let X be a set on which a partial ordering is defined and let A be a subset of X. If $x \in X$ and $x \leq y$ for every $y \in A$, then x is called a *lower bound* of A on X with respect to the partial ordering. If $x \in X$ and $y \leq x$ for every $y \in A$, then x is called an *upper bound* of A on X with respect to the partial ordering. If a particular lower bound of A succeeds (is greater than) any lower bound of A, then it is called the *greatest lower bound*, or *infimum*, of A. If a particular upper bound precedes (is smaller than) every other upper bound of A, then it is called the *least upper bound*, or *supremum*, of A.

A partially ordered set X any two elements of which have a greatest lower bound (also referred to as a *meet*) and a least upper bound (also referred to as a *join*) is called a *lattice*. The meet and join elements of x and y in X are often denoted by $x \wedge y$ and $x \vee y$, respectively. Any lattice on X can thus be defined not only by the pair $\langle X, \leq \rangle$, where $\leq$ is an appropriate partial ordering of X, but also by the triple $\langle X, \wedge, \vee \rangle$, where $\wedge$ and $\vee$ denote the operations of meet and join.

A partially ordered set, any two elements of which have only a greatest lower bound, is called a *lower semilattice* or *meet semilattice*. A partially ordered set, any two elements of which have only a least upper bound, is called an *upper semilattice* or *join semilattice*.

Elements of the power set $\mathcal{P}(X)$ of a universal set X (or any subset of X) can be ordered by the set inclusion $\subseteq$. This ordering, which is only partial, forms a lattice in which the *join* (least upper bound, supremum) and *meet* (greatest lower bound, infimum) of any pair of sets $A, B \in \mathcal{P}(X)$ is given by $A \cup B$ and $A \cap B$, respectively. This lattice is *distributive* (due to the distributive properties of $\cup$ and $\cap$ listed in Table 1.2) and *complemented* (since each set in $\mathcal{P}(X)$ has its complement in $\mathcal{P}(X)$); it is usually called a *Boolean lattice* or a *Boolean algebra*. The connection between the two formulations of this important lattice, $\langle \mathcal{P}(X), \subseteq \rangle$ and $\langle \mathcal{P}(X), \cup, \cap \rangle$, is facilitated by the equivalence

$$A \subseteq B \text{ iff } A \cup B = B \text{ and } A \cap B = A \text{ for any } A, B \in \mathcal{P}(X),$$

where "iff" is a common abbreviation of the phrase "if and only if" or its alter-native "is equivalent to." This convenient abbreviation is used throughout this book.

If $R \subseteq X \times Y$, then we call R a *binary relation* between X and Y. If $\langle x, y \rangle \in R$, then we also write $R(x, y)$ or xRy to signify that x is related to y by R. The *inverse of a binary relation* R on $X \times Y$, which is denoted by R^{-1}, is a binary relation on $Y \times X$ such that

$$\langle y, x \rangle \in R^{-1} \quad \text{iff} \quad \langle x, y \rangle \in R.$$

For any pair of binary relations $R \subseteq X \times Y$ and $Q \subseteq Y \times Z$, the *composition* of R and Q, denoted by $R \circ Q$, is a binary relation on $X \times Z$ defined by the formula

$$R \circ Q = \{\langle x, z \rangle \mid \langle x, y \rangle \in R \text{ and } \langle y, z \rangle \in Q \text{ for some } y\}.$$

If a binary relation on $X \times Y$ is such that each element $x \in X$ is related to exactly one element of $y \in Y$, the relation is called a *function*, and it is usually denoted by a lowercase letter. Given a function f, this unique assignment of one particular element $y \in Y$ to each element $x \in X$ is often expressed as $f(x) = y$. Set X is called a *domain* of f and Y is called its *range*. The domain and range of function f are usually specified in the form $f: X \to Y$; the arrow indicates that function f *maps* elements of set X to elements of set Y; f is called a *completely specified function* iff each element $x \in X$ is included in at least one pair $\langle x, y = f(x) \rangle$ and it is called an *onto function* iff each element $y \in Y$ is included in at least one pair $\langle x, y = f(x) \rangle$. If the domain of a function (and possibly also its range) is a set of functions, then it is common to call such a function a *functional*.

The *inverse of a function* f is another function, f^{-1}, which maps elements of set Y to disjoint subsets of set X. If f is a completely specified and onto function, then f^{-1} maps elements of set Y to blocks of the unique partition, $\pi_f(X)$, that is induced on the set X by function f. This partition consists of $|Y|$ subsets of X,

$$\pi_f(X) = \{X_y \mid y \in Y\},$$

where

$$X_y = \{x \in X \mid f(x) = y\}$$

for each $y \in Y$. Function f^{-1} thus has the form

$$f^{-1}: Y \to \pi_f(X)$$

and is defined by the assignment $f^{-1}(y) = X_y$ for each $y \in Y$.

The notion of a Cartesian product is not restricted to ordered pairs. It may involve ordered n-tuples for any $n \geq 2$. An *n-dimensional Cartesian product* for some particular n is the set of all ordered n-tuples that can be formed from the designated sets in a manner analogous to forming ordered pairs of a 2-dimensional Cartesian product. When the n-tuples are formed from a single set, say, set X, then the n-dimensional Cartesian product is usually denoted by the symbol X^n. For example, if $X = \{0, 1\}$, then $X^3 = \{\langle 0, 0, 0\rangle, \langle 0, 0, 1\rangle, \langle 0, 1, 0\rangle, \langle 0, 1, 1\rangle, \langle 1, 0, 0\rangle, \langle 1, 0, 1\rangle, \langle 1, 1, 0\rangle, \langle 1, 1, 1\rangle\}$. Any subset of a given n-dimensional Cartesian product is called an *n-dimensional relation*.

Several important concepts are associated with n-dimensional relations for any finite $n \geq 2$. For the sake of simplicity, let us define them in terms of a ternary relation $R \subseteq X \times Y \times Z$. Generalizations to $n > 3$ are obvious. A *projection* of R into one of its dimensions, say, dimension X, is the set

$$[R \!\downarrow\! X] = \{x \in X \mid \langle x, y, z\rangle \in R \text{ for some } \langle y, z\rangle \in Y \times Z\}.$$

The symbol $[R \downarrow X]$ indicates that R is projected into dimension X. Projections $[R \downarrow Y]$ and $[R \downarrow Z]$ are defined in a similar way. A projection of R into two of its dimensions, say, $X \times Y$, is the set

$$[R \!\downarrow\! X \times Y] = \{\langle x, y\rangle \in X \times Y \mid \langle x, y, z\rangle \in R \text{ for some } z \in Z\}.$$

Projections $[R \downarrow X \times Z]$ and $[R \downarrow Y \times Z]$ are defined in a similar way.

A *cylindric extension* of projection $[R \downarrow X]$ of a ternary relation $R \subseteq X \times Y \times Z$ with respect to $Y \times Z$ is the set

$$([R \!\downarrow\! X] \!\uparrow\! Y \times Z) = \{\langle x, y, z\rangle \in X \times Y \times Z \mid x \in [R \!\downarrow\! X]\}.$$

Similarly, a cylindric extension of projection $[R \downarrow X \times Y]$ with respect to dimension Z is the set

$$([R \!\downarrow\! X \times Y] \!\uparrow\! Z) = \{\langle x, y, z\rangle \in X \times Y \times Z \mid \langle x, y\rangle \in X \times Y\}.$$

The intersection of cylindric extensions of any given set P of projections of relation R is called a *cylindric closure* of R with respect to projections in P.

For any pair of binary relations $R \subseteq X \times Y$ and $Q \subseteq Y \times Z$, the *join* of R and Q, denoted by $R * Q$, is a ternary relation on $X \times Y \times Z$ defined by the formula

$$R * Q = \{\langle x, y, z\rangle \in X \times Y \times Z \mid \langle x, y\rangle \in R \text{ and } \langle y, z\rangle \in Q\}.$$

Observe that

$$R \circ Q = [R * Q \!\downarrow\! X \times Z].$$

Important subsets of $\mathbb{R}$ are intervals of real numbers. Four types of intervals of real numbers between a and b are distinguished: *Closed intervals*, $[a, b]$, which contain the endpoints a and b; *open intervals*, (a, b), which do not contain the endpoints; *semiopen intervals*, $(a, b]$ (left-open interval) and $[b, a)$ (right-open interval), which do not contain the left-end point and the right-end point, respectively.

An important and frequently used universal set is the set of all points in n-dimensional Euclidean vector space $\mathbb{R}^n$ for some $n \geq 1$ (i.e., all n-tuples of real numbers). Sets defined in terms of $\mathbb{R}^n$ are often required to possess a property referred to as convexity. A subset A of $\mathbb{R}^n$ is called *convex* iff, for every pair of points

$$\mathbf{r} = \langle r_i \,|\, i \in \mathbb{N}_n \rangle \qquad \text{and} \qquad \mathbf{s} = \langle s_i \,|\, i \in \mathbb{N}_n \rangle$$

in A and every real number $\lambda \in [0, 1]$, the point

$$\mathbf{t} = \langle \lambda r_i + (1 - \lambda)s_i \,|\, i \in \mathbb{N}_n \rangle$$

is also in A. In other words, a subset A of $\mathbb{R}^n$ is convex iff, for every pair of points $\mathbf{r}$ and $\mathbf{s}$ in A, all points located on the straight-line segment connecting $\mathbf{r}$ and $\mathbf{s}$ are also in A.

In $\mathbb{R}$, any set defined by a single interval of real numbers is convex; any set defined by more than one interval that does not contain some points between the intervals is not convex. For example, the set $A = [0, 2] \cup [3, 5]$ is not convex, as can be shown by producing one of an infinite number of possible counterexamples: let $r = 1$, $s = 4$, and $\lambda = 0.4$; then, $\lambda r + (1 - \lambda)s = 2.8$ and $2.8 \notin A$.

Let R denote any set of real numbers (i.e., $R \subseteq \mathbb{R}$). If there is a real number r (or a real number s) such that $x \leq r$ (or $x \geq s$, respectively) for every $x \in R$, then r is called an *upper bound* of R (or a *lower bound* of R), and we say that R is *bounded above* by r (or *bounded below* by s).

For any set of real numbers R that is bounded above, a real number r is called the *supremum* of R iff:

(a) r is an upper bound of R.

(b) No number less than r is an upper bound of R.

If r is the supremum of R, we write $r = \sup R$. If R has a maximum, then $\sup R = \max R$. For example, $\sup(0, 1) = \sup[0, 1] = 1$, but only $\max[0, 1] = 1$; maximum of the open interval $(0, 1)$ does not exist.

For any set of real numbers R that is bounded below, a real number s is called the *infimum* of R iff:

(a) s is a lower bound of R.

(b) No number greater than s is a lower bound of R.

If s is the infimum of R, we write $s = \inf R$. If R has a minimum, then $\inf R = \min R$.

Classical sets must satisfy two basic requirements. First, members of each set must be distinguishable from one another; and second, for any given set and any given object, it must be possible to determine whether the object is, or is not, a member of the set.

Fuzzy sets, which play an important role in GIT, differ from classical sets by rejecting the second requirement. Contrary to classical sets, fuzzy sets are not required to have sharp boundaries that distinguish their members from other objects. The membership in a fuzzy set is not a matter of affirmation or denial, as it is in a classical set, but a matter of degree.

Due to their sharp boundaries, classical sets are usually referred to in fuzzy literature as *crisp sets*. This convenient and well-established term is adopted in this book. Also adopted is the usual notation, according to which both crisp and fuzzy sets are denoted by capital letters. This is justified by the fact that crisp sets are special (degenerate) fuzzy sets.

Each fuzzy set is defined in terms of a relevant *crisp universal set* by a function analogous to the characteristic function of crisp sets. This function is called a *membership function*. As explained in Chapter 7, the form of this function depends on the type of fuzzy set that is defined by it. For the most common fuzzy sets, referred to as *standard fuzzy sets*, the membership function used for defining a fuzzy set A on a given universal set X assigns to each element x of X a real number in the unit interval $[0, 1]$. This number is interpreted as the *degree of membership* of x in A. When only the extreme values, 0 and 1, are assigned to each $x \in X$, the membership function becomes formally equivalent to a characteristic function that defines a crisp set. However, there is subtle conceptual difference between the two functions. Contrary to the symbolic role of the numbers in characteristic functions of crisp sets, numbers assigned to objects by membership functions of standard fuzzy sets clearly have a numerical significance. This significance is preserved when crisp sets are viewed (from the standpoint of fuzzy set theory) as special fuzzy sets. For example, when we calculate an average of two or more membership functions, we obtain a membership function that defines a meaningful standard fuzzy set. On the other hand, an average of two or more characteristic functions is not a meaningful characteristic function.

Two distinct notations are most commonly employed in the literature to denote membership functions. In one of them, the membership function of a fuzzy set A is denoted by μ_A, and its form for standard fuzzy sets is

$$\mu_A : X \to [0, 1].$$

For each $x \in X$, the value $\mu_A(x)$ is the degree of membership of x in A. In the second notation, the membership function is denoted by A and, of course, has the same form

$$A : X \to [0, 1].$$

Clearly, $A(x)$ is again the degree of membership of x in A.

According to the first notation, the symbol of the fuzzy set is distinguished from the symbol of its membership function. According to the second notation, this distinction is not made, but no ambiguity results from this double use of the same symbol, since each fuzzy set is uniquely defined by one particular membership function. In this book, the second notation is adopted; it is simpler, and, by and large, more popular in the current literature on fuzzy set theory. Since crisp sets are viewed from the standpoint of fuzzy set theory as special fuzzy sets, the same notation is used for them.

By exploiting degrees of membership, fuzzy sets are capable of expressing gradual transitions from membership to nonmembership. This expressive capability has wide utility. For example, it allows us to capture, at least in a crude way, meanings of expressions in natural language, most of which are inherently vague. Membership degrees in these fuzzy sets express compatibilities of relevant objects with the linguistic expression that the sets attempt to capture. Crisp sets are hopelessly inadequate for this purpose.

Consider the four membership functions whose graphs are shown in Figure 1.4. These functions are defined on the set of nonnegative real numbers. Functions A and B define crisp sets, which are viewed here (from the fuzzy set perspective) as special fuzzy sets. Set A consists of a single object, the number 3; set B consists of all real numbers in the closed interval [2, 4]. Functions C and D define genuine fuzzy sets. Set C captures (in appropriate context) linguistic expressions such as *around* 3, *close to* 3, or *approximately* 3. It may thus be viewed as a fuzzy number. Similarly, fuzzy set D may be viewed as a fuzzy interval.

Observe that the crisp set B in Figure 1.4 also consists of numbers that are around 3. However, the sharp boundaries of the set are at odds with the vague term *around*. The meaning of the term is certainly not captured, for example, by excluding the number 1.999999 while including the number 2. The abrupt transitions from membership to nonmembership make crisp sets virtually unusable for capturing meanings of linguistic terms of natural language.

To explain the role of fuzzy set theory in GIT, an overview of its fundamentals is presented in Chapter 7.

1.5. AN OUTLINE OF THE BOOK

The objective of this book is to survey the current level of development of GIT. The material, which is presented in a textbook-like manner, is organized in the following way.

After setting the stage in this introductory chapter, the actual survey begins with overviews of the two classical uncertainty theories. These are theories based on the notion of *possibility* (Chapter 2) and the notion of *probability* (Chapter 3). Due to extensive coverage of these theories in the literature, especially the one based on probability, only the most fundamental features

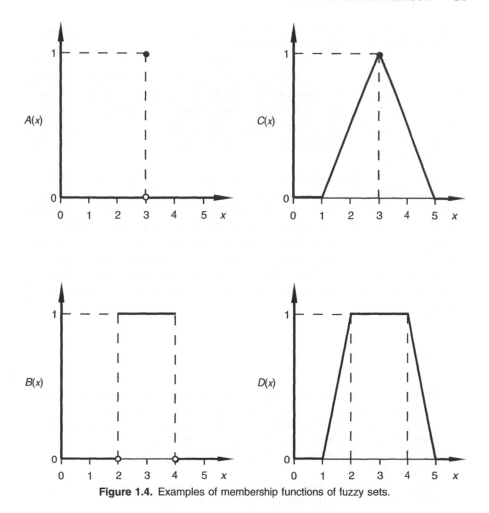

Figure 1.4. Examples of membership functions of fuzzy sets.

of these theories are covered. However, Notes in the two chapters guide the reader through the literature dealing with these classical uncertainty theories.

The next part of the book (Chapters 4–6) is oriented toward introducing some generalizations of the classical probability-based uncertainty theory. These generalizations are obtained by replacing the additivity requirement of probability measures with the weaker requirement of monotonicity of monotone measures, but they are still formalized within the language of classical set theory. These theories may be viewed as theories of imprecise probabilities of various types. While the focus in Chapter 4 is on common properties of uncertainty functions in all these theories, Chapter 5 is concerned with distinctive properties of the individual theories. Covered are only theories that had been well developed when this book was written. Functionals for measuring uncertainty in the introduced theories are examined in Chapter 6. These functionals, which are central to GIT, are shown to be largely invariant with respect to the great diversity of uncertainty theories.

Further generalization of uncertainty theories by extending classical sets to fuzzy sets of various types, which is usually referred to as *fuzzification*, is the subject of Chapter 8. In order to make the book self-contained, relevant concepts of fuzzy set theory are introduced in Chapter 7. This important area of GIT is still largely undeveloped. Only a few uncertainty theories have been fuzzified thus far, and all these developed fuzzifications involve only the standard fuzzy sets.

The survey of GIT is concluded by examining four important principles of uncertainty in Chapter 9. These are methodological principles justified on epistemological grounds. Their applications to a wide variety of practical problems, which are discussed in this chapter, demonstrate the importance of GIT.

The closing chapter of the book (Chapter 10) is devoted to conclusions. GIT is examined in this chapter in both retrospect and prospect. The overall conclusion from this examination is that GIT is still in an early stage of its development, notwithstanding the many important results that are surveyed in this book.

NOTES

1.1. The recent book by Pollack [2003] is recommended as supplementary reading to this chapter. It is a well-written and thorough discussion of the important role of uncertainty in both science and ordinary life. Also recommended is the book by Smithson [1989], in which the many facets of changing attitudes toward uncertainty in science and other areas of human affairs throughout the 20th century are carefully examined.

1.2. The turning point leading to the acceptance of methods based on probability theory in science was the publication of sound mathematical foundations of statistical mechanics by Willard Gibbs [1902].

1.3. Research on a broader conception of uncertainty-based information, liberated from the confines of classical set theory and probability theory, began in the early 1980s [Higashi and Klir, 1983a, b; Höhle, 1982; Yager, 1983]. The name *generalized information theory* (GIT) was coined for this research program by Klir [1991].

1.4. Information measured solely by the reduction of uncertainty is not explicitly concerned with the semantic and pragmatic aspects of information viewed in the broader sense [Cherry, 1957; Jumarie, 1986, 1990; Kornwachs and Jacoby, 1996]. However, these aspects are not ignored, but they are assumed to be addressed prior to each particular application. For example, when dealing with a system of some kind (in general, a set of interrelated variables), we are assumed to understand the language (formalized or natural) in which the system is described (this resolves the semantic aspect of information), and we are also assumed to know the purpose for which the system has been constructed (this resolves the pragmatic aspect). These assumptions certainly restrict the applicability of uncertainty-based information. However, an argument can be made [Dretske, 1981, 1983] that the notion of uncertainty-based information is sufficiently rich as a basis for additional treatment, through which the broader concept of information, pertaining to human communication and cognition, can adequately be formalized.

1.5. The concept of information has also been investigated in terms of the theory of computability. In this approach, which is not covered in this book, the amount of information represented by an object is measured by the length of its shortest description in some standard language (e.g., by the shortest program for the standard Turing machine). Information of this type is usually referred to as *descriptive information* or *algorithmic information* [Kolmogorov, 1965; Chaitin, 1987], and it is connected with the concept of Kolmogorov complexity [Li and Vitányi, 1993].

1.6. Some additional approaches to information have appeared in the literature since the early 1990s. For example, Devlin [1991] formulates and investigates information in terms of logic, while Stonier [1990] views information as a physical property defined as the capacity to organize a system or to maintain it in an organized state. Another physics-based approach to information is known in the literature as Fisher information [Fisher, 1950; Frieden, 1998]. A more recent, measurement-based approach was developed by Harmuth [1992]. Again, these various approaches are not covered in this book.

1.7. A digest of most mathematical concepts that are relevant to the subject of this book is in Section 1.4. A useful reference for strengthening the background in classical set theory, which is suitable for self-study is *Set Theory and Related Topics* by S. Lipschutz (Schaum Series/McGraw-Hill, New York). Basic familiarity with calculus and some aspects of mathematical analysis are also needed for understanding this book. The book *Mathematical Analysis* by T.M. Apostol (Addison-Wesley, Reading, MA) is recommended as a useful reference in this area.

EXERCISES

1.1. For which of the following pairs of sets is $A = B$?

   (a) $A = \{0, 1, 2, 3\}; B = \{1, 3, 2, 0\}$
   (b) $A = \{0, 1, 0, 2, 3\}; B = \{0, 1, 2, 3, 2\}$
   (c) $A = \varnothing; B = \{\varnothing\}$
   (d) $A = \{0\}; B = \{\varnothing\}$

1.2. Which of the following definitions are acceptable as definitions of classical (crisp) sets?

   (a) $A = \{a \mid a$ is a real number$\}$
   (b) $B = \{b \mid b$ is a real number much greater than 1$\}$
   (c) $C = \{c \mid c$ is a living organism$\}$
   (d) $D = \{d \mid d$ is a section in this book$\}$
   (e) $E = \{e \mid e$ is a set$\}$
   (f) $F = \{f \mid f$ is a pretty girl$\}$

1.3. Which of the following statements are correct provided that $X = \{\emptyset, 0, 1, 2, \{0, 1\}\}$?

   (a) $\{0, 1\} \in X$

   (b) $\{0, 1\} \subset X$

   (c) $\{0, 1, 2, \{0, 1\}\} = X$

   (d) $\{\{0, 1\}\} \subset X$

   (e) $\{0, 1, 2\} \in X$

   (f) $\{\emptyset\} \in X$

1.4. Let $A = \mathbb{N}_{40}$, $B = \{b \mid b$ is a natural number divisible by 3, $b \le 30\}$, $C = \{c \mid c$ is an odd natural number, $a \le 50\}$. Determine the following sets:

   (a) $A - B; A - C; C - A, C - B$

   (b) $A \cap B; A \cap C; B \cap C$

   (c) $A \cup B; A \cup C; B \cup C$

   (d) $(A \cap B) \cup C; (A \cup B) \cap C; B \cup (A \cap C)$

1.5. Determine the partition lattices of sets $A = \{1, 2, 3\}$ and $B = \{1, 2, 3, 4\}$.

1.6. How many possible relations can be defined on the following Cartesian products of finite sets?

   (a) $X_1 \times X_2 \times \ldots \times X_n$

   (b) $A \times B^2 \times C^3$

   (c) $\mathcal{P}(A) \times \mathcal{P}(B)$

   (d) $A^2 \times \mathcal{P}^2(B) \times C$

1.7. For each of the following relations, determine whether or not it is reflexive, symmetric, antisymmetric, or transitive:

   (a) $R \subseteq \mathcal{P}(X) \times \mathcal{P}(X)$; $\langle A, B \rangle \in R$ iff $A \subseteq B$ for all $A, B \in \mathcal{P}(X)$.

   (b) $R \subseteq C \times C$, where C denotes the set of courses in a graduate program: $\langle a,b \rangle \in R$ iff course a is a prerequisite of course b.

   (c) $R_n \subseteq \mathbb{N} \times \mathbb{N}$: $\langle a,b \rangle \in R_n$ iff the remainders obtained by diving a and b by n are the same, where n is some specific natural number greater than 1.

   (d) $R \subseteq W \times W$, where W denotes the set of all English words: $\langle a,b \rangle \in R$ iff a is a synonym of b.

   (e) $R \subseteq F \times F$, where F denotes the set of all five-letter English words: $\langle a,b \rangle \in R$ iff a differs from b in at most one position.

   (f) $R \subseteq A \times A$: $\langle a,b \rangle \in R$ iff $f(a) = f(b)$, where f is a function of the form

$$f : A \to A.$$

   (g) $R \subseteq T \times T$, where T denotes all persons included in a family tree: $\langle a,b \rangle \in R$ iff a is an ancestor of b.

(h) $R = \{\langle 0,0 \rangle, \langle 0,1 \rangle, \langle 1,0 \rangle, \langle 1,1 \rangle, \langle 1,2 \rangle, \langle 2,2 \rangle, \langle 0,2 \rangle, \langle 3,3 \rangle\}$

(i) $R = \{\langle 0,0 \rangle, \langle 0,3 \rangle, \langle 1,1 \rangle, \langle 2,2 \rangle, \langle 1,0 \rangle, \langle 0,1 \rangle, \langle 3,1 \rangle, \langle 3,3 \rangle, \langle 3,0 \rangle\}$

1.8. Let $X = \mathbb{N}_4$. Determine which of the following relations on X^2 are equivalence relations, compatibility relations, or partial orderings:

(a) $R_1 = \{\langle x, y \rangle \mid x < y\}$

(b) $R_2 = \{\langle x, y \rangle \mid x \leq y\}$

(c) $R_3 = \{\langle x, y \rangle \mid x = y\}$

(d) $R_4 = \{\langle x, y \rangle \mid 2x = y\}$

(e) $R_5 = \{\langle x, y \rangle \mid x = y - 1\}$

(f) $R_6 = \{\langle x, y \rangle \mid x < y \text{ or } x > y\}$

(g) $R_7 = \{\langle 1, 2 \rangle, \langle 2, 3 \rangle, \langle 3, 2 \rangle, \langle 4, 3 \rangle\}$

1.9. For the binary relations in Exercise 1.8, determine the following compositions and joins:

(a) $R_1 \circ R_2$ and $R_1 * R_2$

(b) $R_1 \circ R_1^{-1}$ and $R_1 * R_2^{-1}$

(c) $R_4 \circ R_5$ and $R_3 * R_7$

(d) $R_6 \circ R_7^{-1}$ and $R_7 * R_6^{-1}$

1.10. For each of the following functions, determine the supremum and infimum, as well as the maximum and minimum (if they exist):

(a) $f(x) = 1 - x$, where $x \in [0, 1)$

(b) $f(x) = \sin x$, where $x \in [0, 2\pi]$

(c) $f(x) = x/(1 - x)$, where $x \in (0, 10)$

(d) $f(x) = \sin x / x$, where $x \in \mathbb{R}$

(f) $f(x) = \max[0, 2x - x^2]$, where $x \in [0, 1]$

1.11. For each of the following binary relations on $\mathbb{R}^2$, determine its 1-dimensional projections, its cylindric extensions, and the cylindric closure of these projections:

(a) $R = \{\langle x, y \rangle \mid x^2 + y^2 \leq 1\}$

(b) $R = \{\langle x, y \rangle \mid 0 \leq y \leq 2x - x^2\}$

(c) $R = \{\langle x, y \rangle \mid |x| + |y| \leq 1\}$

(d) $R = \{\langle x, y \rangle \mid x^2 + 2y^2 \leq 1\}$

2

CLASSICAL POSSIBILITY-BASED UNCERTAINTY THEORY

When you have eliminated the impossible, whatever remains must have been the case, however improbable it may seem to be.

—Sherlock Holmes

2.1. POSSIBILITY AND NECESSITY FUNCTIONS

One of the two classical theories of uncertainty, which is the subject of this chapter, is based on the notion of *possibility* and the associated notion of *necessity*. The other classical theory, which is the subject of Chapter 3, is based on the notion of *probability*. Of the two classical theories, the one based on possibility is simpler, more fundamental, and older. To describe this rather simple theory, let X denote a finite set of mutually exclusive alternatives that are of concern to us (diagnoses, predictions, etc.). This means that in any given situation only one of the alternatives is true. To identify the true alternative, we need to obtain relevant information (e.g., by conducting relevant diagnostic tests). The most elementary and, at the same time, the most fundamental kind of information is a demonstration (based, for example, on outcomes of the conducted diagnostic tests) that some of the alternatives in X are not possible. After excluding these alternatives from X, we obtain a subset E of X. This subset contains only alternatives that, according to the obtained information, are *possible*. We can say that alternatives in E are supported by the evidence.

Let the characteristic function of the set of all possible alternatives, E, be called in this context a *basic possibility function* and be denoted by r_E. Then,

Uncertainty and Information: Foundations of Generalized Information Theory, by George J. Klir
© 2006 by John Wiley & Sons, Inc.

$$r_E(x) = \begin{cases} 1 & \text{when } x \in E \\ 0 & \text{when } x \notin E. \end{cases} \tag{2.1}$$

Using common sense, a possibility function, Pos_E, defined on the power set, $\mathcal{P}(X)$, is then given by the formula

$$Pos_E(A) = \max_{x \in A} r_E(x), \tag{2.2}$$

for all $A \in \mathcal{P}(X)$. It is indeed correct to say that it is possible that the true alternative is in A when A contains at least one possible alternative (an alternative that is also contained in E). It follows immediately that

$$Pos_E(A \cup B) = \max\{Pos_E(A), Pos_E(B)\} \tag{2.3}$$

for any pair of sets $A, B \in \mathcal{P}(X)$.

Given the possibility function Pos_E on the power set of X, it is useful to define another function, Nec_E, to describe for each $A \in \mathcal{P}(X)$ the *necessity* that the true alternative is in A. Clearly, the true alternative is necessarily in A if and only if it is not possible that it is in $\bar{A}$, the complement of A. Hence,

$$Nec_E(A) = 1 - Pos_E(\overline{A}) \tag{2.4}$$

for all $A \in \mathcal{P}(X)$.

2.2. HARTLEY MEASURE OF UNCERTAINTY FOR FINITE SETS

The question of how to measure the amount of uncertainty associated with a finite set E of *possible* alternatives was addressed by Hartley [1928]. He showed that the only meaningful way to measure this amount is to use a functional of the form.

$$c \log_b \sum_{x \in X} r_E(x)$$

or, alternatively,

$$c \log_b |E|,$$

where $|E|$ denotes the cardinality of set E; b and c are positive constants, and it is required that $b \neq 1$. Each choice of values b and c determines the unit in which the uncertainty is measured. Requiring, for example, that

$$c \log_b 2 = 1,$$

which is the most common choice, uncertainty would be measured in *bits*. One bit of uncertainty is equivalent to uncertainty regarding the truth or falsity of one elementary proposition. Conveniently choosing $b = 2$ and $c = 1$ to satisfy the preceding equation, we obtain a unique functional, H, defined for any basic possibility function, r_E, by the formula

$$H(r_E) = \log_2 |E|.$$

This functional is usually called a *Hartley measure* of uncertainty. It is easy to see that uncertainty measured by this functional results from the lack of specificity. Clearly, the larger the set of possible alternatives, the less specific are predictions, diagnoses, and the like. Full specificity is obtained when only one of the considered alternatives is possible. This type of uncertainty is thus well characterized by the term *nonspecificity*.

2.2.1. Simple Derivation of the Hartley Measure

This simple derivation is due to Hartley [1928]. Assume, as before, that from among a finite set X of considered alternatives, the only possible alternatives are those in a nonempty subset E of X. That is, there is evidence that alternatives that are not in set E are not possible.

Given now a particular set E of possible alternatives, sequences of its elements can be formed by successive selections. If s selections are made, then there are $|E|^s$ different potential sequences. The amount of uncertainty in identifying one of the sequences, $H(|E|^s)$, which is equivalent to the amount of information needed to remove this uncertainty, should be proportional to s. That is,

$$H\big(|E|^s\big) = K(|E|) \cdot s,$$

where $K(|E|)$ is some function of $|E|$.

Consider now two nonempty subsets of X, E_1 and E_2, such that $|E_1| \neq |E_2|$. Assume that after making s_1 selections from E_1 and s_2 selections from E_2, the number of sequences in both cases is the same. Then, the amounts of information needed to remove uncertainty associated with the two sets of sequences should be the same as well. That is, if

$$|E_1|^{s_1} = |E_2|^{s_2}, \tag{2.5}$$

then

$$K|E_1| \cdot s_1 = K|E_2| \cdot s_2. \tag{2.6}$$

From Eqs. (2.5) and (2.6), we obtain

$$\frac{s_2}{s_1} = \frac{\log_b |E_1|}{\log_b |E_2|}$$

and

$$\frac{s_2}{s_1} = \frac{K|E_1|}{K|E_2|},$$

respectively. Hence,

$$\frac{K|E_1|}{K|E_2|} = \frac{\log_b|E_1|}{\log_b|E_2|}.$$

This equation can be satisfied only by a function K of the form

$$K(|E|) = c \cdot \log_b|E|,$$

where b, c are positive constants and $b \neq 1$. Each choice of values of the constants b and c determines the unit in which uncertainty is measured. When $b = 2$ and $c = 1$, which is the most common choice, uncertainty is measured in bits, and we obtain

$$H(E) = \log_2|E|, \tag{2.7}$$

This can also be expressed in terms of the basic possibility function r_E as

$$H(r_E) = \log_2 \sum_{x \in X} r_E(x). \tag{2.8}$$

2.2.2. Uniqueness of the Hartley Measure

Hartley's derivation of the measure of possibilistic uncertainty (or nonspecificity) H, which is expressed by either Eq. (2.7) or Eq. (2.8), is certainly convincing. However, it does not explicitly prove that H is the only meaningful functional to measure possiblistic uncertainty in bits. To be meaningful, the functional must satisfy some essential axiomatic requirements. The uniqueness of this functional H was proven on axiomatic grounds by Rényi [1970b].

Since, according to our intuition, the possibilistic uncertainty depends only on the *number of possible alternatives* (the number of elements in the set $E \subseteq X$), Rényi conceptualized the measure of possibilistic uncertainty, H, as a functional of the form:

$$H : \mathbb{N} \to \mathbb{R}^+.$$

Using this form, he characterized the functional by the following axioms:

Axiom (H1) Branching. $H(n \cdot m) = H(n) + H(m)$.

Axiom (H2) Monotonicity. $H(n) \le H(n + 1)$.

Axiom (H3) Normalization. $H(2) = 1$.

Axiom (H1) involves a set with $m \cdot n$ elements, which can be partitioned into n subsets each with m elements. A characterization of an element from the full set requires the amount $H(m \cdot n)$ of information. However, we can also proceed in two steps to characterize the element by taking advantage of the partition of the set. First, we characterize the subset to which the element belongs: the required information is $H(n)$. Then, we characterize the element within the subset: the required information is $H(m)$. These two amounts of information completely characterize an element of the full set and, hence, their sum should equal $H(m \cdot n)$. This is exactly what the axiom requires.

Axiom (H2) expresses an essential and rather obvious requirement: When the number of possible alternatives increases, the amount of information needed to characterize one of them cannot decrease. Axiom (H3) is needed to define the measurement unit. As it is stated by Rényi, the defined measurement unit is the bit.

Using the three axioms, Rényi established that H defined by Eq. (2.7) is the only functional that satisfies these axioms. This is the subject of the following *uniqueness theorem.*

Theorem 2.1. The functional $H(n) = \log_2 n$ is the only functional that satisfies Axioms (H1)–(H3).

Proof. Let n be an integer greater than 2. For each integer i, define the integer $q(i)$ such that

$$2^{q(i)} \le n^i < 2^{q(i)+1}. \tag{2.9}$$

These inequalities can be written as

$$q(i)\log_2 2 \le i\log_2 n < (q(i)+1)\log_2 2.$$

When we divide these inequalities by i and replace $\log_2 2$ with 1, we get

$$\frac{q(i)}{i} \le \log_2 n < \frac{q(i)+1}{i}.$$

Consequently,

$$\lim_{i \to \infty} \frac{q(i)}{i} = \log_2 n. \tag{2.10}$$

Let H denote a functional that satisfies Axioms (H1)–(H3). Then, by Axiom (H2),

$$H(a) \le H(b) \tag{2.11}$$

for $a < b$. Combining Eq. (2.11) and Eq. (2.9), we obtain

$$H(2^{q(i)}) \le H(n^i) \le H(2^{q(i)+1}). \tag{2.12}$$

By Axiom (H1), we obtain

$$H(a^k) = k \cdot H(a).$$

If we apply this to all three terms of Eq. (2.12), we get

$$q(i) \cdot H(2) \le i \cdot H(n) \le (q(i)+1) \cdot H(2).$$

By Axiom (H3), $H(2) = 1$, so these inequalities become

$$q(i) \le i \cdot H(n) \le q(i)+1.$$

Dividing through by i yields

$$\frac{q(i)}{i} \le H(n) \le \frac{q(i)+1}{i},$$

and consequently,

$$\lim_{i \to \infty} \frac{q(i)}{i} = H(n). \tag{2.13}$$

Comparing Eq. (2.10) with Eq. (2.13), we conclude that $H(n) = \log_2 n$ for $n > 2$. Since $\log_2 2 = 1$ and $\log_2 1 = 0$, function H trivially satisfies all the axioms for $n = 1, 2$ as well. ∎

2.2.3. Basic Properties of the Hartley Measure

First, it is easy to see that the Hartley measure defined by Eq. (2.7) satisfies the inequalities

$$0 \le H(E) \le \log_2 |X|$$

for any $E \in \mathcal{P}(X)$. The lower bound is obtained when only one of the considered alternatives is possible, as exemplified by deterministic systems. The upper bound is obtained when all considered alternatives are possible. This expresses the state of total ignorance.

In some applications, it is preferable to use a *normalized Hartley measure*, *NH*, which is defined by the formula

$$NH(E) = \frac{H(E)}{\log_2|X|}. \tag{2.14}$$

The range of NH is clearly the unit interval $[0, 1]$, independent of X and $E \subseteq X$. Moreover, NH is invariant with respect to the choice of measurement units.

Assume now that a given set of possible alternatives, E, is reduced by the outcome of an action to a smaller set $E' \subset E$. Then, the amount of information obtained by the action, $I_H(E, E')$ is measured by the difference $H(E) - H(E')$. That is,

$$I_H(E, E') = \log_2 \frac{|E|}{|E'|}. \tag{2.15}$$

When the action eliminates all alternatives in E except one (i.e., when $|E'| = 1$), we obtain $I_H(E, E') = \log_2|E| = H(E)$. This means that $H(E)$ may also be viewed as the amount of information needed to characterize one element of set E.

Consider now two universal sets, X and Y, and assume that a relation $R \subseteq X \times Y$ describes a set of possible alternatives in some situation of interest. Consider further the sets

$$R_X = \{x \in X \,|\, (x, y) \in R \text{ for some } y \in Y\},$$
$$R_Y = \{y \in Y \,|\, (x, y) \in R \text{ for some } x \in X\},$$

which are usually referred to as *projections* of R on sets X, Y, respectively. Then three distinct Hartley measures are applicable, $H(R_X)$, $H(R_Y)$, and $H(R)$, which are defined on the power sets of X, Y, and $X \times Y$, respectively. The first two,

$$H(R_X) = \log_2|R_X|, \tag{2.16}$$
$$H(R_Y) = \log_2|R_Y|, \tag{2.17}$$

are called *simple* or *marginal Hartley measures*. The third one,

$$H(R) = \log_2|R| \tag{2.18}$$

is called a *joint Hartley measure*.

Two additional Hartley measures are defined,

$$H(R_X \,|\, R_Y) = \log_2 \frac{|R|}{|R_Y|}, \tag{2.19}$$

$$H(R_Y \,|\, R_X) = \log_2 \frac{|R|}{|R_X|}, \tag{2.20}$$

which are called *conditional Hartley measures.* These definitions can be generalized by restricting the set of possible conditions to $R_Y' \subseteq R_Y$ and $R_X' \subseteq R_X$, respectively. The generalized definitions are:

$$H(R_X \mid R_Y') = \log_2 \frac{|R|}{|R_Y'|},\tag{2.21}$$

$$H(R_Y \mid R_X') = \log_2 \frac{|R|}{|R_X'|},\tag{2.22}$$

Observe that the ratio $|R|/|R_Y|$ in Eq. (2.19) represents the average number of elements of R_X that are possible alternatives under the condition that an element of R_Y has already been selected. This means that $H(R_X \mid R_Y)$ measures the average nonspecificity regarding possible choices from R_X for all possible choices from R_Y. Function $H(R_Y \mid R_X)$ defined by Eq. (2.20) clearly has a similar meaning, with the roles of sets R_X and R_Y exchanged. The generalized forms of conditional Hartley measures defined by Eqs. (2.21) and (2.22) obviously have the same meaning under the restricted sets of possible conditions R'_Y and R'_X, respectively.

The marginal, joint, and conditional Hartley measures are related in numerous ways. To describe these various relations generically, it is useful (and a common practice) to identify only the universal sets involved and not the actual subsets of possible alternatives. That is, the generic symbols $H(X), H(Y)$, $H(X \times Y)$, $H(X \mid Y)$, and $H(Y \mid X)$ are used instead of their specific counterparts $H(R_X)$, $H(R_Y)$, $H(R)$, $H(R_X \mid R_Y)$ and $H(R_Y \mid R_X)$, respectively. As is shown later in this book, the generic descriptions of the relations have the same form in every uncertainty theory, even though the related entities are specific to each theory and change from theory to theory.

The equations

$$H(X \mid Y) = H(X \times Y) - H(Y),\tag{2.23}$$

$$H(Y \mid X) = H(X \times Y) - H(X),\tag{2.24}$$

which follow immediately from Eqs. (2.19) and (2.20), express in generic form the relationship between marginal, joint, and conditional Hartley measures. As is demonstrated later in this book, these important equations hold in every uncertainty theory when the Hartley measure is replaced with its counterpart in the other theory.

If possible alternatives from X do not depend on selections from Y, and vice versa, then $R = X \times Y$ and the sets R_X and R_Y are called *noninteractive.* Then, clearly,

$$H(X \mid Y) = H(X),\tag{2.25}$$

$$H(Y \mid X) = H(Y),\tag{2.26}$$

$$H(X \times Y) = H(X) + H(Y).\tag{2.27}$$

In the general case, when sets R_X and R_Y are not necessarily *interactive*, these equations become the inequalities

$$H(X \mid Y) \leq H(X), \tag{2.28}$$

$$H(Y \mid X) \leq H(Y), \tag{2.29}$$

$$H(X \times Y) \leq H(X) + H(Y). \tag{2.30}$$

The following functional, which is usually referred to as *information transmission*, is a useful indicator of the strength of constraint between possible alternatives in sets X and Y:

$$T_H(X,Y) = H(X) + H(Y) - H(X \times Y). \tag{2.31}$$

When the sets are noninteractive, $T_H(X,Y) = 0$; otherwise, $T_H(X,Y) > 0$. Using Eqs. (2.23) and (2.24), $T_H(X,Y)$ can be also expressed in terms of the conditional uncertainties:

$$T_H(X,Y) = H(X) - H(X \mid Y), \tag{2.32}$$

$$T_H(X,Y) = H(Y) - H(Y \mid X). \tag{2.33}$$

The maximum value, $\hat{T}_H(X,Y)$, of information transmission associated with relations $R \subseteq X \times Y$ is obtained when

$$H(X \mid Y) = H(Y \mid X) = 0.$$

This means that

$$H(X \times Y) - H(Y) = 0,$$

$$H(X \times Y) - H(X) = 0,$$

and, hence,

$$H(X \times Y) = H(X) = H(Y).$$

This implies that $|R| = |R_X| = |R_Y|$. These equalities can be satisfied only for $|R| = 1, 2, \ldots, \min\{|X|,|Y|\}$ Clearly, the largest value of information transmission is obtained for

$$|R| = |R_X| = |R_Y| = \min\{|X|,|Y|\}.$$

Hence,

$$\hat{T}_H(X,Y) = \min\{\log_2|X|, \log_2|Y|\}. \tag{2.34}$$

The normalized information transmission, NT_H, is then defined by the formula

$$NT_H(X,Y) = T_H(X,Y)/\hat{T}_H(X,Y). \qquad (2.35)$$

2.2.4. Examples

The meaning of uncertainty measured by the Hartley functional depends on the meaning of the set E. For example, when E is a set of predicted states of a variable (from the set X of all states defined for the variable), $H(E)$ is a measure of *predictive uncertainty*; when E is a set of possible diseases of a patient determined from relevant medical evidence, $H(E)$ is a measure of *diagnostic uncertainty*; when E is a set of possible answers to an unsettled historical question, $H(E)$ is a measure of *retrodictive uncertainty*; when E is a set for possible policies, $H(E)$ is a measure of *prescriptive uncertainty*. The purpose of this section is to illustrate the utility of the Hartley measure on simple examples in some of these application contexts.

EXAMPLE 2.1. Consider a simple dynamic system with four states whose purpose is prediction. Let $S = \{s_1, s_2, s_3, s_4\}$ denote the set of states of the system, and let R denote the state-transition relation on S^2 (the set of possible transitions from present states to next states) that is defined in matrix form by the basic possibility function r_R in Figure 2.1a. Entries in the matrix $\mathbf{M}_R$, are values $r_R(s_i, s_j)$ for all pairs $\langle s_i, s_j \rangle \in S^2$. All possible transitions from present states to next states (for which $r_R(s_i, s_j) = 1$) are also illustrated by the directed arcs (edges) in the diagram in Figure 2.1b. It is assumed that transitions occur only at specified discrete times. The system is clearly nondeterministic, which means that its predictions inevitably involve some nonspecifity. For convenience, let

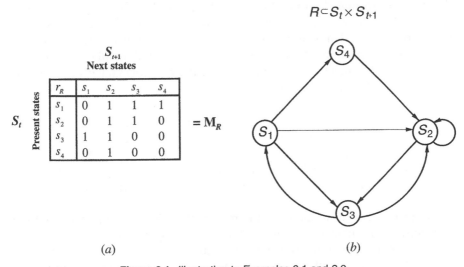

Figure 2.1. Illustration to Examples 2.1 and 2.2.

S_t denote the set of considered states of the system at some specified initial time, and let S_{t+k} for some $k \in \mathbb{N}$ denote the set of considered states of the system at time $t + k$. Clearly, $S_t = S_{t+k} = S$ for any $k \in \mathbb{N}$.

Since the purpose of the system is prediction, it makes sense to ask the system questions regarding possible future states or sequences of future states. For each question, the system provides us with a particular prediction that, in general, is not fully specific. The Hartley measure allows us to calculate the actual amount of nonspecificity in this prediction. The following are a few examples illustrating the use of Hartley measure for this purpose:

(a) Assuming that any of the four states is possible at time t, what is the average nonspecificity in predicting the state at time $t + 1$? Applying Eq. (2.23) for $X \times Y = S_{t+1} \times S_t$, the answer is

$$H(S_{t+1}|S_t) = H(S_t \times S_{t+1}) - H(S_t) = \log_2 8 - \log_2 4 = 1.$$

(b) Assuming that only states s_1 and s_2 are possible at time t, what is the average nonspecificity in predicting the state at time $t + 1$? Applying Eq. (2.23) for $X \times Y = \{\{s_1, s_2\} \times S_{t+1}\}$, the answer is

$$H(S_{t+1} \mid \{s_1, s_2\}) = H(\{s_1, s_2\} \times S_{t+1}) - H(\{s_1, s_2\}) = \log_2 5 - \log_2 2 = 1.32.$$

(c) If any state is possible at time t, what is the average nonspecificity in predicting the sequence of states of length n? For any $n \geq 1$, the answer is given by the formula

$$H(S_{t+1} \times S_{t+2} \times \cdots \times S_{t+n} \mid S_t) = H(S_t \times S_{t+1} \times \cdots \times S_{t+n}) - H(S_t). \quad (2.36)$$

For $n = 1$ this formula becomes the one in Example 2.1a. To apply this formula for any $n \geq 2$, we need to determine the number of possible sequences of the respective lengths. This can be done easily by using the matrix representation of the state-transition relation. For $n = 2$, the total number of possible sequences is obtained by adding all entries in the resulting matrix of the matrix product $\mathbf{M}_R \times \mathbf{M}_R$. In our example,

$$
\begin{bmatrix}
0 & 1 & 1 & 1 \\
0 & 1 & 1 & 0 \\
1 & 1 & 0 & 0 \\
0 & 1 & 0 & 0
\end{bmatrix}
\times
\begin{bmatrix}
0 & 1 & 1 & 1 \\
0 & 1 & 1 & 0 \\
1 & 1 & 0 & 0 \\
0 & 1 & 0 & 0
\end{bmatrix}
=
\begin{bmatrix}
1 & 3 & 1 & 0 \\
1 & 2 & 1 & 0 \\
0 & 2 & 2 & 1 \\
0 & 1 & 1 & 0
\end{bmatrix}.
$$
$$\quad \mathbf{M}_R \qquad\qquad\qquad \mathbf{M}_R \qquad\qquad\quad \mathbf{M}_R \times \mathbf{M}_R$$

By adding all entries in the resulting matrix $\mathbf{M}_R \times \mathbf{M}_R$, we obtain 16, and this is exactly the number of possible sequences of length 2. Moreover, the sums of entries in the individual rows of the resulting matrix are equal to the number of possible sequences of length 2 that begin in states assigned to the respective rows. That is, there are 5, 4, 5, 2 possible sequences of length 2 that begin in states s_1, s_2, s_3, s_4 respectively. Similarly, the sums of the entries in the individual columns of the matrix are equal to the number of possible sequences of length 2 that terminate in states assigned to the respective columns. The same results apply to sequences of lengths 3, 4, and so on, but we need to perform, respectively,

$$(\mathbf{M}_R \times \mathbf{M}_R) \times \mathbf{M}_R, ((\mathbf{M}_R \times \mathbf{M}_R) \times \mathbf{M}_R) \times \mathbf{M}_R,$$

and so on.

Determining the number of possible sequences for $n \in \mathbb{N}_{10}$ and calculating the average predictive nonspecificity for each n in this range by Eq. (2.36), we obtain the following sequence of predictive nonspecificities: 1, 2, 2.95, 4.11, 5.16, 6.21, 7.27, 8.32, 9.37, 10.43. As expected, the predictive nonspecificities increase with n. This means qualitatively that long-term predictions by a nondeterministic system are less specific than short-term predictions by the same system.

Assume now that only one state, s_i, is possible at time t and we want to calculate again the nonspecifity in predicting the sequence of states of length n. In this case,

$$H(S_{t+1} \times S_{t+2} \times \cdots \times S_{t+n} \mid \{s_i\}) = H(\{s_i\} \times S_{t+1} \times S_{t+2} \times \cdots \times S_{t+n}) - H(\{s_i\}).$$

As already mentioned, the number of sequences of states of length n that begin with state s_i, which we need for this calculation, is obtained by adding the entries in the respective row of the matrix resulting from the required chain of $n - 1$ matrix products. For $s_t = s_1$ in our example and $n \in \mathbb{N}_{10}$, we obtain the following predictive nonspecificities: 1.58, 2.32, 3.46, 4.46, 5.55, 6.58, 7.65, 8.70, 9.75, 10.81. As expected from the high initial nonspecificity $H(S_{t+1} \mid \{s_1\})$, all these values are above average. On the other hand, the following values for $s_t = s_4$ are all below average: 0, 1, 2, 3.17, 4.17, 5.25, 6.29, 7.35, 8.40, 9.45.

EXAMPLE 2.2. Consider the same system and the same types of predictions as in Example 2.1. However, let the focus in this example be on the predictive informativeness of the system rather than its predictive nonspecificity. That is, the aim of this example is to calculate the amount of information contained in each prediction of a certain type made by the system. In each case, we need to calculate the maximum amount of predictive nonspecificity, obtained in the face of total ignorance, and the actual amount of predictive nonspecificity associated with the prediction made by the system. The amount of information provided by the system is then defined as the difference between the maximum and actual amounts of predictive nonspecificity.

In general, the distinguishing feature of total ignorance within the classical possibility theory is that all recognized alternatives are possible. In our example, the recognized alternatives are transitions from states to states, each of which is represented by one cell in the matrix in Figure 2.1a. Predictions of states or sequences of states are determined via these transitions. Maximum nonspecificity in each prediction is obtained when all the recognized transitions are possible. When, on the other hand, only one transition from each state is possible, each prediction is fully specific and, hence, the system is deterministic.

The following are examples that illustrate the use of the Hartley measure for calculating informativeness of those types of predictions that are examined in Example 2.1:

(a) Let $H(S_{t+1} \mid S_t)$ have the same meaning as in Example 2.1, and let $\hat{H}(S_{t+1} \mid S_t)$ be the average nonspecificity in predicting the state at time $t + 1$ in the face of total ignorance. Then, the average amount of information, $I_H(S_{t+1} \mid S_t)$, contained in the prediction made by the system (or the informativeness of the system with the respect to predicting the next state) is given by the formula

$$I_H(S_{t+1} \mid S_t) = \hat{H}(S_{t+1} \mid S_t) - H(S_{t+1} \mid S_t).$$

Since there are 16 possible transitions in the face of total ignorance,

$$\hat{H}(S_{t+1} \mid S_t) = \log_2 16 - \log_2 4 = 2.$$

From Example 2.1a, $H(S_{t+1} \mid S_t) = 1$. Hence, $I_H(S_{t+1} \mid S_t) = 2 - 1 = 1$.

(b) In this case,

$$\hat{H}(S_{t+1} \mid \{s_1, s_2\}) = \hat{H}(\{s_1, s_2\} \times S_{t+1}) - H(\{s_1, s_2\}) = \log_2 8 - \log_2 2 = 2.$$

Then, using the result in Example 2.1b, we have

$$I_H(S_{t+1} \mid \{s_1, s_2\}) = \hat{H}(S_{t+1} \mid \{s_1, s_2\}) - H(S_{t+1} \mid \{s_1, s_2\}) = 2 - 1.32 = 0.68.$$

(c) If any state is possible at time t, the number of possible sequences of states of length n in the face of total ignorance is clearly equal to 4^{n+1}. Hence,

$$\hat{H}(S_{t+1} \times S_{t+2} \times \cdots \times S_{t+n} \mid S_t) = \hat{H}(S_t \times S_{t+1} \times \cdots \times S_{t+n}) - H(S_n)$$
$$= \log_2 4^{n+1} - \log_2 4 = 2n,$$

and consequently,

$$I_H(S_{t+1} \times S_{t+2} \times \cdots \times S_{t+n} \mid S_t) = 2n - H(S_{t+1} \times S_{t+2} \times \cdots \times S_{t+n} \mid S_t).$$

Using the values of $H(S_{t+1} \times S_{t+2} \times \cdots \times S_{t+n} | S_t)$ calculated for $n \in \mathbb{N}_{10}$ in Example 2.1c, we readily obtain the corresponding values of $I_H(S_{t+1} \times S_{t+2} \times \cdots \times S_{t+n} | S_t)$: 1, 2, 3.05, 3.89, 4.84, 5.79, 6.73, 7.68, 8.63, 9.57.

When only one state, s_i, is possible at time t, the number of possible sequences of states of length n in the face of total ignorance is equal to 4^n, which means that

$$\hat{H}(S_{t+1} \times S_{t+2} \times \cdots \times S_n | \{s_i\}) = \hat{H}(\{s_i\} \times S_{t+1} \times S_{t+2} \times \cdots \times S_{t+n}) - H(\{s_i\})$$
$$= \hat{H}(\{s_i\} \times S_{t+1} \times \cdots \times S_{t+n})$$
$$= \log 4^n$$
$$= 2n.$$

Hence,

$$I_H(S_{t+1} \times S_{t+2} \times \cdots \times S_n | \{s_i\}) = 2n - H(S_{t+1} \times S_{t+2} \times \cdots \times S_n | \{s_i\}).$$

Using the values of $H(S_{t+1} \times S_{t+2} \times \cdots \times S_n | \{s_i\})$ calculated for $n \in \mathbb{N}_{10}$ in Example 2.1c, we obtain the corresponding values of $I_H(S_{t+1} \times S_{t+2} \times \ldots \times S_n | \{s_i\})$: 0.42, 1.68, 2.54, 4.45, 5.42, 6.35, 7.30, 8.25, 9.19.

EXAMPLE 2.3. Consider a system with four variables, x_1, x_2, x_3, x_4, which take their values from the set $\{0, 1\}$. These variables are constrained via a particular 4-dimensional relation $R \subseteq \{0, 1\}^4$, but this relation is not known. We only know how the following pairs of the four variables are related: $\langle x_1, x_2 \rangle$, $\langle x_1, x_4 \rangle$, $\langle x_2, x_3 \rangle$, $\langle x_3, x_4 \rangle$. Let R_{12}, R_{14}, R_{23}, R_{34} denote, respectively, these partial relations on $\{0, 1\}^2$, and let $P = \{R_{12}, R_{14}, R_{23}, R_{34}\}$. The partial relations are defined in Figure 2.2a. All of the introduced relations can also be represented by their basic possibility functions. Let $r, r_{12}, r_{14}, r_{23}, r_{34}$ denote these functions.

If relation R were known, the four partial relations (or any of the other partial relations) would be uniquely determined as specific projections of R via the max operation of possibility theory. For example, using the labels introduced for all overall states (elements of the Cartesian product $\{0, 1\}^4$) in Figure 2.2c, we have

$$r_{12}(0,0) = \max\{r(s_0), r(s_1), r(s_2), r(s_3)\},$$
$$r_{12}(0,1) = \max\{r(s_4), r(s_5), r(s_6), r(s_7)\},$$
$$r_{12}(1,0) = \max\{r(s_8), r(s_9), r(s_{10}), r(s_{11})\},$$
$$r_{12}(1,1) = \max\{r(s_{12}), r(s_{13}), r(s_{14}), r(s_{15})\}.$$

In our case, R is not known and we want to determine it on the basis of information in the partial relations (projections of R). This inverse problem, illustrated in Figure 2.2b, is usually referred to as *system identification*. In general,

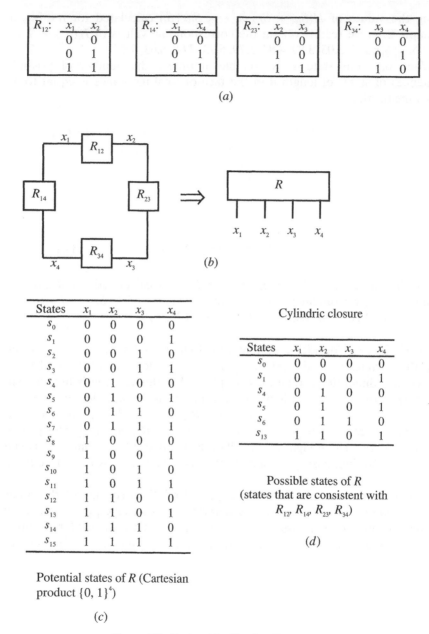

Figure 2.2. System identification (Example 2.3).

R cannot be determined uniquely from its projections. We can only determine a family, $\mathcal{R}_P$, of all relations that are consistent with the given projections in set P. Clearly, $\mathcal{R}_P \subseteq \mathcal{P}(\{0, 1\}^4)$. It is convenient to determine $\mathcal{R}_P$ in two steps. First, we determine the set of all overall states (elements of $\{0, 1\}^4$ in our case) that are *possible* under the given information. These are states that are *consistent* with the given projections. In our case, a particular overall state $\langle \dot{x}_1, \dot{x}_2, \dot{x}_3, \dot{x}_4 \rangle$ is possible if and only if

$$\langle \dot{x}_1, \dot{x}_2 \rangle \in R_{12} \text{ and } \langle \dot{x}_1, \dot{x}_4 \rangle \in R_{14} \text{ and } \langle \dot{x}_2, \dot{x}_3 \rangle \in R_{23} \text{ and } \langle \dot{x}_3, \dot{x}_4 \rangle \in R_{34}.$$

The possibility of each overall state $\langle \dot{x}_1, \dot{x}_2, \dot{x}_3, \dot{x}_4 \rangle \in \{0, 1\}^4$ thus can be determined by the equation

$$r\langle \dot{x}_1, \dot{x}_2, \dot{x}_3, \dot{x}_4 \rangle = \min\{r_{12}(\dot{x}_1, \dot{x}_2), r_{14}(\dot{x}_1, \dot{x}_4), r_{23}(\dot{x}_2, \dot{x}_3), r_{34}(\dot{x}_3, \dot{x}_4)\}.$$

The resulting set of all possible overall states, which is usually called a *cylindric closure* of the given projections, is shown in Figure 2.2d. The term "cylindric closure" emerged from a classical method for determining the set of all possible overall states from given projections. In this method (less efficient than the one described here), the cylindric extension is constructed for each projection with respect to the remaining dimensions and the intersection of all these cylindric extensions is the cylindric closure. The unknown relation R is guaranteed to be a subset of the cylindric closure.

Once the set of all possible overall states (the cylindric closure) is determined, the next step is to determine all its subsets that are *complete* in the sense that they cover all possible states of the given projections. In our example, there are eight such subsets, one of which is the cylindric closure itself:

$$\{s_0, s_1, s_4, s_5, s_6, s_{13}\}$$
$$\{s_0, s_1, s_4, s_6, s_{13}\}$$
$$\{s_0, s_1, s_5, s_6, s_{13}\}$$
$$\{s_0, s_4, s_5, s_6, s_{13}\}$$
$$\{s_1, s_4, s_5, s_6, s_{13}\}$$
$$\{s_0, s_1, s_6, s_{13}\}$$
$$\{s_0, s_5, s_6, s_{13}\}$$
$$\{s_1, s_4, s_6, s_{13}\}$$

Each of these subsets of the Cartesian product $\{0, 1\}^4$ can be the unknown relation R, but we have no basis to decide which one it is. We therefore identified a family, $\mathcal{R}_P$, of all possible overall relations. Each of these relations is both consistent and complete with respect to the given projections in P. The *identification nonspecificity* is given by the Hartley measure

$$H(\mathcal{R}_P) = \log_2|\mathcal{R}_P|$$
$$= \log_2 8 = 3.$$

The identification nonspecificity is of course uniquely determined by the given set P of projections of R. Its maximum is obtained when $P = \varnothing$, which expresses our total ignorance about R. Clearly,

$$H(\mathcal{R}_\varnothing) = \log_2 \left| \mathcal{P}\big(\{0,1\}^4\big) \right|$$
$$= \log_2 2^{16} = 16.$$

The amount of information, $I(P)$, about R contained in the given set P of projections of R is then calculated by the formula

$$I(P) = H(\mathcal{R}_\varnothing) - H(\mathcal{R}_P)$$
$$= 16 - 3 = 13.$$

The dependence of $H(\mathcal{R}_P)$ and $I(P)$ on P is examined in the next example.

EXAMPLE 2.4. Consider a system with three variables, x_1, x_2, x_3. The variables, whose values are in the set $\{0, 1\}$, are constrained by a ternary relation $R \subseteq \{0, 1\}^3$ that is not known. Labels of all potential states of the system, that is, elements of $\{0, 1\}^3$, are introduced in Table 2.1a. Assume that the three binary relations specified Table 2.1b are projections of R. Assume further that we know either all of these projections or only two of them (see Table 2.1c, the first column). Our aim is to identify the unknown relation R from information in each of these four sets of projections, P_1, P_2, P_3, P_4. For each P_i ($i \in \mathbb{N}_4$), we determine first the cylindric closure and all its complete subsets, $\mathcal{R}_{P_i}$, in the same way as in Example 2.3 (the second column in the Table 2.1c). Then, we can calculate for each P_i the identification nonspecificity, $H(\mathcal{R}_{P_i})$, and the information content, $I(\mathcal{R}_{P_i})$, of the projections in P_i (columns 3 and 4 in the Table 2.1c). This example is quite illustrative. It shows that the choice of projections is important. When we know all the projections (P_1), the identification is fully specific and the information content is 8 bits (maximum identification nonspecificity is $\log_2 2^8 = 8$ and the information contained in the three projections reduces it to 0). When we know only projections R_{13} and R_{23} (P_4), the identification is still fully specific. Therefore $I(P_1) = I(P_4)$, which means that adding projection R_{12} to P_4 does not increase the information content. Each of the remaining pairs of projections, P_2 and P_3, identifies seven possible overall relations, so their identification nonspecificity is $\log_2 7 = 2.81$ and their information content is $8 - 2.81 = 5.19$.

EXAMPLE 2.5. The purpose of this example is to illustrate how the various properties of the Hartley measure, expressed by Eqs. (2.16)–(2.35), can be utilized for analyzing n-dimensional relations ($n \geq 2$). A simple system with four variables, x_1, x_2, x_3, x_4, is employed here as an example. The variables take their values in sets X_1, X_2, X_3, X_4, respectively, where $X_1 = X_2 = \{0, 1\}$ and $X_3 = X_4 = \{0, 1, 2\}$. All possible overall states of the system are listed in Table 2.2a. This

Table 2.1. System Identification (Example 2.4)

States	x_1	x_2	x_3	R_{12}:	x_1	x_2	R_{23}:	x_2	x_3
s_0	0	0	0		0	0		0	0
s_1	0	0	1		0	1		0	1
s_2	0	1	0		1	0		1	1
s_3	0	1	1						
s_4	1	0	0						
s_5	1	0	1	R_{13}:	x_1	x_3			
s_6	1	1	0		0	0			
s_7	1	1	1		0	1			
					1	0			

(*a*)

(*b*)

Sets of Known Projections: P_i	Cylindric Closure (CC) and Its Complete Subsets: $\mathcal{R}_{Pi}$	$H(\mathcal{R}_{Pi})$	$I(P_i)$
$P_1 = \{R_{12}, R_{13}, R_{23}\}$	$\{s_0, s_1, s_3, s_4\}$ (CC)	0	8
$P_2 = \{R_{12}, R_{13}\}$	$\{s_0, s_1, s_2, s_3, s_4\}$ (CC) $\{s_0, s_1, s_2, s_4\}$ $\{s_0, s_1, s_3, s_4\}$ $\{s_0, s_2, s_3, s_4\}$ $\{s_1, s_2, s_3, s_4\}$ $\{s_0, s_3, s_4\}$ $\{s_1, s_2, s_4\}$	2.81	5.19
$P_3 = \{R_{12}, R_{23}\}$	$\{s_0, s_1, s_3, s_4, s_5\}$ (CC) $\{s_0, s_1, s_3, s_4\}$ $\{s_0, s_1, s_3, s_5\}$ $\{s_0, s_3, s_4, s_5\}$ $\{s_1, s_3, s_4, s_5\}$ $\{s_0, s_3, s_5\}$ $\{s_1, s_3, s_5\}$	2.81	5.19
$P_4 = \{R_{13}, R_{23}\}$	$\{s_0, s_1, s_3, s_4,\}$ (CC)	0	8

(*c*)

set of overall states is a 4-dimensional relation R on the Cartesian product $\{0, 1\}^2 \times \{0, 1, 2\}^2$. An important way of analyzing such a relation is to search for strong dependencies between various subsets of variables. The capability of measuring conditional uncertainty and information transmission for any two disjoint subsets of variables is essential for conducting such a search in a meaningful way.

Table 2.2. Information Analysis of a 4-Dimensional Relation (Example 2.5)

Relation R			
x_1	x_2	x_3	x_4
0	0	1	0
0	0	1	1
0	1	0	2
0	1	1	1
0	0	2	0
0	0	2	1
0	1	2	2
1	0	0	0
1	0	0	1
1	1	0	1
1	0	1	2
1	1	1	2
1	0	0	2
1	0	1	0
1	0	1	1
1	0	2	0
1	1	2	1
1	1	2	0
1	1	2	1
1	1	2	2

(a)

Conditional uncertainties and information transmissions

$H(X_1|X_2 \times X_3 \times X_4)$ = $4.39 - 3.91 = 0.48$
$H(X_2 \times X_3 \times X_4|X_1)$ = $4.39 - 1 = 3.39$
$T_H(X_1, X_2 \times X_3 \times X_4)$ = $1 + 3.91 - 4.39 = 0.52$
$H(X_2|X_1 \times X_3 \times X_4)$ = $4.39 - 3.91 = 0.48$
$H(X_1 \times X_3 \times X_4|X_2)$ = $4.39 - 1 = 3.39$
$T_H(X_2, X_1 \times X_3 \times X_4)$ = $1 + 3.91 - 4.39 = 0.52$
$H(X_3|X_1 \times X_2 \times X_4)$ = $4.39 - 3.32 = 1.07$
$H(X_1 \times X_2 \times X_4|X_3)$ = $4.39 - 1.58 = 2.81$
$T_H(X_3, X_1 \times X_2 \times X_4)$ = $1.58 + 3.32 - 4.39 = 0.51$
$H(X_4|X_1 \times X_2 \times X_3)$ = $4.39 - 3.46 = 0.93$
$H(X_1 \times X_2 \times X_3|X_4)$ = $4.39 - 1.58 = 2.81$
$T_H(X_4, X_1 \times X_2 \times X_3)$ = $1.58 + 3.46 - 4.39 = 0.65$
$H(X_1 \times X_2|X_3 \times X_4)$ = $4.39 - 3.17 = 1.22$
$H(X_3 \times X_4|X_1 \times X_2)$ = $4.39 - 2 = 2.39$
$T_H(X_1 \times X_2, X_3 \times X_4)$ = $3.17 + 2 - 4.39 = 0.78$
$H(X_1 \times X_3|X_2 \times X_4)$ = $4.39 - 2.85 = 1.54$
$H(X_2 \times X_4|X_1 \times X_3)$ = $4.39 - 2.85 = 1.54$
$T_H(X_1 \times X_3, X_2 \times X_4)$ = $2.85 + 2.85 - 4.39 = 1.31$
$H(X_1 \times X_4|X_2 \times X_3)$ = $4.39 - 2.85 = 1.54$
$H(X_2 \times X_3|X_1 \times X_4)$ = $4.39 - 2.85 = 1.54$
$T_H(X_1 \times X_4, X_2 \times X_3)$ = $2.85 + 2.85 - 4.39 = 1.31$

(b)

Normalized counterparts

$0.48/\log_2 2 = 0.48$
$3.39/\log_2 18 = 0.81$
$0.52/\log_2 2 = 0.52$
$0.48/\log_2 2 = 0.48$
$3.39/\log_2 18 = 0.81$
$0.52/\log_2 2 = 0.52$
$1.07/\log_2 3 = 0.68$
$2.81/\log_2 12 = 0.78$
$0.51/\log_2 3 = 0.32$
$0.93/\log_2 3 = 0.59$
$2.81/\log_2 12 = 0.78$
$0.65/\log_2 3 = 0.41$
$1.22/\log_2 4 = 0.61$
$2.39/\log_2 9 = 0.75$
$0.78/\log_2 6 = 0.39$
$1.54/\log_2 6 = 0.54$
$1.54/\log_2 6 = 0.54$
$1.31/\log_2 6 = 0.51$
$1.54/\log_2 6 = 0.54$
$1.54/\log_2 6 = 0.54$
$1.31/\log_2 6 = 0.51$

(c)

Suppose we want to calculate conditional uncertainties and information transmissions for all partitions of $\{x_1, x_2, x_3, x_4\}$ with two blocks, as listed in Table 2.2b. Due to Eqs. (2.23), (2.24), and (2.31), all these calculations are based on the following values of the Hartley measure, which are obtained directly from the given relation R:

$$H(X_1) = H(X_2) = \log_2 2 = 1,$$
$$H(X_3) = H(X_4) = \log_2 3 = 1.58,$$
$$H(X_1 \times X_2) = \log_2 4 = 2,$$
$$H(X_3 \times X_4) = \log_2 9 = 3.17,$$
$$H(X_1 \times X_3) = H(X_1 \times X_4) = H(X_2 \times X_3) = H(X_2 \times X_4)$$
$$= \log_2 6 = 2.85,$$
$$H(X_1 \times X_2 \times X_3) = \log_2 11 = 3.46,$$
$$H(X_1 \times X_2 \times X_4) = \log_2 10 = 3.32,$$
$$H(X_1 \times X_3 \times X_4) = H(X_2 \times X_3 \times X_4) = \log_2 15 = 3.91,$$
$$H(X_1 \times X_2 \times X_3 \times X_4) = \log_2 21 = 4.39.$$

These values are shown in the calculations in Table 2.2b. Also shown in Table 2.2c are calculations of their normalized counterparts.

The capability of calculating conditional uncertainties and information transmissions between groups of variables, illustrated here by a simple example, is a particularly important tool for analyzing high-dimensional relations. Normally, we need to do these calculations only for some groups of variables that are of interest in each individual application.

2.3. HARTLEY-LIKE MEASURE OF UNCERTAINTY FOR INFINITE SETS

The Hartley measure is applicable only to finite sets. Its counterpart for bounded and convex subsets of the n-dimensional Euclidean space $\mathbb{R}^n$ ($n \geq 1$, finite), which is called a *Hartley-like measure*, emerged only in the mid-1990s (see Note 2.3).

2.3.1. Definition

Let X denote a universal set of concern that is assumed to be a bounded and convex subset of $\mathbb{R}^n$ for some $n \geq 1$, and let HL denote the Hartley-like measure defined on convex subsets of X. Then, HL is a functional of the form

$$HL{:}C \to \mathbb{R}^+,$$

where C denotes the family of all convex subsets of X. Functionals in the class defined for all convex subsets A of X by the formula

$$HL(A) = \min_{t \in T} \left\{ c \log_b \left[\prod_{i=1}^{n} [1 + \mu(A_{i_t})] + \mu(A) - \prod_{i=1}^{n} \mu(A_{i_t}) \right] \right\}, \qquad (2.37)$$

were found to satisfy all axiomatic requirements, stated in Section 2.3.2, that are essential for measuring uncertainty in this case. Symbols in Eq. (2.37) have the following meaning:

- μ denotes the Lebesgue measure;
- T denotes the set of all isometric transformations from one orthogonal coordinate system to another;
- A_{i_t} denotes the ith projection of A in coordinate system t;
- b and c denote positive constants $(b \neq 1)$, whose choice defines a measurement unit.

Equation (2.37) allows us to define a measurement unit for the Hartley-like measure by choosing any positive values b and c except $b = 1$. Let the measurement unit be defined by the requirement that $HL(A) = 1$ when A is a closed interval of real numbers of length 1 in some assumed unit of length, which must be specified in each particular application. That is, we require that

$$c \log_b 2 = 1$$

for the specified unit of length. It is convenient to choose $b = 2$ and $c = 1$ to satisfy this equation. Then,

$$HL(A) = \min_{t \in T} \left\{ \log_2 \left[\prod_{i=1}^{n} [1 + \mu(A_{i_t})] + \mu(A) - \prod_{i=1}^{n} \mu(A_{i_t}) \right] \right\}. \qquad (2.38)$$

The chosen measurement unit is intuitively appealing, since $HL(A) = 1$ when A is a unit interval, $HL(A) = 2$ when A is a unit square, $HL(A) = 3$ when A is a unit cube, and so on.

2.3.2. Required Properties

Let $X = \overset{n}{\underset{i=1}{\times}} X_i$ denote a universal set of concern that is assumed to be a bounded and convex subset of $\mathbb{R}^n$ for some finite $n \geq 1$. Then the Hartley-like measure defined on the family of all convex subsets of X, C, by Eq. (2.38) is expected to satisfy the following axiomatic requirements:

Axiom (HL1) Range. For each $A \in C$, $HL(A) \in [0, \infty)$, where $HL(A) = 0$ if and only if $A = \{x\}$ for some $x \in X$.

Axiom (HL2) Monotonicity. For all $A, B \in C$, if $A \subseteq B$, then $HL(A) \leq HL(B)$.

Axiom (HL3) Subadditivity. For each $A \in C$, $HL(A) \leq \sum_{i=1}^{n} HL(A_i)$, where A_i denotes the 1-dimensional projection of A to dimension i in some coordinate system.

Axiom (HL4) Additivity. For all $A \in C$, such that $A = \underset{i=1}{\overset{n}{\times}} A_i$ where A_i has the same meaning as in Axiom (*HL3*),

$$HL(A) = \sum_{i=1}^{n} HL(A_i).$$

Axiom (HL5) Coordinate Invariance. Functional HL does not change under isometric transformations of the coordinate system.

Axiom (HL6) Continuity. HL is a continuous functional.

It is evident that HL defined by Eq. (2.38) is continuous, invariant with respect to isometric transformations of the coordinate system, and that it satisfies the required range. The monotonicity of the functional follows from the corresponding monotonicity of the Lebesgue measure, and its subadditivity is demonstrated as follows: for any $A \in C$,

$$
\begin{aligned}
HL(A) &= \min_{t \in T}\left\{\log_2\left[\prod_{i=1}^{n}[1 + \mu(A_{i_t})] + \mu(A) - \prod_{i=1}^{n}\mu(A_{i_t})\right]\right\} \\
&\leq \min_{t \in T}\left\{\log_2\left[\prod_{i=1}^{n}[1 + \mu(A_{i_t})]\right]\right\} \\
&= \min_{t \in T}\left\{\sum_{i=1}^{n}\log_2[1 + \mu(A_{i_t})]\right\} \\
&= \sum_{i=1}^{n} HL(A_i).
\end{aligned}
$$

It remains to show that the proposed functional is additive, to be fully justified as a general measure of nonspecificity of convex subsets of X in $\mathbb{R}^n$ for any finite $n \geq 1$.

To prove that the proposed functional is additive, we must prove that

$$HL(A) = \sum_{i=1}^{n} HL(A_i)$$

for any $A \in C$ such that $A = \bigtimes_{i=1}^{n} A_i$. It has already been shown that

$$HL(A) \le \sum_{i=1}^{n} HL(A_i)$$

for any $A \in C$. Hence, it remains to prove that

$$HL(A) \ge \sum_{i=1}^{n} HL(A_i)$$

when $A = \bigtimes_{i=1}^{n} A_i$. This, in turn, amounts to proving that for any rotation of the set A,

$$\prod_{i=1}^{n}[1+\mu(A_{i_t})] + \mu(A) - \prod_{i=1}^{n}\mu(A_{i_t}) \ge \prod_{i=1}^{n}[1+\mu(A_i)].$$

Since $\mu(A) = \prod_{i=1}^{n}\mu(A_i)$, this inequality can be written as

$$\prod_{i=1}^{n}[1+\mu(A_{i_t})] - \prod_{i=1}^{n}\mu(A_{i_t}) \ge \prod_{i=1}^{n}[1+\mu(A_i)] - \prod_{i=1}^{n}\mu(A_i). \qquad (2.39)$$

For $n = 1$, this inequality is trivially satisfied. For $n = 2, 3$, it is proved here by directly examining the effect of rotation of the set $A = \bigtimes_{i=1}^{n} A_i$ $(n = 2, 3)$ on projections. The proof for an arbitrary finite n is not presented here, since it is based on some special results from convexity theory (see Note 2.3).

Generally, in the n-dimensional space, $\mathbb{R}^n$, any rotation can be represented by the orthogonal matrix

$$\mathbf{I} = \begin{bmatrix} \cos\alpha_{11} & \cos\alpha_{12} & \cdots & \cos\alpha_{1n} \\ \cos\alpha_{21} & \cos\alpha_{22} & \cdots & \cos\alpha_{2n} \\ \vdots & \vdots & \ddots & \vdots \\ \cos\alpha_{n1} & \cos\alpha_{n2} & \cdots & \cos\alpha_{nn} \end{bmatrix},$$

where the parameters satisfy the following properties:

1. $\sum_{i=1}^{n}\cos^2\alpha_{ij} = 1, \quad \forall j \in \mathbb{N}_n,$ and $\sum_{j=1}^{n}\cos^2\alpha_{ij} = 1, \quad \forall i \in \mathbb{N}_n.$

2. $\sum_{k=1}^{n}\cos\alpha_{ik}\cos\alpha_{jk} = 0, \quad \forall i,j \in \mathbb{N}_n,$ and
   $\sum_{k=1}^{n}\cos\alpha_{ki}\cos\alpha_{kj} = 0, \quad \forall i,j \in \mathbb{N}_n.$

For each given rotation defined by matrix $\mathbf{I}$, an arbitrary point

$$\mathbf{x} = \langle x_1, x_2, \cdots, x_n \rangle^T$$

in $\mathbb{R}^n$ is transformed to the point

$$\mathbf{x}' = \langle x_1', x_2', \cdots, x_n' \rangle^T$$

by the matrix equation

$$\mathbf{x}' = \mathbf{I}\mathbf{x}.$$

That is,

$$x_1' = x_1 \cos \alpha_{11} + x_2 \cos \alpha_{12} + \cdots + x_n \cos \alpha_{1n}$$
$$x_2' = x_1 \cos \alpha_{21} + x_2 \cos \alpha_{22} + \cdots + x_n \cos \alpha_{2n}$$
$$\vdots$$
$$x_n' = x_1 \cos \alpha_{n1} + x_2 \cos \alpha_{n2} + \cdots + x_n \cos \alpha_{nn}.$$

Let us consider, without any loss of generality, that

$$A = \mathop{\Large\times}\limits_{i=1}^{n} [0, a_1]$$

for some $a_i \in \mathbb{R}$, $i \in \mathbb{N}_n$. Then, the ith projection of this set subjected to rotation defined by the matrix $\mathbf{I}$ is the set

$$A_i = \{ x_i' \mid \mathbf{x}' = \mathbf{I}\mathbf{x}, \forall \mathbf{x} \in A \}$$

for any $i \in \mathbb{N}_n$. The Lebesgue measure of the projection is

$$\mu(A_i) = \max_{\mathbf{x}, \mathbf{y} \in A} \{ |x_i' - y_i'| \}.$$

That is

$$\mu(A_i) = \max_{\mathbf{x}, \mathbf{y} \in A} \left\{ \sum_{j=1}^{n} |(x_j - y_j) \cos \alpha_{ij}| \right\}$$

for any $i \in \mathbb{N}_n$. Since this maximum must be reached by two vertices of the set A, the Lebesgue measure of the projection can be rewritten by

$$\mu(A_i) = \sum_{k=1}^{n} a_k |\cos \alpha_{ik}|$$

for any $i \in \mathbb{N}_n$.

Using the last formula, let us examine some aspects of the proposed function pertaining to its additivity. First let us present the following two basic properties:

(a) $\sum_{i=1}^{n} \mu(A_i) \geq \sum_{i=1}^{n} a_i$. This is because

$$
\begin{aligned}
\sum_{i=1}^{n} \mu(A_i) &= \sum_{i=1}^{n} \sum_{k=1}^{n} a_k |\cos \alpha_{ik}| \\
&= \sum_{k=1}^{n} \sum_{i=1}^{n} a_k |\cos \alpha_{ik}| \\
&\geq \sum_{k=1}^{n} \sum_{i=1}^{n} a_k \cos^2 \alpha_{ik} \\
&= \sum_{k=1}^{n} a_k \sum_{i=1}^{n} \cos^2 \alpha_{ik} \\
&= \sum_{k=1}^{n} a_k.
\end{aligned}
$$

(b) $\prod_{i=1}^{n} \mu(A_i) \geq \prod_{i=1}^{n} a_i$. This is because of the fact that the Lesbesgue measure of the set is less than or equal to the Lesbesgue measure of the Cartesian product of its 1-dimensional projections.

The Two-Dimensional Case   Let set A be a rectangle in the standard coordinate system that is shown in Figure 2.3. Since $A = [0, a_1] \times [0, a_2]$ in this system, we have

$$
\prod_{i=1}^{2} [1 + \mu(A_{i_i})] + \mu(A) - \prod_{i=1}^{2} \mu(A_{i_i}) = (1 + a_1)(1 + a_2).
$$

Now we prove that this is the minimum for all rotations. In the 2-dimensional space, any rotation can be represented by the matrix

$$
\mathbf{I} = \begin{bmatrix} \cos \theta & -\sin \theta \\ \sin \theta & \cos \theta \end{bmatrix}.
$$

Figure 2.3 illustrates a rotated rectangle A and its projections. It is easy to show that

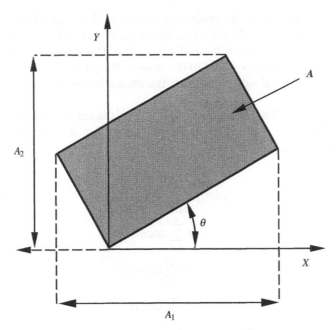

Figure 2.3. Rotation of set A in $\mathbb{R}^2$.

$$\mu(A_1) = a_1|\cos\theta| + a_2|\sin\theta|$$

$$\mu(A_2) = a_1|\sin\theta| + a_2|\cos\theta|.$$

Then under the new coordinate system,

$$
\begin{aligned}
\prod_{i=1}^{2}[1 + \mu(A_{i_t})] &+ \mu(A) - \prod_{i=1}^{2}\mu(A_{i_t}) \\
&= 1 + a_1|\cos\theta| + a_2|\sin\theta| + a_1|\sin\theta| + a_2|\cos\theta| \\
&\quad + (a_1|\cos\theta| + a_2|\sin\theta|)(a_1|\sin\theta| + a_2|\cos\theta|) + a_1a_2 \\
&\geq 1 + a_1\cos^2\theta + a_2\sin^2\theta + a_1\sin^2\theta + a_2\cos^2\theta \\
&\quad + (a_1\cos^2\theta + a_2\sin^2\theta)(a_1\sin^2\theta + a_2\cos^2\theta) + a_1a_2 \\
&\geq (1 + a_1)(1 + a_2).
\end{aligned}
$$

Therefore, in the 2-dimensional case the measure is additive.

The Three-Dimensional Case   To prove the additivity in the 3-dimensional space, we only need to prove that for any rotation of set $A \in C$,

$$\mu(A_1)\mu(A_2) + \mu(A_1)\mu(A_3) + \mu(A_2)\mu(A_3) \geq a_1a_2 + a_1a_3 + a_2a_2.$$

The Cartesian product of the projections includes the original set as a subset, and both of them are cubes. Hence, the area of the surface is a Cartesian product, which is twice the left-hand side of the preceding inequality. Hence, this area is greater than or equal to the area of the surface of the set A, which is twice the right-hand side. Therefore, the inequality holds.

n-Dimensional Case   In addition to $n = 2, 3$, it is easy to prove the additivity for any arbitrary n under the assumption that A is a set with equal edges. To show the proof, let $a_i = a$ for all $i \in \mathbb{N}_n$. Then, we have

$$\mu(A_i) = \sum_{k=1}^{n} a_k |\cos \alpha_{ik}|$$
$$= \sum_{k=1}^{n} a |\cos \alpha_{ik}|$$
$$= a \sum_{k=1}^{n} |\cos \alpha_{ik}|$$
$$\geq a \sum_{k=1}^{n} \cos^2 \alpha_{ik}$$
$$= a,$$

and consequently,

$$\prod_{i=1}^{n} [1 + \mu(A_{i_t})] + \mu(A) - \prod_{i=1}^{n} \mu(A_{i_t}) \geq \prod_{i=1}^{n} [1 + \mu(A_{i_t})] \geq (1 + a)^n.$$

Hence, the additivity holds. For a fully general proof of additivity of the Hartley-like measure, see Note 2.3.

2.3.3. Examples

The purpose of this section is to illustrate by specific examples some subtle issues involved in computing the various forms of the Hartley-like measure (basic, conditional, normalized) as well as information based on the Hartley-like measure. For the sake of simplicity, the examples are restricted to $n = 2$ and it is assumed that the universal set is the Cartesian product of $X \times Y$, where $X = Y = [0, 100]$ in some units of length, which are assumed to be the same in all examples. This means that $\mu(X) = \mu(Y) = 100$ and $\mu(X \times Y) = 100^2$. Due to the additivity of HL, we have

$$HL(X \times Y) = \log_2(101^2 + 100^2 - 100^2)$$
$$= \log_2 101^2 = 13.32.$$

EXAMPLE 2.6. Assume that, according to given evidence, the only possible alternatives (points in $X \times Y$) are located in a square, S, whose side is equal to 8. Due to the additivity of HL, we can readily calculate the nonspecificity of this evidence, the associated conditional nonspecificities, and the information transmission:

$$HL(S) = \log_2(9^2 + 8^2 - 8^2)$$
$$= \log_2 81 = 6.34,$$

$$HL(S_X) = HL(S_Y) = \log_2 9 = 3.17,$$

$$HL(S_X|S_Y) = H(S) - H(S_Y) = 3.17,$$

$$HL(S_Y|S_X) = H(S) - H(S_X) = 3.17,$$

$$T_{HL}(S_X, S_Y) = HL(S_X) + HL(S_Y) - H(S) = 0.$$

Next, we can calculate the amount of information contained in the given evidence:

$$I_{HL}(S) = HL(X \times Y) - HL(S) = 6.98.$$

It may also be desirable to calculate the normalized counterparts of the nonspecificity and information, which are independent of the chosen units:

$$NHL(S) = HL(S)/HL(X \times Y) = 0.48,$$

$$NI_{HL}(S) = I_{HL}(S)/HL(X \times Y) = 0.52.$$

The last result means that the given evidence contains 52% of the total amount of information needed to identify the true alternative. Clearly, $NHL(S) + NI_{HL}(S) = 1$.

EXAMPLE 2.7. In this example, the only possible alternatives are known to be located in a circle, C, with a radius $r = 4$. Clearly, $\mu(C) = \pi \cdot 4^2 = 50.27$. The two projections of C, C_X and C_Y, are in this case invariant with respect to rotations and, clearly, $\mu(C_X) = \mu(C_Y) = 8$. Hence, the calculations of the same quantities as in Example 2.6 is straightforward:

$$HL(C) = \log_2(9^2 + 50.27 - 8^2) = 6.07,$$

$$HL(C_X) = H(C_Y) = \log_2 9 = 3.17,$$

$$HL(C_X | C_Y) = HL(C) - HL(C_Y) = 2.9,$$

$$HL(C_Y | C_X) = HL(C) - HL(C_X) = 2.9,$$

$$T_{HL}(C_X, C_Y) = HL(C_X) + HL(C_Y) - HL(C) = 0.27,$$

$$I_{HL}(C) = HL(X \times Y) - HL(C) = 7.25,$$

$$NHL(C) = HL(C)/HL(X \times Y) = 0.46,$$

$$NI_{HL}(C) = I_{HL}(C)/HL(X \times Y) = 0.54.$$

EXAMPLE 2.8. According to given evidence, the only possible alternatives are located in an equilateral triangle, E, with sides of length a (in some appropriate units of lengths). Assume, without any loss of generality, that one vertex of the triangle is located at the origin of the coordinated system, as shown in Figure 2.4a. When the triangle is rotated around the origin, the projections of E change and, hence, the logarithmic function in Eq. (2.38) changes as well. To calculate $HL(E)$, we need to determine the minimum of this function.

As shown in Figure 2.4a, the position of the triangle can be expressed by the angle α. Due to the symmetry of E, it is sufficient to consider values $\alpha \in [0, 30°]$. Within this range, the dependence of the projections, E_X and E_Y, on α is expressed by the equations

$$E_X(\alpha) = a\cos\alpha,$$

$$E_Y(\alpha) = a\cos(30° - \alpha).$$

Moreover, $\mu(E) = a^2\sqrt{3}/4$ which is the area of the triangle. The logarithmic function in Eq. (2.38) is then a function of α, $f(\alpha)$, expressed by the formula

$$f(\alpha) = \log_2[1 + a\cos\alpha + a\cos(30° - \alpha) + a^2\sqrt{3}/4].$$

A natural way to determine extremes of $f(\alpha)$ is to solve the equation $f'(\alpha) = 0$ for α, where $f'(\alpha)$ denotes the derivative of f with respect to α. That is,

$$f'(\alpha) = \frac{-a\sin\alpha + a\sin(30° - \alpha)}{\ln 2(1 + \cos\alpha + a\cos(30° - \alpha) + a^2\sqrt{3}/4)} = 0.$$

This equation reduces to the simple equation

$$-\sin\alpha + \sin(30° - \alpha) = 0,$$

which is independent of a. The solution is $\alpha = 15°$. However, by plotting the function $f(\alpha)$ for some value of a (or by determining that its second derivative is negative), we can easily find that the function attains its maximum at $\alpha = 15°$. Due to the periodicity of the rotation with the cycle of 30°, maxima of $f(\alpha)$ are also obtained at $\alpha = (15 \pm 30k)°$, for any nonnegative integer k. Three cycles ($k = 0, 1, 2$) are shown are Figure 2.4b for $a = 4$. We can see that the minimum value of $f(\alpha)$ is attained at $\alpha = (0 \pm 30k)°$, which are values at which the function is not differentiable. When $a = 4$, the minimum value of $f(\alpha)$ is 3.944 (Figure 2.4b), which is also the value of $HL(E)$. For an arbitrary a,

$$HL(E) = f(0) = \log_2[1 + a + a\sqrt{3}/2 + a^2\sqrt{3}/4].$$

Moreover, we have

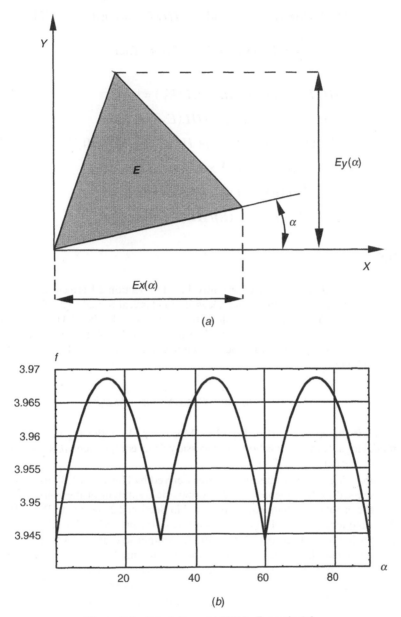

Figure 2.4. Calculation of $HL(E)$ in Example 2.8.

$$HL(E_X) = \log_2[1+a] \quad \text{and} \quad HL(E_Y) = \log_2[1+a\sqrt{3}/2].$$

For $a = 4$, $HL(E_X) = 2.322$ and $HL(E_Y) = 2.158$. Then,

$$HL(E_X \mid E_Y) = HL(E) - HL(E_Y) = 1.786,$$
$$HL(E_Y \mid E_X) = HL(E) - HL(E_X) = 1.622,$$
$$T_{HL}(E_X, E_Y) = HL(E_X) + HL(E_Y) - H(E) = 0.536,$$
$$I_{NL}(E) = HL(X \times Y) - HL(E) = 9.376,$$
$$NHL(E) = HL(E)/HL(X \times Y) = 0.296,$$
$$NI_{HL}(E) = I_{HL}(E)/HL(X \times Y) = 0.704.$$

NOTES

2.1. Possibility and necessity functions introduced in Section 2.1 are closely connected with operators of possibility and necessity in classical modal logic [Chellas, 1980; Hughes and Creswell, 1996]. To show this connection let $Pos_E(A)$ defined by Eq. (2.2) be interpreted for each $A \in \mathcal{P}(X)$ as the truth value of the proposition "Given evidence E, it is possible that the true alternative is in set A." Then, according to modal logic,

$$Pos_E(A \text{ or } B) \quad \text{iff} \quad Pos_E(A) \quad \text{or} \quad Pos_E(B),$$

which is the counterpart of Eq. (2.3). Moreover, the following is a tautology in modal logic: "Any proposition p is necessary iff its complement, $\bar{p}$, is not possible." Its counterpart is Eq. (2.4).

2.2. Possibilistic measure of uncertainty was derived by Hartley [1928]. Its significance is discussed by Kolmogorov [1965]. Axiomatic treatment of the Hartley measure and the proof of its uniqueness presented in Section 2.2 are due to Rényi [1970b]. The uniqueness of the Hartley measure was also proved under different axioms (somewhat less intuitive) and in more complicated ways by the Hungarian mathematician Erdös in 1946 and by the Russian mathematician Fadeev in 1957. References to these historical publications are given in [Rényi, 1970b], including a reference to his own original publication of the proof in 1959.

2.3. The Hartley-like measure, HL, was proposed in a paper by Klir and Yuan [1995b]. The proposed measure was proved in the paper to satisfy all essential require-ments with one exception: the additivity for sets with unequal edges when $n > 3$ remained an open problem. It was posed in *SIAM Review* [**38**(2), 1996, p. 315] in the following form:

Let $A = [0,a_1] \times \cdots \times [0,a_n]$ be a block of $\mathbf{R}^n$, and let A_t be the block obtained by rigid rotation t of A around the origin. For $i = 1, 2, \ldots, n$, let b_i denote the length of the projection of A_t on the ith coordinate axis. Show that

$$\prod_{i=1}^{n} a_i + \prod_{i=1}^{n}(1+b_i) \geq \prod_{i=1}^{n}(1+a_i) + \prod_{i=1}^{n} b_i.$$

The problem has come from an attempt to find a measure for nonspecificity of convex sets, which is a generalization of the Hartley measure for infinite sets. The requirement of additivity leads to the above mentioned inequality.

The problem was solved by Ramer and the solution was presented first in a very concise form in the *SIAM Review* [**39**(3), 1997, p. 516–51]. Its more elaborate version is covered in Ramer and Padet [2001]. A possibility of extending the applicability of the Hartley-like measure to nonconvex sets and some other related issues are also discussed in this paper.

EXERCISES

2.1. Repeat the calculations in Examples 2.1c and 2.2c under the assumption that the possible states at the time t are: (i) $\{s_2\}$; (ii) $\{s_3\}$; (iii) $\{s_1, s_4\}$, (iv) $\{s_2, s_3, s_4\}$.

2.2. Repeat Examples 2.1 and 2.2 for a system with three states, s_1, s_2, s_3, whose state-transition relation R is defined by the following matrix:

r_R	s_1	s_2	s_3	
s_1	1	1	0	
s_2	1	0	1	$= \mathbf{M}_R.$
s_3	0	1	0	

In addition, calculate the nonspecificity in predicting the state at time $t + k$ ($k = 2, 3, 4$).

2.3. Assume that the systems employed in the previous Exercises are used for retrodiction (determining past states or sequences of states) rather than prediction. Repeat, under this assumption, some of the calculations done in the previous Exercises.

2.4. Repeat Example 2.3 for some other sets of projections defined by you. Some projections may be 3-dimensional.

2.5. Consider a system with three variables, x_1, x_2, x_3. The variables, whose values are in the set $\{0, 1\}$, are constrained by a ternary relation R, which is not known. It is only known that each of the three binary projections is the full Cartesian product $\{0, 1\}^2$. Determine the cylindric closure of those projections and all its subsets that are consistent and complete with respect to the projections. Then, calculate the amount of nonspecificity in identifying the overall relation and the amount of information provided by the projections.

2.6. Consider a system with three variables, x_1, x_2, x_3, whose values are in the set $\{0, 1\}$. The variables are constrained by the following ternary relation:

x_1	x_2	x_3
0	0	1
0	1	0
1	0	0
1	1	1

Calculate conditional nonspecificities and information transmissions, as well as their normalized versions, for all pairs of variables and for all two-block partitions of $\{x_1, x_2, x_3\}$.

2.7. Assume that the following messages were received with some missing information. Determine the nonspecificity and informativeness of each of the messages.

(a) A number with 10 decimal digits was received, but k of the digits were not readable ($k \in \mathbb{N}_{10}$).

(b) A coded message with six letters of the English alphabet was received, but k of the letters were not readable ($k \in \mathbb{N}_6$).

2.8. Consider three variables that are related by the equation $v = v_1 + v_2$. Values of v_1 and v_2 are integers from 0 to 100, values of v are integers from 0 to 200. Given a particular value v, what is the nonspecificity in determining the corresponding values of v_1 and v_2, and what is the informativeness of v about v_1 and v_2?

2.9. Consider the equation $d = a + bc$, where a, b, c are input variables whose values are integers in the set $\{0, 1, \ldots, 9\}$, and d is an output variable with values in the set $\{0, 1, \ldots, 90\}$. Assume now that the values of the input variables are not fully specific. We know only that $a \in \{3, 4\}$, $b \in \{1, 2, 3\}$, and $c \in \{7, 8\}$. What are the input and output nonspecificities in this case? What are the amounts of information contained in the input and in the output?

2.10. Assume that 1000 attractive design alternatives are conceived by an engineering designer. After applying requirements r_1, r_2, r_3, r_4, r_5 (in that order), the number of alternatives is reduced to 200, 100, 64, 12, 1 (in the respective order). What are the prescriptive nonspecificities at the individual stages of the design process, and what is the amount of prescriptive information contained in each of the requirements?

2.11. To test a particular digital electronic chip with n inputs and m outputs for correctness means to determine the actual logic function the chip implements at each output solely by manipulating the input variables and observing the output variables. Initially, there are 2^{2^n} possible logic functions at each output and, hence, the diagnostic nonspecificity is 2^n bits. To resolve this nonspecificity and determine that the implemented function is the correct one, 2^n tests must be conducted. If n is large, this is not realistic. However, when less than 100% of the required tests have been carried out, some diagnostic nonspecificity remains (unless a defect in the chip was discovered by one of the tests). As an example, let $n = 30$ and $m = 10$, and assume that only 90% of the required tests have been carried out and they are all positive. Calculate the information obtained by the tests and the remaining diagnostic nonspecificity.

2.12. Consider the 2-dimensional Euclidean space $\mathbb{R}^2$, and let the domain of interest (the universal set) $X \times Y$ be the square $[0, 1000]^2$. This specification is expressed in some chosen units of length. Our aim is to determine the location of an object, which we know must be somewhere in the square $[0, 1000]^2$. From one information source, we know that the object cannot be outside the square area A shown in Figure 2.5. From another source, we know that it cannot be outside the circular area B also shown in the figure. Calculate, assuming that $a = 2$ (in the chosen units of length), the following:

(a) Basic and conditional nonspecificities of A, B, and $A \cap B$;

(b) Normalized versions of these nonspecificities;

(c) Information obtained by source 1, source 2, and both sources taken together and their normalized versions.

2.13. Assume that the chosen unit of length in Exercise 2.12 is a meter. Repeat the calculations by expressing the same length in centimeters.

2.14. Repeat Example 2.8 for the following areas in $\mathbb{R}^2$:

(a) A hexagon with sides equal to 1;

(b) An ellipse with semiaxes $a = 2$ and $b = 1$;

(c) A semicircle with radius $r = 5$.

2.15. Consider the 3-dimensional Euclidean space $\mathbb{R}^3$ within which the domain of interest, $X \times Y \times Z$, is the cube $[0, 100]^3$. For the following convex subsets of possible points in this domain, calculate the various basic and conditional amounts of nonspecificity, and the associated information, as well as values of relevant information transmissions:

(a) A unit cube;

(b) A sphere with radius $r = 2$;

(c) An ellipsoid with semiaxes $a = 4, b = 2, c = 1$;

(d) A regular tetrahedron with sides $s = 2$.

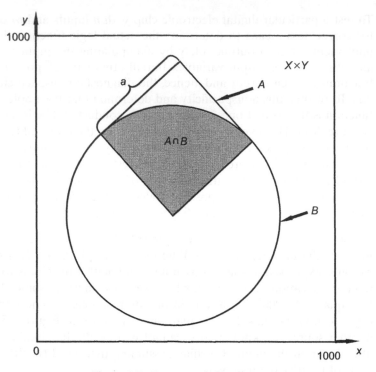

Figure 2.5. Illustration to Exercise 2.12.

2.16. In each of the following algebraic expressions, x and y are input variables whose values are in the interval $[0, 10]$, and z is an output variable whose range is determined by each of the expressions:

(a) $z = x + y$

(b) $z = xy/(x + y)$

(c) $z = x/(x + y) + y/(x + y)$

(d) $z = (x + y)^x$

Assuming that the values of x and y are known only imprecisely, $x \in [\underline{x}, \bar{x}]$ and $y \in [\underline{y}, \bar{y}]$, determine for each expression the input and output nonspecificity and the input informativeness about the output. Consider, for example, $x \in [1, 2]$ and $y \in [5, 7]$, or $x \in [0.1, 1]$ and $y \in [0, 1]$.

2.17. Consider the set A_l of all points on a straight-line segment in the 2-dimensional Euclidean space whose length is $l(l \geq 0)$. Calculate the nonspecificity of A_l.

3

CLASSICAL PROBABILITY-BASED UNCERTAINTY THEORY

Probability is degree of certainty and differs from absolute certainty as the part differs from the whole.

—Jacques Bernoulli

3.1. PROBABILITY FUNCTIONS

Probability-based uncertainty theory, which is the subject of this chapter, is one of the two classical theories of uncertainty. It is based on the notion of probability, which, in turn, is based on the notion of classical (additive) measure. The study of probability has a long history whose outcome is a theory that is now well developed at each of the four fundamental levels (formalization, calculus, measurement, and methodology). The literature on probability theory, including textbooks at various levels, is abundant. Since probability theory is also covered in most academic programs, it is reasonable to assume that the reader is familiar with its fundamentals. Therefore, only a few basic concepts from probability theory are briefly reviewed in this section. These are concepts that are needed for examining in greater depth the various issues regarding the measurement of probabilistic uncertainty (Sections 3.2 and 3.3).

Uncertainty and Information: Foundations of Generalized Information Theory, by George J. Klir
© 2006 by John Wiley & Sons, Inc.

3.1.1. Functions on Finite Sets

As in the case of possibilistic uncertainty, consider first a finite set X of mutually exclusive alternatives (predictions, diagnoses, etc.) that are of concern to us in a given application context. In general, alternatives in set X may be viewed as states of a variable X. Only one of the alternatives is true, but we are not certain which one it is. In probability theory, this uncertainty about the true alternative is expressed by a function

$$p : X \to [0, 1],$$

for which

$$\sum_{x \in X} p(x) = 1. \tag{3.1}$$

This function is called a *probability distribution function*, and the associated tuple of values $p(x)$ for all $x \in X$,

$$\mathbf{p} = \langle p(x) \mid x \in X \rangle,$$

is called a *probability distribution*. For each $x \in X$, the value $p(x)$ expresses the degree of evidential support that x is the true alternative. A variable X whose states $x \in X$ are associated with probabilities $p(x)$ is usually called a *random variable*.

Given a probability distribution function p, the associated probability measure, *Pro*, is obtained for all $A \in \mathcal{P}(X)$ via the formula

$$Pro(A) = \sum_{x \in A} p(x). \tag{3.2}$$

However, it is often not necessary to consider all sets in $\mathcal{P}(X)$. Any family of subsets of X, $C \subseteq \mathcal{P}(X)$, is acceptable provided that it contains X and it is closed under complementation and finite unions. Members of C are called *events*. For any pair of disjoint events A and B,

$$Pro(A \cup B) = Pro(A) + Pro(B). \tag{3.3}$$

This basic property of probability measures is referred to as *additivity*.

Given a probability distribution function p on X and any real-valued function f on X, the functional

$$a(f, p) = \sum_{x \in X} f(x) p(x) \tag{3.4}$$

is called an *expected value* of f. Clearly, $a(f, p)$ is a weighted average of values $f(x)$, in which the weights are probabilities $p(x)$.

Now, consider two sets of alternatives, X and Y, which may be viewed, in general, as sets of states of variables X and Y, respectively. A probability function defined on $X \times Y$ is called a *joint probability distribution function*. The associated *marginal probability distribution functions*, p_X and p_Y, on X and Y, respectively, are determined by the formulas

$$p_X(x) = \sum_{y \in Y} p(x, y), \tag{3.5}$$

for each $x \in X$, and

$$p_Y(y) = \sum_{x \in X} p(x, y), \tag{3.6}$$

for each $y \in Y$. Variables X, Y with marginal probability distribution functions p_X, p_Y, respectively, are called *noninteractive* iff

$$p(x, y) = p_X(x) \cdot p_Y(y) \tag{3.7}$$

for all $x \in X$ and $y \in Y$. *Conditional probability distribution functions*, $p(x|y)$ and $p(y|x)$, are defined for all $x \in X$ and $y \in Y$ such that $p_X(x) \neq 0$ and $p_Y(y) \neq 0$ by the formulas

$$p_{X|Y}(x \mid y) = \frac{p(x, y)}{p_Y(y)}, \tag{3.8}$$

$$p_{Y|X}(y \mid x) = \frac{p(x, y)}{p_X(x)}. \tag{3.9}$$

When $p_{X|Y}(x|y) = p_X(x)$ for all $x \in X$, variable X is said to be *independent* of variable Y. Similarly, when $p_{Y|X}(y|x) = p_Y(y)$ for all y, variable Y is said to be *independent* of variable X.

It is easy to show that the concepts of probabilistic noninteraction and probabilistic independence are equivalent. Given two variables, X and Y, with probability distributions, p_X and p_Y, defined on their states sets, X and Y, assume that they are noninteractive. This means that their joint probability distribution satisfies Eq. (3.7) for all $x \in X$ and $y \in Y$. Then, Eq. (3.8) becomes

$$p_{X|Y}(x \mid y) = \frac{p_X(x) \cdot p_Y(y)}{p_Y(y)}$$
$$= p_X(x)$$

and, similarly, Eq. (3.9) becomes

$$p_{Y|X}(y\,|\,x) = \frac{p_X(x)\cdot p_Y(y)}{p_X(x)}$$
$$= p_Y(y).$$

Hence, noninteraction implies independence.

Assume now that the variables are independent. This means that

$$p_{X|Y}(x\,|\,y) = \frac{p(x,y)}{p_Y(y)}$$
$$= p_X(x)$$

and similarly,

$$p_{X|Y}(x\,|\,y) = \frac{p(x,y)}{p_X(x)}$$
$$= p_Y(y).$$

In both cases, clearly, we obtain Eq. (3.7), which means that independence implies noninteraction. Hence, the two concepts, noninteraction and independence, are equivalent in probability theory. This equivalence does not hold in other theories of uncertainty.

3.1.2. Functions on Infinite Sets

When X is the set of real numbers, $\mathbb{R}$, or a bounded interval of real numbers, $[\underline{x}, \bar{x}]$, the set of alternatives is infinite, and the way in which probability measures are defined for finite X is not applicable. It is not any more meaningful to define probability measures on the full power set $\mathcal{P}(X)$. In each particular application, a relevant family of subsets of X (events), C, must be chosen, which is required to contain X and be closed under complements and countable unions (these requirements imply that C is also closed under countable intersections). Any such family together with the operations of set union, intersection, and complement is usually called a σ-algebra. In many applications, family C consists of all bounded, right-open subintervals of X.

Probability distribution function p cannot be defined for infinite sets in the same way in which it is defined for finite sets. For $X = \mathbb{R}$ or $X = [\underline{x}, \bar{x}]$, function p is defined for all $x \in X$ by the equation

$$p(x) = Pro(\{a \in X \mid a < x\}), \tag{3.10}$$

where Pro denotes, as before, a probability measure. This definition utilizes the ordering of real numbers. Function p is clearly nondecreasing, and it is usually expected to be continuous at each $x \in X$ and differentiable everywhere except at a countable number of points.

Connected with the probability distribution function p is another function, q, defined for all $x \in X$ as the derivative of p. This function is called a *probability density function*. Since p is a nondecreasing function, $q(x) \geq 0$ for all $x \in X$.

Given a σ-algebra defined on a family C, and a probability density function q on X, the probability of any set $A \in C$, $Pro(A)$, can be calculated via the integral

$$Pro(A) = \int_A q(x)\, dx. \tag{3.11}$$

Since it is required that $Pro(X) = 1$ (the true alternative must be in X), the probability density function is constrained by the equation

$$\int_X q(x)\, dx = 1, \tag{3.12}$$

which is the counterpart of Eq. (3.1) for the infinite case.

Given a probability distribution function on X and another real-valued function f on X, the functional

$$a(f, p) = \int_X f(x)\, dp \tag{3.13}$$

is called an *expected value* of f. Clearly, Eq. (3.13) is a counterpart of Eq. (3.4) for the infinite case. Observe that $a(f, p)$ can also be expressed in terms of the probability density function q associated with p as

$$a(f, p) = \int_X f(x)q(x)\, dx. \tag{3.14}$$

When function q is defined on a Cartesian product $X \times Y = [\underline{x}, \bar{x}] \times [\underline{y}, \bar{y}]$, it is called a *joint probability density function*. The associated *marginal probability density functions*, q_X and q_Y, on X and Y, respectively, are defined for each $x \in X$ and each $y \in Y$ by the formulas

$$q_X(x) = \int_Y q(x, y)\, dy, \tag{3.15}$$

$$q_Y(y) = \int_X q(x, y)\, dx. \tag{3.16}$$

Marginal probability density functions are called *noninteractive* iff

$$q(x, y) = q_X(x) \cdot q_Y(y) \tag{3.17}$$

for all $x \in X$ and each $y \in Y$. *Conditional probability density functions*, $q_{X|Y}(x|y)$ and $q_{Y|X}(y|x)$, are defined for all $x \in X$ and all $y \in Y$ such that $p_X(x) \neq 0$ and $p_Y(y) \neq 0$ by the formulas

$$q_{X|Y}(x \mid y) = \frac{q(x, y)}{q_Y(y)}, \tag{3.18}$$

$$q_{Y|X}(y \mid x) = \frac{q(x, y)}{q_X(x)}. \tag{3.19}$$

Clearly, Eqs. (3.15)–(3.19) are counterparts of Eqs. (3.5)–(3.9) for the infinite case. Again, the concepts of probabilistic noninteraction and probabilistic independence are equivalent in this case.

3.1.3. Bayes' Theorem

Consider a σ-algebra with events in a family $C \subseteq \mathcal{P}(X)$ and a probability measure Pro on C. For each pair of sets $A, B \in C$ such that $Pro(B) \neq 0$, the conditional probability of A given B, $Pro(A|B)$, is defined by the formula

$$Pro(A \mid B) = \frac{Pro(A \cap B)}{Pro(B)}. \tag{3.20}$$

Similarly, the conditional probability of B given A, $Pro(B \mid A)$ is defined by the formula

$$Pro(B \mid A) = \frac{Pro(A \cap B)}{Pro(A)}. \tag{3.21}$$

Expressing $Pro(A \cap B)$ from Eqs. (3.20) and (3.21) results in the equation

$$Pro(A \mid B) \cdot Pro(B) = Pro(B \mid A) \cdot Pro(A)$$

that establishes a relationship between the two conditional probabilities. One of the conditional probabilities is then expressed in terms of the other one by the equation

$$Pro(A \mid B) = \frac{Pro(B \mid A) \cdot Pro(A)}{Pro(B)}, \tag{3.22}$$

which is usually referred to as *Bayes' theorem*. Since $Pro(B)$ can be expressed in terms of given elementary, mutually exclusive events $A_i (i \in \mathbb{N}_m)$ of the σ-algebra as

$$Pro(B) = \sum_{i \in \mathbb{N}_m} Pro(A_i \cap B)$$

$$= \sum_{i \in \mathbb{N}_m} Pro(B \mid A_i) \cdot Pro(A_i)$$

when C is finite, Bayes' theorem may also be written in the form

$$Pro(A \mid B) = \frac{Pro(B \mid A) \cdot Pro(A)}{\displaystyle\sum_{i \in \mathbb{N}_m} Pro(B \mid A_i) \cdot Pro(A_i)}. \tag{3.23}$$

For infinite sets, Bayes' theorem must be properly reformulated in terms of probability density functions and the summation in Eq. (3.23) must be replaced with integration.

Bayes' theorem is a simple procedure for updating given probabilities on the basis of new evidence. From prior probabilities $Pro(A)$ and new evidence expressed in terms of conditional probabilities $Pro(B|A)$, we calculate posterior probabilities $Pro(A|B)$. When further evidence becomes available, the posterior probabilities are employed as prior probabilities and the procedure of probability updating is repeated.

EXAMPLE 3.1. Let X denote the population of a given town community. It is known from statistical data that 1% of the town residents have tuberculosis. Using this information, the probability, $Pro(A)$, that a randomly chosen member of the community has tuberculosis (event A) is 0.01. Suppose that this member takes a tuberculosis skin test (TST) and the outcome is positive. On the basis of the information, the prior probability changes to a posterior probability $Pro(A|B)$, where B denotes the event "positive outcome of the TST test." Clearly, the posterior probability depends on the reliability of the TST test. Assume that the following is known about the test: (1) the probability of a positive outcome for a person with tuberculosis, $Pro(B|A)$, is 0.99; and (2) the probability of a positive outcome for a person with no tuberculosis, $Pro(B|\bar{A})$, is 0.04. Using this information regarding the reliability of the TST test, the posterior probability $Pro(A|B)$ that the person has tuberculosis is calculated from the prior probability $Pro(A)$ via Eq. (3.23) as follows:

$$\begin{aligned} Pro(A \mid B) &= \frac{P(B \mid A)P(A)}{Pro(B \mid A)P(A) + Pro(B \mid \bar{A})P(\bar{A})} \\ &= \frac{0.99 \cdot 0.01}{0.99 \cdot 0.01 + 0.04 \cdot 0.99} = 0.2. \end{aligned}$$

The probability that the person has tuberculosis is thus 0.2. Observe that if the test were fully reliable ($Pro(B|A) = 1$ and $Pro(B|\bar{A}) = 0$), we would conclude (as expected) that the person has tuberculosis with probability 1.

3.2. SHANNON MEASURE OF UNCERTAINTY FOR FINITE SETS

The question of how to measure the amount of uncertainty (and the associated information) in classical probability theory was first addressed by

Shannon [1948]. He established that the only meaningful way to measure the amount of uncertainty in evidence expressed by a probability distribution function p on a finite set is to use a functional of the form

$$-c \sum p(x) \log_b p(x),$$

where b and c are positive constants, and $b \neq 1$. Each choice of values b and c determines the unit in which the uncertainty is measured. The most common choice is to define the measurement unit by the requirement that the amount of uncertainty be 1 when $X = \{x_1, x_2\}$ and $p(x_1) = p(x_2) = 1/2$. This requirement, which is usually referred to as a *normalization requirement*, is formally expressed by the equation

$$-c \log_b \frac{1}{2} = 1.$$

It can be conveniently satisfied by choosing $b = 2$ and $c = 1$. The resulting measurement unit is called a *bit*. That is, 1 bit is the amount of uncertainty removed (or information gained) upon learning the answer to a question whose two possible answers were equally likely. The resulting functional,

$$S(p) = -\sum_{x \in X} p(x) \log_2 p(x), \tag{3.24}$$

is called a *Shannon measure of uncertainty* or, more frequently, a *Shannon entropy*.

One way of getting insight into the type of uncertainty measured by the Shannon entropy is to rewrite Eq. (3.24) in the form

$$S(p) = -\sum_{x \in X} p(x) \log_2 \left[1 - \sum_{y \neq x} p(y) \right]. \tag{3.25}$$

The term

$$Con(x) = \sum_{y \neq x} p(\{y\})$$

in Eq. (3.25) represents the total evidential claim pertaining to alternatives that are different from x. That is, $Con(x)$ expresses the sum of all evidential claims that fully conflict with the one focusing on x. Clearly, $Con(x) \in [0, 1]$ for each $x \in X$. The function $-\log_2[1 - Con(x)]$, which is employed in Eq. (3.25), is monotonic increasing with $Con(x)$ and extends its range from $[0, 1]$ to $[0, \infty)$. The choice of the logarithmic function is a result of axiomatic requirements for S, which are discussed later in this chapter. It follows from these facts and from the form of Eq. (3.25) that the Shannon entropy is the mean (expected) value of the *conflict* among evidential claims expressed by each given probability distribution function p.

3.2.1. Simple Derivation of the Shannon Entropy

Suppose that a particular alternative in a finite set X of considered alternatives occurs with the probability $p(x)$. When this probability is very high, say $p(x) = 0.999$, then the occurrence of x is taken almost for granted and, consequently, we are not much surprised when it actually occurs. That is, our uncertainty in anticipating x is quite small and, therefore, our observation that x has actually occurred contains very little information. On the other hand, when the probability is very small, say $p(x) = 0.001$, then we are greatly surprised when x actually occurs. This means, in turn, that we are highly uncertain in our anticipation of x and, hence, the actual observation of x has very large information content. We can conclude from these considerations that the anticipatory uncertainty of x prior to the observation (and the information content of observing x) should be expressed by a decreasing function of the probability $p(x)$: the more likely the occurrence of x, the less information its actual observation contains.

Consider a random experiment with n considered outcomes, $i = 1, 2, \ldots, n$, whose probabilities are $p_1, p_2, \ldots, p_n$, respectively. Assume that $p_i > 0$ for all $i \in \mathbb{N}_n$, which means that no outcomes with zero probabilities are considered. The uncertainty in anticipating a particular outcome i (and the information obtained by actually observing this outcome) should clearly be a function of p_i. Let

$$s : (0, 1] \rightarrow [0, \infty)$$

denote this function. To measure in a meaningful way the anticipatory uncertainty, function s should satisfy the following properties:

(s1) $s(p_i)$ should decrease with increasing p_i.

(s2) $s(1) = 0$.

(s3) s should behave properly when applied to joint outcomes of independent experiments.

To elaborate on property (s3), let r_{ij} denote the joint probabilities of outcomes of two independent experiments. Assume that one of the experiments has n outcomes with probabilities $p_i (i \in \mathbb{N}_n)$ and the other one has m outcomes with probabilities $q_j (j \in \mathbb{N}_m)$. Then, according to the calculus of probability theory,

$$r_{ij} = p_i \cdot q_j \tag{3.26}$$

for all $i \in \mathbb{N}_n$ and all $j \in \mathbb{N}_m$. Since the experiments are independent, the anticipatory uncertainty of a particular joint outcome $\langle i, j \rangle$ should be equal to the sum of anticipatory uncertainties of the individual outcomes i and j. That is, the equation

$$s(r_{ij}) = s(p_i) + s(q_j)$$

should hold when Eq. (3.26) holds. This leads to the functional equation

$$s(p_i \cdot q_j) = s(p_i) + s(q_j),$$

where $p_i, q_j \in (0, 1]$. This is known as one form of the Cauchy equation whose solution is the class of functions defined for each $a \in (0, 1]$ by the equation

$$s(a) = c \log_b a,$$

where c is an arbitrary constant and b is a nonnegative constant distinct from 1. Since s is required by $(s1)$ to be a decreasing function on $(0, 1]$ and the logarithmic function is increasing, c must be negative. Furthermore, defining the measurement unit by the requirement that $s(1/2) = 1$ and choosing conveniently $b = 2$ and $c = -1$, we obtain a unique function s, defined for each $a \in (0, 1]$ by the equation

$$s(a) = -\log_2 a. \tag{3.27}$$

A graph of this function is shown in Figure 3.1.

Now consider a finite set X of considered alternatives with probabilities $p(x)$ for all $x \in X$. Let $S(p)$ denote the expected value of $s[p(x)]$ for all $x \in X$. Then,

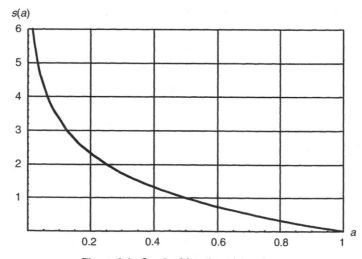

Figure 3.1. Graph of function $s(a) = -\log_2 a$.

$$S(p) = \sum_{x \in X} p(x) \cdot s[p(x)].$$

Substituting for s from Eq. (3.27), we obtain the functional

$$S(p) = -\sum_{x \in X} p(x) \cdot \log_2 p(x),$$

which is the *Shannon entropy* (compare with Eq. (3.24)).

Observe that the term $-p(x) \cdot \log_2 p(x)$ in the formula for $S(p)$ is not defined when $p(x) = 0$. However, employing l'Hospital's rule for indeterminate forms, we can calculate its limit for $p(x) \to 0$:

$$\lim_{p(x) \to 0} (-p(x) \log_2 p(x)) = \lim_{p(x) \to 0} \frac{-\log_2 p(x)}{\dfrac{1}{p(x)}} = \lim_{p(x) \to 0} \frac{\dfrac{-1}{p(x) \ln 2}}{\dfrac{-1}{p^2(x)}} = \lim_{p(x) \to 0} \frac{p(x)}{\ln 2} = 0.$$

When only two alternatives are considered, x_1 and x_2, whose probabilities are $p(x_1) = a$ and $p(x_2) = 1 - a$, the Shannon entropy, $S(p)$, depends only on a in the way illustrated in Figure 3.2a; graphs of the two components of $S(p)$, $s_1(a) = -a \log_2 a$ and $s_2(a) = -(1 - a) \log_2(1 - a)$, are shown in Figure 3.2b.

3.2.2. Uniqueness of the Shannon Entropy

The issue of measuring uncertainty and information in probability theory has also been treated axiomatically in various ways. It has been proved in numerous ways, from several well-justified axiomatic characterizations, that the Shannon entropy is the only meaningful functional for measuring uncertainty and information in probability theory. To survey this more rigorous treatment, assume that $X = \{x_1, x_2 \ldots x_n\}$, and let $p_i = p(x_i)$ for all $i \in \mathbb{N}_n$. In addition, let $\mathcal{P}_n$ denote the set of all probability distributions with n components. That is,

$$\mathcal{P}_n = \{\langle p_1, p_2, \ldots, p_n \rangle\} \mid p_i \in [0, 1] \text{ for all } i \in \mathbb{N}_n \text{ and } \sum_{i=1}^{n} p_i = 1\}.$$

Then, for each integer n, a measure of probabilistic uncertainty is a functional, S_n, of the form

$$S_n : \mathcal{P}_n \to [0, \infty),$$

which satisfies appropriate requirements. For the sake of simplicity, let

$$S_n(p_1, p_2, \ldots, p_n) \text{ be written as } S(p_1, p_2, \ldots, p_n).$$

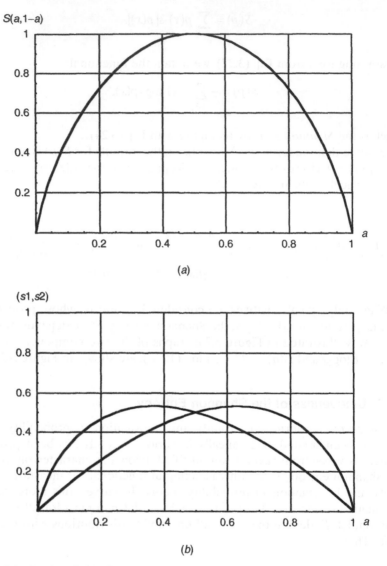

Figure 3.2. Graphs of: (a) $S(a, 1 - a)$; (b) components $s_1(a) = -a\log_2 a$ and $s_2(1 - a) = -(1 - a)\log_2(1 - a)$ of $S(a, 1 - a)$.

Different subsets of the following requirements, which are universally considered essential for a probabilistic measure of uncertainty and information, are usually taken as axioms for characterizing the measure.

Axiom (S1) Expansibility. When a component with zero probability is added to the probability distribution, the uncertainty should not change. Formally,

$$S(p_1, p_2, \ldots, p_n) = S(p_1, p_2, \ldots, p_n, 0)$$

for all $\langle p_1, p_2, \ldots, p_n \rangle \in \mathcal{P}_n$.

Axiom (S2) Symmetry. The uncertainty should be invariant with respect to permutations of probabilities of a given probability distribution. Formally,

$$S(p_1, p_2, \ldots, p_n) = S(\pi(p_1, p_2, \ldots, p_n))$$

for all $\langle p_1, p_2, \ldots, p_n \rangle \in \mathcal{P}_n$ and for all permutations $\pi(p_1, p_2, \ldots, p_n)$.

Axiom (S3) Continuity. Functional S should be continuous in all its arguments $p_1, p_2, \ldots, p_n$. This requirement is often replaced with a weaker requirement: $S(p, 1 - p)$ is a continuous functional of p in the interval $[0, 1]$.

Axiom (S4) Maximum. For each positive integer n, the maximum uncertainty should be obtained when all the probabilities are equal to $1/n$. Formally,

$$S(p_1, p_2, \ldots, p_n) \leq S\left(\frac{1}{n}, \frac{1}{n}, \ldots, \frac{1}{n}\right).$$

Axiom (S5) Subadditivity. The uncertainty of any joint probability distribution should not be greater than the sum of the uncertainties of the corresponding marginal distributions. Formally,

$$S(p_{11}, p_{12}, \ldots, p_{1m}, p_{21}, p_{22}, \ldots, p_{2m}, \ldots, p_{n1}, p_{n2}, \ldots, p_{nm})$$
$$\leq S\left(\sum_{j=1}^{m} p_{1j}, \sum_{j=1}^{m} p_{2j}, \ldots, \sum_{j=1}^{m} p_{nj}\right) + S\left(\sum_{i=1}^{n} p_{i1}, \sum_{i=1}^{n} p_{i2}, \ldots, \sum_{i=1}^{n} p_{im}\right)$$

for any given joint probability distribution $\langle p_{ij} | i \in \mathbb{N}_n, j \in \mathbb{N}_m \rangle$.

Axiom (S6) Additivity. The uncertainty of any joint probability distribution that is noninteractive should be equal to the sum of the uncertainties of the corresponding marginal distributions. Formally,

$$S(p_1 q_1, p_1 q_2, \ldots, p_1 q_m, p_2 q_1, p_2 q_2, \ldots, p_2 q_m, \ldots, p_n q_1, p_n q_2, \ldots, p_n q_m)$$
$$= S(p_1, p_2, \ldots, p_n) + S(q_1, q_2, \ldots, q_m)$$

for any given marginal probability distributions $\langle p_1, p_2, \ldots, p_n \rangle$ and $\langle q_1, q_2, \ldots, q_m \rangle$. This requirement is sometimes replaced with a restricted requirement of *weak additivity* which applies only to uniform marginal probability distributions with $p_i = 1/n$ and $q_j = 1/m$. Formally,

$$S\left(\frac{1}{nm},\frac{1}{nm},\ldots,\frac{1}{nm}\right)=S\left(\frac{1}{n},\frac{1}{n},\ldots,\frac{1}{n}\right)+S\left(\frac{1}{m},\frac{1}{m},\ldots,\frac{1}{m}\right).$$

Introducing a convenient function f such that $f(n)=S(1/n,1/n,\ldots,1/n)$, then the weak additivity can be expressed by the equation

$$f(nm)=f(n)+f(m)$$

for positive integers n and m.

Axiom (S7) Monotonicity. For probability distributions with equal probabilities $1/n$, the uncertainty should increase with increasing n. Formally, for any positive integers m and n, when $m<n$, then $f(m)<f(n)$, where f denotes the function introduced in (S6).

Axiom (S8) Branching. Given a probability distribution $\mathbf{p}=\langle p_i|i\in\mathbb{N}_n\rangle$ on set $X=\{x_i|i\in\mathbb{N}_n\}$ for some integer $n\geq 3$, let X be partitioned into two blocks, $A=\{x_1,x_2,\ldots,x_s\}$ and $B=\{x_{s+1},x_{s+2},\ldots,x_n\}$ for some integer s. Then, the equation

$$S(p_1,p_2,\ldots,p_n)=S(p_A,p_B)+p_A S\left(\frac{p_1}{p_A},\frac{p_2}{p_A},\ldots,\frac{p_s}{p_A}\right)$$
$$+p_B S\left(\frac{p_{s+1}}{p_B},\frac{p_{s+2}}{p_B},\ldots,\frac{p_n}{p_B}\right)$$

should hold, where $p_A=\sum_{i=1}^{s}p_i$ and $p_B=\sum_{i=s+1}^{s}p_i$. On the left-hand side of the equation, the uncertainty is calculated directly; on the right-hand side, it is calculated in two steps, following the calculus of probability theory. In the first step (expressed by the first term), the uncertainty associated with the probability distribution $\langle p_A,p_B\rangle$ on the partition is calculated. In the second step (expressed by the second and third terms), the expected value of uncertainty associated with the conditional probability distributions within the blocks A and B of the partition is calculated. This requirement, which is also called a *grouping requirement* or a *consistency requirement* is sometimes presented in various other forms. For example, one of its weaker forms is given by the formula

$$S(p_1,p_2,p_3)=S(p_1+p_2,p_3)+(p_1+p_2)S\left(\frac{p_1}{p_1+p_2},\frac{p_2}{p_1+p_2}\right).$$

It matters little which of these forms is adopted since they can be derived from one another. (The branching axiom is illustrated later in this chapter by Example 3.6 and Figure 3.6.)

Axiom (S9) Normalization. To ensure (if desirable) that the measurement units of S are bits, it is essential that

$$S\left(\frac{1}{2},\frac{1}{2}\right) = 1.$$

This axiom must be appropriately modified when other measurement units are preferred.

The listed axioms for a probabilistic measure of uncertainty and information are extensively discussed in the abundant literature on classical information theory. The following subsets of these axioms are the best known examples of axiomatic characterization of the probabilistic measure of uncertainty:

1. Continuity, weak additivity, monotonicity, branching, and normalization.
2. Expansibility, continuity, maximum, branching, and normalization.
3. Symmetry, continuity, branching, and normalization.
4. Expansibility, symmetry, continuity, subadditivity, additivity, and normalization.

Any of these collections of axioms (as well as some additional collections) is sufficient to characterize the Shannon entropy uniquely. That is, it has been proven that the Shannon entropy is the only functional that satisfies any of these sets of axioms. To illustrate in detail this important issue of uniqueness, which gives the Shannon entropy its great significance, the uniqueness proof is presented here for the first of the listed sets of axioms.

Theorem 3.1. The only functional that satisfies the axioms of continuity, weak additivity, monotonicity, branching, and normalization is the Shannon entropy.

Proof. (i) First, we prove the proposition $f(n^k) = kf(n)$ for all positive integers n and k by induction on k, where

$$f(n) = S\left(\frac{1}{n},\frac{1}{n},\ldots,\frac{1}{n}\right)$$

is the function that is used in the definition of weak additivity. For $k = 1$, the proposition is trivially true. By the axiom of *weak additivity*, we have

$$f(n^{k+1}) = f(n^k \cdot n) = f(n^k) + f(n).$$

Assume the proposition is true for some $k \in \mathbb{N}$. Then,

$$\begin{aligned} f(n^{k+1}) &= f(n^k) + f(n) \\ &= kf(n) + f(n) \\ &= (k+1)f(n), \end{aligned}$$

which demonstrates that the proposition is true for all $k \in \mathbb{N}$.

(ii) Next, we demonstrate that $f(n) = \log_2 n$. This proof is identical to that of Theorem 2.1, provided that we replace the Hartley measure H with f. Therefore, we do not repeat the derivation here. Observe that the proof requires *weak additivity*, *monotonicity*, and *normalization*.

(iii) We prove now that $S(p, 1-p) = -p\log_2 p - (1-p)\log_2(1-p)$ for rational p. Let $p = r/s$ where $r, s \in \mathbb{N}$. Then

$$\begin{aligned} f(s) &= S\bigg(\underbrace{\frac{1}{s}, \frac{1}{s}, \dots, \frac{1}{s}}_{r}, \underbrace{\frac{1}{s}, \frac{1}{s}, \dots, \frac{1}{s}}_{s-r}\bigg) \\ &= S\bigg(\frac{r}{s}, \frac{s-r}{s}\bigg) + \frac{r}{s}f(r) + \frac{s-r}{s}f(s-r) \end{aligned}$$

by the *branching axiom*. By (ii) and the definition of p we obtain

$$\log_2 s = S(p, 1-p) + p\log_2 r + (1-p)\log_2(s-r).$$

Solving this equation for $S(p, 1-p)$ results in

$$\begin{aligned} S(p, 1-p) &= \log_2 s - p\log_2 r - (1-p)\log_2(s-r) \\ &= p\log_2 s - p\log_2 s + \log_2 s - p\log_2 r - (1-p)\log_2(s-r) \\ &= p\log_2 s - p\log_2 r + (1-p)\log_2 s - (1-p)\log_2(s-r) \\ &= -p\log_2\bigg(\frac{r}{s}\bigg) - (1-p)\log_2\bigg(\frac{s-r}{s}\bigg) \\ &= -p\log_2 p - (1-p)\log_2(1-p). \end{aligned}$$

(iv) We now extend (iii) to the real numbers $p \in [0, 1]$ with the help of the *continuity axiom*. Let p be any number in the unit interval and let p' be a series of rational numbers that approach p as a limit. Then,

$$S(p, 1-p) = \lim_{p' \to p} S(p', 1-p')$$

by the continuity axiom. Moreover,

$$\begin{aligned} \lim_{p' \to p} S(p', 1-p') &= \lim_{p' \to p}[-p'\log_2 p' - (1-p')\log_2(1-p')] \\ &= -p\log_2 p - (1-p)\log_2(1-p), \end{aligned}$$

since all the functions involved are continuous.

(v) We now conclude the proof by showing that

$$S(p_1, p_2, \ldots, p_n) = -\sum_{i=1}^{n} p_i \log_2 p_i.$$

This is accomplished by induction on n. The result is proved in (ii) and (iv) for $n = 1, 2$, respectively. For $n \geq 3$, we may use the *branching axiom* to obtain

$$S(p_1, p_2, \ldots, p_n) = S(p_A, p_n) + p_A S\left(\frac{p_1}{p_A}, \frac{p_2}{p_A}, \ldots, \frac{p_{n-1}}{p_A}\right) + p_n S\left(\frac{p_n}{p_n}\right),$$

where $p_A = \sum_{i=1}^{n-1} p_i$. Since $S(p_n/p_n) = S(1) = 0$ by (ii), we obtain

$$S(p_1, p_2, \ldots, p_n) = S(p_A, p_n) + p_A S\left(\frac{p_1}{p_A}, \frac{p_2}{p_A}, \ldots, \frac{p_{n-1}}{p_A}\right).$$

By (iv) and assuming the proposition we want to prove to be true for $n - 1$, we may rewrite this equation as

$$
\begin{aligned}
S(p_1, p_2, \ldots, p_n) &= -p_A \log_2 p_A - p_n \log_2 p_n - p_A \sum_{i=1}^{n-1} \frac{p_i}{p_A} \log_2 \frac{p_i}{p_A} \\
&= -p_A \log_2 p_A - p_n \log_2 p_n - \sum_{i=1}^{n-1} p_i \log_2 \frac{p_i}{p_A} \\
&= -p_A \log_2 p_A - p_n \log_2 p_n - \sum_{i=1}^{n-1} p_i \log_2 p_i + \sum_{i=1}^{n-1} p_i \log_2 p_A \\
&= -p_A \log_2 p_A - p_n \log_2 p_n - \sum_{i=1}^{n-1} p_i \log_2 p_i + p_A \log_2 p_A \\
&= -\sum_{i=1}^{n} p_i \log_2 p_i. \qquad \blacksquare
\end{aligned}
$$

3.2.3. Basic Properties of the Shannon Entropy

The literature dealing with information theory based on the Shannon entropy is extensive. No attempt is made to give a comprehensive coverage of the theory in this book. However, the most fundamental properties of the Shannon entropy are surveyed.

First, a theorem is presented that plays an important role in classical information theory. This theorem is essential for proving some basic properties of Shannon entropy, as well as introducing some additional important concepts of classical information theory.

Theorem 3.2. The inequality

$$-\sum_{i=1}^{n} p_i \log_2 p_i \le -\sum_{i=1}^{n} p_i \log_2 q_i \qquad (3.28)$$

is satisfied for all probability distributions $\langle p_i \mid i \in \mathbb{N}_n \rangle$ and $\langle q_i \mid i \in \mathbb{N}_n \rangle$ and for all $n \in \mathbb{N}_n$; the equality in (3.28) holds if and only if $p_i = q_i$ for all $i \in \mathbb{N}_n$.

Proof. Consider the function

$$s(p_i, q_i) = p_i(\ln p_i - \ln q_i) - p_i + q_i$$

for $p_i, q_i \in [0, 1]$. This function is finite and differentiable for all values of p_i and q_i except the pair $p_i = 0$ and $q_i \ne 0$. For each fixed $q_i \ne 0$, the partial derivative of s with respect to p_i is

$$\frac{\partial s(p_i, q_i)}{\partial p_i} = \ln p_i - \ln q_i.$$

That is,

$$\frac{\partial s(p_i, q_i)}{\partial p_i} \begin{cases} < 0 & \text{for} \quad p_i < q_i \\ = 0 & \text{for} \quad p_i = q_i \\ > 0 & \text{for} \quad p_i > q_i \end{cases}$$

and, consequently, s is a convex function of p_i, with its minimum at $p_i = q_i$. Hence, for any given i, we have

$$p_i(\ln p_i - \ln q_i) - p_i + q_i \ge 0,$$

where the equality holds if and only if $p_i = q_i$. This inequality is also satisfied for $q_i = 0$, since the expression on its left-hand side is $+\infty$ if $p_i \ne 0$ and $q_i = 0$, and it is zero if $p_i = 0$ and $q_i = 0$. Taking the sum of this inequality for all $i \in \mathbb{N}_n$, we obtain

$$\sum_{i=1}^{n} [p_i \ln p_i - p_i \ln q_i - p_i + q_i] \ge 0,$$

which can be rewritten as

$$\sum_{i=1}^{n} p_i \ln p_i - \sum_{i=1}^{n} p_i \ln q_i - \sum_{i=1}^{n} p_i + \sum_{i=1}^{n} q_i \ge 0.$$

The last two terms on the left-hand side of this inequality cancel each other out, as they both sum up to one. Hence,

$$\sum_{i=1}^{n} p_i \ln p_i - \sum_{i=1}^{n} p_i \ln q_i \geq 0,$$

which is equivalent to Eq. (3.28) when multiplied through by $1/\ln 2$.   ∎

This theorem, sometimes referred to as *Gibbs' theorem*, is quite useful in studying properties of the Shannon entropy. For example, the theorem can be used as follows for proving that the maximum of the Shannon entropy for probability distributions with n elements is $\log_2 n$.

Let $q_i = 1/n$ for all $i \in \mathbb{N}_n$. Then, Eq. (3.28) yields

$$\begin{aligned} S(p_i \,|\, i \in \mathbb{N}_n) &= -\sum_{i=1}^{n} p_i \log_2 p_i \\ &\leq -\sum_{i=1}^{n} p_i \log_2 \frac{1}{n} \\ &= -\log_2 \frac{1}{n} \sum_{i=1}^{n} p_i \\ &= \log_2 n. \end{aligned}$$

Thus, $S(p_i \,|\, i \in \mathbb{N}_n) \leq \log_2 n$. The upper bound is obtained for $p_i = 1/n$ for all $i \in \mathbb{N}_n$.

Let us now examine Shannon entropies of joint, marginal, and conditional probability distributions defined on sets X and Y. In agreement with a common practice in the literature dealing with the Shannon entropy, we simplify the notation in the rest of this section by using $S(X)$ instead of $S(P_X(x) \,|\, x \in X)$ or $S(p_1, p_2, \ldots, p_n)$. Furthermore, assuming $x \in X$ and $y \in Y$ we use symbols $p_X(x)$ and $p_Y(y)$ to denote marginal probabilities on sets X and Y, respectively, the symbol $p(x, y)$ for joint probabilities on $X \times Y$, and the symbols $p(x|y)$ and $p(y|x)$ for the corresponding conditional probabilities. In this simplified notation for conditional probabilities, the meaning of each symbol is uniquely determined by the arguments shown in the parentheses.

Given two sets X and Y which may be viewed, in general, as state sets of random variables X and Y, respectively, we can recognize the following three types of Shannon entropies:

1. A *joint entropy* defined in terms of the joint probability distribution on $X \times Y$,

$$S(X \times Y) = -\sum_{\langle x, y \rangle \in X \times Y} p(x, y) \log_2 p(x, y) \qquad (3.29)$$

2. Two *simple entropies* based on marginal probability distributions:

$$S(X) = -\sum_{x \in X} p_X(x)\log_2 p_X(x), \tag{3.30}$$

$$S(Y) = -\sum_{y \in Y} p_Y(y)\log_2 p_Y(y). \tag{3.31}$$

3. Two *conditional entropies* defined in terms of weighted averages of local conditional probabilities:

$$S(X \mid Y) = -\sum_{y \in Y} p_Y(y) \sum_{x \in X} p(x \mid y)\log_2 p(x \mid y) \tag{3.32}$$

$$S(Y \mid X) = -\sum_{x \in X} p_X(x) \sum_{y \in Y} p(y \mid x)\log_2 p(y \mid x). \tag{3.33}$$

In addition to these three types of Shannon entropies, the functional

$$T_S(X,Y) = S(X) + S(Y) - S(X,Y) \tag{3.34}$$

is often used in the literature as a measure of the strength of the relationship (in the probabilistic sense) between elements of set X and Y. This functional is called an *information transmission*. It is analogous to the functional defined by Eq. (2.31) for the Hartley measure: it can be generalized to more than two sets in the same way.

It remains to examine the relationship among the various types of entropies and the information transmission. The key properties of this relationship are expressed by the next several theorems.

Theorem 3.3

$$S(X \mid Y) = S(X \times Y) - S(Y). \tag{3.35}$$

Proof

$$S(X \mid Y) = -\sum_{y \in Y} p_Y(y) \sum_{x \in X} p(x \mid y)\log_2 p(x \mid y)$$

$$= -\sum_{y \in Y} p_Y(y) \sum_{x \in X} \frac{p(x, y)}{p_Y(y)}\log_2 \frac{p(x, y)}{p_Y(y)}$$

$$= -\sum_{y \in Y} \sum_{x \in X} p(x, y)\log_2 \frac{p(x, y)}{p_Y(y)}$$

$$= -\sum_{y \in Y} \sum_{x \in X} p(x, y)\log_2 p(x, y) + \sum_{y \in Y} \sum_{x \in X} p(x, y)\log_2 p_Y(y)$$

$$= S(X \times Y) + \sum_{y \in Y} \sum_{x \in X} p(x, y)\log_2 p_Y(y)$$

$$= S(X \times Y) + \sum_{y \in Y} \log_2 p_Y(y) \sum_{x \in X} p(x, y)$$

$$= S(X \times Y) + \sum_{y \in Y} p_Y(y)\log_2 p_Y(y)$$

$$= S(X \times Y) - S(Y). \qquad \blacksquare$$

The same theorem can obviously be proved for the conditional entropy of Y given X as well:

$$S(Y \mid X) = S(X \times Y) - S(X). \tag{3.36}$$

The theorem can be generalized to more than two sets. The general form, which can be derived from either Eq. (3.35) or Eq. (3.36), is

$$\begin{aligned} S(X_1 \times X_2 \times \cdots \times X_n) &= S(X_1) + S(X_2 \mid X_1) + S(X_3 \mid X_1 \times X_2) \\ &\quad + \cdots + S(X_n \mid X_1 \times X_2 \times \cdots \times X_{n-1}). \end{aligned} \tag{3.37}$$

This equation is valid for any permutation of the sets involved.

Theorem 3.4

$$S(X \times Y) \le S(X) + S(Y). \tag{3.38}$$

Proof

$$\begin{aligned} S(X) &= -\sum_{x \in X} p_X(x) \log_2 p_X(x) \\ &= -\sum_{x \in X} \sum_{y \in Y} p(x, y) \log_2 \sum_{y \in Y} p(x, y) \\ S(Y) &= -\sum_{y \in Y} p_Y(y) \log_2 p_Y(y) \\ &= -\sum_{y \in Y} \sum_{x \in X} p(x, y) \log_2 \sum_{x \in X} p(x, y) \\ S(X) + S(Y) &= -\sum_{x \in X} \sum_{y \in Y} p(x, y) \left[\log_2 \sum_{y \in Y} p(x, y) + \log_2 \sum_{x \in X} p(x, y) \right] \\ &= -\sum_{\langle x, y \rangle \in X \times Y} p(x, y) [\log_2 p_X(x) + \log_2 p_Y(y)] \\ &= -\sum_{\langle x, y \rangle \in X \times Y} p(x, y) \log_2 [p_X(x) \cdot p_Y(y)]. \end{aligned}$$

By Gibbs' theorem we have

$$\begin{aligned} S(X \times Y) &= -\sum_{\langle x, y \rangle \in X \times Y} p(x, y) \log_2 p(x, y) \\ &\le -\sum_{\langle x, y \rangle \in X \times Y} p(x, y) \log_2 [p_X(x) \cdot p_Y(y)] = S(X) + S(Y). \end{aligned}$$

Hence $S(X \times Y) \leq S(X) = S(Y)$; and furthermore (again by Gibbs' theorem), the equality holds if and only if

$$p(x, y) = p_X(x) \cdot p_Y(y),$$

which means that the random variables whose state sets are X and Y are noninteractive.   ∎

Theorem 3.4 can easily be generalized to more than two sets. Its general form is

$$S(X_1 \times X_2 \times \cdots \times X_n) \leq \sum_{i=1}^{n} S(X_i), \qquad (3.39)$$

which holds for every $n \in \mathbb{N}$.

Theorem 3.5

$$S(X) \geq S(X \mid Y). \qquad (3.40)$$

Proof. From Theorem 3.3,

$$S(X \mid Y) = S(X \times Y) - S(Y),$$

and from Theorem 3.4

$$S(X \times Y) \leq S(X) + S(Y).$$

Hence,

$$S(X \mid Y) + S(Y) \leq S(X) + S(Y),$$

and the inequality

$$S(X \mid Y) \leq S(X)$$

follows immediately.   ∎

Exchanging X and Y in Theorem 3.5, we obtain

$$S(Y) \geq S(Y \mid X).$$

Additional equations expressing the relationships among the various entropies and the information transmission can be obtained by simple formula manipulations with the aid of key properties in Theorems 3.3 through 3.5. For example, when we substitute for $S(X, Y)$ from Eq. (3.35) into Eq. (3.34), we obtain

$$T_S(X,Y) = S(X) - S(X \mid Y); \qquad (3.41)$$

similarly, by substituting Eq. (3.36) into Eq. (3.34), we obtain

$$T_S(X,Y) = S(Y) - S(Y \mid X). \qquad (3.42)$$

By comparing Eqs. (3.41) and (3.42), we also obtain

$$S(X) - S(Y) = S(X \mid Y) - S(Y \mid X). \qquad (3.43)$$

For each type of the Shannon entropy, S, the normalized counterpart, NS, is calculated by dividing the respective entropy by its maximum value. Thus, for example,

$$NS(X) = \frac{S(X)}{\log_2 |X|}, \qquad (3.44)$$

$$NS(X \times Y) = \frac{S(X \times Y)}{\log_2 (|X| \cdot |Y|)}, \qquad (3.45)$$

$$NS(X \mid Y) = \frac{S(X \mid Y)}{\log_2 |X|}. \qquad (3.46)$$

The range of each of these counterparts is, of course, [0, 1]. The maximum value, $\hat{T}_S(X, Y)$, of information transmission associated with joint probability distributions on $X \times Y$ can be derived in a similar way as its possibilistic counterpart (Eq. (2.34)). It is given by the formula

$$\hat{T}_S(X,Y) = \min\{\log_2 |X|, \log_2 |Y|\}. \qquad (3.47)$$

Then,

$$NT_S(X,Y) = \frac{T_S(X,Y)}{\hat{T}_S(X,Y)}. \qquad (3.48)$$

3.2.4. Examples

The purpose of this section is to illustrate the various properties and applications of the Shannon entropy by simple examples, some of which are probabilistic counterparts of examples in Chapter 2.

Table 3.1. Illustration to Example 3.2

x	y	$p(x, y)$
0	0	0.7
0	1	0.2
1	0	0.0
1	1	0.1

(a)

x	$p_X(x)$
0	0.9
1	0.1

y	$p_Y(y)$
0	0.7
1	0.3

(b)

x	y	$p_X(x) \cdot p_Y(y)$
0	0	0.63
0	1	0.27
1	0	0.07
1	1	0.03

(c)

EXAMPLE 3.2. Consider two variables, X and Y, whose states are 0 or 1 and whose joint probabilities $p(x, y)$ on $X \times Y = \{0, 1\}^2$ are specified in Table 3.1a. Uncertainty associated with these joint probabilities is determined by the Shannon entropy

$$S(X \times Y) = -0.7 \log_2 0.7 - 0.2 \log_2 0.2 - 0.1 \log_2 0.1 = 1.16.$$

The marginal probabilities $p_X(x)$ and $p_Y(y)$, calculated by Eqs. (3.5) and (3.6), are shown in Table 3.1b. Their uncertainties are:

$$S(X) = -0.9 \log_2 0.1 - 0.1 \log_2 0.1 = 0.47,$$
$$S(Y) = -0.7 \log_2 0.7 - 0.3 \log_2 0.3 = 0.88.$$

The conditional uncertainties can now be calculated by Eqs. (3.35) and (3.36):

$$S(X \mid Y) = S(X \times Y) - S(Y) = 0.28,$$
$$S(Y \mid X) = S(X \times Y) - S(X) = 0.69.$$

Moreover, the information transmission, which expresses the strength of the relationship between the variables, can be calculated by Eq. (3.34):

$$T_S(X, Y) = S(X) + S(Y) - S(X \times Y) = 0.19.$$

EXAMPLE 3.3. Consider the same variables as in Example 3.2. However, only their marginal probabilities given in Table 3.1b are known. Assume in this example that the variables are independent. Since probabilistic independence is equivalent to probabilistic nonineteraction, as shown in Section 3.1.1, we can calculate their joint probability distribution based on this assumption by Eq. (3.7). This joint distribution is shown in Table 3.1c. The uncertainty, S_{ind}, based on the assumption of independence is thus readily calculated as

$$S_{ind}(X \times Y) = -0.63 \log_2 0.63 - 0.27 \log_2 0.27$$
$$-0.07 \log_2 0.07 - 0.03 \log_2 0.03 = 1.35.$$

Observe that $S_{ind}(X \times Y) - S(X \times Y) = 0.19$. This means that 0.19 bits of information are gained when we know the actual joint probability distribution in Table 3.1a.

EXAMPLE 3.4. Consider three variables, X, Y, Z, whose states are in sets $X = Y = \{0, 1\}$ and $Z = \{0, 1, 2\}$, respectively. The joint probabilities on $X \times Y \times Z$ are given in Table 3.2a. In this case, there are six distinct conditional uncertainties and four distinct information transmissions. To calculate them, we need

Table 3.2. Illustration to Example 3.4

x	y	z	$p(x,y,z)$
0	0	0	0.05
0	1	0	0.10
0	0	2	0.22
1	0	0	0.05
1	0	1	0.20
1	0	2	0.10
1	1	1	0.08
1	1	2	0.20

(a)

x	y	$p_{XY}(x,y)$
0	0	0.27
0	1	0.10
1	0	0.35
1	1	0.28

x	z	$p_{XZ}(x,z)$
0	0	0.15
0	2	0.22
1	0	0.05
1	1	0.28
1	2	0.30

y	z	$p_{YZ}(y,z)$
0	0	0.10
0	1	0.20
0	2	0.32
1	0	0.10
1	1	0.08
1	2	0.20

(b)

x	$p_X(x)$
0	0.37
1	0.63

y	$p_Y(y)$
0	0.62
1	0.38

z	$p_Z(z)$
0	0.20
1	0.28
2	0.52

(c)

$S(X \times Y \times Z) = 2.80$	$S(X \times Y) = 1.89$	$S(X) = 0.95$
	$S(X \times Z) = 2.14$	$S(Y) = 0.96$
	$S(Y \times Z) = 2.41$	$S(Z) = 1.47$

(d)

to determine all two-variable marginal probability distributions (shown in Table 3.2*b*) and all one-variable marginal probability distributions (shown in Table 3.2*c*). Values of the Shannon entropy for all probability distributions in Table 3.2(*a*)–(*c*) are shown in Table 3.2*d*. These values form the basis from which all the conditional uncertainties and information transmissions are calculated as follows:

$$S(X \mid Y \times Z) = S(X \times Y \times Z) - S(Y \times Z) = 0.39$$

$$S(Y \mid X \times Z) = S(X \times Y \times Z) - S(X \times Z) = 0.66$$

$$S(Z \mid X \times Y) = S(X \times Y \times Z) - S(X \times Y) = 0.91$$

$$S(X \times Y \mid Z) = S(X \times Y \times Z) - S(Z) = 1.33$$

$$S(X \times Z \mid Y) = S(X \times Y \times Z) - S(Y) = 1.84$$

$$S(Y \times Z \mid X) = S(X \times Y \times Z) - S(X) = 1.85$$

$$T(X \times Y, Z) = S(X \times Y) + S(Z) - S(X \times Y \times Z) = 0.56$$

$$T(X \times Z, Y) = S(X \times Z) + S(Y) - S(X \times Y \times Z) = 0.30$$

$$T(Y \times Z, X) = S(Y \times Z) + S(X) - S(X \times Y \times Z) = 0.56$$

$$T(X \times Y, Z) = S(X) + S(Y) + S(Z) - S(X \times Y \times Z) = 0.58.$$

EXAMPLE 3.5. This example is in some sense a probabilistic counterpart of the simple nondeterministic dynamic system discussed in possibilistic terms in Examples 2.1 and 2.2. The subject here is a simple probabilistic dynamic system with state set $X = \{x_1, x_2, x_3\}$. State transitions of the system occur only at specified discrete times and are fully determined for each initial probability distribution on X by the conditional probabilities specified by the matrix or the diagram in Figure 3.3*a* and 3.3*b*, respectively.

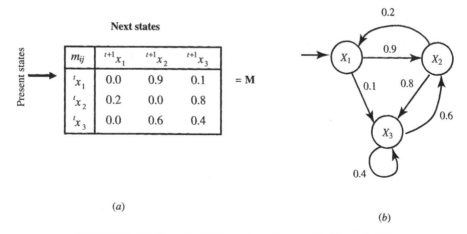

(*a*)

(*b*)

Figure 3.3. Simple probabilistic system discussed in Example 3.5.

To describe how the system behaves, let $t = 1, 2, \ldots$ denote discrete times at which state-transitions occur, let $p({}^{t}x_i)$ denote the probability of state x_i at time t, and let

$$ {}^{t}\mathbf{p} = \langle p({}^{t}x_i) \mid x_i \in X \rangle $$

denote the probability distribution of all states of the system at time t. Furthermore, let $\mathbf{M} = [m_{ij}]$ denote the matrix of conditional probabilities $p({}^{t+1}x_j \mid {}^{t}x_i)$ for all pairs $\langle x_i, x_j \rangle \in X^2$, which are independent of t. That is,

$$ m_{ij} = p({}^{t+1}x_j \mid {}^{t}x_i) $$

for all $i, j \in \mathbb{N}_3$ and all $t \in \mathbb{N}$.

Given the probability distribution ${}^{t}\mathbf{p}$ at some time t, the system is capable of predicting probability distributions at time $t + k$ ($k = 1, 2, \ldots$) or probability distributions of sequences of future states of some lengths. The Shannon entropy of each of these distributions measures the amount of uncertainty in the respective prediction. We can also measure the amount of information contained in each prediction made by the system (predictive informativeness of the system). For each prediction type, this is the difference between the maximum predictive uncertainty allowed by the framework of the system and the actual predictive uncertainty. The maximum predictive uncertainty is obtained for the state-transition matrix, $\hat{M} = [\hat{m}_{ij}]$, in which each row is a uniform probability distribution. In our case $\hat{m}_{ij} = 1/3$ for all $i, j \in \mathbb{N}$.

To illustrate the calculations of predictive uncertainty and predictive informativeness for the various prediction types, let us assume that the system is in state x_1 at time t (as indicated in Figure 3.3 by the arrow pointed at x_1). This is formally expressed as ${}^{t}\mathbf{p} = \langle 1, 0, 0 \rangle$. Maximum and actual uncertainties for some predictions are given in Figure 3.4. The diagram, which contains all sequences of states with nonzero probabilities of length 4 or less, also shows probabilities of individual states at each of the considered times. Each of the arrows under the diagram indicates the time at which the prediction is made and the time span of the prediction. Each of the first four arrows is a prediction about the next-time probability distributions made at different times. The next three arrows indicate predictions made at time t about sequences of states of lengths 2, 3, and 4. The last three arrows indicate predictions made at time t about probability distributions at time $t + 2, t + 3$, and $t + 4$. The two numbers on top of each arrow indicate the two uncertainties needed for calculating the informativeness of the prediction, the maximum one and the actual one. Let us follow in detail the calculation of some of these uncertainties.

Using Figure 3.4 as a guide, the next state prediction made at $t + 2$ is calculated by the formula

$$ {}^{t+3}\mathbf{p} = {}^{t+2}\mathbf{p} \times \mathbf{M}. $$

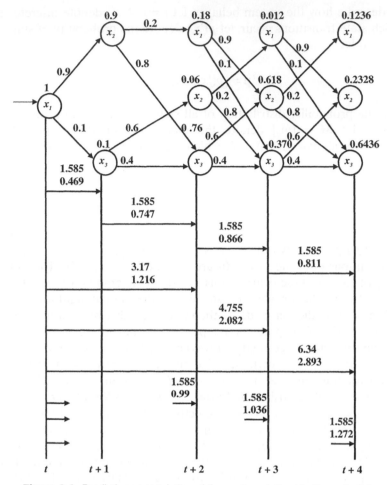

Figure 3.4. Predictive uncertainties of the system defined in Example 3.5.

Substituting for $^{t+2}\mathbf{p}$ and $\mathbf{M}$, we obtain

$$^{t+2}\mathbf{p} = [0.18, 0.06, 0.76] \times \begin{bmatrix} 0.0 & 0.9 & 0.1 \\ 0.2 & 0.0 & 0.8 \\ 0.0 & 0.6 & 0.4 \end{bmatrix}$$
$$= [0.012, 0.618, 0.370].$$

Its uncertainty is measured by the conditional Shannon entropy

$$S[p(^{t+3}x_j \mid {}^{t+2}x_i) \mid i, j \in \mathbb{N}_3] = 0.18 \cdot S(0.0, 0.9, 0.1) + 0.06 \cdot S(0.2, 0.0, 0.8)$$
$$+ 0.76 \cdot S(0.0, 0.6, 0.4)$$
$$= 0.866,$$

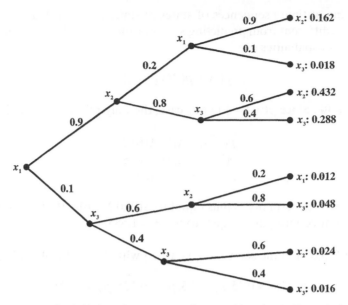

Figure 3.5. Probabilities of sequences of states of length 3 in Example 3.5.

which is calculated here by using Eq. (3.32). Its maximum counterpart, $\hat{S}$, is

$$\hat{S}[\hat{p}(^{t+3}x_j \mid {}^{t+2}x_i) \mid i, j \in \mathbb{N}_3] = 0.18 \cdot S(1/3, 1/3, 1/3) + 0.06 \cdot S(1/3, 1/3, 1/3)$$
$$+ 0.76 \cdot S(1/3, 1/3, 1/3)$$
$$= \log_2 3 = 1.585.$$

Now consider the prediction made at time t of sequences of states of length 3. There are, of course, $3^3 = 27$ such sequences, but only 8 of them have nonzero probabilities; these are shown in Figure 3.5. Probabilities of these sequences are calculated by the formula

$$p(^{t+1}x_i, {}^{t+2}x_j, {}^{t+3}x_k) = p(^tx_1) \times p(^{t+1}x_i \mid {}^tx_1) \times p(^{t+2}x_j \mid {}^{t+1}x_i) \times p(^{t+3}x_k \mid {}^{t+2}x_j)$$

for all $i, j, k \in \mathbb{N}_3$. For example,

$$p(^{t+1}x_2, {}^{t+2}x_3, {}^{t+3}x_2) = 1 \times 0.9 \times 0.8 \times 0.6 = 0.432.$$

The amount of uncertainty in predicting at time t sequences of states at times $t + 1, t + 2, t + 3$ is measured by the Shannon entropy for the probability distribution obtained for the sequences and shown in Figure 3.5. Its value is 2.082. Since there are 27 possible sequences of states of length 3, the associated maximum uncertainty is clearly equal to $\log_2 27 = 4.755$.

Predicting at time t sequences of states at times $t+1, t+2$, and $t+3$ is, of course, very different from predicting states at time $t+3$. The latter prediction is based on probabilities

$$p({}^t x_i) \cdot p({}^{t+2} x_k \mid {}^t x_i)$$

for all $i, k \in \mathbb{N}_3$. Since $p({}^t x_1) = 1$ in our case, the only relevant probabilities are

$$p({}^{t+2} x_1 \mid {}^t x_1) = 0.012,$$
$$p({}^{t+2} x_2 \mid {}^t x_1) = 0.618,$$
$$p({}^{t+2} x_3 \mid {}^t x_1) = 0.370.$$

The predictive uncertainty is thus equal to $S(0.012, 0.618, 0.370) = 1.036$, and the maximum counterpart is equal to $\log_2 3 = 1.585$.

EXAMPLE 3.6. Let the set $X = \{x_1, x_2, x_3, x_4\}$ with the probability distribution

$$\mathbf{p} = \langle p_1 = 0.25, p_2 = 0.5, p_3 = 0.125, p_4 = 0.125 \rangle$$

be given where p_i denotes the probability of x_i for all $i \in \mathbb{N}_4$. Consider the four branching schemes specified in Figure 3.6 for calculating the uncertainty of this probability distribution. Employing the branching property of the Shannon entropy, the resulting uncertainty should be the same regardless of which of the branching schemes we use. Let us perform and compare the four schemes of calculating the uncertainty.

Scheme I. According to this scheme, we calculate the uncertainty directly:

$$S(p) = -0.25 \log_2 0.25 - 0.5 \log_2 0.5 - 2 \times 0.125 \log_2 0.125$$
$$= 0.5 + 0.5 + 0.75 = 1.75.$$

Scheme II

$$S(p) = S(p_A, p_B) + p_A \cdot S(p_1/p_A, p_2/p_A) + p_B \cdot S(p_3/p_B, p_4/p_B)$$
$$= S\left(\frac{3}{4}, \frac{1}{4}\right) + 0.75 \cdot S\left(\frac{1}{3}, \frac{2}{3}\right) + 0.25 \cdot S\left(\frac{1}{2}, \frac{1}{2}\right)$$
$$= 0.811 + 0.689 + 0.25 = 1.75.$$

Scheme III

$$S(p) = S(p_1, p_A) + p_A \cdot S(p_2/p_A, p_3/p_A, p_4/p_A)$$
$$= S\left(\frac{1}{4}, \frac{3}{4}\right) + 0.75 \cdot S\left(\frac{2}{3}, \frac{1}{6}, \frac{1}{6}\right)$$
$$= 0.811 + 0.939 = 1.75.$$

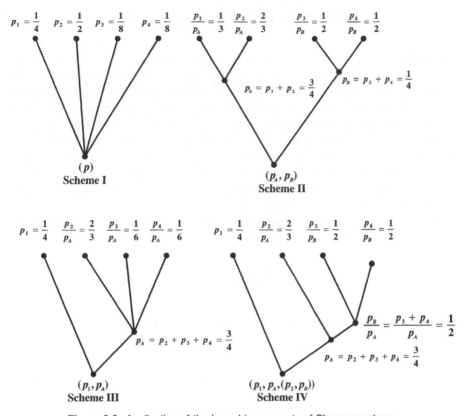

Figure 3.6. Application of the branching property of Shannon entropy.

Scheme IV

$$S(p) = S(p_1, p_A) + p_A \cdot S(p_2/p_A, p_B/p_A) + p_B \cdot S(p_3/p_B, p_4/p_B)$$
$$= S\left(\frac{1}{4}, \frac{3}{4}\right) + 0.75 \cdot S\left(\frac{2}{3}, \frac{1}{3}\right) + 0.25 \cdot S\left(\frac{1}{2}, \frac{1}{2}\right)$$
$$= 0.811 + 0.689 + 0.25 = 1.75.$$

These results thus demonstrate that the uncertainty can be calculated in terms of any branching scheme. There are, of course, many additional branching schemes in this example, each of which can be employed for calculating the uncertainty and each of which must lead to the same result.

3.3. SHANNON-LIKE MEASURE OF UNCERTAINTY FOR INFINITE SETS

One important aspect of the Shannon entropy remains to be discussed. This aspect concerns its restriction to finite sets. Is this restriction necessary? It is suggestive that the function

$$B(q(x) \mid x \in [a, b]) = -\int_a^b q(x) \log_2 q(x)\, dx, \tag{3.49}$$

where q denotes a probability density function on the interval $[a, b]$ of real numbers, could be viewed as the counterpart of the Shannon entropy in the domain of real numbers. Indeed, the form of this functional, usually referred to as a *Boltzmann entropy* or a *differential entropy*, is analogous to the form of the Shannon entropy. The former is obtained from the latter by replacing summation with integration and a probability distribution function with a probability density function. Notwithstanding this analogy, the following question cannot be avoided: Is the Boltzman entropy a genuine counterpart of the Shannon entropy? To answer this nontrivial question, we must establish a connection between the two functionals.

Let q be a probability density function on the interval $[a, b]$ of real numbers. That is, $q(x) \geq 0$ for all $x \in [a, b]$ and,

$$\int_a^b q(x)\, dx = 1. \tag{3.50}$$

Consider a sequence of probability distributions ${}^n\mathbf{p} = \langle {}^n p_1, {}^n p_2, \ldots, {}^n p_n \rangle$ such that

$$ {}^n p_i = \int_{x_{i-1}}^{x_i} q(x)\, dx \tag{3.51}$$

for every $n \in \mathbb{N}_n$, where

$$x_i = a + i \frac{b - a}{n}$$

for each $i \in \mathbb{N}_n$, and $x_0 = a$ by convention. For convenience, let

$$\Delta_n = \frac{b - a}{n}$$

so that

$$x_i = a + i\Delta_n.$$

For each probability distribution ${}^n\mathbf{p} = \langle {}^n p_1, {}^n p_2, \ldots, {}^n p_n \rangle$, let ${}^n\mathbf{d}(x)$ denote a probability density function on $[a, b]$ such that

$$ {}^n\mathbf{d}(x) = \langle {}^n d_i(x) \mid i \in \mathbb{N}_n \rangle,$$

where

$$^nd_i(x) = \frac{^np_i}{\Delta_n} \quad \text{for } x \in [x_{i-1}, x_i) \tag{3.52}$$

for all $i \in \mathbb{N}_n$. Then due to continuity of $q(x)$, the sequence $\langle ^n\mathbf{d}(x)|n \in \mathbb{N}\rangle$ converges to $q(x)$ uniformly on $[a,b]$.

Given the probability distribution $^n\mathbf{p}$ for some $n \in \mathbb{N}$, its Shannon entropy is

$$S(^n\mathbf{p}) = -\sum_{i=1}^{n} {^np_i} \log_2 {^np_i}$$

or, using the introduced probability density function $^n\mathbf{d}$,

$$S(^n\mathbf{p}) = -\sum_{i=1}^{n} {^nd_i(x)}\Delta_n \log_2[{^nd_i(x)}\Delta_n].$$

This equation can be modified as follows:

$$S(^n\mathbf{p}) = -\sum_{i=1}^{n} {^nd_i(x)}\Delta_n \log_2 {^nd_i(x)} - \sum_{i=1}^{n} {^nd_i(x)}\Delta_n \log_2 \Delta_n$$

$$= -\sum_{i=1}^{n} \left[{^nd_i(x)}\log_2 {^nd_i(x)}\right]\Delta_n - \log_2 \Delta_n \sum_{i=1}^{n} {^np_i}.$$

Since probabilities ^np_i of the distribution $^n\mathbf{p}$ must add to one, and by the definition of Δ_n, we obtain

$$S(^n\mathbf{p}) = -\sum_{i=1}^{n} \left[{^nd_i(x)}\log_2 {^nd_i(x)}\right]\Delta_n + \log_2 \frac{n}{b-a}. \tag{3.53}$$

When $n \to \infty$ (or $\Delta_n \to 0$), we have

$$\lim_{n\to\infty} -\sum_{i=1}^{n} \left[{^nd_i(x)}\log_2 {^nd_i(x)}\right]\Delta_n = -\int_a^b q(x)\log_2 q(x)\,dx$$

according to the introduced relation among $^n\mathbf{p}$, $q(x)$, and $^nd_i(x)$, in particular Eqs. (3.51) and (3.52). Equation (3.53) can thus be written for $n \to \infty$ as

$$\lim_{n\to\infty} S(^n\mathbf{p}) = B(q(x)) + \lim_{n\to\infty} \frac{n}{b-a}. \tag{3.54}$$

The last term in this equation clearly diverges. This means that the Boltzmann entropy *is not* a limit of the Shannon entropy for $n \to \infty$ and, consequently, it is not a measure of uncertainty and information.

The discrepancy between the Shannon and Boltzmann entropies can be reconciled in a modified form of the Boltzmann entropy,

$$\hat{B}[q(x),r(x)\,|\,x\in[a,b]] = \int_a^b q(x)\log_2 \frac{q(x)}{r(x)}\,dx, \qquad (3.55)$$

which involves two probability density functions, $q(x)$ and $r(x)$, defined on $[a, b]$. If only q is given, it is convenient to use

$$r(x) = \frac{1}{b-a},$$

which is the probability density function corresponding to the uniform probability distribution on $[a, b]$. The finite counterpart of $\hat{B}$ is the functional

$$\hat{S}[p(x),p'(x)\,|\,x\in X] = \sum_{x\in X} p(x)\log_2 \frac{p(x)}{p'(x)}. \qquad (3.56)$$

This functional, which is known in classical information theory as a *cross-entropy* or a *directed divergence*, measures uncertainty in relative rather than absolute terms. When p' is the uniform probability distribution function $(p'(x) = 1/|X|)$ for all $x \in X$), then

$$\hat{S}[p(x),p'(x)\,|\,x\in X] = \log_2|X| - S[p(x)\,|\,x\in X].$$

In this special form, $\hat{S}$ clearly measures the amount of information carried by function p' with respect to total ignorance (expressed in probabilistic terms).

When $q(x)$ in Eq. (3.55) is replaced with a density function, $q(x,y)$, of a joint probability distribution on $X \times Y$, and $r(x)$ is replaced with the product of density functions of marginal distributions on X and Y, $q_X(x)\cdot q_Y(y)$, $\hat{B}$ becomes the continuous counterpart the information transmission given by Eq. (3.34). This means that the continuous counterpart, T_B, of the information transmission can be expressed as

$$T_B[q(x,y),q_X(x)\cdot q_Y(y)\,|\,x\in[a,b],y\in[c,d]]$$
$$= \int_a^b\int_c^d f(x,y)\log_2 \frac{q(x,y)}{q_X(x)\cdot q_Y(x)}\,dx\,dy. \qquad (3.57)$$

It is well established that this functional is finite when functions q, q_X, and q_Y are continuous; it is always positive, and it is invariant under linear transformations.

NOTES

3.1. As is well described by Hacking [1975], the concept of numerical probability emerged in the mid-17th century. However, its adequate formalization was achieved only in the 20th century by Kolmogorov [1950]. This formalization is based on the classical measure theory [Halmos, 1950]. The literature dealing with probability theory and its applications is copious. Perhaps the most comprehensive study of foundations of probability was made by Fine [1973]. Among the enormous number of other books published on the subject, it makes sense to mention just a few that seem to be significant in various respects: Billingsley [1986], De Finetti [1974, 1975], Feller [1950, 1966], Gnedenko [1962], Jaynes [2003], Jeffreys [1939], Reichenbach [1949], Rényi [1970a, b], Savage [1972].

3.2. A justified way of measuring uncertainty and uncertainty-based information in probability theory was established in a series of papers by Shannon [1948]. These papers, which are also reprinted in the small book by Shannon and Weaver [1949], opened a way for developing the classical probability-based information theory. Among the many books providing general coverage of the theory, particularly notable are classical books by Ash [1965], Billingsley [1965], Csiszár and Körner [1981], Feinstein [1958], Goldman [1953], Guiasu [1977], Jelinek [1968], Jones [1979], Khinchin [1957], Kullback [1959], Martin and England [1981], Reza [1961], and Yaglom and Yaglom [1983], as well as more recent books by Blahut [1987], Cover and Thomas [1991], Gray [1990], Ihara [1993], Kåhre [2002], Mansuripur [1987], and Yeung [2002]. The role of information theory in science is well described in books by Brillouin [1956, 1964] and Watanabe [1969]. Other books focus on more specific areas, such as economics [Batten, 1983; Georgescu-Roegen, 1971; Theil, 1967], engineering [Bell, 1953; Reza, 1961], chemistry [Eckschlager, 1979], biology [Gatlin, 1972], psychology [Attneave, 1959; Garner, 1962; Quastler, 1955; Weltner, 1973], geography [Webber, 1979], and other areas [Hyvärinen, 1968; Kogan, 1988; Moles, 1966; Yu, 1976]. Useful resources to major papers on classical information theory that were published in the 20th century are the books edited by Slepian [1974] and Verdú and McLaughlin [2000]. Claude Shannon's contributions to classical information theory are well documented in [Sloane and Wymer, 1993]. Most current contributions to classical information theory are published in the *IEEE Transactions on Information Theory*. Some additional books on classical information theory, not listed here, are included in Bibliography.

3.3. Various subsets of the axioms for a probabilistic measure of uncertainty that are presented in Section 3.2.2. were shown to be sufficient for providing the *uniqueness of Shannon entropy* by Feinstein [1958], Forte [1975], Khinchin [1957], Rényi [1970b], and others. The uniqueness proof presented as Theorem 3.1 is adopted from a book by Ash [1965]. Excellent overviews of the various axiomatic treatments of Shannon entropy can be found in books by Aczél and Daróczy [1975], Ebanks et al. [1997], and Mathai and Rathie [1975]. All these books are based heavily on the use of functional equations. An excellent and comprehensive monograph on functional equations was prepared by Aczél [1966].

3.4. Several classes of functionals that subsume the Shannon entropy as a special case have been proposed and studied. They include:

1. *Rényi entropies* (also called entropies of degrees α), which are defined for all real numbers $\alpha \neq 1$ by the formula

$$H_\alpha(p_1, p_2, \ldots, p_n) = \frac{1}{1-\alpha} \log_2 \sum_{i=1}^n p_i^\alpha. \qquad (3.58)$$

It is well known that the limit of H for $\alpha \to 1$ is the Shannon entropy. For $\alpha = 0$, we obtain

$$H_\alpha(p_1, p_2, \ldots, p_n) = \log_2 \sum_{i=1}^n p_i^0.$$

This functional represents one of the probabilistic interpretations of Hartley information as a measure that is insensitive to actual values of the given probabilities and distinguishes only between zero and nonzero probabilities. As the name suggests, Rényi entropies were proposed and investigated by Rényi [1970b].

2. *Entropies of order β*, introduced by Daróczy [1970], which have the form

$$H_\beta(p_1, p_2, \ldots, p_n) = \frac{1}{2^{1-\beta} - 1} \left(\sum_{i=1}^n p_i^\beta - 1 \right) \qquad (3.59)$$

for all $\beta \neq 1$. As in the case of Rényi entropies, the limit of H_β for $\beta \to 1$ results in the Shannon entropy.

3. *R-norm entropies*, which are defined for all $R \neq 1$ by the functional

$$H_R(p_1, p_2, \ldots, p_n) = \frac{R}{R-1} \left[1 - \left(\sum_{i=1}^n p_i^R \right)^{1/R} \right]. \qquad (3.60)$$

As in the other two classes of functionals, the limit of H_R for $R \to 1$ is the Shannon entropy. This class of functionals was proposed by Boekee and Van der Lubbe [1980] and was further investigated by Van der Lubbe [1984].

Formulas converting entropies from one class to other classes are well known. Conversion formulas between H_α and H_β were derived by Aczél and Daróczy [1975, p. 185]; formulas for converting H_α and H_β into H_R were derived by Boeke and Van der Lubbe [1980, p.144].

Except for the Shannon entropy, these classes of functionals are not adequate measures of uncertainty, since each of them violates some essential requirement for such a measure. For example, when $\alpha, \beta, R > 0$, Rényi entropies violate subadditivity, entropies of order β violate additivity, and R-norm entropies violate both subadditivity and additivity. The significance of these functions in the context of information theory is thus primarily theoretical, as they help us better understand Shannon entropy as a limiting case in these classes of functions. Strong arguments supporting this claim can be found in papers by Aczél, Forte, and Ng [1974] and Forte [1975].

3.5. Using Theorems 3.3 through 3.5 as a base, numerous theorems regarding the relationship among the information transmission and basic, conditional, and joint Shannon entropies can be derived by simple algebraic manipulations and by mathematical induction to obtain generalizations. Conant [1981] offers some useful ideas in this regard. A good summary of practical theorems for Shannon entropy was prepared by Ashby [1969]. Ashby [1965, 1970, 1972] and Conant [1976, 1988] also demonstrated the utility of these theorems for analyzing complex systems.

3.6. An excellent examination of the difference between *Shannon* and *Boltzmann entropy* is made by Reza [1961]. This issue is also discussed by Ash [1965], Guiasu [1977], Jones [1979], and Ihara [1993]. The origin of the concept of entropy in physics is discussed in detail by Fast [1962]; see also an important early paper by Elsasser [1937].

3.7. Guiasu [1977, Chapter 4] introduced and studied a generalization of the Shannon entropy to the *weighted Shannon entropy*,

$$WS(p(x), w(x) \mid x \in X) = -\sum_{x \in X} w(x) p(x) \log_2 p(x). \qquad (3.61)$$

where $w(x)$ are nonnegative numbers that are called *weights*. For each alternative $x \in X$, the weight $w(x)$ characterizes the *importance* (or *utility*) of the alternative in a given application context. Guiasu showed that the functional WS possesses the following properties:

- $WS(p(x), w(x) \mid x \in X) \geq 0$, where the equality is obtained iff $p(x) = 1$ for one particular $x \in X$.
- WS is subadditive and additive.
- The maximum of WS,

$$WS_{\max} = \lambda + \sum_{x \in X} w^{(1-\lambda)/w(x)} \qquad (3.62)$$

is reached when $p(x) = e^{(1-\lambda)/w(x)}$ for all $x \in X$, where λ is determined by the equation

$$\sum_{x \in X} e^{(1-\lambda)/w(x)} = 1. \qquad (3.63)$$

EXERCISES

3.1. For each of the probability distributions in Table 3.3, calculate the following:

(a) All conditional uncertainties;

(b) Information transmissions for all partitions of the set of variables;

(c) The reduction of uncertainty (information) with respect to the uniform probability distribution;

(d) Normalized counterparts of results obtained in parts (a), (b), (c).

Table 3.3. Probability Distributions in Exercise 3.1

x	y	z	$p_1(x, y, z)$	$p_2(x, y, z)$	$p_3(x, y, z)$	$p_4(x, y, z)$
0	0	0	0.30	0.00	0.10	0.10
0	0	1	0.00	0.00	0.00	0.00
0	1	0	0.00	0.25	0.00	0.20
0	1	1	0.20	0.25	0.30	0.00
1	0	0	0.00	0.25	0.40	0.30
1	0	1	0.10	0.25	0.00	0.00
1	1	0	0.30	0.00	0.00	0.40
1	1	1	0.10	0.00	0.20	0.00

Table 3.4. State-Transition Matrix in Exercise 3.3

	$^{t+1}x_1$	$^{t+1}x_2$	$^{t+1}x_3$	$^{t+1}x_4$
$^{t}x_1$	0.2	0.0	0.8	0.0
$^{t}x_2$	0.0	0.0	0.0	1.0
$^{t}x_3$	0.0	0.9	0.0	0.1
$^{t}x_4$	0.5	0.3	0.2	0.0

3.2. Repeat Example 3.5 using the following assumptions:
   (a) The state at time t is x_2;
   (b) The state at time t is x_3;
   (c) The probability distribution at time t is $^t p(x_1) = 0$, $^t p(x_2) = 0.6$, $^t p(x_3) = 0.4$.

3.3. Consider a system with state set $X = \{x_1, x_2, x_3, x_4\}$ whose transitions from present states to next states are characterized by the state-transition matrix of conditional probabilities specified in Table 3.4. Assuming that the initial probability distribution on the states (at time t) is $^t p(x_1) = 0.5$, $^t p(x_2) = 0.3$, $^t p(x_3) = 0.2$, $^t p(x_4) = 0$, determine the following:
   (a) Uncertainties in predicting states at time $t = 2, 3, 4$;
   (b) Uncertainty in predicting sequences of states of lengths 2, 3, 4;
   (c) Associated normalized uncertainties for parts (a) and (b);
   (d) Associated amounts of information contained in the system with respect to questions regarding predictions in parts (a) and (b).

3.4. Repeat Exercise 3.3 for some other probability distributions on the states at time t.

3.5. Use some additional branching in Example 3.6 to calculate the value of the Shannon entropy for the given probability distributions, for example:
   (a) $p_A = p_1 + p_4$, $p_B = p_2 + p_3$
   (b) $p_A = p_1 + p_2 + p_4$, $p_B = p_3$
   (c) Scheme IV in which p_1 is exchanged with p_4 and p_2 is exchanged with p_3.

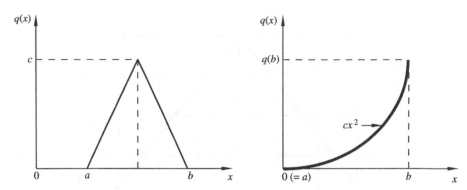

Figure 3.7. Probability density functions in Exercise 3.9.

3.6. Derive the generalized form Eq. (3.37) of Eq. (3.35) in Theorem 3.3.

3.7. Derive the generalized form Eq. (3.39) of Eq. (3.38) in Theorem 3.4.

3.8. The so-called Q-factor, which is defined by the equation

$$Q(X \times Y \times Z) = S(X) + S(Y) + S(Z) - S(X \times Y)$$
$$- S(X \times Z) - S(Y \times Z) + S(X \times Y \times Z),$$

is often used in classical information theory. Express $Q(X \times Y \times Z)$ solely in terms of the various information transmissions.

3.9. For each of the probability density functions $q(x)$ shown in Figure 3.7, calculate the Boltzmann entropy and demonstrate that it is negative, zero, or positive, depending on the values of a, b, c. NOTE: Remember that the condition

$$\int_a^b q(x)\, dx = 1$$

must be satisfied.

3.10. Let the graphs in Figure 3.8a and 3.8b represent, respectively, a probability density function q and the associated probability distribution function p. Determine mathematical definitions of both these functions.

3.11. Consider variables X, Y whose states are in sets X, Y, respectively, and that are characterized by joint probabilities for all $\langle x, y \rangle \in X \times Y$. Show that:

(a) X is independent of Y iff Y is independent of X;

(b) X and Y are noninteractive iff X and Y are independent of one another.

3.12. Calculate the posterior probability $Pro(A \mid B)$ in Example 3.1 under the following assumptions:

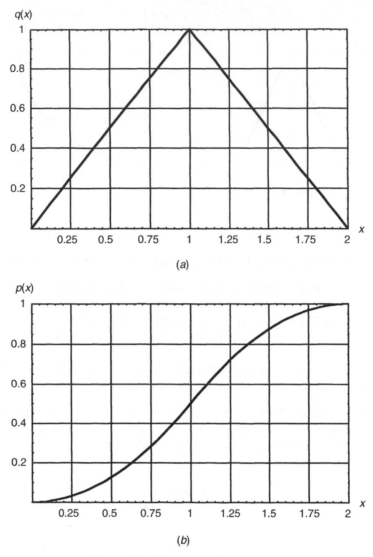

Figure 3.8. Illustration to Exercise 3.10.

(a) The outcome of the TST test is negative;

(b) The reliability of the TST test is higher; $Pro(B|A) = 0.999$ and $Pro(B|\bar{A}) = 0.02$;

(c) the reliability of the TST test is lower: $Pro(B|A) = 0.8$ and $Pro(B|\bar{A}) = 0.15$.

4

GENERALIZED MEASURES AND IMPRECISE PROBABILITIES

An educated mind is satisfied with the degree of precision that the nature of the subject admits and does not seek exactness where only approximation is possible.

—Aristotle

4.1. MONOTONE MEASURE

The term "classical information theory" is used in the literature, by and large, to refer to the theory based on the notion of probability (Chapter 3). Uncertainty functions in this theory are expressed in terms of classical measure theory, which in turn, is formalized in terms of classical set theory. Generalizing the concept of a classical measure is thus one way of enlarging the framework for a broader treatment of the concept of uncertainty and the associated concept of uncertainty-based information. The purpose of this chapter is to discuss this generalization. Further enlargement of the framework, which is discussed in Chapter 8, is obtained by fuzzifications of classical as well as generalized measures. Basic characteristics of both these generalizations are depicted in Figure 4.1.

Given a universal set X and a nonempty family C of subsets of X with an appropriate algebraic structure (e.g., a σ-algebra), a *classical measure*, μ, is a set function of the form

Uncertainty and Information: Foundations of Generalized Information Theory, by George J. Klir
© 2006 by John Wiley & Sons, Inc.

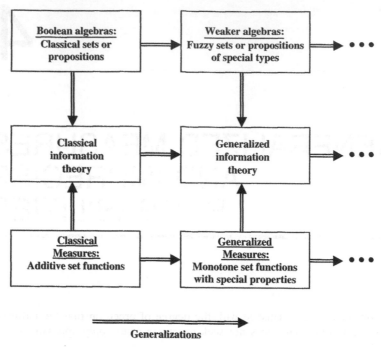

Generalizations

Figure 4.1. Classical information theory and its generalizations.

$$\mu : C \to [0, \infty]$$

that satisfies the following requirements:

(*cm*1) If $\varnothing \in C$ (family C is usually assumed to contain $\varnothing$), then $\mu(\varnothing) = 0$;
(*cm*2) For every sequence $A_1, A_2, \ldots$ of pairwise disjoint sets of C,

$$\text{if } \bigcup_{i=1}^{\infty} A_i \in C \quad \text{then} \quad \mu\!\left(\bigcup_{i=1}^{\infty} A_i\right) = \sum_{i=1}^{\infty} \mu(A_i).$$

Observe that probability is a classical measure such that C is a σ-algebra and $\mu(X) = 1$.

Property (*cm*2), which is the distinguishing feature of classical measures, is called a *countable additivity*. A variant of this property, which is called a *finite additivity*, is defined as follows:

(*cm*2′) for every finite sequence $A_1, A_2, \ldots, A_n$, of pairwise disjoint sets of C,

$$\text{if } \bigcup_{i=1}^{n} A_i \in C \quad \text{then} \quad \mu\!\left(\bigcup_{i=1}^{n} A_i\right) = \sum_{i=1}^{n} \mu(A_i).$$

It is well known that any countable additive measure is also finitely additive, but not the other way around.

The requirement of additivity (countable or finite) of classical measures is based on the assumption that disjoint sets are noninteractive with respect to the measured property. This assumption is too restrictive in some application contexts. Consider, for example, a set of workers in a workshop whose purpose is to manufacture products of a specific type. Assume that the set is partitioned into subsets (working groups) $A_1, A_2, \ldots, A_n$, and let $\mu(A_i)$ denote the number of products made by group $A_i (i \in \mathbb{N}_n)$ within a given unit of time. Then, clearly, any of the following can happen for any two groups A_i, A_j:

- $\mu(A_i \cup A_j) = \mu(A_i) + \mu(A_j)$ when groups A_i and A_j work separately.
- $\mu(A_i \cup A_j) > \mu(A_i) + \mu(A_j)$ when the groups work together and their co-operation is efficient.
- $\mu(A_i \cup A_j) < \mu(A_i) + \mu(A_j)$ when the groups work together and their co-operation is inefficient.

Numerous other examples could be presented to illustrate that the additivity requirement of classical measures severely limits their applicability. Some examples, relevant to the various issues of uncertainty formalization, are discussed later in this chapter.

After recognizing that classical measures are too restrictive, it is not obvious how to generalize them. One possibility is to eliminate the additivity requirement and define generalized measures solely by the requirement (*cm*1). Although this sweeping generalization seems too radical, it has been found useful in some applications. However, its utility for dealing with uncertainty is questionable. Another possibility is to replace the additivity requirement with an appropriate weaker requirement. It is generally recognized that the highest generalization of classical measures that is meaningful for formalizing uncertainty functions is the one that replaces the additivity requirement with a weaker requirement of monotonicity with respect to the subsethood ordering. Generalized measures of this kind are called *monotone measures*. The following is their formal definition.

Given universal set X and nonempty family C of subsets X (usually with an appropriate algebraic structure), a *monotone measure*, μ, on $\langle X, C \rangle$ is a function of the type

$$\mu : C \to [0, \infty]$$

that satisfies the following requirements:

- (μ1) $\mu(\varnothing) = 0$ (*vanishing at the empty set*).
- (μ2) For all $A, B \in C$, if $A \subseteq B$, then $\mu(A) \leq \mu(B)$ (*monotonicity*).
- (μ3) For any increasing sequence $A_1 \subseteq A_2 \subseteq \cdots$ of sets in C,

$$\text{if } \bigcup_{i=1}^{\infty} A_i \in C, \quad \text{then} \quad \lim_{i \to \infty} \mu(A_i) = \mu\left(\bigcup_{i=1}^{\infty} A_i\right) \quad (\textit{continuity from below}).$$

($\mu4$) For any decreasing sequence $A_1 \supseteq A_2 \supseteq \cdots$ of sets in C,

$$\text{if } \bigcap_{i=1}^{\infty} A_i \in C, \quad \text{then} \quad \lim_{i \to \infty} \mu(A_i) = \mu\left(\bigcap_{i=1}^{\infty} A_1\right) \quad (\textit{continuity from above}).$$

Observe that the same symbol, μ, is used for both monotone and additive measures. This does not create any notational confusion since additive measures are contained in the class of monotone measures. It is just required that the meaning of the symbol be stated explicitly when it stands for some special type of monotone measures, such as additive measures.

Functions that satisfy requirements ($\mu1$), ($\mu2$), and either ($\mu3$) or ($\mu4$) are equally important in the theory of monotone measures. In fact, they are essential for formalizing imprecise probabilities (Section 4.3). These functions are called *semicontinuous* from below or above, respectively. When the universal set X is finite, requirements ($\mu3$) and ($\mu4$) are trivially satisfied and may thus be disregarded. If $X \in C$ and $\mu(X) = 1$, μ is called a *regular* monotone measure (or *regular* semicontinuous monotone measure). Uncertainty functions of any type are always regular monotone measures.

Observe that requirement ($\mu2$) defines measures that are actually *monotone increasing*. By changing the inequality $\mu(A) \leq \mu(B)$ in ($\mu2$) to $\mu(A) \geq \mu(B)$, we can define measures that are *monotone decreasing*. Both types of monotone measures are useful, even though monotone increasing measures are more common in dealing with uncertainty. Unless specified otherwise, the term "monotone measure" is used in this book to refer to monotone increasing measures that are regular. The utility of monotone decreasing measures is discussed later in the book.

The following inequalities hold for every monotone measure μ: if A, B, $A \cup B \in C$, then

$$\mu(A \cap B) \leq \min\{\mu(A), \mu(B)\}, \tag{4.1}$$

$$\mu(A \cup B) \geq \max\{\mu(A), \mu(B)\}. \tag{4.2}$$

These inequalities follow from monotonicity of μ and from the facts that $A \cap B \subseteq A$ and $A \cap B \subseteq B$, and similarly, $A \cup B \supseteq A$ and $A \cup B \supseteq B$. If, in addition, either the inequality

$$\mu(A \cup B) \geq \mu(A) + \mu(B) \tag{4.3}$$

or the inequality

$$\mu(A \cup B) \leq \mu(A) + \mu(B) \tag{4.4}$$

holds for all $A, B, A \cup B \in C$ such that $A \cap B = \varnothing$, the monotone measure is called *superadditive* or *subadditive*, respectively.

It is easy to see that additivity implies monotonicity, but not the other way around. For all $A, B, A \cup B \in C$ such that $A \cap B = \varnothing$, a monotone measure μ is capable of capturing any of the following situations:

(a) $\mu(A \cup B) > \mu(A) + \mu(B)$, which expresses a cooperative action or synergy between A and B in terms of the measured property.

(b) $\mu(A \cup B) = \mu(A) + \mu(B)$, which expresses the fact that A and B are noninteractive with respect to the measured property.

(c) $\mu(A \cup B) < \mu(A) + \mu(B)$, which expresses some sort of inhibitory effect or incompatibility between A and B as far as the measured property is concerned.

Observe that probability theory, which is based on classical measure theory, is capable of capturing only situation (b). This demonstrates that the theory of monotone measures provides us with a considerably broader framework than probability theory for formalizing uncertainty. As a consequence, it allows us to capture types of uncertainty that are beyond the scope of probability theory.

The need for monotone measures arises in many problem areas. One example is the area of ordinary measurement in physics. While additivity characterizes well many types of measurement under idealized, error-free conditions, it is not fully adequate to characterize most measurements under real, physical conditions, when measurement errors are unavoidable. To illustrate this claim by an example, consider two disjoint events, A and B, defined in terms of adjoining intervals of real numbers, as shown in Figure 4.2a. Observations in close neighborhoods (within a measurement error) of the end points

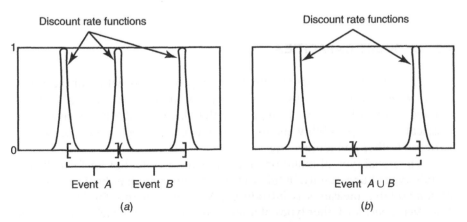

Figure 4.2. An example illustrating the violation of the additivity axiom of probability theory.

of each event are unreliable and should be properly discounted, for example, according to the discount rate functions shown in Figure 4.2a. That is, observations in the neighborhoods of the end points should carry less evidence than those outside them. The closer they are to the end points, the less evidence they should carry. When measurements are taken for the union of the two events, as shown in Figure 4.2b, one of the discount rate functions is not applicable. Hence, the same observations produce more evidence for the single event $A \cup B$ than for the two disjoint events A and B. This implies that the probability of $A \cup B$ should be greater than the sum of the probabilities of A and B. The additivity requirement is thus violated. To properly formalize this situation, we need to use an appropriate monotone measure that is superadditive.

For some historical reasons of little significance, monotone measures are often referred to in literature as *fuzzy measures*. This name is somewhat confusing, since no fuzzy sets are involved in the definition of monotone measures. To avoid this confusion, the term "fuzzy measures" should be reserved to measures (additive or nonadditive) that are defined on families of fuzzy sets.

Since all monotone measures discussed in the rest of this book are regular, it is reasonable to omit the adjective "regular." Therefore, by convention, the term "monotone measure" refers in the rest of this book to regular monotone measures. Moreover, it is assumed, unless it is stated otherwise, that the universal set, X, is finite and that $C = \mathcal{P}(X)$. That is, it is normally assumed that the monotone measures of concern are set functions

$$\mu : \mathcal{P}(X) \to [0, 1],$$

where X is a finite set, that satisfy the following requirements:

($\mu 1'$) $\mu(\varnothing) = 0$ and $\mu(X) = 1$.
($\mu 2'$) For all $A, B \in \mathcal{P}(X)$, if $A \subseteq B$, then $\mu(A) \leq \mu(B)$.

4.2. CHOQUET CAPACITIES

The general notion of a monotone measure provides us with a broad framework, within which various special types of monotone measures can be defined. Among these special types are the classical, additive measures, the classical (crisp) possibility measures and necessity measures, and a great variety of other nonadditive measures.

Each special type of monotone measures has a potential for formalizing a certain type of uncertainty. In this section, an important family of special types of nonadditive measures is introduced. Measures in this family are called *Choquet capacities*. Other types of nonadditive measures, which have been utilized for formalizing imprecise probabilities, are introduced in Chapter 5.

Given a particular integer $k \geq 2$, a *Choquet capacity of order k* (or *k-monotone Choquet capacity*) is a monotone measure μ that satisfies the inequalities

$$\mu\left(\bigcup_{j=1}^{k} A_j\right) \geq \sum_{\substack{K \subseteq N_k \\ K \neq \varnothing}} (-1)^{|K|+1} \mu\left(\bigcap_{j \in K} A_j\right) \tag{4.5}$$

for all families of k subsets of X. For convenience, monotone measures that are not required to satisfy Eq. (4.5) or any other special property are often referred to as Choquet capacities of order 1 (1-monotone). That is, any general monotone measure is also viewed as a Choquet capacity of order 1.

Since sets A_j in Eq. (4.5) are not necessarily distinct, every Choquet capacity of order $k > 2$ is also of order $k' = 2, 3, \ldots, k$. However, a capacity of order k is clearly not a capacity of any higher order ($k + 1, k + 2$, etc.). Hence, capacities of order 2, which satisfy the simple inequalities

$$\mu(A_1 \cup A_2) \geq \mu(A_1) + \mu(A_2) - \mu(A_1 \cap A_2) \tag{4.6}$$

for all pairs of subsets of X, are the most general capacities. The least general ones are those of order k (or k-monotone) for all $k \geq 2$. These are called *Choquet capacities of infinite order* (or *∞-monotone*). They satisfy the inequalities

$$\mu(A_1 \cup A_2 \cup \cdots \cup A_k) \geq \sum_i \mu(A_i) - \sum_{i<j} \mu(A_i \cap A_j) +$$
$$- \cdots + (-1)^{k+1} \mu(A_1 \cap A_2 \cap \cdots A_k) \tag{4.7}$$

for every $k \geq 2$ and every family of k subsets of X. Observe that probability measures are special ∞-monotone Choquet capacities for which all the inequalities in Eq. (4.7) collapse to equalities.

4.2.1. Möbius Representation

It is well known (see Note 4.4) that every set function

$$\mu : \mathcal{P}(X) \to \mathbb{R},$$

where X is a finite set, can be uniquely represented by another set function

$$^{\mu}m : \mathcal{P}(X) \to \mathbb{R},$$

via the formula

$$^{\mu}m(A) = \sum_{B|B \subseteq A} (-1)^{|A-B|} \mu(B) \tag{4.8}$$

for all $A \in P(X)$. This formula is called a *Möbius transform* and function ${}^{\mu}m$ is called a *Möbius representation* of μ (or a *Möbius function*).

The Möbius transform is a one-to-one function and it is thus invertible. Its inverse is defined for all $A \in P(X)$ by the formula

$$\mu(A) = \sum_{B|B\subseteq A} {}^{\mu}m(B). \tag{4.9}$$

EXAMPLE 4.1. A set function $\mu: P(X) \to [0, 1]$, where $X = \{x_1, x_2, x_3\}$, and its Möbius representations ${}^{\mu}m$ are shown in Table 4.1. Subsets of A of X are defined in the table by their characteristic functions. Given values $\mu(A)$ for all $A \in P(X)$, we can calculate the value of ${}^{\mu}m(A)$ for each A by the Möbius transform Eq. (4.8). For example,

$$\begin{aligned}
{}^{\mu}m(\{x_2, x_3\}) &= \mu(\{x_2, x_3\}) - \mu(\{x_2\}) - \mu(\{x_3\}) \\
&= 0.5 - 0 - 0.3 = 0.2,
\end{aligned}$$

$$\begin{aligned}
{}^{\mu}m(\{x_1, x_2, x_3\}) &= \mu(\{x_1, x_2, x_3\}) - \mu(\{x_1, x_2\}) - \mu(\{x_1, x_3\}) - \mu(\{x_2, x_3\}) \\
&\quad + \mu(\{x_1\}) + \mu(\{x_2\}) + \mu(\{x_3\}) \\
&= 1 - 0.4 - 0.3 - 0.5 + 0 + 0 + 0.3 = 0.1.
\end{aligned}$$

Function μ can uniquely be reconstructed from its Möbius representation ${}^{\mu}m$ by the inverse transform defined by Eq. (4.9). For example,

$$\begin{aligned}
\mu(\{x_2, x_3\}) &= {}^{\mu}m(\{x_2, x_3\}) + {}^{\mu}m(\{x_2\}) + {}^{\mu}m(\{x_3\}) \\
&= 0.2 + 0 + 0.3 = 0.5,
\end{aligned}$$

$$\mu(\{x_1, x_2, x_3\}) = \sum_{B \in P(X)} {}^{\mu}m(B) = 1.$$

It can easily be shown that a set function μ is a monotone measure (regular) if and only if its Möbius representation ${}^{\mu}m$ has the following properties:

Table 4.1. Set Function μ and Its Möbius Representation ${}^{\mu}m$

	x_1	x_2	x_3	$\mu(A)$	${}^{\mu}m(A)$
A:	0	0	0	0.0	0.0
	1	0	0	0.0	0.0
	0	1	0	0.0	0.0
	0	0	1	0.3	0.3
	1	1	0	0.4	0.4
	1	0	1	0.3	0.0
	0	1	1	0.5	0.2
	1	1	1	1.0	0.1

($m1$) $^\mu m(\varnothing) = 0$.

($m2$) $\displaystyle\sum_{A \in \mathcal{P}(X)} {}^\mu m(A) = 1$.

($m3$) $\displaystyle\sum_{\{x\} \subseteq B \subseteq A} {}^\mu m(B) \geq 0$ for all $A \in \mathcal{P}(X)$ and all $x \in A$.

Property ($m1$) follows directly from Eq. (4.8) and the requirement $\mu(\varnothing) = 0$ of monotone measures. Property ($m2$) follows from Eq. (4.9) and the requirement $\mu(X)$ of (regular) monotone measures. Property ($m3$) follows from Eq. (4.8) and the fact that monotonicity of μ holds if and only if it holds for pairs A, $A - \{x\}$ of sets ($x \in X$). Observe that property ($m3$) implies that $^\mu m(\{x\}) \geq 0$ for all $x \in X$.

Additional properties of the Möbius representation have been recognized for Choquet capacities of various orders $k \geq 2$. Some of these properties, which are utilized later in this book, are:

($m4$) If μ is a Choquet capacity of order k and $2 \leq |A| \leq k$, then $^\mu m(A) \geq 0$.

($m5$) μ is a Choquet capacity of order ∞ if and only if $^\mu m(A) \geq 0$ for all $A \in \mathcal{P}(X)$.

($m6$) μ is a probability measure if and only if $^\mu m(A) > 0$ when $|A| = 1$ and $^\mu m(A) = 0$ otherwise.

($m7$) μ is a Choquet capacity of order k ($k \geq 2$) if and only if

$$\sum_{C \subseteq B \subseteq A} m(B) \geq 0$$

for all $A \in \mathcal{P}(X)$ and all $C \in \mathcal{P}(X)$ such that $2 \leq |C| \leq k$.

For further information regarding the Möbius representation, including these properties, see Note 4.4.

EXAMPLE 4.2. Four monotone measures defined on $\mathcal{P}(\{x_1, x_2, x_3\})$ and their Möbius representations are specified in Table 4.2. Measure μ_1 is a monotone measure, but it is not a Choquet capacity of any order $k \geq 2$. This follows, for example, from the inequalities

$$\mu_1(\{x_1\} \cup \{x_2\}) < \mu_1(\{x_1\}) + \mu_1(\{x_2\}),$$

$$\mu_1(\{x_2\} \cup \{x_3\}) < \mu_1(\{x_2\}) + \mu_1(\{x_3\}),$$

which violate the required inequalities (4.6) for Choquet capacities of order 2. It also follows from the negative values of $m_1(\{x_1, x_2\})$ and $m_1(\{x_2, x_3\})$, which violates property ($m4$) for $k = 2$. Measure μ_2 is not a Choquet capacity of any order $k \geq 2$ as well; for example,

Table 4.2 Examples of Monotone Measures Discussed in Example 4.2

x_1	x_2	x_3	$\mu_1(A)$	$m_1(A)$	$\mu_2(A)$	$m_2(A)$	$\mu_3(A)$	$m_3(A)$	$\mu_4(A)$	$m_4(A)$
A: 0	0	0	0.0	0.0	0.0	0.0	0.0	0.0	0.0	0.0
1	0	0	0.5	0.5	0.4	0.4	0.4	0.4	0.3	0.3
0	1	0	0.2	0.2	0.1	0.1	0.1	0.1	0.0	0.0
0	0	1	0.4	0.4	0.2	0.2	0.2	0.2	0.2	0.2
1	1	0	0.5	−0.2	0.5	0.0	0.6	0.1	0.3	0.0
1	0	1	0.9	0.0	0.8	0.2	0.7	0.1	0.9	0.4
0	1	1	0.5	−0.1	0.5	0.2	0.5	0.2	0.3	0.1
1	1	1	1.0	0.2	1.0	−0.1	1.0	−0.1	1.0	0.0

$$\mu_2(\{x_1, x_2\} \cup \{x_1, x_3\}) < \mu_2(\{x_1, x_2\}) + \mu_2(\{x_1, x_3\}) - \mu_2(\{x_1\}),$$

which violates the required inequalities (4.6) for 2-monotone measures. However, this cannot be determined by property (*m*4). Measure μ_3 is a Choquet capacity of order 2, as can be easily verified by checking the required inequalities (4.6). However, it is not a Choquet capacity of order 3, since

$$\mu_3(\{x_1, x_2\} \cup \{x_1, x_3\} \cup \{x_2, x_3\}) < \mu_3(\{x_1, x_2\}) + \mu_3(\{x_1, x_3\}) + \mu_3(\{x_2, x_3\})$$
$$- \mu_3(\{x_1\}) - \mu_3(\{x_2\}) - \mu_3(\{x_3\}).$$

Observe, that this measure also violates property (*m*7) for $k = 3$. Measure μ_4 is clearly a Choquet capacity of order ∞ since $m_4(A) \geq 0$ for all $A \subseteq \{x_1, x_2, x_3\}$.

4.3. IMPRECISE PROBABILITIES: GENERAL PRINCIPLES

Classical probability theory requires that probabilities of all recognized alternatives (elementary events) be precise real numbers. Given these numbers, probabilities of the various sets of alternatives are then uniquely determined by the additivity property of probability measures. These, again, are precise real numbers. This requirement of precision is overly restrictive since there are many problem situations in which more than one probability distribution is compatible with given evidence. One such situation is illustrated by the following example.

EXAMPLE 4.3. Consider a universal set $X \times Y$, where $X = \{x_1, x_2\}$ and $Y = \{y_1, y_2\}$ are state sets of random variables X and Y, respectively. Assume that we know the marginal probabilities $p_X(x_1)$, $p_X(x_2) = 1 - p_X(x_1)$, $p_Y(y_1), p_Y(y_2) = 1 - p_Y(p_1)$, and we want to use this information to determine the unknown joint probabilities $p_{ij} = p(x_i, y_j)(i, j \in \{1, 2\})$. According to the calculus of probability theory, the joint probabilities are related to the marginal probabilities by the following equations:

(a) $p_{11} + p_{12} = p_X(x_1)$.
(b) $p_{21} + p_{22} = 1 - p_X(x_1)$.
(c) $p_{11} + p_{21} = p_Y(y_1)$.
(d) $p_{12} + p_{22} = 1 - p_Y(y_1)$.

Only three of these equations are linearly independent. For example, Eq. (d) is a linear combination of the other equations: (d) = (a) + (b) − (c). By excluding it, the remaining three equations are linearly independent. Since they contain four unknowns, one of them must be chosen as a free variable. Choosing, for example, p_{11} as the free variable, we obtain the following solution:

$$p_{12} = p_X(x_1) - p_{11},$$

$$p_{21} = p_Y(y_1) - p_{11},$$

$$p_{22} = 1 - p_X(x_1) - p_Y(y_1) + p_{11}.$$

Since p_{12}, p_{21}, and p_{22} are required to be nonnegative numbers, the free variable p_{11} is constrained by the inequalities

$$\max\{0, p_X(x_1) + p_Y(y_1) - 1\} \le p_{11} \le \min\{p_X(x_1), p_Y(y_1)\}.$$

When a particular value of p_{11} that satisfies these inequalities is chosen, the values of p_{12}, p_{21}, and p_{22} are uniquely determined. The resulting joint probability distribution is consistent with the given marginal distributions. Since values of p_{11} range over a closed interval of real numbers, there is a closed and convex set of joint probability distributions that are consistent with the marginal distributions. According to the given evidence (the known marginal distributions), these are the only possible joint distributions, and the actual one is among them. All the other joint distributions on $X \times Y$ are not possible.

Consider, for example, that $p_X(x_1) = 0.8$ and $p_Y(y_1) = 0.6$ Then, the set of all possible joint probability distribution functions is specified by the following four statements:

$$p_{11} \in [0.4, 0.6],$$

$$p_{12} = 0.8 - p_{11},$$

$$p_{21} = 0.6 - p_{11},$$

$$p_{22} = p_{11} - 0.4.$$

Other types of incomplete information regarding a probability distribution (e.g., knowing only the expected value of a random variable, analyzing statistical data with observation gaps, etc.) result in sets of possible probability distributions as well. To deal with each given incomplete information correctly, we need to recognize and work with the whole set of possible probability distributions, and not to choose only one of them, as required by probability theory. This means, in turn, that we need to work with imprecise probabilities, as shown in the rest of this section.

Sets of probability distribution functions defined on some universal set X are often referred to in the literature as *credal sets* on X. This shorter term is convenient and it is occasionally used in this book.

4.3.1. Lower and Upper Probabilities

Let X denote a finite universal set of concern (a set of elementary events) and let $\mathcal{D}$ denote a given set of probability distribution functions (a credal set), p, on X. Then, the associated *lower probability function*, $^{\mathcal{D}}\underline{\mu}$, is defined for all sets $A \in \mathcal{P}(X)$ by the formula

$$^{\mathcal{D}}\underline{\mu}(A) = \inf_{p \in \mathcal{D}} \sum_{x \in A} p(x). \tag{4.11}$$

Similarly, the associated *upper probability function*, $^{\mathcal{D}}\overline{\mu}$, is defined for all $A \in \mathcal{P}(X)$ by the formula

$$^{\mathcal{D}}\overline{\mu}(A) = \sup_{p \in \mathcal{D}} \sum_{x \in A} p(x). \tag{4.12}$$

It follows directly from Eqs. (4.11) and (4.12) that lower and upper probabilities are monotone measures.

Given a particular set $A \in \mathcal{P}(X)$, let $\dot{p}$ denote one of the probability distribution functions in $\mathcal{D}$ for which the infimum in Eq. (4.11) is obtained. Since

$$\sum_{x \in A} \dot{p}(x) + \sum_{x \notin A} \dot{p}(x) = 1$$

is required by probability theory for each set $A \in \mathcal{P}(X)$, $\dot{p}$ must also be a probability distribution function for which the supremum in Eq. (4.12) is obtained. Hence, the equation

$$^{\mathcal{D}}\overline{\mu}(A) = 1 - {}^{\mathcal{D}}\underline{\mu}(\overline{A}) \tag{4.13}$$

holds for all $A \in \mathcal{P}(X)$. Due to this property, functions $^{\mathcal{D}}\underline{\mu}$ and $^{\mathcal{D}}\overline{\mu}$ are called *dual* (or *conjugate*). One of them is sufficient for capturing information in $\mathcal{D}$; the other one is uniquely determined by Eq. (4.13).

It follows directly from Eqs. (4.11) and (4.12) that

$$^{\mathcal{D}}\underline{\mu}(A) \leq {}^{\mathcal{D}}\overline{\mu}(A) \tag{4.14}$$

for all $A \in \mathcal{P}(X)$ and, in addition,

$$^{\mathcal{D}}\underline{\mu}(\varnothing) = {}^{\mathcal{D}}\overline{\mu}(\varnothing) = 0, \tag{4.15}$$

$$^{\mathcal{D}}\underline{\mu}(X) = {}^{\mathcal{D}}\overline{\mu}(X) = 1. \tag{4.16}$$

Moreover, any lower probability function is superadditive. That is,

$$^{\mathcal{D}}\underline{\mu}(A \cup B) \geq {}^{\mathcal{D}}\underline{\mu}(A) + {}^{\mathcal{D}}\underline{\mu}(B) \tag{4.17}$$

for all $A, B \in \mathcal{P}(X)$ such that $A \cap B = \varnothing$. This follows form the fact that the infima for disjoint sets A and B may be obtained in Eq. (4.11) for distinct probability distribution functions in $\mathcal{D}$, but the infimum for $A \cup B$ must be obtained for a single probability distribution function in $\mathcal{D}$. By the same argument applied to the suprema for A, B, and $A \cup B$ in Eq. (4.12), it follows that the upper probability function is subadditive. That is,

$$^{\mathcal{D}}\overline{\mu}(A \cup B) \leq {}^{\mathcal{D}}\overline{\mu}(A) + {}^{\mathcal{D}}\overline{\mu}(B) \tag{4.18}$$

for all $A, B \in \mathcal{P}(X)$.

Lower probabilities also satisfy the inequality

$$\sum_{x \in X} {}^{\mathcal{D}}\underline{\mu}(\{x\}) \leq 1. \tag{4.19}$$

This can be shown by using Eqs. (4.16) and (4.17), and the associativity of the operation of set union:

$$1 = {}^{\mathcal{D}}\underline{\mu}(X) = {}^{\mathcal{D}}\underline{\mu}\left(\bigcup_{x \in X} \{x\} \right) \geq \sum_{x \in X} {}^{\mathcal{D}}\underline{\mu}(\{x\}).$$

In a similar way, it can be shown that

$$\sum_{x \in X} {}^{\mathcal{D}}\overline{\mu}(\{x\}) \geq 1. \tag{4.20}$$

Due to the inequality (4.14), the lower and upper probabilities form for each set $A \in \mathcal{P}(X)$ a closed interval

$$\left[{}^{\mathcal{D}}\underline{\mu}(A), {}^{\mathcal{D}}\overline{\mu}(A) \right]$$

of possible probabilities of that set. When $^{\mathcal{D}}\underline{\mu}(A) = {}^{\mathcal{D}}\overline{\mu}(A)$ for all $A \in \mathcal{P}(X)$, classical (precise) probabilities are obtained. It is important to realize that a lower probability function (or, alternatively, an upper probability function) can be derived from more than one set of probability distribution functions on a given set X. This possibility is illustrated by the following example.

EXAMPLE 4.4. Consider $X = \{x_1, x_2, x_3\}$ and the following two sets of probability distributions on X:

$$\mathcal{D}_1 = \{\langle p(x_1), p(x_2), p(x_3)\rangle \mid p(x_1) = p(x_2) = a, p(x_3) = 1 - 2a, a \in [0, 0.5]\},$$

$$\mathcal{D}_2 = \{\langle p(x_1), p(x_2), p(x_3)\rangle \mid p(x_1) = a, p(x_2) = b,$$
$$p(x_3) = 1 - a - b, a \in [0, 0.5], b \in [0, 0.5]\}.$$

A geometrical interpretation of these sets is shown in Figure 4.3*a*. It is easy to see that applying Eqs. (4.11) and (4.12) to these sets results in the same lower and upper probability functions given in Figure 4.3*b*. Moreover, any set $\mathcal{D}$ such that $\mathcal{D}_1 \subseteq \mathcal{D} \subseteq \mathcal{D}_2$ is also associated with these functions.

Associated with any given lower probability function μ on $\mathcal{P}(X)$ is the *unique* set, $^\mu\mathcal{D}$, of *all* probability distribution functions p on X that are *consistent* with μ (or *dominate* μ). That is,

$$^\mu\mathcal{D} = \left\{ p \mid \mu(A) \le \sum_{x \in A} p(x) \text{ for all } A \in \mathcal{P}(X) \right\}. \tag{4.21}$$

Clearly, $^\mu\mathcal{D}$ is the largest among those sets of probability distribution functions on X that are associated with μ. Given the lower probability function μ in Example 4.4, the unique set $^\mu\overline{\mathcal{D}}$ defined by Eq. (4.21) is clearly the set $\mathcal{D}_2$.

Alternatively, the set of *all* probability distributions p on X that are *consistent* with a given upper probability function $\bar{\mu}$ on $\mathcal{P}(X)$ (or are *dominated* by $\bar{\mu}$) is defined as follows:

$$^{\bar{\mu}}\mathcal{D} = \left\{ p \mid \bar{\mu}(A) \ge \sum_{x \in A} p(x) \text{ for all } A \in \mathcal{P}(X) \right\}. \tag{4.22}$$

However, $^{\bar{\mu}}\mathcal{D} = ^\mu\mathcal{D}$ due to the duality of μ and $\bar{\mu}$. Introducing the symbol

$$\boldsymbol{\mu} = \langle \mu, \bar{\mu} \rangle$$

for any dual pair of lower and upper probability functions, the symbols $^{\bar{\mu}}\mathcal{D}$ and $^\mu\mathcal{D}$ may conveniently be replaced with one common symbol $^\mu\mathcal{D}$.

It is important to realize that $^\mu\mathcal{D}$ is always a *closed convex set of probability distribution functions* on X. It is the intersection of the closed convex sets of probability distribution functions characterized by the individual inequalities in Eqs. (4.21) or (4.22). When $\boldsymbol{\mu}$ is derived by Eqs. (4.11) and (4.12) from a given set $\mathcal{D}$ that is not convex, then $\mathcal{D}$ is not necessarily contained in $^\mu\mathcal{D}$.

All lower probability functions are superadditive and so are all Choquet capacities of any order $k \ge 2$. These two classes of functions are thus compatible. This means that special types of Choquet capacities (of the various orders $k \ge 2$) represent in a natural way special types of lower probability functions.

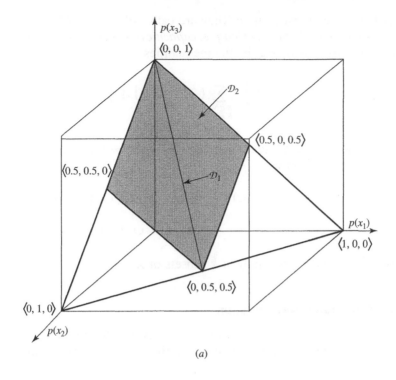

(a)

	x_1	x_2	x_3	$\underline{\mu}(A)$	$\overline{\mu}(A)$
A:	0	0	0	0.0	0.0
	1	0	0	0.0	0.5
	0	1	0	0.0	0.5
	0	0	1	0.0	1.0
	1	1	0	0.0	1.0
	1	0	1	0.5	1.0
	0	1	1	0.5	1.0
	1	1	1	1.0	1.0

(b)

Figure 4.3. Illustration to Example 4.4.

That is, Choquet capacities of each order form a basis for formalizing a particular theory of imprecise probabilities.

4.3.2. Alternating Choquet Capacities

Due to the duality of lower and upper probabilities expressed by Eq. (4.13), each theory of imprecise probabilities may also be formalized in terms of the upper probabilities. In that case, Choquet capacities that are dual to those previously introduced must be used. They are called *alternating Choquet capacities of order k* (or *k-alternating Choquet capacities*) and are defined for all families of k subsets of X ($k \geq 2$) by the inequalities

$$\mu\left(\bigcap_{j=1}^{k} A_j\right) \leq \sum_{\substack{K \subseteq N_k \\ K \neq \varnothing}} (-1)^{|K|+1} \mu\left(\bigcup_{j \in K} A_j\right). \tag{4.23}$$

Moreover, alternating *Choquet capacities of order* ∞ (or ∞-*alternating*) are defined by the inequalities

$$\mu(A_1 \cap A_2 \cap \cdots \cap A_k) \leq \sum_{i} \mu(A_i) - \sum_{i<j} \mu(A_i \cup A_j) + - \cdots$$
$$+ (-1)^{k+1} \mu(A_1 \cup A_2 \cup \cdots \cup A_k) \tag{4.24}$$

for every $k \geq 2$ and every family of k subsets of X.

4.3.3. Interaction Representation

Given a finite universal set $X = \{x_i \mid i \in N_n\}$, let $\widetilde{X}$ denote the set of all $n!$ permutations of X. Denoting for convenience, the set of all permutations of N_n by Π_n, we have

$$\widetilde{X} = \{\mathbf{x}_\pi = \langle x_{\pi(i)} \mid i \in N_n, \pi \in \Pi_n\rangle\}.$$

For each $\pi \in \Pi_n$ and each $k \in N_n$, let

$$A_{\pi,k} = \{x_{\pi(1)}, x_{\pi(2)}, \ldots, x_{\pi(k)}\},$$

and let $A_{\pi,0} = \varnothing$ for any $\pi \in \Pi_n$ by convention. Clearly, the sequence of sets $A_{\pi,k}$ for all $k \in N_{0,n}$ is one particular maximal chain of nested subsets of X in the Boolean lattice $\langle \mathcal{P}(X), \subseteq \rangle$. This chain is uniquely characterized by the chosen permutation π. Given a *regular* monotone measure μ on $\mathcal{P}(X)$, a sequence of values $\mu(A_{\pi,k})$ for $k \in N_{0,n}$ is associated with subsets in the chain. Clearly, $\mu(A_{\pi,0}) = 0$, $\mu(A_{\pi,n}) = 1$,

$$\mu(A_{\pi,k}) - \mu(A_{\pi,k-1}) \in [0, 1]$$

for all $k \in N_n$, and

$$\sum_{k=1}^{n} [\mu(A_{\pi,k}) - \mu(A_{\pi,k-1})] = 1.$$

Hence, probability distribution functions, $^{\mu\pi}p$, defined for each particular $\pi \in \Pi_n$ and all $k \in N_n$ by the formula

$$^{\mu\pi}p\left(x_{\pi(k)}\right) = \mu(A_{\pi,k}) - \mu(A_{\pi,k-1}) \tag{4.25}$$

are induced by μ in the Boolean lattice $\langle \mathcal{P}(X), \subseteq \rangle$. Clearly, functions defined for different permutations are not necessarily distinct. Let $^{\mu}\mathcal{B}$ denote the set of all distinct probability functions defined by Eq. (4.25). Then, clearly, $1 \le |^{\mu}\mathcal{B}| \le n!$.

EXAMPLE 4.5. Two distinct monotone measures, μ_1 and μ_2, are defined in Figures 4.4a and 4.5a. In Figures 4.4b and 4.5b, respectively, these measures are expressed in terms of the Hasse diagrams of the underlying Boolean lattice. Measure μ_1 is additive and, hence, it is completely characterized by a single probability distribution function p whose values are equal to the values of μ_1 for singletons. It can be easily verified that functions $^{\mu_1,\pi}p$ for all $\pi \in \Pi_3$ are equal to p. Measure μ_2 is a Choquet capacity of order 2. Functions $^{\mu_2,\pi}p$ for all $\pi \in \Pi_3$ are shown in Figure 4.5c. Clearly, $^{\mu_2}\mathcal{B}$ consists in this case of four probability distributions, p_1, p_2, p_3, p_4, as is obvious from Figure 4.5c.

The explained representation of a given monotone measure μ by the associated set $^{\mu}\mathcal{B}$ of probability distribution functions is usually called an *interaction representation* in the literature. The term "interaction" refers to the capability of monotone measures to express positive or negative interactions among disjoint sets with respect to the measured property. It turns out that the interactions intrinsic to a monotone measure μ are more explicitly expressed by the associated set $^{\mu}\mathcal{B}$.

The following are some basic properties for the interaction representation of monotone measures that pertain to imprecise probabilities (see Note 4.4):

	x_1	x_2	x_3	$\mu_1(A)$
A:	0	0	0	0.0
	1	0	0	0.5
	0	1	0	0.3
	0	0	1	0.2
	1	1	0	0.8
	1	0	1	0.7
	0	1	1	0.5
	1	1	1	1.0

$\{x_1, x_2, x_3\}$: **1.0**

$\{x_1, x_2\}$: **0.8** $\{x_1, x_3\}$: **0.7** $\{x_2, x_3\}$: **0.5**

$\{x_1\}$: **0.5** $\{x_2\}$: **0.3** $\{x_3\}$: **0.2**

$\varnothing$: **0.0**

(a)   (b)

Figure 4.4. Interaction representation of additive measure μ_1 (Example 4.5).

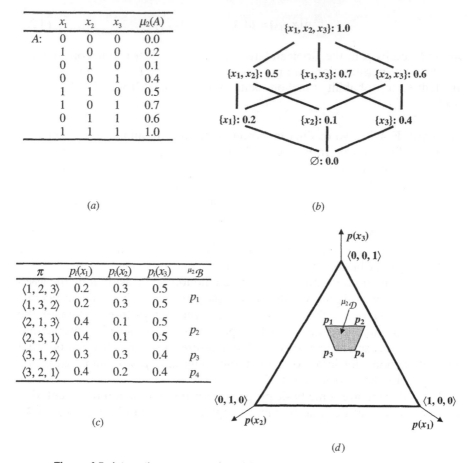

(a)

(b)

(c)

(d)

Figure 4.5. Interaction representation of 2-monotone measure (Example 4.5).

(i1) For any pair of dual monotone measures, $\mu = \langle \underline{\mu}, \overline{\mu} \rangle$, $^\mu B = {}^{\overline{\mu}} B$, but the same probability functions in $^\mu B$ and $^{\overline{\mu}} B$ are obtained for different permutations.

(i2) A monotone measure μ is additive iff $|^\mu B| = 1$.

(i3) Let $\mu = \langle \underline{\mu}, \overline{\mu} \rangle$ be a pair of dual monotone measures, so that $^\mu B = {}^{\overline{\mu}} B = {}^\mu B$. Then, $\underline{\mu}$ is a Choquet capacity of order 2 (and $\overline{\mu}$ is the associated alternating capacity of order 2) iff

$$\underline{\mu}(A) = \min_{p_j \in {}^\mu B} \sum_{x_i \in A} p_j(x_i),$$ (4.26)

$$\overline{\mu}(A) = \max_{p_j \in {}^\mu B} \sum_{x_i \in A} p_j(x_i)$$ (4.27)

for all $A \in \mathcal{P}(X)$. When $\underline{\mu}$ and $\bar{\mu}$ are more general monotone measures, Eqs. (4.26) and (4.27) are not applicable. In that case, $\underline{\mu}$ and $\bar{\mu}$ are determined from $^{\mu}\mathcal{B}$ if the permutations corresponding to each probability distribution function $^{\mu}\mathcal{B}$ are known.

(i4) A given monotone measure μ is a Choquet capacity of order 2 iff $^{\mu}\mathcal{B} \subseteq {}^{\mu}\mathcal{D}$.

(i5) If a given monotone measure μ is a Choquet capacity of order 2, then $^{\mu}\mathcal{B}$ is the set of extreme points of $^{\mu}\mathcal{D}$, which is commonly referred to as the *profile* of $^{\mu}\mathcal{D}$.

Property (i5) is particularly important. It allows us to determine $^{\mu}\mathcal{D}$ directly from the interaction representation $^{\mu}\mathcal{B}$ provided that μ is a Choquet capacity of order 2.

EXAMPLE 4.6. Consider the lower and upper probability functions $\underline{\mu}$ and $\bar{\mu}$ defined in Figure 4.3b. Hasse diagrams of the Boolean lattice with values $\underline{\mu}$ (A) and $\bar{\mu}(A)$ are shown in Figure 4.6a and 4.6b, respectively. Probability distribution functions $^{\mu,\pi}p$ and $^{\bar{\mu},\pi}p$ for all permutations $\pi \in \Pi_3$ are shown in Figure 4.6c and 4.6d, respectively. We can see that $^{\mu}\mathcal{B} = {}^{\bar{\mu}}\mathcal{B} = {}^{\mu}\mathcal{B}$ is the set of the extreme points of $^{\mu}\mathcal{D}$, which are shown in Figure 4.3a.

EXAMPLE 4.7. The lower probability function μ_2 defined in Figrue 4.5a is a Choquet capacity of order 2. According to property (i5) of the interaction representation, $^{\mu}\mathcal{B} = \{p_1, p_2, p_3, p_4\}$ (given in Figure 4.5c) is the set of extreme points of $^{\mu}\mathcal{D}$. Hence, $^{\mu}\mathcal{D}$ is characterized by the linear combination of these points. Locations of the extreme points in the probabilistic simplex and the set of all points in $^{\mu}\mathcal{D}$ are shown in Figure 4.5d.

4.3.4. Möbius Representation

When the Möbius transform in Eq. (4.8) is applied to lower and upper probability functions, $\underline{\mu}$ and $\bar{\mu}$, distinct functions, $\underline{m}$ and $\bar{m}$, are obtained respectively. By applying the inverse transform in Eq. (4.9) to $\underline{m}$ and $\bar{m}$, we obtain $\underline{\mu}$ and $\bar{\mu}$, respectively. Since the functions $\underline{\mu}$ and $\bar{\mu}$ are dual, the corresponding functions $\underline{m}$ and $\bar{m}$ are dual as well. It is established that the duality of the latter functions is expressed for all $A \in \mathcal{P}(X)$ by the equation

$$\bar{m}(A) = (-1)^{|A|+1} \sum_{B|B \supseteq A} \underline{m}(B). \tag{4.28}$$

For more information, see Note 4.4.

EXAMPLE 4.8. Lower and upper probability functions, $\underline{\mu}$ and $\bar{\mu}$, and their Möbius representations, $\underline{m}$ and $\bar{m}$, are given in Table 4.3. Since functions $\underline{\mu}$ and

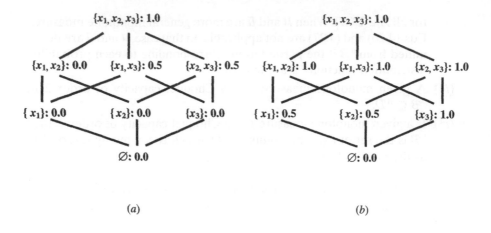

(a)   (b)

π	$p_j(x_1)$	$p_j(x_2)$	$p_j(x_3)$	$\underline{\mu}_{\mathcal{B}}$
$\langle 1,2,3 \rangle$	0.0	0.0	1.0	p_1
$\langle 2,1,3 \rangle$	0.0	0.0	1.0	
$\langle 1,3,2 \rangle$	0.0	0.5	0.5	p_2
$\langle 2,3,1 \rangle$	0.5	0.0	0.5	p_3
$\langle 3,1,2 \rangle$	0.5	0.5	0.0	p_4
$\langle 3,2,1 \rangle$	0.5	0.5	0.0	

π	$p_j(x_1)$	$p_j(x_2)$	$p_j(x_3)$	$\overline{\mu}_{\mathcal{B}}$
$\langle 1,2,3 \rangle$	0.5	0.5	0.0	p_1
$\langle 2,1,3 \rangle$	0.5	0.5	0.0	
$\langle 1,3,2 \rangle$	0.5	0.0	0.5	p_2
$\langle 2,3,1 \rangle$	0.0	0.5	0.5	p_3
$\langle 3,1,2 \rangle$	0.0	0.0	1.0	p_4
$\langle 3,2,1 \rangle$	0.0	0.0	1.0	

(c)   (d)

Figure 4.6. Illustration to Example 4.6.

Table 4.3. Duality of Möbius Representations of Lower and Upper Probability
Functions (Example 4.8)

	x_1	x_2	x_3	$\underline{\mu}(A)$	$\underline{m}(A)$	$\overline{\mu}(A)$	$\overline{m}(A)$
A:	0	0	0	0.0	0.0	0.0	0.0
	1	0	0	0.1	0.1	0.4	0.4
	0	1	0	0.3	0.3	0.7	0.7
	0	0	1	0.2	0.2	0.5	0.5
	1	1	0	0.5	0.1	0.8	−0.3
	1	0	1	0.3	0.0	0.7	−0.2
	0	1	1	0.6	0.1	0.9	−0.3
	1	1	1	1.0	0.2	1.0	0.2

$\bar{\mu}$ are dual, $\underline{m}$ and $\bar{m}$ are dual as well. This means that they uniquely determine each other via Eq. (4.28). For example,

$$\bar{m}(\{x_1, x_2\}) = -[\underline{m}(\{x_1, x_2\}) + \underline{m}(\{x_1, x_2, x_3\})]$$
$$= -[0.1 + 0.2] = -0.3,$$

$$\bar{m}(\{x_3\}) = \underline{m}(\{x_3\}) + \underline{m}(\{x_1, x_3\}) + \underline{m}(\{x_2, x_3\}) + \underline{m}(\{x_1, x_2, x_3\})$$
$$= 0.2 + 0.0 + 0.1 + 0.2 = 0.5,$$

$$\bar{m}(\{x_2, x_3\}) = -[\underline{m}(\{x_2, x_3\}) + \underline{m}(\{x_1, x_2, x_3\})]$$
$$= -[0.1 + 0.2] = -0.3.$$

4.3.5. Joint and Marginal Imprecise Probabilities

Let $\mathcal{D}$, $\underline{\mu}$, $\bar{\mu}$ and m denote the four basic representations of joint imprecise probabilities on a Cartesian product $X \times Y$, and let $\mathcal{D}_X$, $\mathcal{D}_Y$, $\underline{\mu}_X$, $\underline{\mu}_Y$, $\bar{\mu}_X$, $\bar{\mu}_Y$, m_X, and m_Y denote their marginal counterparts. For each of the four basic representations of the joint imprecise probabilities, its marginal counterparts are determined by appropriate rules of projection that are defined as follows.

- *Marginal sets of probability distributions*:

$$\mathcal{D}_X = \left\{ p_X \mid p_X(x) = \sum_{y \in Y} p(x, y) \text{ for some } p \in \mathcal{D} \right\}, \tag{4.29}$$

$$\mathcal{D}_Y = \left\{ p_Y \mid p_Y(y) = \sum_{x \in X} p(x, y) \text{ for some } p \in \mathcal{D} \right\}. \tag{4.30}$$

- *Marginal lower probabilities*:

$$\underline{\mu}_X(A) = \underline{\mu}(A \times Y) \text{ for all } A \in \mathcal{P}(X), \tag{4.31}$$

$$\underline{\mu}_Y(B) = \underline{\mu}(X \times B) \text{ for all } B \in \mathcal{P}(Y). \tag{4.32}$$

- *Marginal upper probabilities*:

$$\bar{\mu}_X(A) = \bar{\mu}(A \times Y) \text{ for all } A \in \mathcal{P}(X), \tag{4.33}$$

$$\bar{\mu}_Y(B) = \bar{\mu}(X \times B) \text{ for all } B \in \mathcal{P}(Y). \tag{4.34}$$

- *Marginal Möbius functions*:

$$m_X(A) = \sum_{R \mid A = R_X} m(R) \text{ for all } A \in \mathcal{P}(X), \tag{4.35}$$

$$m_Y(B) = \sum_{R|B=R_Y} m(R) \text{ for all } B \in \mathcal{P}(Y), \qquad (4.36)$$

where

$$R_X = \{x \in X \mid \langle x, y \rangle \in R \text{ for some } y \in Y\},$$
$$R_Y = \{y \in Y \mid \langle x, y \rangle \in R \text{ for some } x \in X\}.$$

It is well established that marginal lower and upper probability functions calculated by these formulas are measures of the same type as the given joint lower and upper probability functions.

4.3.6. Conditional Imprecise Probabilities

Given a lower probability function $\underline{\mu}$ or an upper probability function $\bar{\mu}$ on $\mathcal{P}(X)$, a natural way of defining conditional lower probabilities $\underline{\mu}(A \mid B)$ for any subsets A, B of X is to employ the associated convex set of probability distribution functions $\mathcal{D}$ (defined by Eq. (4.21) or Eq. (4.22), respectively). For each $p \in \mathcal{D}$, the conditional probability is defined in the classical way,

$$Pro(A \mid B) = \frac{Pro(A \cap B)}{Pro(B)},$$

and Eqs. (4.11) and (4.12) are modified for the conditional probabilities to calculate the lower and upper conditional probabilities. That is,

$$\underline{\mu}(A \mid B) = \inf_{p \in \mathcal{D}} \frac{\sum_{x \in A \cap B} p(x)}{\sum_{x \in B} p(x)}, \qquad (4.37)$$

$$\bar{\mu}(A \mid B) = \sup_{p \in \mathcal{D}} \frac{\sum_{x \in A \cap B} p(x)}{\sum_{x \in B} p(x)}, \qquad (4.38)$$

for all $A, B \in \mathcal{P}(X)$.

Consider now joint lower and upper probability functions $\underline{\mu}$ and $\bar{\mu}$ on $\mathcal{P}(X \times Y)$ and the associated set $\mathcal{D}$ of joint probability distribution functions p that dominate $\underline{\mu}$ and are dominated by $\bar{\mu}$. Then, for all $A \in X, B \in Y$ and $p \in \mathcal{D}$,

$$Pro(A \mid B) = \frac{Pro(A \times B)}{Pro(X \times B)} = \frac{\sum_{\langle x,y \rangle \in A \times B} p(x, y)}{\sum_{\langle x,y \rangle \in X \times B} p(x, y)},$$

and $\underline{\mu}(A \mid B)$ or $\bar{\mu}(A \mid B)$ are obtained, respectively, by taking the infimum or supremum of $Pro(A \mid B)$ for all $p \in \mathcal{D}$. Similarly,

$$Pro(B \mid A) = \frac{Pro(A \times B)}{Pro(A \times Y)} = \frac{\displaystyle\sum_{\langle x,y \rangle \in A \times B} p(x, y)}{\displaystyle\sum_{\langle x,y \rangle \in A \times Y} p(x, y)}$$

and $\underline{\mu}(B \mid A)$ or $\bar{\mu}(B \mid A)$ are obtained, respectively, by taking the infimum or supremum of $Pro(B \mid A)$.

4.3.7. Noninteraction of Imprecise Probabilities

It remains to address the following question: Given marginal imprecise probabilities on X and Y, how to define the associated joint imprecise probabilities under the assumption that the marginal ones are noninteractive? The answer depends somewhat on the type of monotone measures by which the marginal imprecise probabilities are formalized, as is discussed in Chapters 5 and 8. However, when operating at the general level of Choquet capacities of order 2, the question is adequately answered via the convex sets of probabilities distributions associated with the given marginal lower and upper probabilities as follows:

(a) Given marginal lower and upper probability functions, $\mu_X = \langle \underline{\mu}_X, \bar{\mu}_X \rangle$ and $\mu_Y = \langle \underline{\mu}_Y, \bar{\mu}_Y \rangle$, on $\mathcal{P}(X)$ and $\mathcal{P}(Y)$, respectively, determine the associated convex sets of marginal probability distributions, $\mathcal{D}_X$ and $\mathcal{D}_Y$, on X and Y.

(b) Assuming that μ_X and μ_Y are noninteractive, apply the notion of noninteraction in classical probability theory, expressed by Eq. (3.7), to probability distributions in set $\mathcal{D}_X$ and $\mathcal{D}_Y$ to define a unique set of joint probability distributions, $\mathcal{D}$, on set $X \times Y$.

(c) Apply Eqs. (4.11) and (4.12) to $\mathcal{D}$ to determine the lower and upper probability functions $\mu = \langle {}^{\mathcal{D}}\underline{\mu}, {}^{\mathcal{D}}\bar{\mu} \rangle$. These are, *by definition*, the unique joint lower and upper probability functions that correspond to *noninteractive* marginal lower and upper probability functions μ_X and μ_Y.

It is known (see Note 4.5) that the following properties hold under this definition for all $A \in \mathcal{P}(X)$ and all $B \in \mathcal{P}(Y)$:

$$\underline{\mu}(A \times B) = \underline{\mu}_X(A) \cdot \underline{\mu}_Y(B), \tag{4.39}$$

$$\bar{\mu}(A \times B) = \bar{\mu}_X(A) \cdot \bar{\mu}_Y(B) \tag{4.40}$$

$$\underline{\mu}[(A \times Y) \cup (X \times B)] = \underline{\mu}_X(A) + \underline{\mu}_Y(B) - \underline{\mu}_X(A) \cdot \underline{\mu}_Y(B), \tag{4.41}$$

$$\bar{\mu}[(A \times Y) \cup (X \times B)] = \bar{\mu}_X(A) + \bar{\mu}_Y(B) - \bar{\mu}_X(A) \cdot \bar{\mu}_Y(B). \quad (4.42)$$

Moreover, it is guaranteed that the marginal measures of $\mu = \langle \underline{\mu}, \bar{\mu} \rangle$ are again the given marginal measures $\mu_X = \langle \underline{\mu}_X, \bar{\mu}_X \rangle$ and $\mu_Y = \langle \underline{\mu}_Y, \bar{\mu}_Y \rangle$.

EXAMPLE 4.9. Consider the marginal lower and upper probability functions $\mu_X = \langle \underline{\mu}_X, \bar{\mu}_X \rangle$ and $\mu_Y = \langle \underline{\mu}_Y, \bar{\mu}_Y \rangle$ on sets $X = \{x_1, x_2\}$ and $Y = \{y_1, y_2\}$, respectively, which are given in Table 4.4a. Assuming that these marginal probability functions are noninteractive, the corresponding joint lower and upper probabilities $\mu = \langle \underline{\mu}, \bar{\mu} \rangle$ are uniquely determined by the introduced definition of noninteraction. One way to compute them is to directly follow the definition of noninteraction. Another, more convenient way is to use Eqs. (4.39)–(4.42) for all subsets of $X \times Y$ for which the equations are applicable and use the direct method only for the remaining subsets.

Following the definition of noninteraction, we need to determine first the convex sets $\mathcal{D}_X$ and $\mathcal{D}_Y$ of marginal probability distributions. From Table 4.4a,

$$p_X(x_1) \in [0.6, 0.8], \, p_X(x_2) \in [0.2, 0.4].$$
$$p_Y(y_1) \in [0.5, 0.9], \, p_Y(y_2) \in [0.1, 0.5].$$

Hence, $\mathcal{D}_X$ is the convex combination of the extreme distributions,

$$\langle p_X(x_1) = 0.6, p_X(x_2) = 0.4 \rangle \quad \text{and} \quad \langle p_X(x_1) = 0.8, p_X(x_2) = 0.2 \rangle,$$

that dominate the lower probabilities (Figure 4.7a) or, alternatively, are dominated by the upper probabilities (Figure 4.7c). That is,

$$p_X(x_1) \in \{0.6\lambda_X + 0.8(1 - \lambda_X)| \, \lambda_X \in [0,1]\}$$
$$= \{0.8 - 0.2\lambda_X \,| \, \lambda_X \in [0,1]\}$$
$$p_X(x_2) \in \{0.4\lambda_X + 0.2(1 - \lambda_X)| \, \lambda_X \in [0,1]\}$$
$$= \{0.2 + 0.2\lambda_X \,| \, \lambda_X \in [0,1]\}$$

and

$$\mathcal{D}_X = \{\langle 0.8 - 0.2\lambda_X, 0.2 + 0.2\lambda_X \rangle | \, \lambda_X \in [0,1]\}.$$

Similarly,

$$p_Y(y_1) \in \{0.5\lambda_Y + 0.9(1 - \lambda_Y)| \, \lambda_Y \in [0,1]\}$$
$$= \{0.9 - 0.4\lambda_Y \,| \, \lambda_Y \in [0,1]\}$$
$$p_Y(y_2) \in \{0.5\lambda_Y + 0.1(1 - \lambda_Y)| \, \lambda_Y \in [0,1]\}$$
$$= \{0.1 + 0.4\lambda_Y \,| \, \lambda_Y \in [0,1]\},$$

Table 4.4. Joint Lower and Upper Probability Functions Based on Noninteractive Marginal Lower and Upper Probability Functions (Example 4.9)

	x_1	x_2	$\underline{\mu}_X(A)$	$\bar{\mu}_X(A)$
A:	0	0	0.0	0.0
	1	0	0.6	0.8
	0	1	0.2	0.4
	1	1	1.0	1.0

	y_1	y_2	$\underline{\mu}_Y(B)$	$\bar{\mu}_Y(B)$
B:	0	0	0.0	0.0
	1	0	0.5	0.9
	0	1	0.1	0.5
	1	1	1.0	1.0

(a)

	x_1 y_1	x_1 y_2	x_2 y_1	x_2 y_2	$\underline{\mu}(C)$	$\bar{\mu}(C)$	$m(C)$
C:	0	0	0	0	0.00	0.00	0.00
	1	0	0	0	0.30	0.72	0.30
	0	1	0	0	0.06	0.40	0.06
	0	0	1	0	0.10	0.36	0.10
	0	0	0	1	0.02	0.20	0.02
	1	1	0	0	0.60	0.80	0.24
	1	0	1	0	0.50	0.90	0.10
	1	0	0	1	0.50	0.74	0.18
	0	1	1	0	0.26	0.50	0.10
	0	1	0	1	0.10	0.50	0.02
	0	0	1	1	0.20	0.40	0.08
	1	1	1	0	0.80	0.98	−0.10
	1	1	0	1	0.64	0.90	−0.18
	1	0	1	1	0.60	0.94	−0.18
	0	1	1	1	0.28	0.70	−0.10
	1	1	1	1	1.00	1.00	0.36

(b)

and

$$\mathcal{D}_Y = \{\langle 0.9 - 0.4\lambda_Y, 0.1 + 0.4\lambda_Y \rangle \mid \lambda_X \in [0,1]\},$$

as can be derived with the help of Figure 4.7b (or alternatively, Figure 4.7d). Values of λ_X and λ_Y are independent of each other.

Now applying the definition of noninteraction, we determine the set $\mathcal{D}$ of joint probability distributions $\mathbf{p}$ by taking pairwise products of components of the marginal probability distributions. That is,

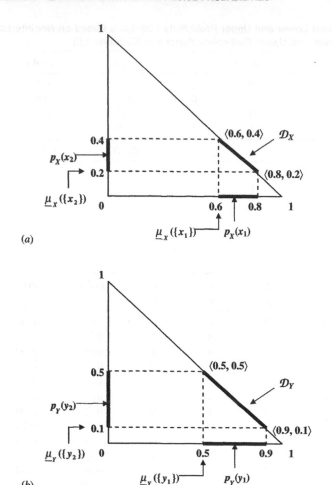

Figure 4.7. Illustration to Example 4.9. (a) $\mathcal{D}_X$ via lower probability $\underline{\mu}_X$. (b) $\mathcal{D}_Y$ via lower probability $\underline{\mu}_Y$.

$$\mathcal{D} = \{\langle p(x_i, y_j) = p_X(x_1) \cdot p_Y(y_j) | \ x_i \in X, y_j \in Y\rangle\}.$$

For example,

$$p(x_1, y_1) \in \{(0.8 - 0.2\lambda_X)(0.9 - 0.4\lambda_Y) | \ \lambda_X, \lambda_Y \in [0,1]\}.$$

Clearly, the minimum $p(x_1, y_1)$ is 0.3, and it is obtained for $\lambda_X = \lambda_Y = 1$; the maximum is 0.72, and it is obtained for $\lambda_X = \lambda_Y = 0$. Hence,

$$p(x_1, y_1) \in [0.3, 0.72],$$

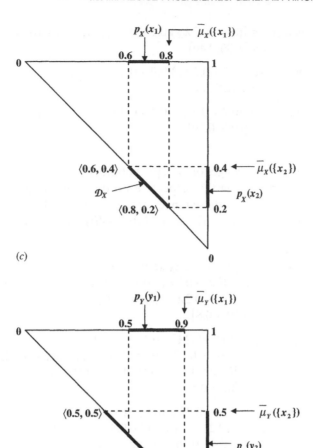

Figure 4.7. (c) $\mathcal{D}_X$ via upper probability $\bar{\mu}_X$. (d) $\mathcal{D}_Y$ via upper probability $\bar{\mu}_Y$.

and, consequently,

$$\underline{\mu}(\{x_1, y_1\}) = 0.3 \quad \text{and} \quad \bar{\mu}(\{x_1, y_1\}) = 0.72.$$

Similarly, we can calculate the ranges of joint probabilities $p(x_i, y_i)$ for the other pairs of $x_i \in X$ and $y_j \in Y$ whose minima and maxima are the joint lower and upper probabilities of singleton sets, which are given in Table 4.4b:

$$p(x_1, y_2) \in \{(0.8 - 0.2\lambda_X)(0.1 + 0.4\lambda_Y)| \; \lambda_X, \lambda_Y \in [0,1]\}$$
$$= [0.06, 0.40],$$

$$p(x_2, y_1) \in \{(0.2 + 0.2\lambda_X)(0.9 - 0.4\lambda_Y)| \; \lambda_X, \lambda_Y \in [0,1]\}$$
$$= [0.1, 0.36],$$

$$p(x_2, y_2) \in \{(0.2 + 0.2\lambda_X)(0.1 + 0.4\lambda_Y)| \; \lambda_X, \lambda_Y \in [0,1]\}$$
$$= [0.02, 0.20].$$

We can also calculate ranges of joint probability measures, *Pro*, for the other subsets of $X \times Y$ by adding the respective products of marginal probability. The minima and maxima of these ranges are, respectively, the lower and upper probabilities of these subsets (shown in Table 4.4*b*). For example,

$$Pro(\{\langle x_1, y_1 \rangle, \langle x_1, y_2 \rangle\}) \in \{(0.8 - 0.2\lambda_X)(0.9 - 0.4\lambda_Y)$$
$$+ (0.8 - 0.2\lambda_X)(0.1 + 0.4\lambda_Y)| \; \lambda_X, \lambda_Y \in [0,1]\}$$
$$= \{0.8 - 0.2\lambda_X \; | \; \lambda_X \in [0,1]\}$$
$$= [0.6, 0.8].$$

$$Pro(\{\langle x_1, y_2 \rangle, \langle x_2, y_1 \rangle\}) \in \{(0.8 - 0.2\lambda_X)(0.1 + 0.4\lambda_Y)$$
$$+ (0.2 + 0.2\lambda_X)(0.9 - 0.4\lambda_Y)| \; \lambda_X, \lambda_Y \in [0,1]\}$$
$$= \{0.26 + 0.16\lambda_X + 0.24\lambda_Y - 0.16\lambda_X\lambda_Y \; | \; \lambda_X, \lambda_Y \in [0,1]\}$$
$$= [0.26, 0.50].$$

$$Pro(\{\langle x_1, y_1 \rangle, \langle x_2, y_2 \rangle\}) \in \{(0.8 - 0.2\lambda_X)(0.9 - 0.4\lambda_Y)$$
$$+ (0.2 + 0.2\lambda_X)(0.1 + 0.4\lambda_Y)| \; \lambda_X, \lambda_Y \in [0,1]\}$$
$$= \{0.74 - 0.16\lambda_X - 0.24\lambda_Y + 0.16\lambda_X\lambda_Y \; | \; \lambda_X, \lambda_Y \in [0,1]\}$$
$$= [0.50, 0.74].$$

$$Pro(\{\langle x_1, y_1 \rangle, \langle x_1, y_2 \rangle, \langle x_2, y_1 \rangle\}) \in \{(0.8 - 0.2\lambda_X)(0.9 - 0.4\lambda_Y)$$
$$+ (0.8 - 0.2\lambda_X)(0.1 + 0.4\lambda_Y)$$
$$+ (0.2 + 0.2\lambda_X)(0.9 - 0.4\lambda_Y)| \; \lambda_X, \lambda_Y \in [0,1]\}$$
$$= \{0.98 - 0.02\lambda_X - 0.08\lambda_Y$$
$$- 0.08\lambda_X\lambda_Y \; | \; \lambda_X, \lambda_Y \in [0,1]\}$$
$$= [0.80, 0.98].$$

Using Eqs. (4.39)–(4.42), the joint lower and upper probabilities can be calculated more conveniently for all subsets of $X \times Y$ except the subsets $\{\langle x_1, y_1 \rangle, \langle x_2, y_2 \rangle\}$, and $\{\langle x_1, y_2 \rangle, \langle x_2, y_1 \rangle\}$. For example,

$$\underline{\mu}(\{\langle x_1, y_1 \rangle\}) = \underline{\mu}_X(\{x_1\}) \cdot \underline{\mu}_Y(\{y_1\})$$
$$= 0.6 \cdot 0.5 = 0.3,$$

$$\overline{\mu}(\{\langle x_1, y_1 \rangle\}) = \overline{\mu}_X(\{x_1\}) \cdot \overline{\mu}_Y(\{y_1\})$$
$$= 0.8 \cdot 0.9 = 0.72,$$

$$\underline{\mu}(\{\langle x_1, y_1 \rangle, \langle x_1, y_2 \rangle\}) = \underline{\mu}_X(\{x_1\}) \cdot \underline{\mu}_Y(Y)$$
$$= 0.6 \cdot 1 = 0.6,$$

$$\overline{\mu}(\{\langle x_1, y_1 \rangle, \langle x_1, y_2 \rangle\}) = \overline{\mu}_X(\{x_1\}) \cdot \overline{\mu}_Y(Y)$$
$$= 0.8 \cdot 1 = 0.8,$$

$$\underline{\mu}(\{\langle x_1, y_1 \rangle, \langle x_1, y_2 \rangle, \langle x_2, y_1 \rangle\}) = \underline{\mu}[(\{x_1\} \times Y) \cup (X \times \{y_1\})]$$
$$= \underline{\mu}_X(\{x_1\}) + \underline{\mu}_Y(\{y_1\}) - \underline{\mu}_X(\{x_1\}) \cdot \underline{\mu}_Y(\{y_1\})$$
$$= 0.6 + 0.5 - 0.6 \cdot 0.5 = 0.8,$$

$$\overline{\mu}(\{\langle x_1, y_1 \rangle, \langle x_1, y_2 \rangle, \langle x_2, y_1 \rangle\}) = \overline{\mu}[(\{x_1\} \times Y) \cup (X \times \{y_1\})]$$
$$= \overline{\mu}_X(\{x_1\}) + \overline{\mu}_Y(\{y_1\}) - \overline{\mu}_X(\{x_1\}) \cdot \overline{\mu}_Y(\{y_1\})$$
$$= 0.8 + 0.9 - 0.8 \cdot 0.9 = 0.98.$$

4.4. ARGUMENTS FOR IMPRECISE PROBABILITIES

The need for enlarging the framework of classical probability theory by allowing imprecision in probabilities has been discussed quite extensively in the literature, and many arguments for imprecise probabilities have been put forward. The following are some of the most common of these arguments.

1. Situations in which given information implies a *set of probability distributions* on a given set X are quite common. One type of such situations is illustrated by Example 4.3. All these situations can be properly formalized by imprecise probabilities, as is shown in Section 4.3, but not by classical, precise probabilities. Choosing any particular probability distribution from the given set, as is required by classical probability theory, is utterly arbitrary. Regardless how this choice is justified within classical probability theory, it is not supported by available information. The extreme case is reached in a situation of *total ignorance*—a situation in which no information about probability distributions is available. This situation, in which all probability distributions on X are possible, is properly formalized by special lower and upper probability functions that are defined for all $A \in \mathcal{P}(X)$ by the formulas

$$\underline{\mu}(A) = \begin{cases} 0 \text{ when } A \neq X \\ 1 \text{ when } A = X. \end{cases} \tag{4.43}$$

$$\overline{\mu}(A) = \begin{cases} 0 \text{ when } A = \varnothing \\ 1 \text{ when } A \neq \varnothing. \end{cases} \tag{4.44}$$

These lower and upper probabilities define, in turn, the following ranges of probabilities, *Pro*, that are maximally imprecise:

$$Pro(A) \in [0,1] \text{ for all } A \in \mathcal{P}(X) - \{\varnothing, X\},$$

and, of course, $Pro(\varnothing) = 0$ and $Pro(X) = 1$. These maximally imprecise probabilities are usually called *vacuous probabilities*. It is obvious that they are associated with the set of all probability distributions on X.

2. Imprecision of probabilities is needed to reflect the amount of statistical information on which they are based. The precision should increase with amount of statistical information. Imprecise probabilities allow us to utilize this sensible principle methodologically. As a simple example, let X denote a finite set of states of a variable and let the variable be observed at discrete times. Assume that in a sequence of N observations of the variable, each state $x \in X$ was observed $n(x)$-times. According to classical probability theory, probabilities of individual states, $p(x)$, are estimated by the ratios $n(x)/N$ for all $x \in X$. While these estimates are usually acceptable when N is sufficiently large relative to the number of all possible states, they are questionable when N is small. An alternative is to estimate lower and upper probabilities, $\underline{\mu}(x)$ and $\overline{\mu}(x)$, in such a way that we start with the maximum imprecision ($\underline{\mu}(x) = 0$ and $\overline{\mu}(x) = 1$ for all $x \in X$) when $N = 0$ (total ignorance) and define the imprecision (expressed by the differences $\underline{\mu}(x) - \overline{\mu}(x)$) by a function that is monotone decreasing with N. This can be done, for example, by using for each $x \in X$ the functions (estimators)

$$\underline{\mu}(x) = \frac{n(x)}{N + c}, \tag{4.45}$$

$$\overline{\mu}(x) = \frac{n(x) + c}{N + c}, \tag{4.46}$$

where $c \geq 1$ is a coefficient that expresses how quickly the imprecision in estimated probabilities decreases with the amount of statistical information, which is expressed by the value of N. The chosen value of c expresses the caution in estimating the probabilities. The larger the value, the more cautious the estimators are. As a simple example, let $X = \{0, 1\}$ be a set of states of a single variable v, whose observations, $v(t)$, at discrete times $t \in \mathbb{N}_{50}$ are given in Figure 4.8a. These observations were actually randomly generated with probabilities $p(0) = 0.39$ and $p(1) = 0.61$. Figure 4.8b shows lower and upper probabilities of $x = 0$ estimated for each $N \in \mathbb{N}_{50}$ by Eqs. (4.45) and (4.46), respectively, with $c = 4$. Figure 4.8c shows the same for $x = 1$.

3. Classical probability theory requires that $Pro(A) + Pro(\overline{A}) = 1$. This means, for example, that a little evidence supporting A implies a large amount of evidence supporting $\overline{A}$. However, in many real-life situations, we have a little evidence supporting A and a little evidence supporting $\overline{A}$ as well. Suppose, for

$v(t) = $ 0111001010100111010111010
1101110111011110101011011

(a)

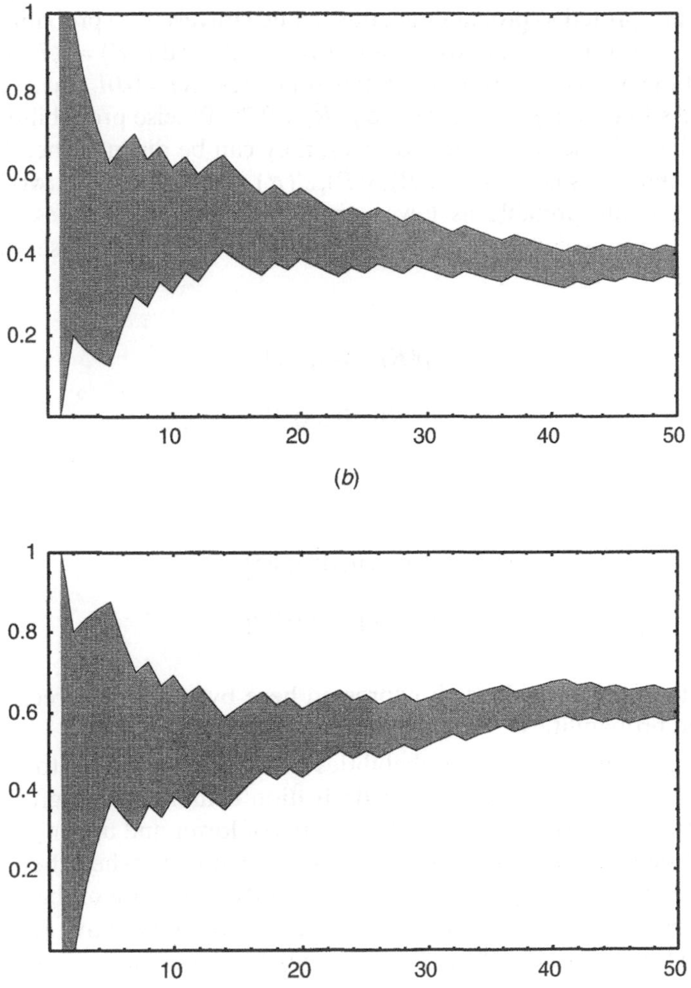

(b)

(c)

Figure 4.8. Example of the decrease of imprecision in estimated probabilities with the amount of statistical information.

example, that an old painting was discovered that somewhat resembles paintings of Raphael. Such a discovery is likely to generate various questions regarding the status of the painting. The most obvious questions are:

(a) Is the discovered painting a genuine painting by Raphael?

(b) It the discovered painting a counterfeit?

According to these questions, the discovered painting belongs either to the set of Raphael's paintings, R, or to its complement, $\bar{R}$. Evidential support for R and $\bar{R}$, assessed by an expert after examining the painting, may be expressed by numbers $s(R)$ and $s(\bar{R})$ in $[0, 1]$. Assume that the assessment is rather small for both R and $\bar{R}$, say $s(R) = 0.1$ and $s(\bar{R}) = 0.3$. These numbers cannot be directly interpreted as probabilities, but can be converted to probabilities $p(R)$ and $p(\bar{R})$ by normalization. We obtain $p(R) = 0.25$ and $p(\bar{R}) = 0.75$. Assuming now that the assessment were ten times smaller, $s'(R) = 0.01$ and $s'(\bar{R}) = 0.03$, we would still obtain $p(R) = 0.25$ and $p(\bar{R}) = 0.75$. Precise probabilities cannot thus distinguish the two cases. However, they can be distinguished by imprecise probabilities, since values $s(R)$, $s(\bar{R})$, $s'(R)$, and $s'(\bar{R})$ in the two examples are directly interpretable as lower probabilities. In both cases, we obtain only imprecise assessments of the probabilities. In the first case, the assessments are

$$p(R) \in [0.1, 0.7],$$

$$p(\bar{R}) \in [0.3, 0.9],$$

while in the second case they are

$$p'(R) \in [0.01, 0.97],$$

$$p'(\bar{R}) \in [0.03, 0.99].$$

Clearly, the weaker evidence is expressed here by greater imprecision of the estimated probabilities.

4. In general, imprecise probabilities are not only easier to elicit from experts than precise ones, but their derivation is more meaningful. Allowing experts to express their assessments in terms of lower and upper probabilities does not deter them from making precise assessments, when they feel comfortable doing so, but gives them more flexibility. Imprecise probabilities, contrary to precise ones, can also be constructed from qualitative judgments of various types that are easy to understand and justify in some applications. Moreover, we may not be able to elicit precise probabilities in practice, even if that is possible in principle. This may be due to lack of time or the computational resources needed for a thorough analysis of a complex body of evidence.

5. Precise probabilistic assessments regarding some decision-making situation that are obtained from several sources (sensors, experts, individuals of a group in a group decision) are often inconsistent. In order to be able to combine information from the individual sources, their inconsistencies must be resolved. A natural way to do that is to use imprecise probabilities. The degree of imprecision in the combined probabilities is a manifestation of the

extent to which the sources are inconsistent. This interesting methodological issue is discussed in Chapter 9.

6. Assume that a certain monetary unit, u, is available to bet on the occurrence of the various events $A \in \mathcal{P}(X)$. If we know the probability, $Pro(A)$, of event A, the highest acceptable rate (a fraction of u) for betting on the occurrence of A is $Pro(A)$. This means that we are willing to bet no more than $u \cdot Pro(A)$ to receive u when A occurs and to receive nothing when A does not occur. We are also willing to bet no more than $u(1 - Pro(A))$ against the occurrence of A. Imprecise probabilities are obviously needed in any betting situation in which relevant information is not sufficient to determine a unique probability measure on recognized events. In such situations, the lower probability of any event A is the highest acceptable rate for betting on the occurrence of A, while the upper probability of A is the highest acceptable rate for betting against the occurrence of A. If a utility function f defined on X is also involved, the betting behavior is guided by the lower and upper expected values of f (Section 4.5).

4.5. CHOQUET INTEGRAL

When working with imprecise probabilities, the classical notion of expected value of a given function is generalized. Instead of one expected value, we have now a range of expected values. This range is captured by lower and upper expected values, which are based on lower and upper probability functions, respectively. Since lower and upper probability functions are monotone measures, which in general are nonadditive, the classical Lebesgue integral, which is applicable only to additive measures, must be appropriately generalized to calculate an expected value of a function with respect to a monotone measure. It is generally recognized in the literature that a proper generalization of the Lebesgue integral to monotone measures is a functional that is called a *Choquet integral.*

Given a real-valued function f (a utility function) on set X, a monotone measure μ on an appropriate family C of subsets of X that contains $\varnothing$ and X, and a particular set $A \in C$, the Choquet integral of f with respect to μ on A, $(C)\int_A f d\mu$, is a functional defined by the equation

$$(C)\int_A f \, du = \int_A \mu(A \cap {}^{\alpha}F) \, d\alpha, \tag{4.47}$$

where ${}^{\alpha}F = \{x \mid f(x) \geq \alpha\}$. Observe that the Choquet integral is defined via a special Lebesgue integral (the one on the right-hand side of this equation).

The Choquet integral possesses some properties that make it a meaningful generalization of the Lebesgue integral for monotone measures. They include (see Note 4.6):

- When μ is an additive measure, the Choquet integral coincides with the Lebesgue integral:

$$(C)\int_A f \, d\mu = \int_A f \, d\mu.$$

- If $f_1(x) \le f_2(x)$ for all $x \in A$, then

$$(C)\int_A f_1 \, d\mu \le (C)\int_A f_2 \, d\mu.$$

- If $\mu_1(B) \le \mu_2(B)$ for all $B \in P(A)$, then

$$(C)\int_A f \, d\mu_1 \le (C)\int_A f \, d\mu_2.$$

- For any nonnegative constant c,

$$(C)\int_A (c \cdot f) \, d\mu = c(C)\int_A f \, d\mu,$$

$$(C)\int_A (c + f) \, d\mu = c\mu(A) + (C)\int_A f \, d\mu.$$

- When f is a nonnegative function, the Choquet integral $(C)\int_A f d\mu$ produces a measure that preserves the following properties of the given monotone measure μ: monotonicity, superadditivity, subadditivity, continuity from above, and continuity from below.

Applying the Choquet integral of function f on X to relevant lower and upper probability functions, $\underline{\mu}$ and $\bar{\mu}$, we obtain the following lower and upper expected values of f:

$$\underline{a}(f,\underline{\mu}) = (C)\int_X \underline{\mu}(X \cap {}^\alpha F) \, d\alpha, \tag{4.48}$$

$$\bar{a}(f,\bar{\mu}) = (C)\int_X \bar{\mu}(X \cap {}^\alpha F) \, d\alpha, \tag{4.49}$$

Hence, the expected value is expressed imprecisely by the interval $[\underline{a}(f, \underline{\mu}), \bar{a}(f, \bar{\mu})]$.

When X is a finite set and f is a nonnegative function, it is convenient to introduce the following special notation to simplify the computation of the Choquet integral.

Let $X = \{x_1, x_2, \ldots, x_n\}$ and $f(x_k) = f_k$. Assume that elements of X are permuted in such a way that $f_1 \ge f_2 \ge \ldots \ge f_n \ge 0$ and let $f_{n+1} = 0$ by convention. For each $x_k \in X$, let $A_k = \{x_1, x_2, \ldots, x_k\}$. Then, given a monotone measure μ on

X, the Choquet integral of f on X with respect to μ can be expressed by the simple formula

$$(C)\int_X f\, d\mu = \sum_{k=1}^{n} (f_k - f_{k+1})\mu(A_i),\qquad(4.50)$$

where $f_{n+1} = 0$ by convention.

EXAMPLE 4.10. To illustrate the use of Eq. (4.50), consider the joint lower and upper probability functions on $X \times Y$ in Table 4.4b, and assume that the following function is defined on the same set $X \times Y$: $f(\langle x_1, y_1 \rangle) = 5$, $f(\langle x_1, y_2 \rangle) = 3$, $f(\langle x_2, y_1 \rangle) = 2$, $f(\langle x_2, y_2 \rangle) = 3$. Let $Z = \{z_1, z_2, z_3, z_4\} = X \times Y$, where $z_1 = \langle x_1, y_1 \rangle$, $z_2 = \langle x_1, y_2 \rangle$, $z_3 = \langle x_2, y_2 \rangle$, $z_4 = \langle x_2, y_1 \rangle$, and let $f(z_k) = f_k$. Then, $f_1 \geq f_2 \geq f_3 \geq f_4$ and Eq. (4.50) is applicable for calculating the lower and upper expected values of f:

$$\underline{a}(f,\underline{\mu}) = \sum_{k=1}^{4}(f_k - f_{k+1})\underline{\mu}(\{z_1, z_2, \cdots, z_k\})$$
$$= 2 \cdot 0.30 + 1 \cdot 0.60 + 0 \cdot 0.64 + 2 \cdot 1 = 3.20$$

$$\overline{a}(f,\overline{\mu}) = \sum_{k=1}^{4}(f_k - f_{k+1})\overline{\mu}(\{z_1, z_2, \cdots, z_k\})$$
$$= 2 \cdot 0.72 + 1 \cdot 0.80 + 0 \cdot 0.90 + 2 \cdot 1 = 4.24.$$

Hence, the expected value of f is in the interval $[3.20, 4.24]$.

4.6. UNIFYING FEATURES OF IMPRECISE PROBABILITIES

Imprecise probabilities can be formalized in terms of classical set theory in many different ways, each based on using a monotone measure of some particular type. An important class of types of monotone measures are the Choquet capacities of orders $2, 3, \ldots$, which are introduced in Section 4.2. This class plays an important role in the area of imprecise probabilities, primarily due to its large scope and the natural ordering of the described types of monotone measure by their generalities. Notwithstanding the significance of imprecise probabilities based on the Choquet capacities of various orders, other types of monotone measures have increasingly been recognized as useful for formalizing imprecise probabilities. Some of the most prominent among them are examined in Chapter 5.

Formalizing imprecise probabilities is thus associated with considerable diversity and this, in turn, results in high methodological diversity. Fortunately, this ever increasing diversification of the area of imprecise probabilities can be countered, at least to some extent, by the various common, and thus unifying, features of imprecise probabilities that are discussed in the preceding

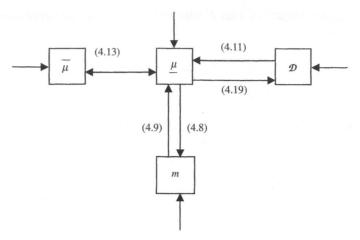

Figure 4.9. Unifying features of imprecise probabilities.

sections of this chapter. Before embarking on the examination of diverse theories of imprecise probabilities in Chapter 5, their unifying features are summarized as follows. For convenience, the summary is facilitated via Figure 4.9.

Perhaps the principal feature that unifies all imprecise probabilities is that each is invariably associated with a set of classical probability measures. This set may often be the initial representation of imprecise probabilities. It consists of all probability measures pertaining to a given situation that are consistent with available information. Clearly, its structure reflects the nature of the given information.

The set of probability measures derived from given information is usually, but not always, convex. If it is convex, then it can be converted to other, methodologically more convenient representations. These representations include lower or upper probability functions and their Möbius as well as interaction representations, all introduced in Section 4.3. Given any of these representations, it can be converted to the other ones as needed. Using symbols introduced in Section 4.3, the conversions are summarized in Figure 4.9. When applicable, references are made in the figure to equations that define the individual conversions.

If the set of probability measures derived from given information is not convex, the conversions are not applicable. This means that the imprecisions in probabilities have to be dealt with directly in terms of the given set. Alternatively, the nonconvex set can be approximated by the smallest convex set that contains it (its convex hull). However, the methodology for dealing with nonconvex sets of probability measures has not been adequately developed as yet.

A set of probability measures is not always the initial representation. Depending on the application context, any of the other representations may be a natural initial representation. In many applications, for example, we begin with lower or upper probabilities assessed by human experts and the set of probability measures consistent with them can be derived by the conversion equation when needed. In other applications, we begin with the Möbius representation or the interaction representation and convert them to the other representations as needed.

NOTES

4.1. *Classical measure theory* has been recognized as an important area of mathematics since the late 19th century. For a self-study, the classic text by Halmos [1950] is recommended. Among many other books, only a few are mentioned here, which are significant in various respects. The book by Caratheodory [1963], whose original German version was published in 1956, is one of the earliest and most influential books on classical measure theory. Books by Temple [1971] and Weir [1973] provide pedagogically excellent introductions to classical measure theory and require only basic knowledge of calculus and algebra as prerequisites. The books by Billingsley [1986] and Kingman and Taylor [1966] focus on the connections between classical measure theory and probability theory. The history of classical measure theory and Lebesgue integral is carefully traced in a fascinating book by Hawkins [1975]. He describes how modern mathematical concepts regarding these theories (involving concepts such as a function, continuity, convergence, measure, integration, and the like) developed (primarily in the 19th century and the early 20th century) through the work of many mathematicians, including Cauchy, Fourier, Borel, Riemann, Cantor, Dirichlet, Hankel, Jordan, Weierstrass, Volterra, Peano, Lebesgue, and Radon.

4.2. The idea of *nonadditive measures* is due to Gustave Choquet, a distinguished French mathematician. He conceived of this idea in 1953, during his one-year residence at the University of Kansas at Lawrence, and used it for developing his *theory of capacities*. The theory was published initially as a large research report at the University of Kansas and shortly afterwards in the open literature [Choquet, 1953–54]. In spite of this important publication, the idea of nonadditive measures was almost completely ignored by the scientific community for many years. It seems that is was recognized only in the early 1970s by Huber in his effort to develop a robust statistics (see [Huber, 1981]). It began to attract more attention only in the 1980s, primarily under the influence of work by Michio Sugeno on *monotone measures*, which he introduced in his doctoral dissertation [Sugeno, 1977] under the name "fuzzy measures." Although this name is ill-conceived (there is no fuzziness in these measures), it is unfortunately still widely used in the literature. Since the late 1970s, research on the theory of monotone (fuzzy) measures has been steadily growing. The outcome of this research, which is covered in too many papers to be cited here, is adequately represented by the following books: a textbook by Wang and Klir [1992], monographs by Denneberg [1994] and Pap [1995], and a book edited by Grabisch et al. [2000]. Additional useful sources include: a survey article by Sims and Wang [1990], a Special Issue of *Fuzzy Sets and Systems* on Fuzzy Measures and Integrals [**92**(2), 1997, pp. 137–264] and, above all, an extensive handbook edited by Pap [2002].

4.3. According to an interesting historical study made by Shafer [1978], some aspects of nonadditive (and, hence, imprecise) probabilities are recognizable in the work of Bernoulli and Lambert in the 17th and 18th centuries. However, these traces of nonadditivity were lost in the course of history and were rediscovered only in the second half of the 20th century. Perhaps the first thorough investigation of imprecise probabilities was made by Dempster [1967a,b], even though it was preceded by a few earlier, but narrower investigations. Since the publication of Dempster's papers, the number of publications on imprecise probabilities has been rapidly growing. However, most of these publications deal with various special types of imprecise probabilities. Some notable early exceptions are papers by Walley and Fine [1979, 1982] and Kyburg [1987]. Since the early 1990s, a greater emphasis can be observed in the literature on studying imprecise probabilities from various highly general perspectives. It is likely that this trend was influenced by the publication of an important book by Walley [1991]. Employing simple, but very fundamental principles of *avoiding sure loss, coherence*, and *natural extension*, Walley develops a general theory of imprecise probabilities in this book. In addition to its profound contribution to the area of imprecise probabilities, the book is also a comprehensive guide to the literature and history of this area. Short versions of the material in the book and some new ideas are presented in [Walley, 1996, 2000]. An important resource for researchers in the area of imprecise probabilities is a Web site dedicated to the "Imprecise Probabilities Project." The purpose of the project is "to help advance the theory and applications of imprecise probabilities, mainly by the dissemination of relevant information." The Web site, whose address is http://ippserv.rug.ac.be, contains a bibliography, information about people working in the field, abstracts of recent papers, and other relevant information. Associated with the Imprecise Probability Project are biennial International Symposia on Imprecise Probabilities and Their Applications (ISIPTA), which were initiated in 1999.

4.4. While the Möbius transform is well established in combinatorics, the interaction representation of monotone measures has its roots in the theory of cooperative games. These connections are discussed in several papers by Grabisch [1997*a–c*, 2000]. Mathematical properties of the Möbius representation and the interaction representation as well as conversions between these representations are thoroughly investigated in [Grabisch, 1997a]. In particular, this paper contains a derivation of Eq. (4.28). Properties $(m1)$–$(m7)$ of the Möbius representation of Choquet capacities, which are listed in Section 4.2.1, are proven in [Chateauneuf and Jaffray, 1989]. Properties $(i1)$–$(i5)$ of the interaction representation, which are listed in Section 4.3.3, are proven in [De Campos and Bolaños [1989] as well as in [Chateauneuf and Jaffray, 1989]. Miranda et al. [2003] investigates a generalization of the interaction representation to infinite sets.

4.5. The concept of noninteraction for lower and upper probabilities is investigated in [De Campos and Huete, 1993]. This paper contains proofs of Eqs. (4.39)–(4.42). The concept of conditional monotone measure is investigated in [De Campos et al., 1990]. Various algorithms for dealing with imprecise probabilities represented by convex sets of probability measures are presented in [Cano and Moral, 2000].

4.6. The Choquet integral, as well as other integrals based on monotone measures, are covered quite extensively in the literature. An excellent tutorial on this subject was written by Murofushi and Sugeno [2000]. Some other notable references dealing with various aspects of the Choquet integral include [Weber, 1984;

Murofushi and Sugeno, 1989, 1993; De Campos and Bolaños, 1992; Wang et al., 1996; Benvenuti and Mesiar, 2000; Dennenberg, 1994; and Pap, 1995, 2002]. Murofushi et al. [1994] show that the Choquet integral is applicable to set functions that are not monotone as well.

4.7. The various issues of constructing monotone measures from available evidence are discussed by Klir et al. [1997]. Reche and Salmerón [2000] develop a procedure for constructing lower and upper monotone measures that are compatible with coherent partial information. Coletti and Scozzafava [2002] deal with probability (imprecise in general) in terms for coherent partial assessments and coherent extensions.

4.8. It seems that the term "credal sets," which is fairly routinely used in the literature for closed and convex sets of probability distributions, emerged from some writings by Levi [1980, 1984, 1986, 1996, 1997].

4.9. The interaction representation is studied in papers by De Campos and Bolaños [1987] and Grabisch [1997a,b, 2000]. Grabisch shows that this representation is closely connected with the interaction between players, as represented in the mathematical theory of non-atomic games studied by Shapley [1971] and Auman and Shapley [1974].

4.10. Imprecise probabilities can also be studied within the mathematics of general interval structures. Relevant references in this regard are [Aubin and Frankowska, 1990] and [Wong et al., 1995]. Another way of studying imprecise probabilities is to look at them as deviations from precise (additive) probabilities [Ban and Gal, 2002].

EXERCISES

4.1. Determine which of the following set functions μ is a monotone measure or a semicontinuous monotone measure:

(a) Given $\langle X, C \rangle, \mu$ is defined for all $A \in C$ by the formula

$$\mu(A) = \begin{cases} 1 & \text{when } x_0 \in A \\ 0 & \text{when } x_0 \notin A, \end{cases}$$

where x_0 is a fixed element of X.

(b) Let $X = \{x_1, x_2, x_3\}$, $C = \mathcal{P}(X)$, and

$$\mu(A) = \begin{cases} 1 & \text{when } A = X \\ 0 & \text{when } A = \varnothing \\ 1/2 & \text{otherwise.} \end{cases}$$

(c) Let $X = \{1, 2, \ldots, n\}$, $C = \mathcal{P}(X)$, and $\mu(A) = |A|/n$ for all $A \in C$.

(d) Let X be the set of all positive integers, $C = \mathcal{P}(X)$ and $\mu(A) = \Sigma_{i \in A} i$ for all $A \in C$.

(e) Let $X = [0, 1]$, $C = \mathcal{P}(X)$, and $\mu(A) = \inf_{x \in A} f(x)$ for all $A \in C$, where f is a real-valued function on X.

(f) Repeat part (e) for $\mu(A) = \sup_{x \in A} f(x)$.

4.2. Determine the Möbius representations of the set functions defined in Exercise 4.1b and c.

4.3. Show that every additive measure is also monotone, but not the other way around.

4.4. Let μ be a regular monotone measure on $\langle X, C \rangle$ that is continuous from above, where C is a Boolean algebra of subsets of X. Show that the set function v defined on $\langle X, C \rangle$ by the equation

$$v(E) = 1 - \mu(\bar{E})$$

is a regular monotone measure that is continuous from below.

4.5. Construct some monotone measures on $\langle X = \{x_1, x_2, x_3\}, \mathcal{P}(X) \rangle$ that are:

(a) Superadditive

(b) Subadditive

(c) Neither superadditive nor subadditive

4.6. Determine for each monotone measure constructed in Exercise 4.5 its dual measure and the Möbius representations of the resulting pair of dual measures. Then, answer the question: Is one of the dual measures a Choquet capacity of order 2?

4.7. Show that every Choquet capacity of order k, where $k > 2$, is also a Choquet capacity of order $k - 1$, but not the other way around.

4.8. Consider the set of joint probability distribution functions defined at the end of Example 4.3. Determine the corresponding lower and upper probabilities, their Möbius representations, and whether or not the measure representing the lower probabilities is:

(a) Superadditive

(b) A Choquet capacity of order 2

Repeat the exercise symbolically for any values of the marginal probabilities (assume for convenience that $p_X(x_1) \geq p_Y(y_1)$).

4.9. Determine the interaction representation of the following monotone measures:

(a) Measures μ_3 and μ_4 in Table 4.2;

(b) The lower and upper probability measures in Table 4.3

(c) The monotone measures defined in Table 4.5.

Table 4.5. Monotone Measures in Exercises 4.9–4.14

| | x_1 | x_1 | x_2 | x_2 | $\mu_1(C)$ | $\mu_2(C)$ | $\mu_3(C)$ |
	y_1	y_2	y_1	y_2			
C:	0	0	0	0	0.00	0.00	0.0
	1	0	0	0	0.26	0.08	0.3
	0	1	0	0	0.26	0.00	0.4
	0	0	1	0	0.26	0.00	0.4
	0	0	0	1	0.00	0.00	0.2
	1	1	0	0	0.59	0.20	0.7
	1	0	1	0	0.53	0.40	0.7
	1	0	0	1	0.27	0.08	0.4
	0	1	1	0	0.53	0.00	0.8
	0	1	0	1	0.27	0.00	0.6
	0	0	1	1	0.27	0.00	0.6
	1	1	1	0	0.87	0.52	1.0
	1	1	0	1	0.61	0.20	0.7
	1	0	1	1	0.55	0.40	0.7
	0	1	1	1	0.55	0.00	0.9
	1	1	1	1	1.00	1.00	1.0

4.10. For each of the measures defined in Table 4.5, determine the following:
   (a) The dual measure;
   (b) Möbius representations of the measure and its dual;
   (c) The interaction representation.

4.11. For each of the measures defined in Table 4.5, determine the highest order $k(k \geq 1)$ for which the measure is a Choquet capacity.

4.12. For each of the measures defined in Table 4.5, determine the corresponding marginal measures.

4.13. For each of the marginal measures determined in Exercise 4.12, construct the joint measure based on the assumption of noninteraction.

4.14. For each of the marginal measures in Table 4.5 and their duals, calculate the Choquet integral of the following functions on $\{x_1, x_2\} \times \{y_1, y_2\}$ (for convenience, let $f(x_i, y_j) = f_{ij}$):
   (a) $f_{11} = 0.5, f_{12} = 1.0, f_{21} = 0.7, f_{22} = 0.3$
   (b) $f_{11} = 250, f_{12} = 120, f_{21} = 500, f_{22} = 750$
   (c) $f_{11} = 1, f_{12} = 3, f_{21} = 3, f_{22} = 1$

4.15. Perform computer experiments by generating two values, 0 and 1, of a single random variable with probabilities $p(0)$ and $p(1) = 1 - p(0)$. Apply Eqs. (4.45) and (4.46) to generated sequences of length $N \geq 0$ and observe the convergence to the given probabilities as a function N and c.

4.16. Generalize Eqs. (4.45) and (4.46) to more values and to more variables.

5

SPECIAL THEORIES OF IMPRECISE PROBABILITIES

I think it wiser to avoid the use of a probability model when we do not have the necessary data than to fill the gaps arbitrarily; arbitrary assumptions yield arbitrary conclusions

—Terrence L. Fine

5.1. AN OVERVIEW

This chapter is in some sense complementary to Chapter 4. While the focus of Chapter 4 is on the examination of the unifying features of theories of imprecise probabilities, the purpose of this chapter is to explore the great diversity of these theories. Clearly, this diversity results from the many ways in which monotone measures can be constrained by special requirements.

The theories of imprecise probabilities that are well developed and have been proved useful in some application contexts are covered in detail. Other theories, which have not been sufficiently developed or tested in applications as yet, are introduced only as themes for future research.

Among the theories of imprecise probabilities that are examined in this chapter are those based on the Choquet capacities of various orders, which are already known from Chapter 4. In particular, the one based on capacities of order ∞ is covered in detail. This theory, which is usually referred to in the literature as the *Dempster–Shafer theory*, is already quite well developed and has been utilized in many applications. Two more special theories, both sub-

Uncertainty and Information: Foundations of Generalized Information Theory, by George J. Klir
© 2006 by John Wiley & Sons, Inc.

sumed under the Dempster–Shafer theory, are also examined in detail: a theory based on *graded possibilities* and a theory based on special monotone measures that are called *Sugeno λ-measures* (or just *λ-measures*). One additional theory, which is based on monotone measures derived from *interval-valued probability distributions*, is covered in this chapter in detail. This theory is not comparable with the Dempster–Shafer theory, but it is subsumed under the theory based on Choquet capacities of order 2.

Ordering of the mentioned theories by levels of their generality is shown in Figure 5.1. Each arrow $T \rightarrow T'$ in the figure means that theory T' is more general than theory T. The presentation in this chapter follows these arrows, starting with the two least general theories of imprecise probabilities shown in the figure. One of them is a simple generalization of classical possibility theory, in which possibilities are graded. The other one is a simple generalization of classical probability theory, which is based on λ-measures. The presentation then proceeds to the Dempster–Shafer theory, and the theory based on interval-valued probability distributions. The chapter concludes with a survey of other types of monotone measures that can be used for formalizing imprecise probabilities.

5.2. GRADED POSSIBILITIES

The theory examined in this section is a generalization of the classical possibility theory, which is reviewed in Chapter 2. Instead of distinguishing only between possibility and impossibility, as in the classical possibility theory, the generalized possibility theory is designed to distinguish grades (or degrees) of possibility. It is thus appropriate to view it as a *theory of graded possibilities*.

In analogy with the classical possibility theory, its generalized counterpart is based on two dual monotone measures: a *possibility measure* and a *necessity measure*. Contrary to the classical possibility and necessity measures, whose values are in the set {0, 1}, the values of their generalized counterparts cover the whole unit interval [0, 1].

As in the classical case, it is convenient to formalize the generalized possibility theory in terms of generalized possibility measures, which are appropriate monotone measures that characterize graded possibilities. For each given generalized possibility measure, *Pos*, its dual generalized possibility measure, *Nec*, is then defined for each recognized set A by the duality equation

$$Nec(A) = 1 - Pos(\overline{A}), \tag{5.1}$$

which is a generalization of Eq. (2.4).

Family C on which generalized possibility measures are defined is required to be an *ample field*. This is a family of subsets of X that are closed under arbitrary unions and intersections, and under complementation in X. When X is finite, C is usually the whole power set of X.

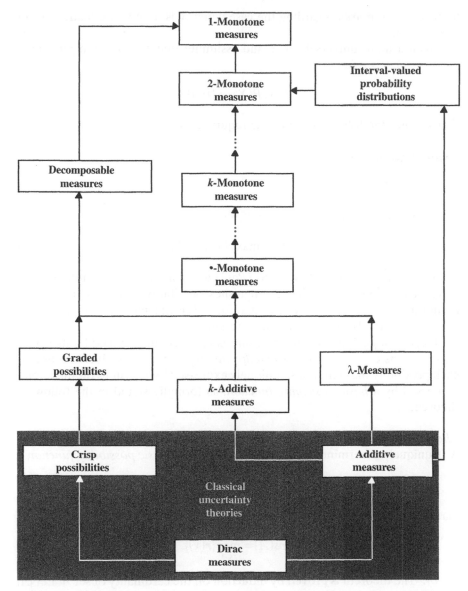

Figure 5.1. Ordering of monotone measures used for representing imprecise probabilities by their levels of generality. (Dirac measures are defined in Note 5.12.)

Since the generalized possibility theory subsumes the classical one as a special case and classical possibility, and necessity measures are special cases of their graded counterparts, it is sensible to omit the adjectives "generalized" and "graded" from now on.

For the sake of clarity, the following formalization of possibility measures is based on the assumption that the set of all considered alternatives, X, is finite. A few remarks regarding the case when X is infinite are made later in this section.

Given a finite universal set X and assuming that $C = \mathcal{P}(X)$, a *possibility measure*, *Pos*, is a function

$$Pos:\mathcal{P}(X)\to[0,1]$$

that satisfies the following axiomatic requirements:

Axiom (Pos1). $Pos(\varnothing) = 0$.

Axiom (Pos2). $Pos(X) = 1$.

Axiom (Pos3). For any sets $A, B \in \mathcal{P}(X)$,

$$Pos(A \cup B) = \max\{Pos(A), Pos(B)\}. \tag{5.2}$$

Observe that axioms (*Pos1*) and (*Pos2*) are shared by all monotone measures. It is axiom (*Pos3*) that distinguishes possibility measures from other monotone measures. Observe also that Eq. (5.2) is the limiting case of inequality (4.2) that holds for all monotone measures.

Recall that each probability measure is uniquely determined by its values on singletons, expressed by a *probability distribution* function. It turns our that each possibility measure is also uniquely expressed by its values on singletons, expressed by a *basic possibility function*, as formally stated in the following theorem.

Theorem 5.1. Every possibility measure, *Pos*, defined on subsets of a finite set X is uniquely determined for each $A \in \mathcal{P}(X)$ by a *basic possibility function*

$$r:X \to [0,1]$$

via the formula

$$Pos(A) = \max_{x \in A}\{r(x)\}. \tag{5.3}$$

Proof. The theorem is proved by induction on the cardinality of set A. Let $|A| = 1$. Then, $A = \{x\}$ for some $x \in X$, and Eq. (5.3) is trivially satisfied. Assume now that Eq. (5.3) is satisfied for $|A| = n - 1$ and let $A = \{x_1, x_1, \ldots, x_n\}$. Then, by Eq. (5.2),

$$\begin{aligned}
Pos(A) &= \max\{Pos(\{x_1, x_2, \ldots x_{n-1}\}), Pos(\{x_n\})\}\\
&= \max\left\{\max_{i \in \mathbb{N}_{n-1}}\{Pos(\{x_i\})\}, Pos(\{x_n\})\right\}\\
&= \max_{i \in \mathbb{N}_n}\{Pos(\{x_i\})\}\\
&= \max_{i \in \mathbb{N}_n}\{r(x_i)\}.
\end{aligned}$$

■

Although it can be easily shown that Eq. (5.2) implies monotonicity, this is even more transparent from Eq. (5.3). Observe also that

$$\max_{x \in X}\{r(x)\} = 1 \tag{5.4}$$

due to Axiom (*Pos2*). This property is usually called a *possibilistic normalization.*

In the literature function *r* is often called a possibility distribution function. In spite of its common use, this term is avoided in this book since it is misleading. The term "distribution function" makes perfectly good sense in probability theory, where any probability distribution function *p* is required to satisfy the equation

$$\sum_{x \in X} p(x) = 1.$$

That is, each probability distribution function actually *distributes* the fixed value of 1 among the individual elements of *X*. On the contrary, a basic possibility function *r does not distribute* any fixed value among the elements of *X*, as is obvious from the following inequalities

$$1 \le \sum_{x \in X} r(x) \le |X|.$$

Basic possibility functions are thus non-distributive and it is misleading to call them distribution functions.

The property that distinguishes necessity measures from other monotone measures can be determined from the duality between possibility and necessity measures, as expressed by the following theorem.

Theorem 5.2. Let *Pos* denote a particular possibility measure defined on subsets of a finite set *X*. If *Nec* is a particular necessity measure that is dual to *Pos* via Eq. (5.1), then for any sets $A, B \in \mathcal{P}(X)$

$$Nec(A \cap B) = \min\{Nec(A), Nec(B)\}. \tag{5.5}$$

Proof

$$\begin{aligned}
Nec(A \cap B) &= 1 - Pos\overline{(A \cap B)} \\
&= 1 - Pos(\overline{A} \cup \overline{B}) \\
&= 1 - \max\{Pos(\overline{A}), Pos(\overline{B})\} \\
&= 1 - \max\{1 - Nec(A), 1 - Nec(B)\} \\
&= 1 - [1 - \min\{Nec(A), Nec(B)\}] \\
&= \min\{Nec(A), Nec(B)\}. \qquad \blacksquare
\end{aligned}$$

Observe that the distinguishing property Eq. (5.5) of necessity measures is the limiting case of the inequality (4.1) that holds for all monotone measures. Observe also that possibility theory may equally well be formalized via axioms of necessity measures and possibility measures defined via the duality equation. The following is this alternative formalization.

Given a finite universal set X, a *necessity measure*, *Nec*, is a function

$$Nec:\mathcal{P}(X)\to[0,1]$$

that satisfies the following axiomatic requirements:

Axiom (Nec1). $Nec(\varnothing) = 0$.

Axiom (Nec2). $Nec(X) = 1$.

Axiom (Nec3). For any sets $A, B \in \mathcal{P}(X)$,

$$Nec(A \cap B) = \min\{Nec(A), Nec(B)\}.$$

Necessity measures do not possess a natural counterpart of the property in Eq. (5.3), which allows any possibility measure to be fully determined by its values on singletons. Therefore, the formalization of possibility theory in terms of possibility measures is more transparent and offers some methodological advantages.

In addition to their axiomatic properties, Eqs. (5.2) and (5.5), possibility and necessity measures must also conform to the properties in Eqs. (4.1) and (4.2) of general monotone measures. This means that

$$Pos(A \cap B) \leq \min\{Pos(A), Pos(B)\} \tag{5.6}$$

and

$$Nec(A \cup B) \geq \max\{Nec(A), Nec(B)\} \tag{5.7}$$

for all $A, B \in \mathcal{P}(X)$. Moreover,

$$\max\{Pos(A), Pos(\overline{A})\} = 1 \tag{5.8}$$

and

$$\min\{Nec(A), Nec(\overline{A})\} = 0. \tag{5.9}$$

These equations follow directly from the axioms of possibility and necessity measures, respectively. A direct consequence of Eq. (5.8) is the inequality

$$Pos(A) + Pos(\overline{A}) \geq 1, \tag{5.10}$$

and applying to it the duality equation, Eq. (5.1), we obtain the equation

$$Nec(A) + Nec(\overline{A}) \leq 1. \tag{5.11}$$

It can also easily be shown that

$$Nec(A) \leq Pos(A) \tag{5.12}$$

for all $A \in \mathcal{P}(X)$. Clearly, if $Pos(A) < 1$, then $Pos(\overline{A}) = 1$ by Eq. (5.8) and $Nec(A) = 0$ by Eq. (5.1). Hence, $Nec(A) \leq Pos(A)$. If $Pos(A) = 1$, then trivially $Nec(A) \leq Pos(A)$.

Due to their properties, necessity and possibility measures can be interpreted as lower and upper probabilities, respectively. For any $A \in \mathcal{P}(X)$, the probability interval has the form

$$[Nec(A), \ Pos(A)] = \begin{cases} [0, Pos(A)] & \text{when } Pos(A) < 1 \\ [Nec(A), 1] & \text{when } Nec(A) > 0. \end{cases}$$

5.2.1. Möbius Representation

In order to investigate the Möbius representation in possibility theory, it is convenient to introduce the following special notation. It is usually assumed that the Möbius representation employed is the one based on the lower probability function (i.e., the necessity measure in this case).

Given a basic possibility function on $X = \{x_1, x_1, \ldots, x_n\}$, where $n \geq 1$, it is assumed that the elements of X are reordered in such a way that

$$r(x_i) \geq r(x_{i+1})$$

for all $i \in \mathbb{N}_{n-1}$. Let $r(x_i) = r_i$ and let the n-tuple

$$\mathbf{r} = \langle r_i \mid i \in \mathbb{N}_n \rangle$$

be called a *possibility profile of length n*. Clearly, $r_1 = 1$ due to possibilistic normalization. Let

$$A_i = \{x_1, x_2, \ldots, x_i\}$$

for all $i \in \mathbb{N}_n$, and let m denote the Möbius function based on the necessity measure. Due to the ordered possibility profile, $m(A) \neq 0$ only when $A = A_i$ for some $i \in \mathbb{N}_n$. Let $m(A_i) = m_i$ for all $i \in \mathbb{N}_n$, and let the n-tuple

$$\mathbf{m} = \langle m_i \mid i \in \mathbb{N}_n \rangle$$

denote the Möbius representation of the possibility profile. Then,

$$r_i = \sum_{k=i}^{n} m_k \qquad (5.13)$$

for all $i \in \mathbb{N}_n$. These equations have following form:

$$
\begin{aligned}
r_1 &= m_1 + m_2 + m_3 + \cdots + m_i + m_{i+1} + \cdots + m_n \\
r_2 &= \phantom{m_1 + {}} m_2 + m_3 + \cdots + m_i + m_{i+1} + \cdots + m_n \\
&\vdots \\
r_i &= \phantom{m_1 + m_2 + m_3 + \cdots + {}} m_i + m_{i+1} + \cdots + m_n \\
&\vdots \\
r_n &= \phantom{m_1 + m_2 + m_3 + \cdots + m_i + m_{i+1} + \cdots + {}} m_n.
\end{aligned}
$$

Solving them for $m_i (i \in \mathbb{N}_n)$, we obtain

$$m_i = r_i - r_{i+1}, \qquad (5.14)$$

where $r_{n+1} = 0$ by convention. Now,

$$Nec(A_i) = \sum_{k=1}^{i} m_k. \qquad (5.15)$$

Substituting m_k from Eq. (5.14) and recognizing that $r_1 = 1$, we obtain

$$Nec(A_i) = 1 - r_{i+1} \qquad (5.16)$$

for all $i \in \mathbb{N}_n$ (with $r_{n+1} = 0$). Clearly, $Nec(A) = 0$ when $A \neq A_i$ for all $i \in \mathbb{N}_n$. Moreover, $Pos(A_i) = 1$ and $Pos(A_n - A_{i-1}) = r_i$ (where $A_0 = \varnothing$ by convention and $A_n = X$) for all $i \in \mathbb{N}_n$.

The special notation introduced in this section and the calculation of $m_i = m(A_i)$, $Nec(A_i)$, $Pos(A_i)$, and $Pos(A_n - A_i)$ for a given possibility profile on $X = \{x_1, x_1, \ldots, x_n\}$ is illustrated in Figure 5.2.

EXAMPLE 5.1. A specific example illustrating the special notation and the various calculations is shown in Figure 5.3. Assuming that the possibility profile $\langle r_1, r_2, r_3, r_4 \rangle = \langle 1.0, 0.8, 0.4, 0.3 \rangle$ is given, all values of $m(A)$, $Nec(A)$, and $Pos(A)$ shown in the figure can be calculated for all $A \subseteq \{x_1, x_2, x_3, x_4\}$ in two different ways:

(a) Calculate $Pos(A)$ for all A by Eq. (5.3), calculate $Nec(A)$ for all A by Eq. (5.1), and calculate $m(A)$ for all A by the Möbius transform Eq. (4.8).
(b) Calculate $m(A)$ for all A by the formula

$$m(A) = \begin{cases} r_i - r_{i+1} & \text{when } A = A_i \\ 0 & \text{otherwise.} \end{cases}$$

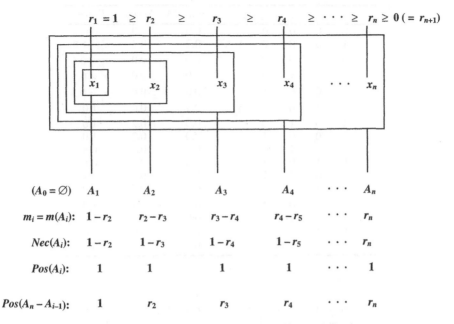

Figure 5.2. Illustration of the special notation for possibility theory.

Calculate then $Nec(A)$ for all A by the inverse Möbius transform, Eq. (4.9), and calculate $Pos(A)$ for all A by Eq. (5.1).

5.2.2. Ordering of Possibility Profiles

Possibility profiles of the same length can be partially ordered in the following way: given any two possibility profiles of length n,

$$^j\mathbf{r} = \langle {}^j r_1, {}^j r_2, \ldots, {}^j r_n \rangle,$$

$$^k\mathbf{r} = \langle {}^k r_1, {}^k r_2, \ldots, {}^k r_n \rangle,$$

we define

$$^j\mathbf{r} \leq {}^k\mathbf{r} \quad \text{iff} \quad {}^j r_i \leq {}^k r_i$$

for all $i \in \mathbb{N}_n$. This partial ordering forms a lattice, $\mathcal{R}_n$, on the set of all possibility profiles of a particular length n. Its join, $\vee$, and meet, $\wedge$, are defined, respectively, as

$$^j\mathbf{r} \vee {}^k\mathbf{r} = \langle \max\{ {}^j r_i, {}^k r_i \} \mid i \in \mathbb{N}_n \rangle,$$

$$^j\mathbf{r} \wedge {}^k\mathbf{r} = \langle \min\{ {}^j r_i, {}^k r_i \} \mid i \in \mathbb{N}_n \rangle,$$

	x_1	x_2	x_3	x_4	A_i	$m_i = m(A_i)$	$Nec(A)$	$Pos(A)$
A:	0	0	0	0		0.0	0.0	0.0
	1	0	0	0	A_1	0.2	0.2	$r_1 = 1.0$
	0	1	0	0		0.0	0.0	$r_2 = 0.8$
	0	0	1	0		0.0	0.0	$r_3 = 0.4$
	0	0	0	1		0.0	0.0	$r_4 = 0.3$
	1	1	0	0	A_2	0.4	0.6	1.0
	1	0	1	0		0.0	0.2	1.0
	1	0	0	1		0.0	0.2	1.0
	0	1	1	0		0.0	0.0	0.8
	0	1	0	1		0.0	0.0	0.8
	0	0	1	1		0.0	0.0	0.4
	1	1	1	0	A_3	0.1	0.7	1.0
	1	1	0	1		0.0	0.6	1.0
	1	0	1	1		0.0	0.2	1.0
	0	1	1	1		0.0	0.0	0.8
	1	1	1	1	A_4	0.3	1.0	1.0

(a)

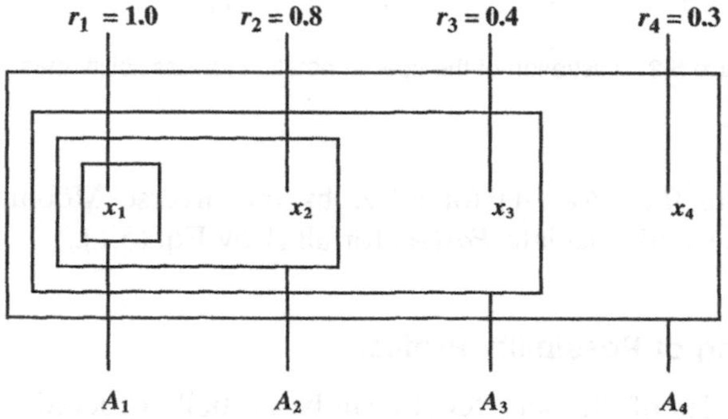

(b)

Figure 5.3. Illustration to Example 5.1.

for all pairs of possibility profiles of the same length. The smallest possibility profile, which expresses no uncertainty, is

$$\langle 1, 0, \ldots, 0 \rangle;$$

the greatest one, which expresses total ignorance, is

$$\langle 1, 1, \ldots, 1 \rangle.$$

For any $^i\mathbf{r}, {}^k\mathbf{r} \in \mathcal{R}_n$, if $^i\mathbf{r} \le {}^k\mathbf{r}$, then $^k\mathbf{r}$ represents greater uncertainty than $^i\mathbf{r}$ or, in other words, $^i\mathbf{r}$ contains more information then $^k\mathbf{r}$.

Observe that there exists a one-to-one correspondence between possibility profiles

$$\mathbf{r} = \langle r_i \mid i \in \mathbb{N}_n \rangle$$

and their Möbius representations

$$\mathbf{m} = \langle m_i \mid i \in \mathbb{N}_n \rangle$$

that is expressed by Eqs. (5.13) and (5.14). The natural ordering of possibility profiles thus induces an ordering of the associated Möbius representations. In this sense, the smallest and the largest Möbius representations are, respectively,

$$\langle 1,0,0,\ldots,0 \rangle \quad \text{and} \quad \langle 0,0,\ldots,0,1 \rangle$$

5.2.3. Joint and Marginal Possibilities

When a basic possibility function r is defined on a Cartesian product $X \times Y$, its values $r(x, y)$ for all $x \in X$ and $y \in Y$ are called *joint possibilities*. The associated marginal possibility functions, r_X and r_Y, are defined by the formulas

$$r_X(x) = \max_{y \in Y}\{r(x,y)\} \tag{5.17}$$

for all $x \in X$, and

$$r_Y(y) = \max_{x \in X}\{r(x,y)\} \tag{5.18}$$

for all $y \in Y$. These formulas follow directly from Eq. (5.3). To see this, note that

$$Pos_X(\{x\}) = Pos(\{x\} \times Y)$$

holds for each $x \in X$. Using Eq. (5.3), this equation can be rewritten as Eq. (5.17). Equation (5.18) can be derived in a similar way.

Alternatives in sets X and Y may be viewed as states of associated variables $\mathcal{X}$ and $\mathcal{Y}$, respectively. The marginal possibility functions r_X and r_Y, then, describe information regarding the individual variables within the theory of graded possibilities. Similarly, the joint possibility function r describes (within this theory) information regarding a relation between the variables.

Now assume that the marginal possibility functions r_X and r_Y are given. These functions are clearly not sufficient to determine the joint possibility function r. However, when the variables do not interact, the joint possibility function r is defined for all $x \in X$ and $y \in Y$ by the equation

$$r(x, y) = \min\{r_X(x), r_Y(y)\}. \tag{5.19}$$

This definition is justified in the following way:

(a) For each $x \in X$,

$$Pos(\{x\} \times Y) = Pos_X(\{x\}),$$

and for each $y \in Y$

$$Pos(X \times \{y\}) = Pos_Y(\{y\}).$$

(b) For each pair $\langle x, y \rangle \in X \times Y$,

$$\begin{aligned} Pos(\{\langle x, y \rangle\}) &= Pos((\{x\} \times Y) \cap (X \times \{y\})) \\ &\le \min\{Pos(\{x\} \times Y), Pos(X \times \{y\})\} \\ &= \min\{Pos_X(\{x\}), Pos_Y(\{y\})\}. \end{aligned}$$

Hence,

$$r(x, y) \le \min\{r_X(x), r_Y(y)\}.$$

(c) Under the assumption of noninteraction, the joint possibilities must express the minimal constraint, which means that for each $x \in X$ and each $y \in Y$ the values $r(x, y)$ must be the largest acceptable values. This immediately results in Eq. (5.19).

EXAMPLE 5.2. Figure 5.4*a* illustrates the situation in which the joint possibilities $r(x, y)$ are given, and we want to determine the marginal possibilities $r_X(x)$ and $r_Y(y)$. Using Eq. (5.17), for example,

$$\begin{aligned} r_X(1) &= \max\{r(1,0), r(1,1)\} \\ &= \max\{0.6, 0.4\} = 0.6. \end{aligned}$$

Figure 5.4*b* illustrates a different situation, in which the marginal possibilities $r_X(x)$ and $r_Y(y)$ are given, and we want to determine the joint possibilities $r'(x, y)$ under the assumption of noninteraction of the marginals. Using Eq. (5.19), we get

$$\begin{aligned} r(1,1) &= \min\{r_X(1), r_Y(1)\} \\ &= \min\{0.6, 0.8\} = 0.6, \end{aligned}$$

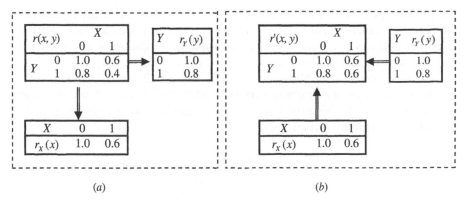

Figure 5.4. Illustration to Example 5.2.

for example. Observe that the marginal possibilities in the two cases in Figure 5.4 are the same, but the joint possibilities are different. In fact, among all joint possibility profiles that are consistent with the marginal possibilities in Figure 5.4, the one based on the assumption of noninteractive marginals, $\langle r'(x, y) \mid x \in X, y \in Y \rangle$, is the largest one.

5.2.4. Conditional Possibilities

Again consider marginal possibility functions r_X and r_Y that describe information regarding variables X and Y. As before, state sets of the variables are X and Y, respectively. When the variables do not interact, their joint possibility function r is defined by Eq. (5.19), as is explained in Section 5.2.3. When they do interact, we need to employ appropriate conditional possibility functions, $r_{X|Y}$ and $r_{Y|X}$, to account for the interaction. In general, the joint possibility function can be expressed via either of the following equations:

$$r(x, y) = \min\{r_Y(y), r_{X|Y}(x \mid y)\}, \tag{5.20}$$

$$r(x, y) = \min\{r_X(x), r_{Y|X}(y \mid x)\}, \tag{5.21}$$

The conditional possibility functions are implicitly defined by these functional equations. By solving the equations for $r_{X|Y}$ and $r_{Y|X}$, we obtain for each $\langle x, y \rangle \in X \times Y$ the following formulas:

$$r_{X|Y}(x \mid y) = \begin{cases} r(x, y) & \text{when } r_Y(y) > r(x, y) \\ [r(x, y), 1] & \text{when } r_Y(y) = r(x, y) \end{cases} \tag{5.22}$$

$$r_{Y|X}(y \mid x) = \begin{cases} r(x, y) & \text{when } r_X(x) > r(x, y) \\ [r(x, y), 1] & \text{when } r_X(x) = r(x, y). \end{cases} \tag{5.23}$$

Conditional possibilities are thus interval-valued. However, to satisfy the possibilistic normalization, at least one of the intervals must be replaced with its maximum value of 1. Since there is no rationale for choosing any particular interval for this replacement, the usual way is to replace each of them with 1. This unique choice from all the intervals in Eqs. (5.22) and (5.23) is justified by the principle of maximum uncertainty, which is introduced and discussed in Chapter 9. Using this principle, we obtain the following definitions of the conditional possibility functions:

$$r_{X|Y}(x \mid y) = \begin{cases} r(x, y) & \text{when } r_Y(y) > r(x, y) \\ 1 & \text{when } r_Y(y) = r(x, y) \end{cases} \qquad (5.24)$$

$$r_{Y|X}(y \mid x) = \begin{cases} r(x, y) & \text{when } r_X(x) > r(x, y) \\ 1 & \text{when } r_X(x) = r(x, y). \end{cases} \qquad (5.25)$$

EXAMPLE 5.3. An example of joint and marginal possibility functions on $X \times Y$ is given in Figure 5.5. The associated conditional possibility functions $r_{X|Y}$ and $r_{Y|X}$ are shown in Figure 5.5b and 5.5c, respectively. For the sake of completeness, intervals determined by Eqs. (5.22) and (5.23) are shown in these figures. However, according to Eqs. (5.24) and (5.25), each of these intervals is normally replaced with 1, as required by the possibilistic normalization and the principle of maximum uncertainty.

Conditional possibilities can be used for defining possibilistic independence. Variable X is said to be *independent of variable* Y (in the possibilistic sense) iff

$$r_{X|Y}(x \mid y) = r_X(x) \qquad (5.26)$$

for all $\langle x, y \rangle \in X \times Y$. Similarly, variable Y is said to be *independent of variable* X iff

$$r_{Y|X}(y \mid x) = r_Y(y) \qquad (5.27)$$

for all $\langle x, y \rangle \in X \times Y$.

Assume now that Eq. (5.26) holds. Then, by replacing $r_{X|Y}$ in Eq. (5.20) with r_X, we obtain Eq. (5.19). Similarly, assuming that Eq. (5.27) holds and replacing $r_{Y|X}$ in Eq. (5.21) with r_Y, we obtain Eq. (5.19). Hence, *possibilistic independence implies possibilistic noninteraction.* The converse, however, is not true. To show this, assume that Eq. (5.19) holds. Then, Eq. (5.20) can be written as

$$\min\{r_X(x), r_Y(y)\} = \min\{r_Y(y), r_{X|Y}(x \mid y)\}.$$

$r(x_i, y_j)$		X					
		x_1	x_2	x_3	x_4	x_5	x_6
Y	y_1	1.0	0.8	0.5	0.4	0.3	0.0
	y_2	0.7	0.5	0.4	0.4	0.3	0.0
	y_3	0.2	0.6	0.5	0.4	0.0	0.0
	y_4	0.0	0.0	0.0	0.2	0.2	0.1

Y	$r_Y(y_j)$
y_1	1.0
y_2	0.7
y_3	0.6
y_4	0.2

X	x_1	x_2	x_3	x_4	x_5	x_6
$r_X(x_i)$	1.0	0.8	0.5	0.4	0.3	0.1

(a)

$r_{X|Y}(x_i | y_j)$		X					
		x_1	x_2	x_3	x_4	x_5	x_6
Y	y_1	1.0	0.8	0.5	0.4	0.3	0.0
	y_2	[0.7, 1]	0.5	0.4	0.4	0.3	0.0
	y_3	0.2	[0.6, 1]	0.5	0.4	0.0	0.0
	y_4	0.0	0.0	0.0	[0.2, 1]	[0.2, 1]	0.1

(b)

$r_{Y|X}(y_j | x_i)$		X					
		x_1	x_2	x_3	x_4	x_5	x_6
Y	y_1	1.0	[0.8, 1]	[0.5, 1]	[0.4, 1]	[0.3, 1]	0.0
	y_2	0.7	0.5	0.4	[0.4, 1]	[0.3, 1]	0.0
	y_3	0.2	0.6	[0.5, 1]	[0.4, 1]	0.0	0.0
	y_4	0.0	0.0	0.0	0.2	0.2	[0.1, 1]

(c)

Figure 5.5. Illustration to Example 5.3.

Solving this equation for $r_{X|Y}(x \mid y)$, we obtain

$$r_{X|Y}(x \mid y) = \begin{cases} r_X(x) & \text{when } r_X(x) < r_Y(y) \\ [r_Y(y), 1] & \text{when } r_Y(y) \leq r_X(x), \end{cases}$$

and not Eq. (5.26). Similarly, Eq. (5.21) can be written as

$$\min\{r_X(x), r_Y(y)\} = \min\{r_X(x), r_{Y|X}(y \mid x)\}$$

and, by solving it for $r_{Y|X}(y \mid x)$, we obtain

$$r_{Y|X}(y \mid x) = \begin{cases} r_Y(y) & \text{when } r_Y(y) < r_X(x) \\ [r_X(x), 1] & \text{when } r_X(x) \leq r_Y(y), \end{cases}$$

and not Eq. (5.27).

Observing that both probability measures and possibility measures are uniquely represented by a function on X, it is interesting to compare the various properties of probability theory and possibility theory. A summary of this comparison is in Table 5.1.

5.2.5. Possibilities on Infinite Sets

When X is an infinite set and C is an ample field of subsets of X, a possibility measure, Pos, is a function

$$Pos:C \rightarrow [0,\ 1]$$

that satisfies the following axiomatic requirements

Axiom (Pos1). $Pos(\emptyset) = 0$.

Axiom (Pos2). $Pos(X) = 1$.

Axiom (Pos3). For any family $C = \{A_i \mid i \in I\}$, where I is an arbitrary index set,

$$Pos\left(\bigcup_{i \in I} A_i\right) = \sup_{i \in I} Pos(A_i).$$

In analogy with the finite case, every possibility measure is uniquely determined by a basic possibility function, r, via the formula

$$Pos(A) = \sup_{x \in A} r(x)$$

for each $A \in C$. Clearly, r is required to satisfy the possibilistic normalization

$$\sup_{x \in X} r(x) = 1.$$

Alternatively, the theory can be formalized by the following axiomatic requirements of the dual necessity measure, Nec:

Axiom (Nec1). $Nec(\emptyset) = 0$.

Axiom (Nec2). $Nec(X) = 1$.

Axiom (Nec3). For any family $C = \{A_i \mid i \in I\}$, where I is an arbitrary index set,

$$Nec\left(\bigcap_{i \in I} A_i\right) = \inf_{i \in I} Nec(A_i).$$

It is well established that the possibility measures and necessity measures are semicontinuous from below and from above, respectively.

Table 5.1. Probability Theory Versus Possibility Theory: Comparison of Mathematical Properties for Finite Sets

Probability Theory	Possibility Theory
Based on measures of *one type*: probability measures, *Pro*	Based on measures of *two types*: possibility measures, *Pos*, and necessity measures, *Nec*
Body of evidence consists of *singletons*	Body of evidence consists of a *family of nested subsets*
Unique representation of *Pro* by a *probability* distribution function $p: X \to [0, 1]$ via the formula $$Pro(A) = \sum_{x \in A} p(x)$$	Unique representation of *Pro* by a *basic possibility function* $r: X \to [0, 1]$ via the formula $$Pos(A) = \max_{x \in A} r(x)$$
Normalization: $$\sum_{x \in X} p(x) = 1$$	Normalization: $$\max_{x \in X} r(x) = 1$$
Additivity: $Pro(A \cup B) = Pro(A) + Pro(B) - Pro(A \cap B)$	Max/Min rules: $Pos(A \cup B) = \max \{Pos(A), Pos(B)\}$ $Pos(A \cap B) \leq \min \{Pos(A), Pos(B)\}$ $Nec(A \cap B) = \min \{Nec(A), Nec(B)\}$ $Nec(A \cup B) \geq \max \{Nec(A), Nec(B)\}$
Not applicable	Duality: $Nec(A) = 1 - Pos(\bar{A})$ $Pos(A) < 1 \Rightarrow Nec(A) = 0$ $Nec(A) > 0 \Rightarrow Pos(A) = 1$
$Pro(A) + Pro(\bar{A}) = 1$	$Pos(A) + Pos(\bar{A}) \geq 1$ $Nec(A) + Nec(\bar{A}) \leq 1$ $\max \{Pos(A), Pos(\bar{A})\} = 1$ $\min \{Nec(A), Nec(\bar{A})\} = 0$
Total ignorance: $p(x) = 1/|X|$ for all $x \in X$	Total ignorance: $r(x) = 1$ for all $x \in X$
Conditional probabilities: $$p_{X|Y}(x|y) = \frac{p(x,y)}{p_Y(y)}$$ $$p_{X|Y}(y|x) = \frac{p(x,y)}{p_X(x)}$$	Conditional possibilities: $r_{X|Y}(X|Y) = \begin{cases} r(x,y) & \text{when } r_Y(y) > r(x,y) \\ [r(x,y),1] & \text{when } r_Y(y) = r(x,y) \end{cases}$ $r_{Y|X}(Y|X) = \begin{cases} r(x,y) & \text{when } r_X(x) > r(x,y) \\ [r(x,y),1] & \text{when } r_X(y) = r(x,y) \end{cases}$
Probabilistic noninteraction: $p(x, y) = p_X(x) \cdot p_Y(y)$ (a)	Possibilistic noninteraction: $r(x, y) = \min \{r_X(x), r_Y(y)\}$ (α)
Probabilistic independence: $p_{X|Y}(x|y) = p_X(x)$ $p_{Y|X}(y|x) = p_Y(y)$ (b) (a) $\Rightarrow$ (b) and (b) $\Rightarrow$ (a)	Possibilistic independence: $r_{X|Y}(x|y) = r_X(x)$ $r_{Y|X}(y|x) = r_Y(y)$ (β) $(\beta) \Rightarrow (\alpha)$ but $(\alpha) \nRightarrow (\beta)$

5.2.6. Some Interpretations of Graded Possibilities

Viewing dual pairs of necessity and possibility measures as *imprecise probabilities* is just one interpretation of possibility theory. Other interpretations, totally devoid of any connection to probability theory, appear to be even more fundamental.

Perhaps the most important interpretation of possibility theory is based on defining possibility grades in terms of grades of membership of relevant fuzzy sets. This *fuzzy-set interpretation* of possibility theory is examined in Chapter 8.

Another important interpretation of possibility theory is based on the concept of *similarity*. In this interpretation, the possibility $r(x)$ reflects the degree of similarity between x and an ideal prototype, x_i, for which the possibility degree is 1. That is, $r(x)$ is expressed by a suitable distance between x and x_i defined in terms of relevant attributes of the elements involved. The closer x is to x_i according to the chosen distance, the more possible we consider it in this interpretation of possibility theory. In some cases, the closeness may be determined objectively by a defined measurement procedure. In other cases, it may be based on a subjective judgment of a person (e.g., an expert in the application area involved).

A quite common interpretation of possibility theory is founded on special orderings, $\leq_{Pos}$, defined on the power set $\mathcal{P}(X)$. For any $A, B \in \mathcal{P}(X), A \leq_{Pos} B$ means that B is *at least as possible as* A. This phrase "at least as possible as" may, of course, have various special interpretations, such as, for example, "at least as easy to achieve" or "at most constrained as." When $\leq_{Pos}$ satisfies the requirement

$$A \leq_{Pos} B \Rightarrow A \cup C \leq_{Pos} B \cup C$$

for all $A, B, C, \in \mathcal{P}(X)$, it is called a *comparative possibility relation*. It is known that the only measures that conform to comparative possibility orderings are possibility measures. It is also known that for each ordering $\leq_{Pos}$ there exists a dual ordering, $\leq_{Nec}$, defined by the equivalence

$$A \leq_{Pos} B \Leftrightarrow \overline{A} \leq_{Nec} \overline{B}.$$

These dual orderings are called *comparative necessity relations*; the only measures that conform to them are necessity measures.

5.3. SUGENO λ-MEASURES

Sugeno λ-measures (or just *λ-measures*) are special monotone measures, $^\lambda\mu$, that are characterized by the following axiomatic requirement: for all A, $B \in \mathcal{P}(X)$, if $A \cap B = \varnothing$, then

$$^{\lambda}\mu(A \cup B) = {}^{\lambda}\mu(A) + {}^{\lambda}\mu(B) + \lambda {}^{\lambda}\mu(A) {}^{\lambda}\mu(B), \tag{5.28}$$

where $\lambda > -1$ is a parameter by which different λ-measures are distinguished. Equation (5.28) is usually called a λ-*rule*.

When X is a finite set and values $^{\lambda}\mu(\{x\})$ are given for all $x \in X$, then it is obvious that the value $^{\lambda}\mu(A)$ for any $A \in \mathcal{P}(X)$ can be determined from these values on singletons by a repeated application of the λ-rule. This value can be expressed succinctly as

$$^{\lambda}\mu(A) = \frac{1}{\lambda} \left[\prod_{x \in A} (1 + \lambda {}^{\lambda}\mu(\{x\})) - 1 \right]. \tag{5.29}$$

Observe that, given values $^{\lambda}\mu(\{x\})$ for all $x \in X$, the value of λ can be determined by the requirement that $^{\lambda}\mu(X) = 1$. Applying this requirement to Eq. (5.29) results in the equation

$$1 + \lambda = \prod_{x \in X} (1 + \lambda {}^{\lambda}\mu(\{x\})) \tag{5.30}$$

for λ. This equation determines the parameter uniquely under the conditions stated in the following theorem.

Theorem 5.3. Let $^{\lambda}\mu(\{x\}) < 1$ for all $x \in X$ and let $^{\lambda}\mu(\{x\}) > 0$ for at least two elements of X. Then, Eq. (5.30) determines the parameter λ uniquely as follows:

(a) If $\sum_{x \in X} {}^{\lambda}\mu(\{x\}) < 1$, then λ is equal to the *unique* root of the equation in the interval $(0, \infty)$.

(b) If $\sum_{x \in X} {}^{\lambda}\mu(\{x\}) = 1$, then $\lambda = 0$, which is the only root of the equation.

(c) If $\sum_{x \in X} {}^{\lambda}\mu(\{x\}) > 1$, then λ is equal to the *unique* root of the equation in the interval $(-1, 0)$.

Proof. [Wang and Klir, 1992, pp. 46–47].

Any λ-measure is thus completely determined by its values on all singletons, $^{\lambda}\mu(\{x\})$, for all $x \in X$. Given values $^{\lambda}\mu(\{x\})$ for all $x \in X$, the value of λ is determined via Eq. (5.30), and values $^{\lambda}\mu(A)$ for all subsets of X are then determined by Eq. (5.29). According to Theorem 5.3, three situations are distinguished:

(a) When $\sum_{x \in X} {}^{\lambda}\mu(\{x\}) < 1$, which means that $^{\lambda}\mu$ qualifies as a lower probability, $\lambda > 0$.

(b) When $\sum_{x \in X} {}^{\lambda}\mu(\{x\}) = 1$, which means that ${}^{\lambda}\mu$ is a classical probability measure, $\lambda = 0$.

(c) When $\sum_{x \in X} {}^{\lambda}\mu(\{x\}) > 1$, which means that ${}^{\lambda}\mu$ qualifies as an upper probability, $\lambda < 0$.   ∎

EXAMPLE 5.4. Let $X = \{x_1, x_2, x_3\}$ and let ${}^{\lambda}\mu(\{x_1\}) = 0.2$, ${}^{\lambda}\mu(\{x_2\}) = 0$, and ${}^{\lambda}\mu(\{x_3\}) = 0.5$. The equation for λ has the form

$$1 + \lambda = (1 + 0.2\lambda)(1 + 0.5\lambda).$$

Its only solution is $\lambda = 3$. The remaining values now can be readily determined by the λ-rule: ${}^{\lambda}\mu(\{x_1, x_2\}) = 0.2$, ${}^{\lambda}\mu(\{x_1, x_3\}) = 1$, ${}^{\lambda}\mu(\{x_2, x_3\}) = 0.5$, ${}^{\lambda}\mu(X) = 1$.

When ${}^{\lambda}\mu$ is a joint λ-measure defined on subsets of $X \times Y$ and ${}^{\lambda}\mu_X$ and ${}^{\lambda}\mu_Y$ are the associated marginal λ-measures, the general equations

$$ {}^{\lambda}\mu_X(A) = {}^{\lambda}\mu_X(A \times Y), \tag{5.31}$$

$$ {}^{\lambda}\mu_Y(B) = {}^{\lambda}\mu_X(X \times B) \tag{5.32}$$

hold for any $A \subseteq X$ and any $B \subseteq Y$. However, using Eq. (5.29), the relationship between the joint and marginal λ-measures can also be expressed for all $x \in X$ and all $y \in Y$ by the formulas

$$ {}^{\lambda}\mu_X(\{x\}) = \frac{1}{\lambda}\left[\prod_{y \in Y}(1 + \lambda^{\lambda}\mu(\{x, y\})) - 1\right] \tag{5.33}$$

$$ {}^{\lambda}\mu_Y(\{y\}) = \frac{1}{\lambda}\left[\prod_{x \in X}(1 + \lambda^{\lambda}\mu(\{x, y\})) - 1\right]. \tag{5.34}$$

Let ${}^{\lambda}\mu$ and ${}^{\bar{\lambda}}\mu$ denote λ-measures that represent, respectively, a lower probability and the corresponding upper probability defined for each $A \in \mathcal{P}(X)$ by the duality equation

$$ {}^{\bar{\lambda}}\mu(A) = 1 - {}^{\underline{\lambda}}\mu(\bar{A}). \tag{5.35}$$

To investigate the relationship between $\underline{\lambda}$ and $\bar{\lambda}$, the meaning of ${}^{\lambda}\mu(\bar{A})$ for any λ-measure must be determined first. This can be done by applying the λ-rule to sets A and $\bar{A}$:

$$ {}^{\lambda}\mu(A \cup \bar{A}) = {}^{\lambda}\mu(A) + {}^{\lambda}\mu(\bar{A}) + \lambda^{\lambda}\mu(A){}^{\lambda}\mu(\bar{A}).$$

Recognizing that ${}^{\lambda}\mu(A \cup \bar{A}) = {}^{\lambda}\mu(X) = 1$ and solving this equation for ${}^{\lambda}\mu(\bar{A})$, we readily obtain

$$^{\lambda}\mu(\overline{A}) = \frac{1 - {}^{\lambda}\mu(A)}{1 - \lambda\mu(A)} \tag{5.36}$$

and

$$1 - {}^{\lambda}\mu(\overline{A}) = \frac{(1 + \lambda)^{\lambda}\mu(A)}{1 + \lambda^{\lambda}\mu(A)}. \tag{5.37}$$

The last equation makes it possible to prove the following theorem.

Theorem 5.4. For any pair of dual λ-measures ${}^{\underline{\lambda}}\mu$ and ${}^{\overline{\lambda}}\mu$

$$\overline{\lambda} = -\frac{\underline{\lambda}}{1 + \underline{\lambda}}. \tag{5.38}$$

Proof. For any sets $A, B \in \mathcal{P}(X)$ such that $A \cap B = \emptyset$, we have

$$
\begin{aligned}
{}^{\overline{\lambda}}\mu(A \cup B) &= {}^{\overline{\lambda}}\mu(A) + {}^{\overline{\lambda}}\mu(B) - \frac{\underline{\lambda}}{1 + \underline{\lambda}} {}^{\overline{\lambda}}\mu(A)^{\overline{\lambda}}\mu(B) \\
&= 1 - {}^{\underline{\lambda}}\mu(\overline{A}) + 1 - {}^{\underline{\lambda}}\mu(\overline{B}) - \frac{\underline{\lambda}}{1 + \underline{\lambda}}[(1 - {}^{\underline{\lambda}}\mu(\overline{A}))(1 - {}^{\underline{\lambda}}\mu(\overline{B}))] \\
&= \frac{(1 + \underline{\lambda})^{\underline{\lambda}}\mu(A)}{1 + \underline{\lambda}^{\underline{\lambda}}\mu(A)} + \frac{(1 + \underline{\lambda})^{\underline{\lambda}}\mu(B)}{1 + \underline{\lambda}^{\underline{\lambda}}\mu(B)} - \underline{\lambda}\frac{(1 + \underline{\lambda})^{\underline{\lambda}}\mu(A)^{\underline{\lambda}}\mu(B)}{[1 + \underline{\lambda}^{\underline{\lambda}}\mu(A)][1 + \underline{\lambda}^{\underline{\lambda}}\mu(B)]} \\
&= \frac{(1 + \underline{\lambda})[{}^{\underline{\lambda}}\mu(A) + {}^{\underline{\lambda}}\mu(B) + \underline{\lambda}^{\underline{\lambda}}\mu(A)^{\underline{\lambda}}\mu(B)]}{[1 + \underline{\lambda}^{\underline{\lambda}}\mu(A)][1 + \underline{\lambda}^{\underline{\lambda}}\mu(B)]} \\
&= \frac{(1 + \underline{\lambda})^{\underline{\lambda}}\mu(A \cup B)}{1 + \underline{\lambda}^{\underline{\lambda}}\mu(A \cup B)} \\
&= 1 - {}^{\underline{\lambda}}\mu(\overline{A \cup B}) \\
&= {}^{\overline{\lambda}}\mu(A \cup B).
\end{aligned}
$$
∎

EXAMPLE 5.5. The λ-measure in Example 5.4 represents clearly a lower probability function and, therefore, it should be denoted as ${}^{\underline{\lambda}}\mu$. Its values (calculated in Example 5.4) are shown in Table 5.2. Also shown in the table are values of the dual λ-measure ${}^{\overline{\lambda}}\mu$, which represents an upper probability function. The value of $\overline{\lambda}$ can be calculated either by Eq. (5.30) or by Eq. (5.38). Equation (5.30) has the form

$$1 + \overline{\lambda} = (1 + 0.5\overline{\lambda})(1 + 0.8\overline{\lambda}),$$

and its unique solution is $\overline{\lambda} = -0.75$. Since $\underline{\lambda} = 3$, we obtain

$$\overline{\lambda} = -\frac{\underline{\lambda}}{1 + \underline{\lambda}} = 0.75$$

by Eq. (5.38). As expected, these results are the same.

Table 5.2. Illustration to Examples 5.4 and 5.5

	x_1	x_2	x_3	$^{\underline{\lambda}}\mu(A)$	$^{\bar{\lambda}}\mu(A)$
A:	0	0	0	0.0	0.0
	1	0	0	0.2	0.5
	0	1	0	0.0	0.0
	0	0	1	0.5	0.8
	1	1	0	0.2	0.5
	1	0	1	1.0	1.0
	0	1	1	0.5	0.8
	1	1	1	1.0	1.0

(a)

	$^{\underline{\lambda}}\mu(\{x,y\})$	X	
		0	1
Y	0	0.2	0.2
	1	0.1	0.0

Y	$^{\underline{\lambda}}\mu_Y(\{y\})$
0	0.6
1	0.1

X	0	1
$^{\underline{\lambda}}\mu_X(\{x\})$	0.4	0.2

(b)

	$\langle x, y \rangle$				$^{\underline{\lambda}}\mu(A)$	$^{\bar{\lambda}}\mu(A)$
	00	01	10	11		
A:	0	0	0	0	0.0	0.0
	1	0	0	0	0.2	0.6
	0	1	0	0	0.2	0.6
	0	0	1	0	0.1	0.4
	0	0	0	1	0.0	0.0
	1	1	0	0	0.6	0.9
	1	0	1	0	0.4	0.8
	1	0	0	1	0.2	0.6
	0	1	1	0	0.4	0.8
	0	1	0	1	0.2	0.6
	0	0	1	1	0.1	0.4
	1	1	1	0	1.0	1.0
	1	1	0	1	0.6	0.9
	1	0	1	1	0.4	0.8
	0	1	1	1	0.4	0.8
	1	1	1	1	1.0	1.0

Figure 5.6. Illustration to Example 5.6.

EXAMPLE 5.6. Consider the values of the joint λ-measure on singletons that are given in Figure 5.6a. Since

$$\sum_{\langle x,y\rangle \in X \times Y} {}^{\lambda}\mu(\{\langle x,y\rangle\}) < 1,$$

the λ-measure represents a lower probability, which is characterized by a parameter $\underline{\lambda} > 0$. The value of $\underline{\lambda}$ is determined by the unique positive root of the equation

$$1 + \underline{\lambda} = (1 + 0.2\underline{\lambda})(1 + 0.2\underline{\lambda})(1 + 0.1\underline{\lambda}),$$

which is $\underline{\lambda} = 5$. The marginal lower probability functions ${}^{\lambda}\mu_X$ and ${}^{\lambda}\mu_Y$, which are also shown in Figure 5.6a, are readily determined from ${}^{\lambda}\mu$ by the λ-rule. For example,

$$ {}^{\lambda}\mu_X(\{0\}) = {}^{\lambda}\mu(\{\langle 0,0 \rangle\}) + {}^{\lambda}\mu(\{\langle 0,1 \rangle\}) + \lambda {}^{\lambda}\mu(\{\langle 0,0 \rangle\}) {}^{\lambda}\mu(\{\langle 0,1 \rangle\}) = 0.4. $$

The joint lower probabilities ${}^{\lambda}\mu(A)$ are determined for all $A \subseteq X \times Y$ from the values ${}^{\lambda}\mu(\{\langle x, y \rangle\})$ by the λ-rule or by Eq. (5.29). The joint upper probabilities ${}^{\bar{\lambda}}\mu(A)$ are determined from the lower probabilities by the duality equation, Eq. (5.35). The results are shown in Figure 5.6.

5.3.1. Möbius Representation

The Möbius representation of λ-measure ${}^{\lambda}\mu$ $(\underline{\lambda} > 0)$ can be expressed by a special formula, which is a consequence of the λ-rule. This formula is a subject of the following theorem.

Theorem 5.5. The Möbius representation, ${}^{\lambda}m$, of a λ-measure ${}^{\lambda}\mu$ $(\underline{\lambda} > 0)$ defined on subsets of a finite set X can be expressed for all $A \in \mathcal{P}(X)$ by the formula

$$ {}^{\lambda}m(A) = \begin{cases} \underline{\lambda}^{|A|-1} \displaystyle\prod_{x \in A} {}^{\lambda}\mu(\{x\}) & \text{when } A \neq \varnothing \\ 0 & \text{when } A = \varnothing. \end{cases} \tag{5.39} $$

Proof. Using Eq. (5.29), we have

$$
\begin{aligned}
{}^{\lambda}\mu(A) &= \frac{1}{\underline{\lambda}}\left[\prod_{x \in A}\left(1 + \underline{\lambda}\,{}^{\lambda}\mu(\{x\})\right) - 1 \right] \\
&= \frac{1}{\underline{\lambda}}\left[\sum_{\varnothing \subset B \subseteq A}\left[\underline{\lambda}^{|B|} \prod_{x \in B} {}^{\lambda}\mu(\{x\}) \right] \right] \\
&= \sum_{\varnothing \subset B \subseteq A}\left[\underline{\lambda}^{|B|-1} \prod_{x \in B} {}^{\lambda}\mu(\{x\}) \right] \\
&= \sum_{B \subseteq A} {}^{\lambda}m(B).
\end{aligned}
$$

Since ${}^{\lambda}\mu(X) = 1$, $\displaystyle\sum_{B \subseteq X} {}^{\lambda}m(B) = 1$. Hence, ${}^{\lambda}m$, defined by Eq. (5.29) is the Möbius representation of ${}^{\lambda}\mu$. ∎

It is obvious from Eq. (5.39) that ${}^{\lambda}m(A) \geq 0$ for all $A \in \mathcal{P}(X)$. Hence, ${}^{\lambda}\mu$ is a Choquet capacity of order infinity and its dual, ${}^{\bar{\lambda}}\mu$, is an alternate capacity of order infinity. This means that imprecise probabilities formalized in terms of λ-measures are also subject to the rules of those formalized in terms of capacities of order ∞, which are examined in Section 5.4.

5.4. BELIEF AND PLAUSIBILITY MEASURES

According to a generally accepted terminology in the literature, *belief measures* are Choquet capacities of order ∞ and *plausibility measures* are alternating capacities of order ∞. A theory based on dual pairs of these measures is usually referred to as *evidence theory* or *Dempster–Shafer theory* (DST). The latter name, which seems more common in recent publications, is adopted in this book.

Given a universal set X, assumed here to be finite, a *belief measure*, *Bel*, is a function

$$Bel:\mathcal{P}(X)\to[0,1]$$

such that $Bel(\varnothing) = 0$, $Bel(X) = 1$, and

$$Bel(A_1 \cup A_2 \cup \cdots \cup A_n) \geq \sum_j Bel(A_j) - \sum_{j<k} Bel(A_j \cap A_k)$$
$$+\cdots+(-1)^{n+1} Bel(A_1 \cap A_2 \cap \cdots \cap A_n) \quad (5.40)$$

for all possible families of subsets of X. When X is infinite, the domain of *Bel* is a family C of subsets of X with an appropriate algebraic structure (σ-algebra, etc.) and it is required that *Bel* be *semicontinuous from above*.

For each $A \in \mathcal{P}(X)$, $Bel(A)$ is interpreted as the *degree of belief* (based on available evidence) that the true alternative of X (prediction, diagnosis, etc.) belongs to the set A. We may also view the various subsets of X as answers to a particular question. We assume that some of the answers are correct, but we do not know with full certainty which ones they are. In DST, X is often called a *frame of discernment*. When the sets $A_1, A_2, \ldots, A_n$ in Eq. (5.40) are pairwise disjoint, the inequality requires that the degree of belief associated with the union of the sets is not smaller than the sum of the degrees of belief pertaining to the individual sets. This basic property of belief measures is thus a weaker version of the additivity property of probability measures. This implies that probability measures are special cases of belief measures for which the equality in Eq. (5.40) is always satisfied.

Let $A_1 = A$ and $A_2 = \bar{A}$ in Eq. (5.40) for $n = 2$. Since $A \cup \bar{A} = X$, $A \cap \bar{A} = \varnothing$, and it is required that $Bel(X) = 1$ and $Bel(\varnothing) = 0$, we can immediately derive from Eq. (5.40) the following fundamental property of belief measures:

$$Bel(A)+ Bel(\bar{A}) \leq 1. \quad (5.41)$$

A plausibility measure is a function

$$Pl:\mathcal{P}(X)\to[0,1]$$

such that $Pl(\varnothing) = 0$, $Pl(X) = 1$, and

$$Pl(A_1 \cap A_2 \cap \cdots \cap A_n) \le \sum_j Pl(A_j) - \sum_{j<k} Pl(A_j \cup A_k)$$
$$+ \cdots + (-1)^{n+1} Pl(A_1 \cup A_2 \cup \cdots \cup A_n) \quad (5.42)$$

for all possible families of subsets of X. When X is infinite, function Pl is also required to be *semicontinuous from below*.

Let $n = 2$, $A_1 = A$ and $A_2 = \bar{A}$ in Eq. (5.42). Since $A \cup \bar{A} = X$, $A \cap \bar{A} = \varnothing$, and it is required that $Pl(X) = 1$ and $Pl(\varnothing) = 0$, we immediately obtain the following basic inequality of plausibility measures from Eq. (5.42):

$$Pl(A) + Pl(\bar{A}) \ge 1. \quad (5.43)$$

Belief measures and plausibility measures are dual in the usual sense. That is,

$$Pl(A) = 1 - Bel(\bar{A}) \quad (5.44)$$

for all $A \in \mathcal{P}(X)$. Pairs of these dual measures form the basis of DST.

Any belief measure, Bel, can of course be represented by its Möbius representation, m, which is obtained for each $A \in \mathcal{P}(X)$ by the usual formula

$$m(A) = \sum_{B|B \subseteq A} (-1)^{|A-B|} Bel(B). \quad (5.45)$$

In DST, it is guaranteed that $m(A) \ge 0$. Due to this special property, function m is usually called a *basic probability assignment* in DST. For each set $A \in \mathcal{P}(X)$, the value $m(A)$ expresses the proportion to which all available and relevant evidence supports the claim that a particular element of X, whose characterization in terms of relevant attributes is deficient, belongs to the set A. This value, $m(A)$, pertains solely to one set, set A; it does not imply any additional claims regarding subsets of A. If there is some additional evidence supporting the claim that the element belongs to a subset of A, say $B \subset A$, it must be expressed by another value $m(B)$.

Since values $m(A)$ are positive and add to 1 for all $A \in \mathcal{P}(X)$, function m resembles a probability distribution function. However, there is a fundamental difference between probability distribution functions in probability theory and basic probability assignments in DST: the former are defined on X, while the latter are defined on $\mathcal{P}(X)$. Observe also that none of the properties of monotone measures are required for function m. It is thus not a measure. To obtain a monotone measure, elementary pieces of evidence expressed by values $m(A)$ must be properly aggregated. Two obvious aggregations, which result in a belief measure and a plausibility measure, are expressed for all $A \in \mathcal{P}(X)$ by the formulas

$$Bel(A) = \sum_{B|B \subseteq A} m(B), \tag{5.46}$$

$$Pl(A) = \sum_{B|A \cap B \neq \varnothing} m(B). \tag{5.47}$$

Equation (5.46) is, of course, the inverse of the Möbius transform in Eq. (5.45) [compare with Eqs. (4.9) and (4.8)], and Eq. (5.47) follows from the duality equation, Eq. (5.44). Clearly,

$$
\begin{aligned}
Pl(A) &= 1 - Bel(\overline{A}) \\
&= 1 - \sum_{B|B \subseteq \overline{A}} m(B) \\
&= \sum_{B|B \not\subseteq \overline{A}} m(B) \\
&= \sum_{B|A \cap B \neq \varnothing} m(B).
\end{aligned}
$$

The relationship between $m(A)$ and $Bel(A)$, expressed by Eq. (5.46), has the following meaning. The value of $m(A)$ characterizes the degree of evidence that the true alternative is in set A, but is does not take into account any additional evidence for the various subsets of A. The associated value $Bel(A)$ represents the total evidence (or belief) that the true alternative is in set A, which is obtained by adding degrees of evidence for the set itself, as well as for any of its subsets. The value $Pl(A)$, as expressed by Eq. (5.47), has a different meaning. It represents not only the total evidence that the true alternative is in set A, but also partial (or plausible) evidence for the set that is associated with any set that overlaps with A. From Eqs. (5.46) and (5.47), clearly,

$$Pl(A) \leq Bel(A) \tag{5.48}$$

for all $A \in \mathcal{P}(X)$.

Given a basic probability assignment m on $\mathcal{P}(X)$, every set A for which $m(A) > 0$ is called a *focal set* (or *focal element*). The pair $\langle \mathcal{F}, m \rangle$, where $\mathcal{F}$ denotes the family of all focal sets induced by m, is called a *body of evidence*.

Total ignorance is expressed in terms of the basic probability assignment by $m(X) = 1$ and $m(A) = 0$ for all $A \neq X$. That is, we know that the element is in the universal set, but we have no evidence about its location in any subset of X. It follows from Eq. (5.46) that the expression of total ignorance in terms of the corresponding belief measure is exactly the same: $Bel(X) = 1$ and $Bel(A) = 0$ for all $A \neq X$. However, the expression of total ignorance in terms of the associated plausibility measure (obtained from the belief measure by Eq. (5.44) is quite different: $Pl(\varnothing) = 0$ and $Pl(A) = 1$ for all $A \neq 0$. This expression follows directly from Eq. (5.47).

Table 5.3. Conversion Formulas in the Dempster–Shafer Theory

	m	Bel	Pl	Q
$m(A) =$	$m(A)$	$\displaystyle\sum_{B\subseteq A}(-1)^{\|A-B\|}Bel(B)$	$\displaystyle\sum_{B\subseteq A}(-1)^{\|A-B\|}\left[1-Pl(\overline{B})\right]$	$\displaystyle\sum_{A\subseteq B}(-1)^{\|A-B\|}Q(B)$
$Bel(A) =$	$\displaystyle\sum_{B\subseteq A}m(B)$	$Bel(A)$	$1-Pl(\overline{A})$	$\displaystyle\sum_{B\subseteq\overline{A}}(1)^{\|B\|}Q(B)$
$Pl(A) =$	$\displaystyle\sum_{B\cap A\neq\varnothing}m(B)$	$1-Bel(\overline{A})$	$Pl(A)$	$\displaystyle\sum_{\varnothing\neq B\subseteq\overline{A}}(-1)^{B+1}Q(B)$
$Q(A) =$	$\displaystyle\sum_{A\subseteq B}m(B)$	$\displaystyle\sum_{B\subseteq A}(-1)^{\|B\|}Bel(\overline{B})$	$\displaystyle\sum_{\varnothing\neq B\subseteq A}(-1)^{\|B\|+1}Pl(B)$	$Q(A)$

In addition to the basic probability assignment, belief measure, and plausibility measure, it is also sometimes convenient to use the function

$$Q(A)=\sum_{B\mid A\subseteq B}m(B), \tag{5.49}$$

which is called a *commonality function*. For each $A\in\mathcal{P}(X)$, the value $Q(A)$ represents the total portion of belief that can move freely to every point of A. Function Q is an example of a monotone decreasing measure. It can be used for representing given evidence, since it is uniquely convertible to functions m, Bel, and Pl. All conversion formulas between the four functions are given in Table 5.3.

5.4.1. Joint and Marginal Bodies of Evidence

Given joint belief and plausibility measures, Bel and Pl, defined on subsets of $X\times Y$, the associated marginal measures are determined for each $A\in X$ and $B\in Y$ by the formulas

$$Bel_X(A)=Bel(A\times Y), \tag{5.50}$$

$$Bel_Y(B)=Bel(X\times B), \tag{5.51}$$

$$Pl_X(A)=Pl(A\times Y), \tag{5.52}$$

$$Pl_Y(B)=Pl(X\times B). \tag{5.53}$$

However, the relationship between the joint and marginal bodies of evidence can also be expressed in terms of the basic probability assignment. To do that, we need to define its projections for each subset C of $X\times Y$,

$$C_X = \{x \in X \mid \langle x, y \rangle \in C \text{ for some } y \in Y\},$$
$$C_Y = \{y \in Y \mid \langle x, y \rangle \in C \text{ for some } x \in X\},$$

on X and Y, respectively. Then,

$$m_X(A) = \sum_{C \mid A = C_X} m(C) \text{ for all } A \in \mathcal{P}(X), \qquad (5.54)$$

and

$$m_Y(B) = \sum_{C \mid A = C_Y} m(C) \text{ for all } B \in \mathcal{P}(Y). \qquad (5.55)$$

Given marginal bodies of evidence, $\langle \mathcal{F}_X, m_X \rangle$ and $\langle \mathcal{F}_Y, m_Y \rangle$, are said to be *noninteractive* if and only if for all $A \in \mathcal{F}_X$, $B \in \mathcal{F}_Y$, and $C \in \mathcal{P}(X \times Y)$

$$m(C) = \begin{cases} m_X(A) \cdot m_Y(B) & \text{when } C = A \times B \\ 0 & \text{otherwise.} \end{cases} \qquad (5.56)$$

That is, if marginal bodies of evidence are noninteractive, then the only focal sets of the joint body of evidence are Cartesian products of the marginal focal sets.

EXAMPLE 5.7. Given the marginal bodies of evidence in Table 5.4a, the joint body of evidence in Table 5.4b has been calculated under the assumption that the marginal ones are noninteractive. Using Eq. (5.56), we obtain, for example, for the first and seventh joint focal sets in Table 5.4b:

$$m(\{x_2, x_3\} \times \{y_2, y_3\}) = m_X(\{x_2, x_3\}) \cdot m_Y(\{y_2, y_3\}) = 0.25 \times 0.25 = 0.0625$$
$$m(\{x_1, x_3\} \times \{y_1\}) = m_X(\{x_1, x_3\}) \cdot m_Y(\{y_1\}) = 0.15 \times 0.25 = 0.0375.$$

5.4.2. Rules of Combination

Evidence obtained in the same context from two independent sources (for example, from two experts in the field of inquiry or from two sensors) and expressed by two basic probability assignments m_1 and m_2 on some power set $\mathcal{P}(X)$ must be appropriately aggregated to obtain the combined basic probability assignment $m_{1,2}$. In general, evidence can be combined in various ways, some of which can take into consideration the reliability of the sources and other relevant aspects. The standard way of combing evidence is expressed by the formula

$$m_{1,2}(A) = \frac{\displaystyle\sum_{B \cap C = A} m_1(B) \cdot m_2(C)}{1 - c} \qquad (5.57)$$

Table 5.4. Illustration of Noninteractive Marginal Bodies of Evidence (Example 5.7)

(a) Given Marginal Bodies of Evidence

	x_1	x_2	x_3	$m_X(A)$		y_1	y_2	y_3	$m_Y(B)$
A:	0	1	1	0.25	B:	0	1	1	0.25
	1	0	1	0.15		1	0	0	0.25
	1	1	0	0.30		1	1	1	0.50
	1	1	1	0.30					

(b) Joint Body of Evidence Under the Assumption That the Given Marginal Bodies of Evidence Are Noninteractive

	x_1 y_1	x_1 y_2	x_1 y_3	x_2 y_1	x_2 y_2	x_2 y_3	x_3 y_1	x_3 y_2	x_3 y_3	$m(C)$
C:	0	0	0	0	1	1	0	1	1	0.0625
	0	0	0	1	0	0	1	0	0	0.0625
	0	0	0	1	1	1	1	1	1	0.1250
	0	1	1	0	0	0	0	1	1	0.0375
	0	1	1	0	1	1	0	0	0	0.0750
	0	1	1	0	1	1	0	1	1	0.0750
	1	0	0	0	0	0	1	0	0	0.0375
	1	0	0	1	0	0	0	0	0	0.0750
	1	0	0	1	0	0	1	0	0	0.0750
	1	1	1	0	0	0	1	1	1	0.0750
	1	1	1	1	1	1	0	0	0	0.1500
	1	1	1	1	1	1	1	1	1	0.1500

for all $A \neq \varnothing$, and $m_{1,2}(\varnothing) = 0$, where

$$c = \sum_{B \cap C = \varnothing} m_1(B) \cdot m_2(C). \tag{5.58}$$

Equation (5.57) is referred to as *Dempster's rule of combination*. According to this rule, the degree of evidence $m_1(B)$ from the first source that focuses on set $B \in \mathcal{P}(X)$ and the degree of evidence $m_2(C)$ from the second source that focuses on set $C \in \mathcal{P}(X)$ are combined by taking the product $m_1(B) \cdot m_2(C)$, which focuses on the intersection $B \cap C$. This is exactly the same way in which the joint probability distribution is calculated from two independent marginal distributions; consequently, it is justified on the same grounds. However, since some intersections of focal sets from the first and second source may result in the same set A, we must add the corresponding products to obtain $m_{1,2}(A)$. Moreover, some of the intersections may be empty. Since it is required that $m_{1,2}(\emptyset) = 0$, the value c expressed by Eq. (5.58) is not included in the definition of the joint basic probability assignment $m_{1,2}$. This means that the sum of products $m_1(B) \cdot m_2(C)$ for all focal sets B of m_1 and all focal sets C of m_2 such that $B \cap C \neq \emptyset$ is equal to $1 - c$. To obtain a normalized basic probability assignment $m_{1,2}$, as required, we must divide each of these products by this factor $1 - c$, as indicated in Eq. (5.57).

EXAMPLE 5.8. Assume that an old painting was discovered that strongly resembles paintings of Raphael. Such a discovery is likely to generate various questions regarding the status of the painting. Assume the following three questions:

1. Is the discovered painting a genuine painting by Raphael?
2. Is the discovered painting a product of one of Raphael's many disciples?
3. Is the discovered painting a counterfeit?

Let R, D, and C denote subsets of our universal set X—the set of all paintings—which contain the set of all paintings by Raphael, the set of all paintings by disciples of Raphael, and the set of all counterfeits of Raphael's paintings, respectively.

Assume now that two experts performed careful examinations of the painting independently of each other and subsequently provided us with basic probability assignments m_1 and m_2, specified in Table 5.5. These are the degrees of evidence that each expert obtained by the examination and that support the various claims that the painting belongs to one of the sets of our concern. For example, $m_1(R \cup D) = 0.15$ is the degree of evidence obtained by the first expert that the painting was done by Raphael himself or that the painting was done by one of this disciples. Using Eq. (5.46), we can easily calculate the total evidence, Bel_1 and Bel_2, in each set, as shown in Table 5.5.

Table 5.5. Illustration of the Dempster Rule of Combination (Example 5.8)

Focal Sets	Expert 1		Expert 2		Combined evidence	
	m_1	Bel_1	m_2	Bel_2	$m_{1,2}$	$Bel_{1,2}$
R	0.05	0.05	0.15	0.15	0.21	0.21
D	0.00	0.00	0.00	0.00	0.01	0.01
C	0.05	0.05	0.05	0.05	0.09	0.09
$R \cup D$	0.15	0.20	0.05	0.20	0.12	0.34
$R \cup C$	0.10	0.20	0.20	0.40	0.20	0.50
$D \cup C$	0.05	0.10	0.05	0.10	0.06	0.16
$R \cup D \cup C$	0.60	1.00	0.50	1.00	0.31	1.00

Applying Dempster's rule of combination to m_1 and m_2, we obtain the joint basic assignment $m_{1,2}$, which is also shown in Table 5.5. To determine the values of $m_{1,2}$, we calculate the normalization factor $1 - c$ first. Applying Eq. (5.58), we obtain

$$
\begin{aligned}
c = {}& m_1(R)\cdot m_2(D) + m_1(R)\cdot m_2(C) + m_1(R)\cdot m_2(D\cup C) + m_1(D)\cdot m_2(R) \\
& + m_1(D)\cdot m_2(C) + m_1(D)\cdot m_2(R\cup C) + m_1(C)\cdot m_2(R) + m_1(C)\cdot m_2(D) \\
& + m_1(C)\cdot m_2(R\cup D) + m_1(R\cup D)\cdot m_2(C) + m_1(R\cup C)\cdot m_2(D) \\
& + m_1(D\cup C)\cdot m_2(R) \\
= {}& 0.03.
\end{aligned}
$$

The normalization factor is then $1 - c = 0.97$. Values of $m_{1,2}$ are calculated by Eq. (5.57). For example,

$$
\begin{aligned}
m_{1,2}(R) = {}& [m_1(R)\cdot m_2(R) + m_1(R)\cdot m_2(R\cup D) + m_1(R)\cdot m_2(R\cup C) \\
& + m_1(R)\cdot m_2(R\cup D\cup C) + m_1(R\cup D)\cdot m_2(R) \\
& + m_1(R\cup D)\cdot m_2(D\cup C) + m_1(R\cup C)\cdot m_2(R) \\
& + m_1(R\cup C)\cdot m_2(R\cup D) + m_1(R\cup D\cup C)\cdot m_2(R)]/0.97 \\
= {}& 0.21,
\end{aligned}
$$

$$
\begin{aligned}
m_{1,2}(D) = {}& [m_1(D)\cdot m_2(D) + m_1(D)\cdot m_2(R\cup D) + m_1(D)\cdot m_2(D\cup C) \\
& + m_1(D)\cdot m_2(R\cup D\cup C) + m_1(R\cup D)\cdot m_2(D) \\
& + m_1(R\cup D)\cdot m_2(D\cup C) + m_1(D\cup C)\cdot m_2(D) \\
& + m_1(D\cup C)\cdot m_2(R\cup D) + m_1(R\cup D\cup C)\cdot m_2(D)]/0.97 \\
= {}& 0.01,
\end{aligned}
$$

$$
\begin{aligned}
m_{1,2}(R\cup C) = {}& [m_1(R\cup C)\cdot m_2(R\cup C) + m_1(R\cup C)\cdot m_2(R\cup D\cup C) \\
& + m_1(R\cup D\cup C)\cdot m_2(R\cup C)/0.97 \\
= {}& 0.2,
\end{aligned}
$$

$$
\begin{aligned}
m_{1,2}(R\cup C\cup C) = {}& [m_1(R\cup D\cup C)\cdot m_2(R\cup D\cup C)/0.97 \\
= {}& 0.31,
\end{aligned}
$$

and similarly for the remaining focal sets, C, $R\cup D$, and $D\cup C$. The joint basic assignment can now be used to calculate the joint belief $Bel_{1,2}$ (Table 5.5) and $Pl_{1,2}$.

The Dempster rule of combination is well justified when $c = 0$ in Eq. (5.57). However, it is controversial when $c \neq 0$ (see Note 5.6). This happens when evidence obtained from distinct sources is conflicting. In fact, the value of c can be viewed as the degree of this conflict. An *alternative rule of combination*, which is epistemologically sound and, hence, eliminates the controversies of the Dempster rule, works as follows:

$$
m_{1,2}(A) = \begin{cases}
\displaystyle\sum_{B\cap C=A} m_1(B)\cdot m_2(C) & \text{when } A\neq\varnothing \quad \text{and} \quad A\neq X \\
\displaystyle m_1(X)\cdot m_2(X) + \sum_{B\cap C=\varnothing} m_1(B)\cdot m_2(C) & \text{when } A = X \\
0 & \text{when } A = \varnothing.
\end{cases} \tag{5.59}
$$

Table 5.6. Comparison of the Introduced Rules of Combination (Example 5.9)

Focal Sets	Source 1 m_1	Source 2 m_2	Eqs. (5.57) and (5.58) $m_{1,2}$	Eq. (5.59) $m_{1,2}$
A	0.01	0.90	0.646	0.181
$\bar{A}$	0.80	0.00	0.286	0.080
X	0.19	0.10	0.068	0.739

According to this alternative rule of combination, $m_{1,2}$ is normalized by moving c to $m_{1,2}(X)$. This means that the conflict between the two sources of evidence is not hidden, but it is explicitly recognized as a contributor to our ignorance.

EXAMPLE 5.9. Given some universal set X, assume that we are only interested in finding whether the true alternative is in set A or its complement $\bar{A}$. Assume further that we have evidence from two independent sources shown in Table 5.6. Also shown in the table is the combined evidence obtained by the Dempster rule of combination, expressed by Eqs. (5.57) and (5.58), and by the alternative rule of combination, expressed by Eq. (5.59). In both cases, we first calculate

$$c = m_1(A) \cdot m_2(\bar{A}) + m_1(\bar{A}) \cdot m_2(A) = 0.72.$$

Using the alternative rule, we then calculate

$$m_{1,2}(A) = m_1(A) \cdot m_2(A) + m_1(A) \cdot m_2(X) + m_1(X) \cdot m_2(A) = 0.181,$$
$$m_{1,2}(\bar{A}) = m_1(\bar{A}) \cdot m_2(\bar{A}) + m_1(\bar{A}) \cdot m_2(X) + m_1(X) \cdot m_2(\bar{A}) = 0.180.$$

Finally, $m_{1,2}(X) = 1 - m_{1,2}(A) - m_{1,2}(\bar{A}) = 0.739$. This value can be calculated directly as

$$m_{1,2}(X) = m_1(X) \cdot m_2(X) + c = 0.739.$$

Using the Dempster rule, we normalize the previously obtained values of $m_{1,2}(A)$ and $m_{1,2}(\bar{A})$ by dividing them by $1 - c = 0.28$. This results in the values given in Table 5.6.

5.4.3. Special Classes of Bodies of Evidence

Consider a body of evidence $\langle \mathcal{F}, m \rangle$ in the sense of DST. If the associated belief measure is also additive, then it is a classical probability measure. The following theorem establishes necessary and sufficient conditions for probabilistic bodies of evidence.

Theorem 5.6. A belief measure *Bel* on a finite power set $\mathcal{P}(X)$ is a probability measure if and only if its basic assignment m is given by $m(\{x\}) = Bel(\{x\})$ and $m(A) = 0$ for all subsets A or X that are not singletons.

Proof. Assume that *Bel* is a probability measure. For the empty set $\varnothing$, the theorem trivially holds, since $m(\varnothing) = 0$ by the definition of m. Let $A \neq \varnothing$ and assume $A = \{x_1, x_2, \ldots, x_n\}$. Then by repeated application of additivity, we obtain

$$
\begin{aligned}
Bel(A) &= Bel(\{x_1\}) + Bel(\{x_2, x_3, \ldots, x_n\}) \\
&= Bel(\{x_1\}) + Bel(\{x_2\}) + Bel(\{x_3, x_4, \ldots, x_n\}) \\
&\ \ \vdots \\
&= Bel(\{x_1\}) + Bel(\{x_2\}) + \cdots = Bel(\{x_n\}).
\end{aligned}
$$

Since $Bel(\{x\}) = m(\{x\})$ for any $x \in X$, by Eq. (5.46), we have

$$
Bel(A) = \sum_{i=1}^{n} m(\{x_i\}).
$$

Hence, *Bel* is defined in terms of a basic probability assignment that focuses only on singletons.

Assume now that a basic probability assignment m is given such that

$$
\sum_{x \in X} m(\{x\}) = 1.
$$

Then for any sets $A, B \in \mathcal{P}(X)$ such that $A \cap B = \varnothing$, we have

$$
\begin{aligned}
Bel(A) + Bel(B) &= \sum_{x \in A} m(\{x\}) + \sum_{x \in B} m(\{x\}) \\
&= \sum_{x \in A \cup B} m(\{x\}) = Bel(A \cup B)
\end{aligned}
$$

and, consequently, *Bel* is a probability measure. ∎

Given a body of evidence $\langle \mathcal{F}, m \rangle$, let $\mathcal{F} = \{A_1, A_2, \ldots, A_n\}$. When

$$
A_{\pi(1)} \subset A_{\pi(2)} \subset \ldots \subset A_{\pi(n)}
$$

for some permutation π of $\mathbb{N}_n$ (i.e., when $\mathcal{F}$ is a nested family of focal sets), the body of evidence is called *consonant*. This name appropriately reflects the fact that the degrees of evidence allocated to focal sets that are nested do conflict with one another in a minimal way. Belief and plausibility measures associated with consonant bodies of evidence have special properties that are characterized by the following theorem.

Theorem 5.7. Given a consonant body of evidence $\langle \mathcal{F}, m \rangle$, the associated consonant belief and plausibility measures possess the following properties:

(i) $Bel(A \cap B) = \min\{Bel(A), Bel(B)\}$ for all $A, B \in \mathcal{P}(X)$;

(ii) $Pl(A \cup B) = \max\{Pl(A), Pl(B)\}$ for all $A, B \in \mathcal{P}(X)$.

Proof. (i) Since the focal elements in $\mathcal{F}$ are nested, they may be linearly ordered by the subset relation. Let $\mathcal{F} = \{A_1, A_2, \ldots, A_n\}$ and assume that $A_i \subset A_j$ whenever $i < j$. Now consider arbitrary subsets A and B of X. Let i_1 be the largest integer i such that $A_i \subset A$, and let i_2 be the largest integer i such that $A_i \subset B$. Then $A_i \subset A$ and $A_i \subset B$ if and only if $i \leq i_1$ and $i \leq i_2$, respectively. Moreover, $A_i \subset A \cap B$ if and only if $i \leq \min\{i_1, i_2\}$. Hence,

$$Bel(A \cap B) = \sum_{i=1}^{\min\{i_1, i_2\}} m(A_i)$$
$$= \min\left\{ \sum_{i=1}^{i_1} m(A_i), \sum_{i=1}^{i_2} m(A_i) \right\}$$
$$= \min\{Bel(A), Bel(B)\}.$$

(ii) Assume that (i) holds. Then by Eq. (5.44),

$$Pl(A \cup B) = 1 - Bel\left(\overline{A \cup B}\right)$$
$$= 1 - Bel(\overline{A} \cap \overline{B})$$
$$= 1 - \min\{Bel(\overline{A}), Bel(\overline{B})\}$$
$$= \max\{1 - Bel(\overline{A}), 1 - Bel(\overline{B})\}$$
$$= \max\{Pl(A), Pl(B)\}.$$

for all $A, B \in \mathcal{P}(X)$. ∎

It follows from this theorem that consonant belief and plausibility measures are the same functions as necessity and possibility measures, respectively.

Necessity and possibility measures are thus special belief and plausibility measures, respectively, which are characterized by nested bodies of evidence. Given any nested bodies of evidence, they can be manipulated either by the calculus of possibility theory or by calculus of DST. In the former case, the resulting bodies of evidence are again nested, and hence, we remain within the domain of possibility theory. In the latter case, they may not be nested, which means that we leave the domain of possibility theory. To operate within possibility theory thus requires the use of rules of possibilistic calculus, which in general are different from the rules of the calculus of DST. Possibility theory is thus a special branch of DST only at the level of representation, but not at the calculus level. This is illustrated by the following example.

Table 5.7. Noninteraction in Possibility Theory Versus Noninteraction in DST (Example 5.10)

	$X \times Y$			Eq. (5.19)			Eq. (5.56)		
x_1 y_1	x_1 y_2	x_2 y_1	x_2 y_2	$m(A)$	$Bel(A)$	$Pl(A)$	$m(A)$	$Bel(A)$	$Pl(A)$
A: 0	0	0	0	0.0	0.0	0.0	0.00	0.00	0.00
1	0	0	0	0.2	0.2	1.0	0.08	0.08	1.00
0	1	0	0	0.0	0.0	0.6	0.00	0.00	0.60
0	0	1	0	0.0	0.0	0.8	0.00	0.00	0.80
0	0	0	1	0.0	0.0	0.6	0.00	0.00	0.48
1	1	0	0	0.0	0.2	1.0	0.12	0.20	1.00
1	0	1	0	0.2	0.4	1.0	0.32	0.40	1.00
1	0	0	1	0.0	0.2	1.0	0.00	0.08	1.00
0	1	1	0	0.0	0.0	0.8	0.00	0.00	0.92
0	1	0	1	0.0	0.0	0.6	0.00	0.00	0.60
0	0	1	1	0.0	0.0	0.8	0.00	0.00	0.80
1	1	1	0	0.0	0.4	1.0	0.00	0.52	1.00
1	1	0	1	0.0	0.2	1.0	0.00	0.20	1.00
1	0	1	1	0.0	0.4	1.0	0.00	0.40	1.00
0	1	1	1	0.0	0.0	0.8	0.00	0.00	0.92
1	1	1	1	0.6	1.0	1.0	0.48	1.00	1.00

EXAMPLE 5.10. Consider the following marginal basic probability assignments, m_X and m_Y, on $X = \{x_1, x_2\}$ and $Y = \{y_1, y_2\}$:

$$m_X(\{x_1\}) = 0.2, \quad m_X(X) = 0.8,$$
$$m_Y(\{y_1\}) = 0.4, \quad m_Y(Y) = 0.6.$$

Clearly, both bodies of evidence induced by m_X and m_Y are nested. Assuming their noninteraction, we can calculate the associated joint body of evidence by applying either the rule of possibility theory, expressed by Eq. (5.19), or the rule of DST, expressed by Eq. (5.56). The two results, which are very different, are shown in Table 5.7. While using Eq. (5.19) results in the nested family of focal sets,

$$\{\{\langle x_1, y_1 \rangle\}, \{\langle x_1, y_1 \rangle, \langle x_2, y_1 \rangle\}, X\},$$

the family of focal sets obtained by Eq. (5.56) is not nested, as it contains two focal sets,

$$\{\langle x_1, y_1 \rangle, \langle x_1, y_2 \rangle\} \quad \text{and} \quad \{\langle x_1, y_1 \rangle, \langle x_2, y_1 \rangle\},$$

neither of which is a subset of the other one.

Another special class of bodies of evidence in DST consists of those that are based on λ-measures. Again, the rules of λ-measures must be followed in order to remain within this special class.

5.5. REACHABLE INTERVAL-VALUED PROBABILITY DISTRIBUTIONS

In the theory of imprecise probabilities based on λ-measures (Section 5.3), the dual lower and upper probability measures are fully characterized by *either* the lower probability function *or* the upper probability function on singletons. This means that $|X|$ values of one of the dual measures are sufficient to determine all other values of both measures. In the theory of imprecise probabilities discussed in this section, the lower and upper probability functions are determined by *both the lower and upper probability functions on singletons*. That is, $2 \times |X|$ values (all for singletons) are needed to determine all other values of both measures. In other words, the lower and upper probability measures are determined in this case by probability distribution functions that are interval-valued.

Given a finite set $X = \{x_1, x_2, \ldots, x_n\}$ of considered alternatives, let

$$I = \langle\, [l_i, u_i] \mid i \in \mathbb{N}_n \rangle$$

denote an n-tuple of probability intervals defined on individual alternatives $x_i \in X$. This n-tuple may be viewed as an interval-valued probability distribution on X provided that the inequalities

$$0 \le l_i \le u_i \le 1$$

are satisfied for all $i \in \mathbb{N}_n$. Associated with I is a convex set

$$\mathcal{D} = \left\{ p \mid l_i \le p(x_i) \le u_i, \ i \in \mathbb{N}_n, \ \sum_{x_i \in X} p(x_i) = 1 \right\} \tag{5.60}$$

of probability distribution functions p on X whose values are bounded for each $x_i \in X$ by values l_i and u_i. Clearly, $\mathcal{D} \ne \varnothing$ if and only if

$$\sum_{i=1}^{n} l_i \le 1, \tag{5.61}$$

and

$$\sum_{i=1}^{n} u_i \ge 1. \tag{5.62}$$

Tuples of probability intervals that satisfy these inequalities are called *proper* (or *reasonable*). These are obviously the only meaningful interval-valued probability distributions to represent imprecise probabilities.

Set $\mathcal{D}$ defined by Eq. (5.60) forms always an $(n - 1)$-dimensional polyhedron. Any point (a probability distribution) in this polyhedron can be expressed as a linear combination of its extreme points (vertices, corners). It is known that the number c of these extreme points is bounded by the inequalities

$$n \le c \le n(n-1). \tag{5.63}$$

From the probability distributions in set $\mathcal{D}$, the lower and upper probability measures, l and u, are defined for all $A \in \mathcal{P}(X)$ in the usual way (recall Eqs. (4.11) and (4.12)):

$$l(A) = \inf_{p \in \mathcal{D}} \sum_{x_i \in A} p(x_i), \tag{5.64}$$

$$u(A) = \sup_{p \in \mathcal{D}} \sum_{x_i \in A} p(x_i). \tag{5.65}$$

Due to the definition of set $\mathcal{D}$,

$$p(\{x_i\}) \ge l_i \quad \text{and} \quad p(\{x_i\}) \le u_i \tag{5.66}$$

for all $i \in \mathbb{N}_n$. For consistency of the two representations of imprecise probabilities, one based on I and one based on $\mathcal{D}$, it is desirable, if possible, to modify the lower and upper bounds l_i and u_i to obtain the equalities in Eq. (5.66) without changing the set $\mathcal{D}$. It is easy to see that the equalities in Eq. (5.66) are obtained if and only if the n-tuple I satisfies the inequalities

$$\sum_{j \ne i} l_j + u_i \le 1, \tag{5.67}$$

$$\sum_{j \ne i} u_j + l_i \ge 1, \tag{5.68}$$

for all $i \in \mathbb{N}_n$. These conditions guarantee for each $i \in \mathbb{N}_n$ the existence of probability distribution functions ${}^i p$ and ${}^i q$ in $\mathcal{D}$ such that

$${}^i p(x_i) = u_i \quad \text{and} \quad l_j \le {}^i p(x_i) \le u_j \quad \text{for all } j \ne i,$$

$${}^i q(x_i) = l_i \quad \text{and} \quad l_j \le {}^i q(x_i) \ge u_j \quad \text{for all } j \ne i.$$

These functions guarantee, in turn, that the equalities in Eq. (5.66) are reached. Tuples I for which the inequalities (5.67) and (5.68) are satisfied are thus called *reachable* (or *feasible*).

It is well established in the literature (see Note 5.8) that any given tuple of probability intervals I can be converted to its reachable counterpart,

$$I' = \{[l_i', u_i'] \mid i \in \mathbb{N}_n\},$$

via the formulas

$$l_i' = \max\left\{l_i, 1 - \sum_{j \neq i} u_j\right\}, \tag{5.69}$$

$$u_i' = \min\left\{u_i, 1 - \sum_{j \neq i} u_j\right\}. \tag{5.70}$$

for all $i \in \mathbb{N}_n$. Moreover, the sets of probability distributions associated with I and I' are the same.

Observe that l_i' defined by Eq. (5.69) is either equal to or greater than l_i. Similarly, u_i' defined by Eq. (5.70) is either equal to or smaller than u_i. This means that the probability intervals in I' are equal to or narrower than the corresponding intervals in I. The reachable tuples of probability intervals thus provide us with a more accurate representation of imprecise probabilities than those that are not reachable. It is thus reasonable to limit ourselves to only reachable probability intervals. If the given probability intervals are not reachable, they always can be converted to the reachable ones by Eqs. (5.69) and (5.70).

Assuming now that a given n-tuple of I of probability intervals is reachable, the lower and upper probability measures, l and u, are calculated for all $A \in \mathcal{P}(X)$ by the formulas

$$l(A) = \max\left\{\sum_{x_i \in A} l_i, 1 - \sum_{x_i \notin A} u_i\right\}, \tag{5.71}$$

$$u(A) = \min\left\{\sum_{x_i \in A} u_i, 1 - \sum_{x_i \notin A} l_i\right\}. \tag{5.72}$$

Values $l(A)$ and $u(A)$ for all $A \in \mathcal{P}(X)$ are thus determined by $2n$ values l_i, u_i, $i \in \mathbb{N}_n$.

Imprecise probabilities based on reachable probability intervals are known to belong to the class of imprecise probabilities that are based on Choquet capacities of order 2. They are also known to be incomparable with imprecise probabilities represented by Choquet capacities of order ∞ (i.e., belief and plausibility measures of DST).

EXAMPLE 5.11. Let $X = \{x_1, x_2, x_3\}$ and assume that the interval-valued probability distribution

$$I = \langle [0.3, 0.6], [0.2, 0.5], [0.1, 0.4] \rangle$$

on X is given (elicited from an expert or obtained in some other way). This distribution is proper, since it satisfies the inequalities (5.61) and (5.62). To show that it is also reachable, the inequalities (5.67) and (5.68) must be checked for $i = 1, 2, 3$:

$$l_1 + l_2 + u_3 = 0.3 + 0.2 + 0.4 = 0.9 \leq 1,$$

$$l_1 + l_3 + u_2 = 0.3 + 0.1 + 0.5 = 0.9 \leq 1,$$

$$l_2 + l_3 + u_1 = 0.2 + 0.1 + 0.6 = 0.9 \leq 1,$$

$$u_1 + u_2 + l_3 = 0.6 + 0.5 + 0.1 = 1.2 \geq 1,$$

$$u_1 + u_3 + l_2 = 0.6 + 0.4 + 0.2 = 1.2 \geq 1,$$

$$u_2 + u_3 + l_1 = 0.5 + 0.4 + 0.3 = 1.2 \geq 1.$$

All the required inequalities are satisfied, and hence, the given I is reachable. Now, we can calculate $l(A)$ and $u(A)$ for subsets of X that are not singletons by Eqs. (5.71) and (5.72). The results are shown in Figure 5.7a. For example:

$$l(\{x_2, x_3\}) = \max\{l_2 + l_3, 1 - u_1\} = \max\{0.3, 0.4\} = 0.4$$

$$u(\{x_2, x_3\}) = \min\{u_2 + u_3, 1 - l_1\} = \min\{0.9, 0.7\} = 0.7.$$

One of the two measures may, of course, be calculated from the other one by the duality equation as well.

A geometrical interpretation of the three probability intervals in this example and the associated set $\mathcal{D}$ of probability distributions is shown in Figure 5.7c. We can see that this set is convex and any of its elements can be expressed as a linear combination of the six extreme points shown in the figure. According to Eq. (5.63), this is the maximum number of extreme points for $n = 3$. Since l is a Choquet capacity of order 2, these extreme points are also the probability distributions in the interaction representation of l (introduced in Section 4.3.3). These latter probability distributions are readily obtained from the Hasse diagram of the underlying Boolean lattice shown in Figure 5.7b. We can see that the probability distributions associated with the individual maximal chains in the lattice are exactly the same as the extreme points in Figure 5.7c.

EXAMPLE 5.12. Let $X = \{x_1, x_2, x_3\}$ and consider the interval-valued probability distribution

$$I = \langle [0.2, 0.5], [0.3, 0.4], [0.1, 0.2] \rangle$$

	x_1	x_2	x_3	$l(A)$	$u(A)$
A:	0	0	0	0.0	0.0
	1	0	0	0.3	0.6
	0	1	0	0.2	0.5
	0	0	1	0.1	0.4
	1	1	0	0.6	0.9
	1	0	1	0.5	0.8
	0	1	1	0.4	0.7
	1	1	1	1.0	1.0

(a)

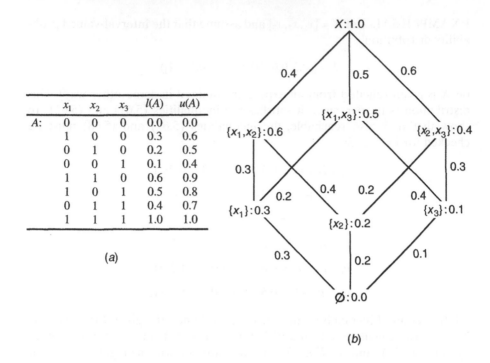

(b)

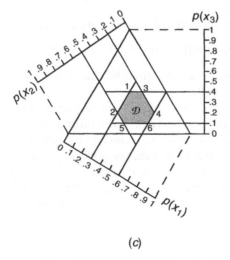

(c)

Figure 5.7. Illustration to Example 5.11.

on X. This distribution is clearly proper, but it is not reachable. For example,

$$u_2 + u_3 + l_1 = 0.8 < 1$$

violates Eq. (5.68). The distribution can be converted to its reachable counterpart, I', by Eqs. (5.69) and (5.70). For example,

$$l_1' = \max\{l_1, 1 - u_2 - u_3\} = \max\{0.2, 0.4\} = 0.4.$$

By calculating values l_i' and u_i' for all $i = 1, 2, 3$, we obtain

$$I' = \langle [0.4, 0.5], [0.3, 0.4], [0.1, 0.2] \rangle.$$

5.5.1. Joint and Marginal Interval-Valued Probability Distributions

Consider sets $X = \{x_1, x_2, \ldots, x_n\}$ and $Y = \{y_1, y_2, \ldots, y_m\}$ of states of two random variables and let a joint interval-valued probability distribution

$$I = \langle [l_{ij}, u_{ij}] \mid i \in \mathbb{N}_n, j \in \mathbb{N}_m \rangle$$

on $\mathcal{P}(X \times Y)$ be given, which is assumed to be reachable. The joint lower and upper probability measures l and u, are determined from I via Eqs. (5.71) and (5.72). The associated marginal measures are defined in the usual way: for every $A \in X$ and every $B \in Y$,

$$l_X(A) = l(A \times Y), \qquad u_X(A) = u(A \times Y), \tag{5.73}$$

$$l_Y(B) = l(X \times B), \qquad u_Y(B) = u(X \times B). \tag{5.74}$$

It is well known that the marginal measures obtained by these equations are exactly the same as those obtained by marginalization of the convex set of probability distributions associated with I. Moreover, the marginal tuples of probability intervals,

$$I_X = \langle [^X l_i, {}^X u_i] \mid i \in \mathbb{N}_n \rangle$$

$$I_Y = \langle [^X l_i, {}^X u_j] \mid j \in \mathbb{N}_n \rangle$$

are determined from I by the following equations ($i \in \mathbb{N}_n, j \in \mathbb{N}_m$):

$$^X l_i = \max\left\{ \sum_{j=1}^{m} l_{ij}, 1 - \sum_{k \neq i} \sum_{j=1}^{m} u_{kj} \right\} \tag{5.75}$$

$$^X u_i = \min\left\{ \sum_{j=1}^{m} u_{ij}, 1 - \sum_{k \neq i} \sum_{j=1}^{m} l_{kj} \right\} \tag{5.76}$$

$$^Y l_j = \max\left\{ \sum_{i=1}^{n} l_{ij}, 1 - \sum_{k \neq j} \sum_{i=1}^{n} u_{ik} \right\} \tag{5.77}$$

$$^{Y}u_j = \min\left\{\sum_{i=1}^{n} u_{ij}, 1 - \sum_{k \neq j} \sum_{i=1}^{n} u_{ik}\right\}. \tag{5.78}$$

By comparing the forms of these equations with the forms of Eqs. (5.69) and (5.70), we can conclude that I_X and I_Y are reachable.

For calculating conditional lower and upper probabilities, the general concept of conditional monotone measures introduced in Section 4.3.6 can be applied. See also Note 5.8.

EXAMPLE 5.13. Joint imprecise probabilities on $X \times Y = \{x_1, x_2\} \times \{y_1, y_2\}$ are defined in this example by the interval-valued probability distribution shown by the bold numbers in Table 5.8a. Since the given intervals are reachable, the remaining values of lower and upper probabilities in the table, $l(C)$ and $u(C)$, are calculated from the four intervals by Eqs. (5.71) and (5.72). The associated marginal lower and upper probabilities are shown in Table 5.8b, 5.8c. They can be calculated either by Eqs. (5.73) and (5.74) or by Eqs. (5.75)–(5.78). For example,

$$l_X(\{x_1\}) = l(\{x_1\} \times Y) = l(\{\langle x_1, y_1 \rangle, \langle x_1, y_2 \rangle\}) = 0.3,$$

$$u_Y(\{y_2\}) = u(X \times \{y_2\}) = u(\{\langle x_1, y_2 \rangle, \langle x_2, y_2 \rangle\}) = 0.8$$

Table 5.8. Illustration to Example 5.13

	x_1	x_1	x_2	x_2	$l(C)$	$u(C)$
	y_1	y_2	y_1	y_2		
C:	0	0	0	0	0.0	0.0
	1	0	0	0	**0.1**	**0.3**
	0	1	0	0	**0.2**	**0.4**
	0	0	1	0	**0.1**	**0.2**
	0	0	0	1	**0.3**	**0.5**
	1	1	0	0	0.3	0.6
	1	0	1	0	0.2	0.5
	1	0	0	1	0.4	0.7
	0	1	1	0	0.3	0.6
	0	1	0	1	0.5	0.8
	0	0	1	1	0.4	0.7
	1	1	1	0	0.5	0.7
	1	1	0	1	0.8	0.9
	1	0	1	1	0.6	0.8
	0	1	1	1	0.7	0.9
	1	1	1	1	1.0	1.0

(a)

	x_1	x_2	$l_X(A)$	$u_X(A)$
A:	0	0	0.0	0.0
	1	0	0.3	0.6
	0	1	0.4	0.7
	1	1	1.0	1.0

(b)

	y_1	y_2	$l_Y(B)$	$u_Y(B)$
B:	0	0	0.0	0.0
	1	0	0.2	0.5
	0	1	0.5	0.8
	1	1	1.0	1.0

(c)

by using Eqs. (5.73) and (5.74), respectively. Similarly,

$$^{X}l_1 = l_X(\{x_1\}) = \max\{l_{11} + l_{12}, 1 - u_{21} - u_{22}\} = 0.3$$

by Eq. (5.75), and

$$^{Y}u_2 = u_Y(\{y_2\}) = \min\{u_{12} + u_{22}, 1 - l_{11} - l_{21}\} = 0.8$$

by Eq. (5.78).

5.6. OTHER TYPES OF MONOTONE MEASURES

The Choquet capacities of various orders introduced in Chapter 4 and the special types of monotone measures introduced so far in this chapter are the main types of set-valued functions that have been utilized for representing imprecise probabilities. The purpose of this section is to survey some additional types of monotone measures that have been discussed in the literature and are potentially useful for representing imprecise probabilities as well.

A large class of monotone measures, referred to as *decomposable measures*, is based on the mathematical concepts of triangular norms (or *t*-norms) and triangular conorms (or *t*-conorms). These are pairs of dual binary operations on the unit interval that are commutative, associative, monotone, and subject to appropriate boundary conditions. They play an important role in fuzzy set theory as intersection and union operations on standard fuzzy sets. In this book, they are introduced in Section 7.3.2.

A monotone measure, $^{u}\mu$, defined on subsets of a finite set X, is called *decomposable with respect to a given t-conorm u* if and only if

$$^{u}\mu(A \cup B) = u[\,^{u}\mu(A),\,^{u}\mu(B)] \tag{5.79}$$

for each pair of disjoint sets $A, B \in \mathcal{P}(X)$. In decomposable measures, the degree of uncertainty of the union of any pair of disjoint sets is thus dependent solely on the degrees of uncertainty of the individual sets.

An important consequence of Eq. (5.79) and associativity of *t*-conorms is that any decomposable measure is uniquely determined by its values on all singletons. That is,

$$^{u}\mu(A) = {}^{u}\mu\left(\bigcup_{x \in A} \{x\}\right) = u_{x \in A}[\,^{u}\mu(x)] \tag{5.80}$$

for all $A \in \mathcal{P}(X)$. This is similar to possibility measures and λ-measures, which are in fact special decomposable measures. For possibility measures, the

t-conorm is the maximum operation; for λ-measures, the t-conorm for each particular value of λ is the function

$$\min\{1, a + b + \lambda ab\}.$$

There are, of course, many other classes of t-conorms or classes of t-norms, some of which are introduced in Section 7.2.2. Thus, decomposable measures form a very broad class of monotone measures. Their characteristic feature is that each measure in this class is fully determined by its values on singletons. Decomposable measures are thus attractive from the standpoint of computational complexity.

Another important class of monotone measures, which are called k-*additive measures*, has been discussed in the literature. A monotone measure, μ, is said to be k-additive ($k \geq 1$) if its Möbius representation, m, satisfies the following requirement: $m(A) = 0$ for all sets $A \in \mathcal{P}(X)$ such that $|A| > k$ and there exists at least one set A with k elements for which $m(A) \neq 0$.

Similar to decomposable measures, k-additive measures were introduced in an attempt to reduce the computational complexity in dealing with monotone measures. In general, to define a monotone measure μ on $\mathcal{P}(X)$ requires that its values be specified for all subsets of X except $\varnothing$ and X (since $m(\varnothing) = 0$ and $m(X) = 1$ by axioms of monotone measures). Hence, $2^{|X|} - 2$ values must be specified in general to characterize a monotone measure. For k-additive measures, the number of required values is reduced to $\sum_{i=1}^{k} \binom{|X|}{i}$, and for decomposable measures it is reduced to $|X|$. Decomposable measures and k-additive measures are thus suitable for approximating monotone measures that are difficult to handle. General principles for dealing with these and other approximation problems are discussed in Chapter 9.

NOTES

5.1. The idea of graded possibilities was originally introduced in the context of fuzzy sets by Zadeh [1978a]. However, the need for a theory of graded possibilities in economics was perceived by the British economist George Shackle already in the 1940s [Shackle, 1949]. He formalized his ideas in terms of a monotone decreasing measure called a *potential surprise* [Shackle, 1955, 1961, 1979]. However, this formalization can be reformulated in terms of standard possibility theory, surveyed in Section 5.2 [Klir, 2002a]. The notion of potential surprise is also examined in [Prade and Yager, 1994] and in some writings by Levi [1984, 1986, 1996].

5.2. The literature on possibility theory is now quite extensive. An early book by Dubois and Prade [1988a] is a classic in this area. More recent developments in possibility are covered in a text by Kruse et al. [1994] and in monographs by Wolkenhauer [1998] and Borgelt and Kruse [2002]. Important sources are also edited books by De Cooman et al. [1995] and Yager [1982a]. A sequence of three papers by De Cooman [1997] is perhaps the most comprehensive and general treatment of possibility theory. Thorough surveys of possibility theory with extensive bibliographies were written by Dubois et al. [1998, 2000]. A sequence of two papers by De Campos and Huete [1999] and a paper by Vejnarová [2000] deal in a comprehensive way with the various issues of independence in possibility theory. Shafer [1985] discusses possibility measures in the broader context of Dempster–Shafer theory. An interesting idea of using second-order possibility distributions in statistical reasoning is explored by Walley [1997]. Constructing possibility profiles from interval data is addressed by Joslyn [1994, 1997].

5.3. As is explained in Section 5.2, possibility theory is based on pairs of dual monotone measures, each consisting of a necessity measure, Nec, and a possibility measure, Pos, that are connected via Eq. (5.1). These two measures, whose range is [0, 1], can be converted by a one-to-one transformation to a single combined function, C, whose range is [−1, 1]. For each $A \in \mathcal{P}(X)$,

$$C(A) = Nec(a) + Pos(A) - 1, \qquad (5.81)$$

and, conversely,

$$Nec(A) = \begin{cases} 0 & \text{when } C(A) \leq 0 \\ C(A) & \text{when } C(A) > 0, \end{cases} \qquad (5.82)$$

$$Pos(A) = \begin{cases} C(A)+1 & \text{when } C(A) \leq 0 \\ 1 & \text{when } C(A) > 0. \end{cases} \qquad (5.83)$$

Positive values of $C(A)$ indicate the *degree of confirmation* of A by the available evidence, while its negative values express the *degree of disconfirmation* of A by the evidence.

5.4. Sugeno λ-measures were introduced by Sugeno [1974, 1977] and were further investigated by Banon [1981], Wierzchoń [1982, 1983], Kruse [1982a,b], and Wang and Klir [1992]. Due to the small number of parameters that characterize any λ-measure (values of the measure on singletons), λ-measures are relatively easy (compared to more general measures) to construct from data. They are often constructed with the help of neural networks [Wang and Wang, 1997] or genetic algorithms [Wang et al., 1998].

5.5. Mathematical theory based on belief and plausibility measures, usually known as *evidence theory* or *Dempster–Shafer theory*, was originated and developed by Glenn Shafer [1976a]. It was motivated by previous work on lower and upper probabilities by Dempster [1967a,b, 1968a,b], as well as by Shafer's historical reflection upon the concept of probability [1978] and his critical examination of the Bayesian approach to evidence [Shafer, 1981]. Although the book by Shafer [1976a] is still the best introduction to the theory, other books devoted to the theory, which are more up-to-date, were written by Guan and Bell [1991–92], Kohlas and Monney [1995], Kramosil [2001], and edited by Yager et al. [1994]. There are too many articles on the theory to be listed here, but some are included in Bibliography. Most of them can be found in reference lists of the mentioned books or in the *Special Issue of the International Journal of Intelligent Systems on the Dempster–Shafer Theory of Evidence* (**18**(1), 2003, pp. 1–148). A broad discussion of the theory is in papers by Shafer [1981, 1982, 1990], Dubois and Prade

[1986a], Smets [1988, 1998], and Smets and Kennes [1994]. Papers by Delmotte [2001] and Wilson [2000] deal with algorithmic aspects of the theory. Although most literature on the Dempster–Shafer theory is restricted to finite sets, this restriction is not necessary, as is shown by Shafer [1979] and Kramosil [2001].

5.6. The Dempster rule of combination is an important ingredient of DST. It is in some sense a generalization of the Bayes rule in classical probability theory. Both rules allow us, within their respective domains, to change any prior expression of uncertainty in the light of new evidence. As the name suggests, the Dempster rule was first proposed by Dempster [1967a], and it has played a prominent role in the development of DST by Shafer [1976a]. The rule was critically examined by Zadeh [1986], and formally investigated by Dubois and Prade [1986b], Harmanec [1997], Norton [1988], Klawonn and Schwecke [1992], and Hájek [1993]. The alternative rule of combination expressed by Eq. (5.59) was proposed by Yager [1987b].

5.7. An interesting connection between modal logic and the various uncertainty theories is suggested and examined in papers by Resconi et al. [1992, 1993]. Modal logic interpretations of belief and plausibility measures on finite sets is studied in detail by Harmanec et al. [1994] and Tsiporkova et al. [1999], and on infinite sets by Harmanec et al. [1996]. A modal logic interpretation of possibility theory is established in a paper by Klir and Harmanec [1994].

5.8. Key references in the area of reachable interval-valued probability distributions are papers by De Campos et al. [1994], Tanaka et al. [2004], and Weichselberger [2000], and a book by Weichselberger and Pöhlman [1990]. These references contain proofs of all propositions in Section 5.5. Conditional interval-valued probability distributions are examined in detail by De Campos et al. [1994]. Bayesian inference based on interval-valued prior probability distributions and likelihoods is developed in paper by Pan and Klir [1997]. Sgarro [1997] demonstrates that the theory based on reachable interval-valued probability distributions is not comparable with DST; neither of these two theories is more general than the other one.

5.9. Decomposable measures were introduced and studied by Dubois and Prade [1982b]. They have been further investigated by various authors, among them Weber [1984] and Pap [1997]. The basis of these measures are triangular norms, whose most comprehensive coverage is in the monograph by Klement et al. [2000].

5.10. k-Additive measures were introduced by Grabisch [1997a]. In a survey paper [Grabisch, 2000], which contains additional references to this subject, he also discusses the applicability of k-additivity to possibility measures (called k-possibility measures) and the issue of approximating monotone measures by k-additive measures.

5.11. Pairs of nondecreasing functions, $\underline{p}$ and $\bar{p}$, from $\mathbb{R}$ to $[0, 1]$ that represent, respectively, lower and upper bounds on the unknown probability distribution functions of random variables on $\mathbb{R}$ were introduced by Williamson and Downs [1990], and further developed under the name "probability boxes" or "p-boxes" by Ferson [2002]. Different methods for constructing p-boxes and their discrete approximations in terms of DST are discussed in [Ferson et al., 2003]. Among other references dealing with p-boxes, the most notable are [Ferson and Hajagos, 2004] and [Regan et al., 2004]. An example of a probability box is shown in Figure 5.8.

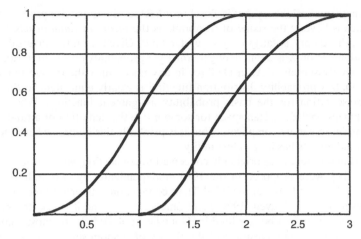

Figure 5.8. Example of a probability box.

5.12. A Dirac measure (see Figure 5.1) is a function $\mu \colon \mathcal{P}(X) \to \{0, 1\}$ defined for each $A \in \mathcal{P}(X)$ by the formula

$$\mu(A) = \begin{cases} 1 & \text{when } x_0 \in A \\ 0 & \text{otherwise,} \end{cases}$$

where x_0 is a particular (given) element of X. In any situation with no uncertainty, possibilistic and probabilistic representations collapse to a Dirac measure: $r(x_0) = p(x_0) = \mu(\{x_0\}) = 1$ and $r(x) = p(x) = \mu(\{x\}) = 0$ for all $x \neq x_0$.

5.13. Walley and De Cooman [1999] investigate conditional possibilities in terms of the behavioral interpretation of possibility measures. They show that the largest (or the least informative) conditional possibilities within this interpretation are defined by the formulas

$$r_{X|Y}(x \mid y) = \frac{r(x, y)}{r(x, y) + 1 - \max\{r(x, y), a(y)\}},$$

$$r_{Y|X}(y \mid x) = \frac{r(x, y)}{r(x, y) + 1 - \max\{r(x, y), b(x)\}},$$

where

$$a(y) = \max\{r_Y(z) \mid z \in Y, z \notin y\},$$

$$b(x) = \max\{r_X(z) \mid z \in X, z \notin x\},$$

This paper also reviews literature dealing with conditional possibility measures and contains all relevant references.

5.14. A mathematical theory that is closely connected with imprecise probabilities, but which is beyond the scope of this book, is the *theory of random sets*. For finite sets, the theory is fairly easy to understand. Given a finite universal set X, a random set on X is defined, in general, by a probability distribution function on nonempty subsets of X. In DST (or in any uncertainty theory subsumed under DST), this probability distribution function is clearly equivalent to the Möbius representation (or the basic probability assignment function). When X is an n-dimensional Euclidean space for some $n \geq 1$, the definition of a random set is considerably more complicated, but it is primarily this domain where the theory of random sets has its greatest utility.

Random sets were originally conceived in connection with stochastic geometry. They were proposed in the 1970s, independently by two authors, Kendall [1973, 1974] and Matheron [1975]. Their connection with belief functions of DST is examined by Nguyen [1978b] and Smets [1992a], and it is also the subject of several papers in a book edited by Goutsias et al. [1997]. The most comprehensive and up-to-date coverage of the theory of random sets is a recent book by Molchanov [2005].

EXERCISES

5.1. Determine which of the Möbius representations in Table 5.9 characterize special monotone measures such as necessity measures, probability measures, λ-measures, or belief measures.

5.2. Assume that the Möbius representations in Table 5.9 characterize joint bodies of evidence on $X \times Y$, where $X = \{x_1, x_2\}$ and $Y = \{y_1, y_2\}$, and let $a = \langle x_1, y_1 \rangle$, $b = \langle x_1, y_2 \rangle$, $c = \langle x_2, y_1 \rangle$, and $d = \langle x_2, y_2 \rangle$. Determine the associated marginal bodies of evidence.

5.3. Given noninteractive marginal possibility profiles $\mathbf{r}_X = \langle 1, 0.8, 0.5 \rangle$ and $\mathbf{r}_Y = \langle 1, 0.7 \rangle$ on sets $X = \{x_1, x_2, x_3\}$ and $Y = \{y_1, y_2\}$, determine the joint body of evidence in two different ways:

(a) By the rules of possibility theory;

(b) By the rules of DST.

Show visually (as in Figure 5.2) the marginal and joint bodies of evidence.

5.4. Repeat Exercise 5.3 for the following noninteractive marginal possibility profiles:

(a) $\mathbf{r}_X = \langle 1, 0.7, 0.2 \rangle$ on $X = \{x_1, x_2, x_3\}$ and $\mathbf{r}_Y \langle 1, 1, 0, 0.4 \rangle$ on $Y = \{y_1, y_2, y_3, y_4\}$

(b) $\mathbf{r}_X = \langle 1, 0.7, 0.6, 0.2 \rangle$ on $X = \{x_1, x_2, x_3, x_4\}$ and $\mathbf{r}_Y = \langle 1, 0.6 \rangle$ on $Y = \{y_1, y_2\}$

Table 5.9. Möbius Representations Employed in Exercises

	a	b	c	d	m_1	m_2	m_3	m_4	m_5	m_6	m_7	m_8	m_9	m_{10}
A:	0	0	0	0	0.0	0.0	0.00	0.00	0.00	0.0	0.0	0.0	0.0	0.0
	1	0	0	0	0.0	0.0	0.00	0.40	0.40	0.0	0.1	0.0	0.0	0.0
	0	1	0	0	0.4	0.2	0.10	0.07	0.20	0.0	0.1	0.0	0.0	0.0
	0	0	1	0	0.1	0.0	0.20	0.00	0.00	0.0	0.1	0.0	0.0	0.0
	0	0	0	1	0.5	0.0	0.48	0.10	0.00	0.0	0.1	0.6	0.0	0.0
	1	1	0	0	0.0	0.3	0.08	0.13	0.20	0.0	0.1	0.0	0.2	0.0
	1	0	1	0	0.0	0.0	0.00	0.00	0.20	0.0	0.1	0.0	0.1	0.0
	1	0	0	1	0.0	0.0	0.00	0.20	0.00	0.4	0.1	0.0	0.0	0.5
	0	1	1	0	0.0	0.0	0.08	0.00	0.00	0.0	0.1	0.0	0.3	0.5
	0	1	0	1	0.0	0.0	0.02	0.03	0.20	0.0	0.1	0.0	0.0	0.0
	0	0	1	1	0.0	0.0	0.00	0.00	0.20	0.0	0.1	0.1	0.0	0.0
	1	1	1	0	0.0	0.0	0.02	0.00	-0.20	0.0	0.0	0.0	0.0	0.0
	1	1	0	1	0.0	0.4	0.02	0.07	-0.20	0.5	0.0	0.0	0.1	0.0
	1	0	1	1	0.0	0.0	0.02	0.00	-0.20	0.0	0.0	0.2	0.0	0.0
	0	1	1	1	0.0	0.0	0.02	0.00	-0.20	0.0	0.0	0.2	0.3	0.0
	1	1	1	1	0.0	0.1	-0.04	0.00	0.40	0.1	0.0	0.1	0.0	0.0

5.5. Determine the Möbius representation, possibility measure, and necessity measure for each of the following possibility profiles defined on $X = \{x_i \mid i \in \mathbb{N}_n\}$ for appropriate values of n:

(a) $\mathbf{r} = \langle 1, 0.8, 0.5, 0.2 \rangle$

(b) $\mathbf{r} = \langle 1, 1, 1, 0.7, 0.7, 0.7 \rangle$

(c) $\mathbf{r} = \langle 1, 0.9, 0.8, 0.6, 0.3, 0.3 \rangle$

(d) $\mathbf{r} = \langle 1, 0.5, 0.4, 0.3, 0.2, 0.1 \rangle$

(e) $\mathbf{r} = \langle 1, 1, 0.8, 0.8, 0.5, 0.5, 0.2 \rangle$

Assume in each case that $r_i = r(x_i), i \in \mathbb{N}_n$.

5.6. Determine the possibility profile, possibility measure, and necessity measure for each of the following Möbius representations defined on $X = \{x_i \mid i \in \mathbb{N}_n\}$ for appropriate values of n:

(a) $\mathbf{m} = \langle 0, 0, 03, 02, 0, 0.4, 0.1 \rangle$

(b) $\mathbf{m} = \langle 0.1, 0.1, 0.1, 0, 0.1, 0.2, 0.2, 0.2 \rangle$

(c) $\mathbf{m} = \langle 0, 0, 0, 0, 0.5, 0.5 \rangle$

(d) $\mathbf{m} = \langle 0, 0.2, 0.2, 0, 0.3, 0, 0.3 \rangle$

(e) $\mathbf{m} = \langle 0.1, 0.2, 0.3, 0.4 \rangle$

(f) $\mathbf{m} = \langle 0.4, 0.3, 0.2, 0.1 \rangle$

Assume in each case that $m_i = m(\{x_1, x_2, \ldots, x_i\}), i \in \mathbb{N}_n$.

5.7. Convert some of the possibility and necessity measures in Exercises 5.5 and 5.6 to the combined function C defined by Eq. (5.81) and convert them back by Eqs. (5.82) and (5.83).

5.8. Let $X = \{a, b, c, d, e\}$ and $Y = \mathbb{N}_8$. Using a joint possibility profile on $X \times Y$ given in terms of the matrix

$$
\begin{bmatrix}
1 & 0 & 0 & .3 & .5 & .2 & .4 & .1 \\
0 & .7 & 0 & .6 & 1 & 0 & .4 & .3 \\
0 & .5 & 0 & 0 & 1 & 0 & 1 & .5 \\
1 & 1 & 1 & .5 & 0 & 0 & 1 & .4 \\
.8 & 0 & .9 & 0 & 1 & .7 & 1 & .2
\end{bmatrix}
$$

where rows are assigned to elements a, b, c, d, e and columns are assigned to numbers $1, 2, \ldots, 8$, determine:

(a) Marginal possibility profiles;

(b) Joint and marginal Möbius representations;

(c) Both conditional possibilities given by Eqs. (5.24) and (5.25);

(d) Hypothetical joint possibly profile based on the assumption of noninteraction.

5.9. Determine the λ-measure μ_λ on $X = \{x_1, x_2\}$ from each of the following values of μ_λ on singletons:

(a) $\mu_\lambda (\{x_1\}) = 0.6, \mu_\lambda(\{x_2\}) = 0.1$

(b) $\mu_\lambda (\{x_1\}) = 0.4, \mu_\lambda(\{x_2\}) = 0.2$

(c) $\mu_\lambda (\{x_1\}) = 0.6, \mu_\lambda(\{x_2\}) = 0.5$

(d) $\mu_\lambda (\{x_1\}) = 0.7, \mu_\lambda(\{x_2\}) = 0.4$

(e) $\mu_\lambda (\{x_1\}) = 0.5, \mu_\lambda(\{x_2\}) = 0.1$

(f) $\mu_\lambda (\{x_1\}) = 0.8, \mu_\lambda(\{x_2\}) = 0.7$

5.10. Determine the joint λ-measure μ_λ on $X \times Y = \{x_1, x_2\} \times \{y_1, y_2\}$ from each of the following values of μ_λ on singletons, which are given in the order $\mu_\lambda(\{\langle x_1, y_1 \rangle\}), \mu_\lambda(\{\langle x_1, y_2 \rangle\}), \mu_\lambda(\{\langle x_2, y_1 \rangle\}), \mu_\lambda(\{\langle x_2, y_2 \rangle\})$:

(a) $0.4; 0.6\overline{6}; 0.0; 0.1$

(b) $0.2; 0.2; 0.1; 0.0$

(c) $0.1153; 0.2; 0.1; 0.0$

(d) $0.4; 0.5696; 0.7; 0.0$

(e) $0.5; 0.0; 0.0; 0.4$

(f) $0.5; 0.0343; 0.0; 0.3$

5.11. For some of the λ-measures in Exercises 5.9 and 5.10, calculate their dual measures by:

(a) The duality equation, Eq. (5.35)

(b) Equation (5.38)

5.12. In Exercise 5.10 determine the associated marginal λ-measures for each of the joint λ-measures.

5.13. In Exercise 5.12 determine for each pair of marginal λ-measures the set of joint probability measures that dominate (if the marginal λ-measures are lower probability measures) or are dominated (if they are upper probability measures) by the marginal λ-measures.

5.14. In Table 5.9 determine which of the Möbius representations characterize belief measures and convert each of these belief measures to the corresponding plausibility measures and commonality functions.

5.15. Let a, b, c, d in Table 5.9 denote, respectively, elements $\langle x_1, y_1 \rangle, \langle x_1, y_2 \rangle$, $\langle x_2, y_1 \rangle, \langle x_2, y_2 \rangle$ of the Cartesian product $\{x_1, x_2\} \times \{y_1, y_2\}$. For each of the Möbius representations in the Table 5.9, determine the associated lower probability function and its marginal lower probability functions.

5.16. For each pair $^i m_X, ^i m_Y$ of marginal bodies of evidence defined in Table 5.10, determine the joint body of evidence under the assumption that the marginal bodies are noninteractive.

Table 5.10. Bodies of Evidence in Exercises 5.16 and 5.17

	x_1	x_2	x_3	^1m_X	^2m_X	^3m_X		y_1	y_2	y_3	^1m_Y	^2m_Y	^3m_Y
A:	0	0	0	0.0	0.0	0.0	B:	0	0	0	0.0	0.0	0.0
	1	0	0	0.0	0.3	0.1		1	0	0	0.3	0.0	0.1
	0	1	0	0.0	0.0	0.2		0	1	0	0.0	0.5	0.1
	0	0	1	0.0	0.1	0.3		0	0	1	0.0	0.0	0.1
	1	1	0	0.4	0.2	0.0		1	1	0	0.1	0.0	0.2
	1	0	1	0.1	0.2	0.2		1	0	1	0.1	0.4	0.2
	0	1	1	0.5	0.2	0.0		0	1	1	0.1	0.0	0.2
	1	1	1	0.0	0.0	0.2		1	1	1	0.4	0.1	0.1

5.17. Assume that the basic probability assignments ^1m_X and ^2m_Y in Table 5.10 represent evidence obtained from two independent sources. Determine the combined (aggregated) evidence by using:

(a) The Dempster rule of combination;

(b) The alternative rule of combination expressed by Eq. (5.59).

5.18. Repeat Exercise 5.17 for other pairs of basic probability assignments in Table 5.10.

5.19. Assume that $^1m_X, {}^2m_X, {}^3m_X$, and alternatively, $^1m_Y, {}^2m_Y, {}^3m_Y$ in Table 5.10 represent evidence obtained from three independent sources. Determine the combined evidence by using:

(a) The Dempster rule of combination;

(b) The alternative rule.

5.20. Show that the Dempster rule of combination is associative so that the combined evidence does not depend on the order in which the sources are used.

5.21. Show that the alternative rule of combination expressed by Eq. (5.59) is not associative so that the combined result depends on the order in which the sources are used. Can the alternative rule be generalized to more than two sources to be independent of the order in which the sources are used?

5.22. Consider the following triples of probability intervals on $X = \{x_1, x_2, x_3\}$:

$I_1 = \langle[0.3, 0.4], [0.3, 0.5], [0.3, 0.5]\rangle$

$I_2 = \langle[0.2, 0.4], [0.4, 0.6], [0.1, 0.2]\rangle$

$I_3 = \langle[0.5, 0.7], [0.2, 0.4], [0.1, 0.2]\rangle$

$I_4 = \langle[0.4, 0.6], [0.5, 0.7], [0.2, 0.4]\rangle$

$I_5 = \langle[0.1, 0.2], [0.3, 0.4], [0.0, 0.3]\rangle$

$I_6 = \langle[0.2, 0.6], [0.2, 0.6], [0.2, 0.6]\rangle$

$I_7 = \langle [0.4, 0.5], [0.3, 0.4], [0.2, 0.5] \rangle$

$I_8 = \langle [0.1, 0.4], [0.1, 0.6], [0.2, 0.5] \rangle$

For each of these triples, do the following:

(a) Check if the triple is proper to qualify as an interval-valued probability distribution;

(b) Check if the triple is reachable; if it is not reachable, convert it to its reachable counterpart.

(c) Determine the lower and upper probabilities for each $A \in \mathcal{P}(X)$.

(d) Determine the interaction representation of lower and upper probabilities and the convex set of probability distributions that are consistent with them.

5.23. Repeat Exercise 5.22 for the following 4-tuples of probability intervals on $X = \{x_1, x_2, x_3, x_4\}$:

$I_1 = \langle [0, 0.3], [0.1, 0.2], [0.3, 0.4], [0.1, 0.4] \rangle$

$I_2 = \langle [0, 0.1], [0.1, 0.3], [0.2, 0.3], [0.48, 0.52] \rangle$

$I_3 = \langle [0, 0.3], [0.1, 0.4], [0.2, 0.5], [0.3, 0.6] \rangle$

$I_4 = \langle [0.2, 0.6], [0.2, 0.5], [0.2, 0.4], [0.2, 0.3] \rangle$

5.24. Repeat Exercise 5.8c for the conditional possibilities defined in Note 5.13.

6

MEASURES OF UNCERTAINTY AND INFORMATION

The mathematical theory of information had come into being when it was realized that the flow of information can be expressed numerically in the same way as distance, mass, temperature, etc.

—Alfréd Rényi

6.1. GENERAL DISCUSSION

Each particular formal theory of uncertainty is based on uncertainty functions, u, that satisfy appropriate axiomatic requirements. A common property of all uncertainty functions is that they are monotone measures. Their additional properties are specific to each theory. Uncertainty functions of each theory share the same special properties. It is useful to represent uncertainty functions pertaining to an uncertainty theory by several distinct forms. These representations are required to be equivalent in the sense that they be uniquely convertible to one another. The most common representations are the various lower and upper probability functions and their Möbius representations, which are extensively discussed in Chapters 4 and 5.

A measure of uncertainty of some type in a particular uncertainty theory is a functional, U, which for each function u in the theory measures the amount of uncertainty of the considered type that is embedded in u. Two distinct types of uncertainty, which emerge from the two classical uncertainty theories, are *nonspecificity* and *conflict*. In the classical theories, they are measured,

Uncertainty and Information: Foundations of Generalized Information Theory, by George J. Klir
© 2006 by John Wiley & Sons, Inc.

respectively, by the Hartley and Shannon functionals. In the various generalizations of the classical uncertainty theories, these two types of uncertainty coexist and, therefore, we need to determine justifiable generalizations of both the Hartley and Shannon functionals in each of these generalized uncertainty theories.

Each functional for measuring uncertainty in a given uncertainty theory must satisfy some essential requirements expressed in terms of the calculus of the theory involved. However, due to conceptual similarities between these various formal expressions, the requirements can be described informally in a generic form, independent of the various uncertainty calculi.

The following requirements, each expressed here in a generic form, are commonly those from which axioms for characterizing measures of uncertainty in the various uncertainty theories are drawn:

1. *Subadditivity*: The amount of uncertainty in a joint representation of evidence (defined on a Cartesian product) cannot be greater than the sum of the amounts of uncertainty in the associated marginal representations of evidence.

2. *Additivity*: The amount of uncertainty in a joint representation of evidence is equal to the sum of the amounts of uncertainty in the associated marginal representations of evidence if and only if the marginal representations are noninteractive according to the rules of the uncertainty calculus involved.

3. *Monotonicity*: When evidence can be ordered in the uncertainty theory employed (as in possibility theory), the relevant uncertainty measure must preserve this ordering.

4. *Continuity*: Any measure of uncertainty must be a continuous functional.

5. *Expansibility*: Expanding the universal set by alternatives that are not supported by evidence must not affect the amount of uncertainty.

6. *Symmetry*: The amount of uncertainty does not change when elements of the universal set are rearranged.

7. *Range*: The range of uncertainty is $[0, M]$, where 0 must be assigned to the unique uncertainty function that describes full certainty and M depends on the size of the universal set involved and on the chosen unit of measurement (normalization).

8. *Branching/Consistency*: When uncertainty can be computed in multiple ways, all acceptable within the calculus of the uncertainty theory involved, the results must be the same (consistent).

9. *Normalization*: A measurement unit defined by specifying what the amount of uncertainty should be for a particular (and usually very simple) uncertainty function.

10. *Coordinate Invariance*: When evidence is described within the n-dimensional Euclidean space ($n \geq 1$), the relevant uncertainty measure must not change under isometric transformations of the coordinate system.

These common names of the requirements are used consistently in this book for specific forms of requirements in the various uncertainty theories. When distinct types of uncertainty coexist in a given uncertainty theory, it is not necessary that these requirements be satisfied by each uncertainty type. However, they must be satisfied by an overall uncertainty measure, which appropriately aggregates measures of the individual uncertainty types.

The strongest justification of a functional as a meaningful measure of the amount of uncertainty of a considered type in a given uncertainty theory is obtained when we can prove that it is the only functional that satisfies the relevant axiomatic requirements and measures the amount of uncertainty in some specific measurement unit. Only some of the listed requirements are usually needed to uniquely characterize the functional. For example, only three of the listed requirements (branching, monotonicity, and normalization) are needed to prove the uniqueness of the Hartley measure (recall Theorem 2.1).

The purpose of this chapter is to examine generalizations of the Hartley and Shannon measures of uncertainty in the various theories of imprecise probabilities introduced in Chapters 4 and 5. The presentation follows, by and large, chronologically the emergence of ideas and results pertaining to this area since the early 1980s. Various unifying features of uncertainty measures are then emphasized in Section 6.11. The presentation is not interrupted by the many relevant historical and bibliographical remarks, which are covered in the Notes Section to this chapter.

6.2. GENERALIZED HARTLEY MEASURE FOR GRADED POSSIBILITIES

It is not surprising that the Hartley measure was first generalized from classical (crisp) possibilities to graded possibilities. The generalized Hartley measure for graded possibilities is usually denoted in the literature by U and it is called *U-uncertainty*.

The U-uncertainty can be expressed in various forms. A simple form, easy to understand, is based on the special notation introduced for graded possibilities in Section 5.2.1. In this notation, $X = \{x_1, x_2, \ldots, x_n\}$ and r_i denotes for each $i \in \mathbb{N}_n$ the possibility of x_i. Elements of X are assumed to be appropriately rearranged so that the possibility profile

$$\mathbf{r} = \langle r_1, r_2, \ldots, r_n \rangle$$

is ordered in such a way that

$$1 = r_1 \geq r_2 \geq \cdots \geq r_n$$

and $r_{n+1} = 0$ by convention. Moreover, set $A_i = \{x_1, x_2, \ldots, x_i\}$ is defined for each $i \in \mathbb{N}_n$.

Using this simple notation, the U-uncertainty is expressed for each given possibility profile $\mathbf{r}$ by the formula

$$U(\mathbf{r}) = \sum_{i=1}^{n} (r_i - r_{i+1}) \log_2 |A_i|. \tag{6.1}$$

Since, clearly,

$$\sum_{i=1}^{n} (r_i - r_{i+1}) = 1,$$

the U-uncertainty is a weighted average of the Hartley measure for sets A_i, $i \in \mathbb{N}_n$, where the weights are the associated differences $r_i - r_{i+1}$ in the given possibility profile. These differences are, of course, values of the basic probability assignment function for sets A_i.

Since $|A_i| = i$ and $\log_2 |A_1| = \log_2 1 = 0$, Eq. (6.1) can be rewritten in the simpler form

$$U(\mathbf{r}) = \sum_{i=2}^{n} (r_i - r_{i+1}) i \tag{6.2}$$

or, alternatively, in the form

$$U(\mathbf{r}) = \sum_{i=2}^{n} r_i \log_2 \left(\frac{i}{i-1} \right). \tag{6.3}$$

It follows directly from Eq. (6.3) that $U(^1\mathbf{r}) \leq U(^2\mathbf{r})$ for any pair of possibility profiles defined on the same set and such that $^1\mathbf{r} \leq {}^2\mathbf{r}$. The U-uncertainty thus preserves the ordering of possibility profiles defined on the same set. Moreover,

$$0 \leq U(\mathbf{r}) \leq \log_2 |X| \tag{6.4}$$

for any possibility profile $\mathbf{r}$ on X. The lower and upper bounds are obtained, respectively, for the smallest and the largest possibility profiles, $\langle 1, 0, \ldots, 0 \rangle$ and $\langle 1, 1, \ldots, 1 \rangle$.

Although Eqs. (6.1)–(6.3) are convenient for intuitive understanding of the U-uncertainty, they are based on the assumption that the given possibility profile is ordered. This assumption makes it difficult to describe the relationship between joint, marginal, and conditional U-uncertainties. Fortunately, the U-uncertainty can also be defined in the following alternative way that does not require that the given possibility profile be ordered.

Let $\mathbf{r} = \langle r(x)| \ x \in X \rangle$ denote a possibility profile on X that is not required to be ordered and let

$$^{\alpha}r = \{x \in X \mid r(x) \geq \alpha\} \tag{6.5}$$

for each $\alpha \in [0, 1]$. Then,

$$U(\mathbf{r}) = \int_0^1 \log_2 |{}^{\alpha}r| \, d\alpha. \tag{6.6}$$

EXAMPLE 6.1. Consider the possibility profile on $\mathbb{N}_{15}$ that is defined by the dots in the diagram in Figure 6.1. Applying Eq. (6.6) to this possibility profile, we obtain

$$\int_0^1 \log_2 |{}^{\alpha}r| \, d\alpha = \int_0^{0.1} \log_2 15 \, d\alpha + \int_{0.1}^{0.3} \log_2 12 \, d\alpha + \int_{0.3}^{0.4} \log_2 11 \, d\alpha$$
$$+ \int_{0.4}^{0.6} \log_2 9 \, d\alpha + \int_{0.6}^{0.7} \log_2 7 \, d\alpha + + \int_{0.7}^{0.9} \log_2 5 \, d\alpha + \int_{0.9}^{0.1} \log_2 3 \, d\alpha$$
$$= 0.1 \ \log_2 15 + 0.2 \log_2 12 + 0.1 \log_2 11 + 0.2 \log_2 9$$
$$+ 0.1 \log_2 7 + 0.2 \log_2 5 + 0.1 \log_2 3$$
$$= 0.39 + 0.72 + 0.35 + 0.63 + 0.28 + 0.46 + 0.16 = 2.99.$$

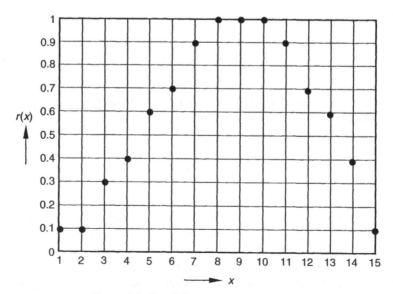

Figure 6.1. Possibility profile in Example 6.1.

The same result can be obtained by ordering the given possibility profile and applying Eq. (6.2). The ordered possibility profile is

$$\mathbf{r} = \langle 1, 1, 1, 0.9, 0.9, 0.7, 0.7, 0.6, 0.6, 0.4, 0.4, 0.3, 0.1, 0.1, 0.1 \rangle.$$

When these values are applied to Eq. (6.2), we obtain the following nonzero terms:

$$\begin{aligned} U(\mathbf{r}) &= 0.1 \log_2 3 + 0.2 \log_2 5 + 0.1 \log_2 7 + 0.2 \log_2 9 \\ &\quad + 0.1 \log_2 11 + 0.2 \log_2 12 + 0.1 \log_2 15 \\ &= 2.99. \end{aligned}$$

6.2.1. Joint and Marginal U-Uncertainties

Assume now that a possibility profile $\mathbf{r}$ is defined on a Cartesian product $X \times Y$, and let

$$^{\alpha}r = \{\langle x, y \rangle \in X \times Y \mid r(x, y) \geq \alpha\}. \tag{6.7}$$

Then,

$$U(\mathbf{r}) = \int_0^1 \log_2 |^{\alpha}r| d\alpha, \tag{6.8}$$

and the U-uncertainties of the marginal possibility profiles, $\mathbf{r}_X$ and $\mathbf{r}_Y$, are

$$U(\mathbf{r}_X) = \int_0^1 \log_2 |^{\alpha}r_X| d\alpha, \tag{6.9}$$

and

$$U(\mathbf{r}_Y) = \int_0^1 \log_2 |^{\alpha}r_Y| d\alpha, \tag{6.10}$$

where

$$^{\alpha}r_X = \{x \in X \mid r_X(x) \geq \alpha\} \tag{6.11}$$

and

$$^{\alpha}r_Y = \{y \in Y \mid r_Y(y) \geq \alpha\} \tag{6.12}$$

for all $\alpha \in [0, 1]$. It is now fairly easy to show that the U-uncertainty is both subadditive and additive.

Theorem 6.1. Let **r** be a possibility profile on $X \times Y$, and let $\mathbf{r}_X$ and $\mathbf{r}_Y$ be the associated marginal possibility profiles on X and Y, respectively. Then,

$$U(\mathbf{r}) \le U(\mathbf{r}_X) + U(\mathbf{r}_X), \tag{6.13}$$

where the equality is obtained if and only if $\mathbf{r}_X$ and $\mathbf{r}_Y$ are noninteractive.

Proof

$$
\begin{aligned}
U(\mathbf{r}) &= \int_0^1 \log_2 |{}^\alpha r| \, d\alpha \\
&\le \int_0^1 \log_2 |{}^\alpha r_X \times {}^\alpha r_Y| \, d\alpha \\
&= \int_0^1 \log_2 |{}^\alpha r_X| \cdot |{}^\alpha r_Y| \, d\alpha \\
&= \int_0^1 \log_2 |{}^\alpha r_X| \, d\alpha + \int_0^1 \log_2 |{}^\alpha r_Y| \, d\alpha \\
&= U(\mathbf{r}_X) + U(\mathbf{r}_X).
\end{aligned}
$$

It is clear that the equality is obtained if and only if ${}^\alpha r = {}^\alpha r_X \times {}^\alpha r_Y$ for all $\alpha \in [0, 1]$, which is the case when marginal possibility profiles are noninteractive. ∎

EXAMPLE 6.2. In Figure 6.2a, a joint possibility profile **r** on $X \times Y = \{x_1, x_2, x_3\} \times \{y_1, y_2, y_3\}$ is given and the associated marginal possibility profiles $\mathbf{r}_X$ and $\mathbf{r}_Y$ are derived from it. In Figure 6.2b, a joint possibility profile, **r′**, is derived

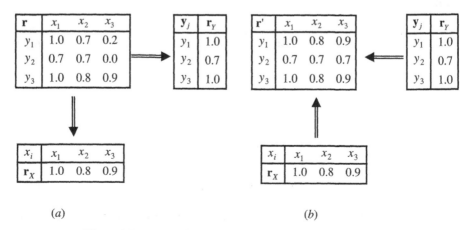

(a)   (b)

Figure 6.2. Joint and marginal possibility profiles (Example 6.2).

from $\mathbf{r}_X$ and $\mathbf{r}_Y$ under the assumption that they are noninteractive. Ordering the individual profiles and applying Eq. (6.2), we obtain

$$U(\mathbf{r}) = 0.1 \log_2 2 + 0.1 \log_2 3 + 0.1 \log_2 4 + 0.5 \log_2 7 + 0.2 \log_2 8 = 2.46,$$

$$U(\mathbf{r}') = 0.1 \log_2 2 + 0.1 \log_2 4 + 0.1 \log_2 6 + 0.7 \log_2 9 = 2.78,$$

$$U(\mathbf{r}_X) = 0.1 \log_2 2 + 0.8 \log_2 3 = 1.37,$$

$$U(\mathbf{r}_Y) = 0.3 \log_2 2 + 0.7 \log_2 3 = 1.41.$$

We can see that $U(\mathbf{r}) < U(\mathbf{r}') = U(\mathbf{r}_X) + U(\mathbf{r}_X)$, which verifies the subadditivity and additivity of the U-uncertainty.

6.2.2. Conditional U-Uncertainty

Applying the definitions of conditional Hartley measures, expressed by Eqs. (2.19) and (2.20), to sets $^\alpha r$, $^\alpha r_X$, and $^\alpha r_Y$, we readily obtain their generalized counterparts for the U-uncertainty:

$$U(\mathbf{r}_X \mid \mathbf{r}_Y) = \int_0^1 \log_2 \frac{|^\alpha r|}{|^\alpha r_Y|} \, d\alpha, \tag{6.14}$$

$$U(\mathbf{r}_Y \mid \mathbf{r}_X) = \int_0^1 \log_2 \frac{|^\alpha r|}{|^\alpha r_X|} \, d\alpha. \tag{6.15}$$

It is obvious that these equations can also be rewritten as

$$U(\mathbf{r}_X \mid \mathbf{r}_Y) = U(\mathbf{r}) - U(\mathbf{r}_Y), \tag{6.16}$$

$$U(\mathbf{r}_Y \mid \mathbf{r}_X) = U(\mathbf{r}) - U(\mathbf{r}_X). \tag{6.17}$$

Let the generic symbols $U(X)$, $U(Y)$, $U(X \times Y)$, $U(X \mid Y)$, and $U(Y \mid X)$ now be used instead of their specific counterparts $U(\mathbf{r}_X)$, $U(\mathbf{r}_Y)$, $U(\mathbf{r}_X)$, $U(\mathbf{r}_X|\mathbf{r}_Y)$, and $U(\mathbf{r}_Y | \mathbf{r}_X)$, respectively. This is a common practice in the literature, and is used in Chapter 2 for the marginal, joint, and conditional Hartley measures and then again in Chapter 3 for the Shannon entropies. This notation makes it easier to recognize that marginal, joint, and conditional uncertainties based on distinct measures of uncertainty are related in analogous ways.

Using the generic symbols, Eqs. (6.16) and (6.17) are rewritten as

$$U(X \mid Y) = U(X \times Y) - U(Y), \tag{6.18}$$

$$U(Y \mid X) = U(Y \times X) - U(X). \tag{6.19}$$

Observe that these equations are analogous to Eqs. (2.23) and (2.24) for the Hartley measure, as well as to Eqs. (3.35) and (3.36) for the Shannon entropy.

Observe also that the conditional U-uncertainties can be calculated without using conditional possibilities.

U-uncertainty analogies of other equations that hold for the Hartley (and for the Shannon entropy) can be easily derived as well. For example, when the marginal possibility profiles are noninteractive, we have the equation

$$U(X \times Y) = U(X) + U(Y) \tag{6.20}$$

by Theorem 6.1, which is analogous to Eq. (2.27) for the Hartley measure. When we combine this equation with Eqs. (6.18) and (6.19), we readily obtain the equations

$$U(X \mid Y) = U(X), \tag{6.21a}$$

$$U(Y \mid X) = U(Y), \tag{6.21b}$$

which are analogous to Eqs. (2.25) and (2.26) for the Hartley measure. Similarly, the inequality

$$U(X \times Y) \leq U(X) + U(Y), \tag{6.22}$$

which is a generic form of Eq. (6.13), is analogous to Eq. (2.30) for Hartley measure and to Eq. (3.38) for the Shannon entropy. We can also define information transmission, $T_U(X, Y)$, for the U-uncertainty by the equation

$$T_U(X, Y) = U(X) + U(Y) - U(X \times Y), \tag{6.23}$$

in analogy with Eq. (2.31) for the Hartley measure and Eq. (3.34) for the Shannon entropy.

EXAMPLE 6.3. Consider the joint and marginal possibility profiles in Figure 6.2a. Applying Eqs. (6.18) and (6.19) and utilizing the joint and marginal U-uncertainties calculated in Example 6.2, we have

$$U(X \mid Y) = 2.46 - 1.41 = 1.05,$$

$$U(Y \mid X) = 2.46 - 1.37 = 1.09,$$

$$T_U(X, Y) = 1.37 + 1.41 - 2.46 = 0.32.$$

Similarly, for the possibility profiles in Figure 6.2b, we obtain

$$U(X \mid Y) = 2.78 - 1.41 = 1.37[= U(X)],$$

$$U(Y \mid X) = 2.78 - 1.38 = 1.41[= U(Y)],$$

$$T_U(X, Y) = 2.78 - 1.37 - 1.41 = 0,$$

as is expected under the assumption of noninteraction.

6.2.3. Axiomatic Requirements for the *U*-Uncertainty

As any measure of uncertainty, the *U*-uncertainty must satisfy the axiomatic requirements listed in Section 6.1. In this case, the requirements must be expressed in terms of the calculus of graded possibilities. If we consider only finite sets, then the requirement of coordinate invariance is not relevant. Hence, we have the following nine specific requirements:

Axiom (U1) Subadditivity. For any joint possibility profile $\mathbf{r}$ on $X \times Y$, $U(\mathbf{r})$ $\leq U(\mathbf{r}_X) + U(\mathbf{r}_Y)$.

Axiom (U2) Additivity. When a joint possibility profile $\mathbf{r}$ on $X \times Y$ is derived from noninteractive marginal possibility profiles $\mathbf{r}_X$ and $\mathbf{r}_Y$, then $U(\mathbf{r}) = U(\mathbf{r}_X) + U(\mathbf{r}_Y)$.

Axiom (U3) Monotonicity. For any pair ${}^1\mathbf{r}$, ${}^2\mathbf{r}$ of possibility profiles of the same length, if ${}^1\mathbf{r} \leq {}^2\mathbf{r}$, then $U({}^1\mathbf{r}) \leq U({}^2\mathbf{r})$.

Axiom (U4) Continuity. U is a continuous functional.

Axiom (U5) Expansibility. When components of zero are added to a given possibility profile, the value of U does not change.

Axiom (U6) Symmetry. If $\mathbf{r}$ is a possibility profile and $\mathbf{r}'$ is a permutation of $\mathbf{r}$, then $U(\mathbf{r}) = U(\mathbf{r}')$.

Axiom (U7) Range. $U(\mathbf{r}) \in [0, M]$, where the minimum and maximum are obtained for $\mathbf{r}_{\min} = \langle 1, 0, \ldots, 0 \rangle$ and $\mathbf{r}_{\max} = \langle 1, 1, \ldots, 1 \rangle$, respectively, and the value M depends on the cardinality of the universal set and the chosen unit of measurement (normalization).

Axiom (U8) Branching/Consistency. For every possibility profile $\mathbf{r} = \langle r_1, r_2, \ldots, r_n \rangle$ of any length n,

$$
\begin{aligned}
U(r_1, r_2, \ldots, r_n) = {}&U(r_1, r_2, \ldots, r_{k-2}, r_k, r_k, r_{k+1}, \ldots, r_n) \\
&+ (r_{k-2} - r_k)U\left(\mathbf{1}_{k-2}, \frac{r_{k-1} - r_k}{r_{k-2} - r_k}, \mathbf{0}_{n-k+1}\right) \\
&- (r_{k-2} - r_k)U(\mathbf{1}_{k-2}, \mathbf{0}_{n-k+2})
\end{aligned}
\tag{6.24}
$$

for each $k = 3, 4, \ldots, n$, where, for any given integer i, $\mathbf{1}_i$ denotes a sequence of i ones and $\mathbf{0}_i$ denotes a sequence of i zeroes.

Axiom (U9) Normalization. To define bits as measurement units, it is required that $U(1, 1) = 1$.

The branching requirement, which can also be formulated in other forms, needs some explanation. This requirement basically states that the *U*-uncertainty must have the capability of measuring possibilistic nonspecificity in two ways. It can be measured either directly for the given possibility profile or indirectly by adding U-uncertainties associated with a combination of possibility profiles that reflect a two-stage measuring process. In the first stage of measurement, the distinction between two neighboring components, r_{k-1} and r_k, is ignored (r_{k-1} is replaced with r_k) and the U-uncertainty of the resulting, less refined possibility profile is calculated. In the second stage, the U-uncertainty is calculated in a local frame of reference, which is defined by a possibility profile that distinguishes only between the two neighboring possibility values that are not distinguished in the first stage of measurement. The U-uncertainty calculated in the local frame must be scaled back to the original frame by a suitable weighting factor. The sum of the two U-uncertainties obtained by the two stages of measurement is equal to the total U-uncertainty of the given possibility profile.

The first term on the right-hand side of Eq. (6.24) represents the U-uncertainty obtained in the first stage of measurement. The remaining two terms represent the second stage, associated with the local U-uncertainty. The first of these two terms expresses the loss of uncertainty caused by ignoring the component r_{k-1} in the given possibility profile, but it introduces some additional U-uncertainty equivalent to the uncertainty of a crisp possibility profile with $k - 2$ components. This additional U-uncertainty is excluded by the last term in Eq. (6.24). That is, the local uncertainty is expressed by the last two terms in Eq. (6.24).

The meaning of the branching property of the U-uncertainty is illustrated by Figure 6.3. Four hypothetical possibility profiles involved in Eq. (6.24) are shown in the figure, and in each of them the local frame involved in the branching is indicated.

It turns out that five of the nine axiomatic requirements are sufficient to prove the uniqueness of the U-uncertainty: additivity, monotonicity, expansibility, branching, and normalization. The proof, which is quite intricate, is covered in Appendix A.

6.2.4. U-Uncertainty for Infinite Sets

U-uncertainty for infinite sets is a generalization of the Hartley-like measure HL (introduced in Section 2.3) for graded possibilities. It is thus reasonable to denote it by UL. This generalization is obtained directly from Eq. (6.6) by replacing $\log_2 |^\alpha r|$ with $HL(^\alpha r)$. Hence,

$$UL(\mathbf{r}) = \int_0^1 HL(^\alpha r)\, d\alpha \qquad (6.25)$$

or, more specifically,

$$UL(\mathbf{r}) = \int_0^1 \min_{t \in T}\left\{\log_2\left[\prod_{i=1}^n (1 + \mu(^\alpha r_{i_t})) + \mu(^\alpha r) - \prod_{i=1}^n (\mu(^\alpha r_{i_t})\right]\right\} d\alpha, \qquad (6.26)$$

where $^\alpha r$ is defined in terms of $\mathbf{r}$ by Eq. (6.5).

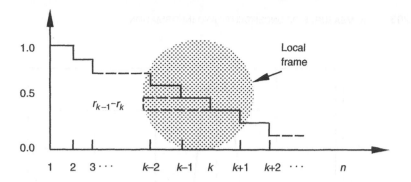

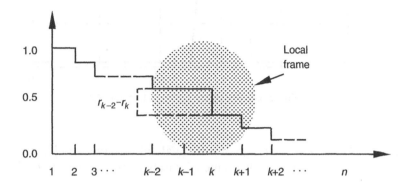

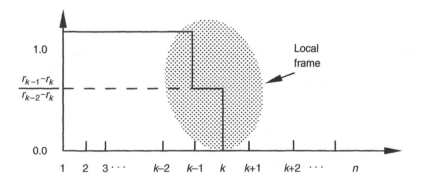

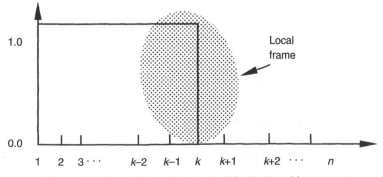

Figure 6.3. Possibility profiles involved in the branching property.

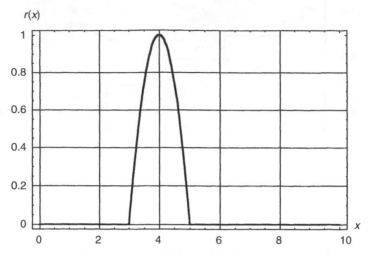

Figure 6.4. Possibility profile in Example 6.4.

EXAMPLE 6.4. Let the range of a real-valued variable x be $[0, 10]$. Assume that we are able to predict the value of x only approximately in terms of the possibility profile

$$r(x) = \max\{0, 2(x-3)-(x-3)^2\}$$

for each $x \in [1, 10]$. The graph of this possibility profile is shown in Figure 6.4. For each $\alpha \in [0, 1]$, $^\alpha r$ is in this case a closed interval $[\underline{x}(\alpha), \bar{x}(\alpha)]$ of real numbers. The endpoints of these intervals for all $\alpha \in [0, 1]$ are functions of α, which can be obtained by solving the equation

$$2(x-3)-(x-3)^2 = \alpha$$

for x. The solution is

$$x_{1,2} = 4 \pm \sqrt{1-\alpha}.$$

Clearly, $\underline{x}(\alpha) = 4 - \sqrt{1-\alpha}$ and $\bar{x}(\alpha) = 4 + \sqrt{1-\alpha}$. Since we deal with a 1-dimensional variable, Eq. (6.26) assumes the form

$$UL(\mathbf{r}) = \int_0^1 \log_2[1 + \mu(^\alpha r)]d\alpha,$$

where $^\alpha r = \bar{x}(\alpha) - \underline{x}(\alpha) = 2\sqrt{1-\alpha}$. Hence,

$$UL(\mathbf{r}) = \int_0^1 \log_2[1 + 2\sqrt{1-\alpha}]d\alpha = 1.1887.$$

6.3. GENERALIZED HARTLEY MEASURE IN DEMPSTER–SHAFER THEORY

Once the U-uncertainty was well established as a generalized Hartley measure for graded possibilities, its further generalization to Dempster–Shafer theory (DST) became conceptually fairly straightforward. It emerged quite naturally from two simple facts: (1) the U-uncertainty is a weighted average of the Hartley measure for all focal subsets; and (2) the weights in this average, which are expressed by the differences $r_i - r_{i+1}$ in ordered possibility profiles, are values of the basic probability assignment function. Although the focal subsets are always nested in the theory of graded possibilities, the concept of the weighted average of the Hartley measure is applicable to any family of focal subsets. The *generalized Hartley measure*, GH, in DST is thus defined by the functional

$$GH(m) = \sum_{A \in \mathcal{F}} m(A) \log_2 |A|, \qquad (6.27)$$

where $\langle \mathcal{F}, m \rangle$ is any arbitrary body of evidence in the sense of DST.

Observe that the functional GH is defined in terms of m while the functional U is defined in terms of r. However, the difference $r_i - r_{i+1}$, in Eq. (6.1) is clearly equal to $m(A_i)$, which means that the U-uncertainty can be defined in terms of the basic probability assignment as well.

It is obvious that the GH measure defined by Eq. (6.27) is a *continuous* functional, which satisfies the *expansibility requirement* and whose *range* is

$$0 \leq GH(m) \leq \log_2 |X|. \qquad (6.28)$$

The lower bound is obtained when all focal subsets are singletons, which means that m is actually a probability distribution on X. This implies that $GH(m) = 0$ for all probability measures. That is, probability measures are fully specific. We can also see that $GH(m) = 1$ when $m(A) = 1$ and $|A| = 2$, which means that the units in which the functional GH measures nonspecificity are bits. For characterizing the functional, this property must be required as a *normalization*.

The GH measure is also invariant with respect to permutations of values of the basic probability assignment function within each group of subsets of X that have equal cardinalities. This invariance is, in fact, the meaning of the requirement of *symmetry* in DST.

The issue of the uniqueness of the GH measure in DST is addressed in Appendix B.

6.3.1. Joint and Marginal Generalized Hartley Measures

The functional GH defined by Eq. (6.27) is also subadditive and additive, as is established by the following two theorems.

Theorem 6.2. For any joint basic probability assignment function m on $X \times Y$ and its associated marginal functions m_X and m_Y,

$$GH(m) \leq GH(m_X) + GH(m_Y).\qquad(6.29)$$

Proof. Recalling Eqs. (5.54) and (5.55), we have

$$
\begin{aligned}
GH(m_X) &= \sum_{A \subseteq X} m_X(A) \log_2 |A| \\
&= \sum_{A \subseteq X} \sum_{C|A=C_X} m(C) \log_2 |A| \\
&= \sum_{A \subseteq X} \sum_{C|A=C_X} m(C) \log_2 |C_X| \\
&= \sum_{C \subseteq X \times Y} m(C) \log_2 |C_X|.
\end{aligned}
$$

Similarly,

$$GH(m_Y) = \sum_{C \subseteq X \times Y} m(C) \log_2 |C_Y|.$$

Hence,

$$
\begin{aligned}
GH(m_X) + GH(m_Y) &= \sum_{C \subseteq X \times Y} m(C)(\log_2 |C_X| + \log_2 |C_Y|) \\
&= \sum_{C \subseteq X \times Y} m(C) \log_2 (|C_X| \cdot |C_Y|) \\
&\geq \sum_{C \subseteq X \times Y} m(C) \log_2 (C) \\
&= GH(m). \qquad\blacksquare
\end{aligned}
$$

Theorem 6.3. Let m_X and m_Y be noninteractive basic probability assignment functions on subsets of X and Y, respectively, and let m be the associated joint basic probability assignment function. Then,

$$GH(m) = GH(m_X) + GH(m_Y).\qquad(6.30)$$

Proof. Recalling Eq. (5.56), we have

$$
\begin{aligned}
GH(m) &= \sum_{C \subseteq X \times Y} m(C) \log_2 |C| \\
&= \sum_{A \subseteq X} \sum_{B \subseteq Y} m_X(A) \cdot m_Y(B) \log_2 |A \times B| \\
&= \sum_{A \subseteq X} \sum_{B \subseteq Y} m_X(A) \cdot m_Y(B) \log_2 (|A| \cdot |B|)
\end{aligned}
$$

$$= \sum_{A \subseteq X} \sum_{B \subseteq Y} m_X(A) \cdot m_Y(B)(\log_2|A| + \log_2|B|)$$

$$= \sum_{B \subseteq Y} m_Y(B) \sum_{A \subseteq X} m_X(A) \cdot \log_2|A| + \sum_{A \subseteq X} m_X(A) \sum_{B \subseteq Y} m_Y(B) \cdot \log_2|B|$$

$$= \sum_{A \subseteq X} m_X(A) \cdot \log_2|A| + \sum_{B \subseteq Y} m_Y(B) \cdot \log_2|B|$$

$$= GH(m_X) + GH(m_Y). \qquad \blacksquare$$

According to the common notational convention in the literature, it is convenient to replace the specific symbols $GH(m)$, $GH(m_X)$, and $GH(m_Y)$ with their generic counterparts $GH(X \times Y)$, $GH(X)$, and $GH(Y)$. The conditional GH measures, addressed in Section 6.3.3, are then denoted as $GH(X|Y)$ and $GH(Y|X)$.

6.3.2. Monotonicity of the Generalized Hartley Measure

Bodies of evidence in DST can be partially ordered on the basis of set inclusion. This ordering is then used for defining the requirement of monotonicity for the GH measure. For any pair $\langle \mathcal{F}_1, m_1 \rangle$ and $\langle \mathcal{F}_2, m_2 \rangle$ of bodies of evidence on X, $\langle \mathcal{F}_1, m_1 \rangle$ is said to be smaller than or equal to $\langle \mathcal{F}_2, m_2 \rangle$, which is written as

$$\langle \mathcal{F}_1, m_1 \rangle \leq \langle \mathcal{F}_2, m_2 \rangle,$$

if and only if the following requirements are satisfied:

(a) For each $A \in \mathcal{F}_1$, there exists some $B \in \mathcal{F}_2$ such that $A \subseteq B$;
(b) For each $B \in \mathcal{F}_2$ there exists some $A \in \mathcal{F}_1$ such that $A \subseteq B$;
(c) There exists a function $f : \mathcal{P}(X) \times \mathcal{P}(X) \to [0, 1]$ such that $f(A, B) > 0$ implies $A \subseteq B$ and

$$m_1(A) = \sum_{B|A \subseteq B} f(A, B) \qquad \text{for each } A \subseteq X, \qquad (6.31)$$

$$m_2(B) = \sum_{A|A \subseteq B} f(A, B) \qquad \text{for each } B \subseteq X. \qquad (6.32)$$

EXAMPLE 6.5. To illustrate the definition of ordering of bodies of evidence, especially requirement (c) in the definition, consider the following two bodies of evidence on $X = \{x_1, x_2, x_3\}$:

$$\langle \mathcal{F}_1, m_1 \rangle : \mathcal{F}_1 = \{\{x_2\}, \{x_1, x_2\}, \{x_1, x_3\}\}$$
$$m_1(\{x_2\}) = 0.2, \quad m_1(\{x_1, x_2\}) = 0.2, \quad m_1(\{x_1, x_3\}) = 0.6$$
$$\langle \mathcal{F}_2, m_2 \rangle : \mathcal{F}_2 = \{\{x_1, x_3\}, \{x_2, x_3\}, X\}$$
$$m_2(\{x_1, x_3\}) = 0.5, \quad m_2(\{x_2, x_3\}) = 0.1, \quad m_2(X) = 0.4.$$

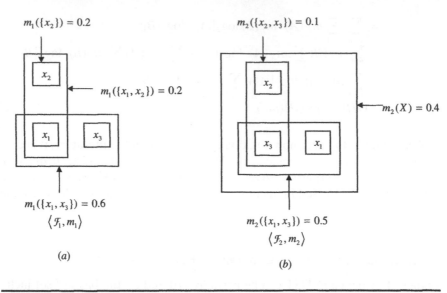

$m_1(\{x_2\}) = 0.2$

$m_1(\{x_1, x_2\}) = 0.2$

$m_1(\{x_1, x_3\}) = 0.6$

$\langle \mathcal{F}_1, m_1 \rangle$

(a)

$m_2(\{x_2, x_3\}) = 0.1$

$m_2(X) = 0.4$

$m_2(\{x_1, x_3\}) = 0.5$

$\langle \mathcal{F}_2, m_2 \rangle$

(b)

A	$\varnothing$	$\{x_1\}$	$\{x_2\}$	$\{x_3\}$	$\{x_1, x_2\}$	$\{x_1, x_3\}$	$\{x_2, x_3\}$	X	B	$m_2(B)$
$m_1(A)$			0.2		0.2	0.6				
$f(A,B)$									$\varnothing$	
									$\{x_1\}$	
									$\{x_2\}$	
									$\{x_3\}$	
									$\{x_1, x_2\}$	
					0.5				$\{x_1, x_3\}$	0.5
			0.1						$\{x_2, x_3\}$	0.1
			0.1		0.2	0.1			X	0.4

(c)

Figure 6.5. Illustrations of ordered bodies of evidence in DST (Example 6.5).

These bodies of evidence are also shown graphically in Figure 6.5a and 6.5b, respectively. To determine whether $\langle \mathcal{F}_1, m_1 \rangle \leq \langle \mathcal{F}_2, m_2 \rangle$, the three requirements, (a)–(c), for ordering bodies of evidence in DST must be checked. Requirement (a) is trivially satisfied, since $X \in \mathcal{F}_2$. Requirement (b) also satisfied for $\{x_1, x_3\} \in \mathcal{F}_2, \{x_1, x_3\} \in \mathcal{F}_1$; for $\{x_2, x_3\} \in \mathcal{F}_2, \{x_2\} \in \mathcal{F}_1$; for $X \in \mathcal{F}_2$, each set in $\mathcal{F}_1$ satisfies the condition. Requirement (c) is satisfied by constructing a function f that satisfies Eqs. (6.31) and (6.32). For the given bodies of evidence, such a function is shown in the table in Figure 6.5c. For clarity, all zero entries in the table are left blank.

Using the introduced ordering of bodies of evidence in DST, the *monotonicity requirement* for the *GH* measure can be stated as follows:

$$\text{if } \langle \mathcal{F}_1, m_1 \rangle \leq \langle \mathcal{F}_2, m_2 \rangle, \text{ then } GH(m_1) \leq GH(m_2).$$

As is stated by the following theorem, the functional GH defined by Eq. (6.27) satisfies this monotonicity.

Theorem 6.4. For any pair of bodies of evidence in DST, $\langle \mathcal{F}_1, m_1 \rangle$ and $\langle \mathcal{F}_2, m_2 \rangle$, and for the functional GH defined by Eq. (6.27),

$$\langle \mathcal{F}_1, m_1 \rangle \leq \langle \mathcal{F}_2, m_2 \rangle \Rightarrow GH(m_1) \leq GH(m_2). \tag{6.33}$$

Proof

$$
\begin{aligned}
\sum_{A \subseteq X} m_1(A) \log_2 |A| &= \sum_{A \subseteq X} \sum_{B | A \subseteq B} f(A, B) \log_2 |A| \\
&\leq \sum_{A \subseteq X} \sum_{B | A \subseteq B} f(A, B) \log_2 |B| \\
&= \sum_{B \subseteq X} \sum_{A | A \subseteq B} f(A, B) \log_2 |B| \\
&= \sum_{B \subseteq X} m_2(B) \log_2 |B|.
\end{aligned}
$$
∎

6.3.3. Conditional Generalized Hartley Measures

The GH measure defined by Eq. (6.27) is basically a weighted average of the Hartley measure. Its conditional counterparts should thus be weighted averages of the respective conditional Hartley measures. It is thus meaningful to define $GH(X|Y)$ and $GH(Y|X)$ by the formulas

$$GH(X|Y) = \sum_{C \subseteq X \times Y} m(C) \log_2 \frac{|C|}{|B|}, \tag{6.34}$$

$$GH(Y|X) = \sum_{C \subseteq X \times Y} m(C) \log_2 \frac{|C|}{|A|}, \tag{6.35}$$

where $B \subseteq Y$ and $A \subseteq X$.

Now, Eq. (6.34) can be rewritten as

$$
\begin{aligned}
GH(X|Y) &= \sum_{C \subseteq X \times Y} m(C) \log_2 |C| - \sum_{C \subseteq X \times Y} m(C) \log_2 |B| \\
&= GH(X \times Y) - \sum_{C \subseteq X \times Y} m(C) \log_2 |B| \\
&= GH(X \times Y) - \sum_{B \subseteq Y} \sum_{C \subseteq X \times Y} m(C) \log_2 |B| \\
&= GH(X \times Y) - \sum_{B \subseteq Y} m_Y(B) \log_2 |B|.
\end{aligned}
$$

Hence,

$$GH(X|Y) = GH(X \times Y) - GH(Y), \tag{6.36}$$

and by a similar derivation,

$$GH(Y|X) = GH(X \times Y) - GH(X). \tag{6.37}$$

These equations verify once more that the relationship among joint, marginal, and conditional uncertainties is an invariant property of the uncertainty theories.

6.4. GENERALIZED HARTLEY MEASURE FOR CONVEX SETS OF PROBABILITY DISTRIBUTIONS

It is desirable to try to generalize the Hartley functional in DST to arbitrary convex sets of probability distributions (credal sets). This generalized version of the functional would be defined by the formula

$$GH(\mathcal{D}) = \sum_{A \subseteq X} m_{\mathcal{D}}(A) \log_2 |A|, \tag{6.38}$$

where $\mathcal{D}$ is a given set of probability distributions on X. Function $m_{\mathcal{D}}$ in this formula is obtained by the Möbius transformation applied to the lower probability function that is derived from $\mathcal{D}$ via Eq. (4.11).

It is easy to show that this functional GH is additive. That is,

$$GH(\mathcal{D}) = GH_X(\mathcal{D}_X) + GH_Y(\mathcal{D}_Y), \tag{6.39}$$

when $\mathcal{D}$ is derived from $\mathcal{D}_X$ and $\mathcal{D}_Y$ that are noninteractive, which means that for all $A \in \mathcal{P}(X)$ and all $B \in \mathcal{P}(Y)$,

$$m_{\mathcal{D}}(A \times B) = m_{\mathcal{D}_X}(A) \cdot m_{\mathcal{D}_Y}(B) \tag{6.40}$$

and $m_{\mathcal{D}}(C) = 0$ for all $C \neq A \times B$. The proof of additivity of functional GH in this generalized case is virtually the same as the proof of additivity of GH in DST (the proof of Theorem 6.3). Although values $m_{\mathcal{D}}(A)$ in Eq. (6.38) may be negative for some subsets of X, the equation

$$\sum_{A \subseteq X} m_{\mathcal{D}}(A) = 1,$$

upon which the proof is based, still holds.

It is clear that the GH functional has the proper range, $[0, \log_2|X|]$, when the units of measurement are bits: 0 is obtained when $\mathcal{D}$ contains a single probability; $\log_2|X|$ is obtained when $\mathcal{D}$ contains all possible probability distributions on X, and thus represents total ignorance. The functional GH is also continuous, symmetric (invariant with respect to permutations of the probability distributions in $\mathcal{D}$), and expansible (it does not change when components with zero probabilities are added to the probability distributions in $\mathcal{D}$).

One additional property of the generalized Hartley functional, which is significant when we deal with credal sets, is its monotonicity with respect to subsethood relationship between credal sets. This means that for every pair of credal sets on X, $^i\mathcal{D}$ and $^j\mathcal{D}$, if $^i\mathcal{D} \subseteq {}^j\mathcal{D}$ then $GH(^i\mathcal{D}) \le GH(^j\mathcal{D})$. This property, whose proof is available in the literature (Note 6.6), is illustrated by the following example.

EXAMPLE 6.6. Consider six convex sets of probability distributions on $X = \{x_1, x_2, x_3\}$, which are denoted by $^i\mathcal{D}(i \in \mathbb{N}_6)$ and are defined geometrically in Figure 6.6. Clearly, $^1\mathcal{D} \supseteq {}^2\mathcal{D} \supseteq {}^3\mathcal{D} \supseteq {}^4\mathcal{D}$ and also $^3\mathcal{D} \supseteq {}^5\mathcal{D}$. However, $^6\mathcal{D}$ is neither a subset nor a superset of any of the other sets. For each set $^i\mathcal{D}(i \in \mathbb{N}_6)$, the associated lower probability function, $^i\underline{\mu}$, its Möbius representation, ^im, and the value $GH(^i\mathcal{D})$ of the GH measure are shown in Table 6.1. In conformity with the monotonicity of GH,

$$GH(^1\mathcal{D}) \ge GH(^2\mathcal{D}) \ge GH(^3\mathcal{D}) \ge GH(^4\mathcal{D})$$

and also $GH(^3\mathcal{D}) \ge GH(^5\mathcal{D})$. Moreover, $GH(^6\mathcal{D}) \ge GH(^i\mathcal{D})$ for all $i \in \mathbb{N}_5$, which illustrates that nonspecificity $GH(\mathcal{D})$ does not express the size of $\mathcal{D}$.

While subadditivity of the generalized Hartley functional has been proven for all uncertainty theories that are subsumed under DST (Theorem 6.2), the following example demonstrates that the functional is not subadditive for arbitrary convex sets of probability distributions.

EXAMPLE 6.7. Let $X = \{x_1, x_2\}$ and $Y = \{y_1, y_2\}$, and let $z_{ij} = \langle x_i, y_j \rangle$ for all $i, j \in \{1, 2\}$. Furthermore, let $\mathbf{p} = \langle p_{11}, p_{12}, p_{21}, p_{22} \rangle$ denote joint probability distributions on $X \times Y$, where $p_{ij} = p(z_{ij})$. Given the set $\mathcal{D}$ of all convex combinations of probability distributions $\mathbf{p}_A = \langle 0.4, 0.4, 0.2, 0 \rangle$ and $\mathbf{p}_B = \langle 0.6, 0.2, 0, 0.2 \rangle$, we obtain the associated sets $\mathcal{D}_X = \{\langle 0.8, 0.2 \rangle\}$ and $\mathcal{D}_Y = \{\langle 0.6, 0.4 \rangle\}$ of marginal probability distributions. Clearly, $GH(\mathcal{D}_X) + GH(\mathcal{D}_Y) = 0$. The lower probability function, $\underline{\mu}_\mathcal{D}$ associated with $\mathcal{D}$ and its Möbius representation, $m_\mathcal{D}$, are shown in Table 6.2. Clearly, $GH(\mathcal{D}) = 0.332$. Hence, GH is not subadditive in this example.

It is easy to determine that the lower probability function $\underline{\mu}_\mathcal{D}$ in Example 6.7 is not 2-monotone. It contains eight violations of 2-monotonicity. One of them is the violation of the inequality

$$\underline{\mu}_\mathcal{D}(\{z_{11}, z_{12}, z_{21}\}) \ge \underline{\mu}_\mathcal{D}(\{z_{11}, z_{12}\}) + \underline{\mu}_\mathcal{D}(\{z_{11}, z_{21}\}) - \underline{\mu}_\mathcal{D}(\{z_{11}\}).$$

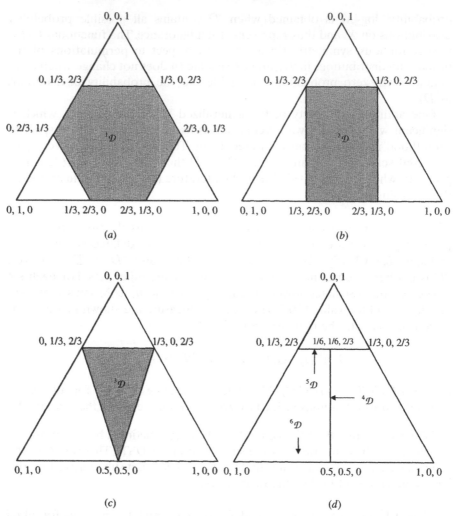

Figure 6.6. Closed convex sets of probability distributions on $X = \{x_1, x_2, x_3\}$ discussed in Example 6.6.

Indeed, $0.8 \not\geq 0.8 + 0.6 - 0.4$. Whether GH is subadditive or not for some less general credal sets is an open question. However, this question looses its significance within the context discussed in Section 6.8.

6.5. GENERALIZED SHANNON MEASURE IN DEMPSTER–SHAFER THEORY

The issue of how to generalize the Shannon entropy from probability theory to DST has been discussed in the literature since the early 1980s. By and large,

Table 6.1 Lower Probability Functions and Their Möbius Representations for the Convex Sets of Probability Distributions on $X = \{x_1, x_2, x_3\}$ Defined in Figure 6.6 and the Associated Values of the Generalized Hartley Measure GH (Example 6.6)

	x_1	x_2	x_3	$^1\mathcal{D}$		$^2\mathcal{D}$		$^3\mathcal{D}$		$^4\mathcal{D}$		$^5\mathcal{D}$		$^6\mathcal{D}$	
				$^1\mu$	^1m	$^2\mu$	^2m	$^3\mu$	^3m	$^4\mu$	^4m	$^5\mu$	^5m	$^6\mu$	^6m
A:	0	0	0	0	0	0	0	0	0	0	0	0	0	0	0
	1	0	0	0	0	0	0	0	0	1/6	1/6	0	0	0	0
	0	1	0	0	0	0	0	0	0	1/6	1/6	0	0	0	0
	0	0	1	0	0	0	0	0	0	0	0	2/3	2/3	0	0
	1	1	0	1/3	1/3	1/3	1/3	1/3	1/3	1/3	0	1/3	1/3	1	1
	1	0	1	1/3	1/3	1/3	1/3	0.5	0.5	0.5	1/3	2/3	0	0	0
	0	1	1	1/3	1/3	1/3	1/3	0.5	0.5	0.5	1/3	2/3	0	0	0
	1	1	1	1	0	1	0	1	-1/3	1	0	1	0	1	0
$GH(^i\mathcal{D})$				1		1		0.805		2/3		1/3		1	

217

Table 6.2. Lower Probability Function $\underline{\mu}_D$ in Examples 6.7 and 6.13

	z_{11}	z_{12}	z_{21}	z_{22}	$\underline{\mu}_D(A)$	$m_D(A)$
A:	0	0	0	0	0.0	0.0
	1	0	0	0	0.4	0.4
	0	1	0	0	0.2	0.2
	0	0	1	0	0.0	0.0
	0	0	0	1	0.0	0.0
	1	1	0	0	0.8	0.2
	1	0	1	0	0.6	0.2
	1	0	0	1	0.4	0.0
	0	1	1	0	0.2	0.0
	0	1	0	1	0.4	0.2
	0	0	1	1	0.2	0.2
	1	1	1	0	0.8	−0.2
	1	1	0	1	0.8	−0.2
	1	0	1	1	0.6	−0.2
	0	1	1	1	0.4	−0.2
	1	1	1	1	1.0	0.4

the prospective generalized Shannon entropy was viewed in these discussions as a measure of conflict among evidential claims in each given body of evidence in DST. This view was inspired by the recognition that the Shannon entropy itself is a measure of conflict associated with each given probability distribution (recall the discussion of Eq. (3.25) in Chapter 3).

Although several intuitively promising functionals have been proposed as candidates for the generalized Shannon entropy in DST, each of them was found upon closer scrutiny to violate some of the essential requirements for uncertainty measures. In most cases, it was the requirement of subadditivity, perhaps the most fundamental requirement, that was violated. The efforts to determine a justifiable generalization of the Shannon entropy in DST have thus been unsuccessful. The reasons for this failure, which are now understood, are explained later in this section.

Although none of the functionals proposed as a prospective generalization of the Shannon entropy in DST is acceptable on mathematical grounds, it seems useful to present their overview for at least two reasons: (1) ideas are always better understood when we are familiar with the history of their development; and (2) knowing which of the promising candidates have failed will avoid their "reinventing."

In the following overview of the unsuccessful attempts to generalize the Shannon entropy to DST, no references are made. The relevant references are all given in Note 6.7.

Two of the candidates for the generalized Shannon entropy in DST were proposed in the early 1980s. One of them is functional E defined by the formula

$$E(m) = -\sum_{A \in \mathcal{F}} m(A)\log_2 Pl(A), \tag{6.41}$$

which is usually called a measure of *dissonance*. The other one is functional C defined by the formula

$$C(m) = -\sum_{A \in \mathcal{F}} m(A)\log_2 Bel(A), \tag{6.42}$$

which is referred to as a measure of *confusion*. It is obvious that both of these functionals collapse into the Shannon entropy when m defines a probability measure.

To decide if either of the two functionals is an appropriate generalization of the Shannon entropy in DST, we have to determine what these functionals actually measure. From Eq. (5.47) and the general property of basic probability assignments (satisfied for every $A \in \mathcal{P}(X)$),

$$\sum_{A \cap B = \emptyset} m(B) + \sum_{A \cap B \neq \emptyset} m(B) = 1, \tag{6.43}$$

we obtain

$$E(m) = -\sum_{A \in \mathcal{F}} m(A)\log_2\left(1 - \sum_{A \cap B = \emptyset} m(B)\right). \tag{6.44}$$

The term

$$K(A) = \sum_{A \cap B = \emptyset} m(B)$$

in Eq. (6.44) represents the total evidential claim pertaining to focal elements that are disjoint with the set A. That is, $K(A)$ expresses the sum of all evidential claims that fully conflict with the one focusing on the set A. Clearly, $K(A) \in [0, 1]$. The function

$$-\log_2[1 - K(A)],$$

which is employed in Eq. (6.44), is monotonic increasing with $K(A)$, and extends its range from $[0, 1]$ to $[0, \infty)$. The choice of the logarithmic function is motivated in the same way as in the classical case of the Shannon entropy.

It follows from these facts and the form of Eq. (6.44) that $E(m)$ is the mean (expected) value of the conflict among evidential claims within a given body of evidence $\langle \mathcal{F}, m \rangle$; it measures the conflict in bits and its range is $[0, \log_2|X|]$.

Functional E is not fully satisfactory since we feel intuitively that $m(B)$ may conflict with $m(A)$ not only when $B \cap A = \emptyset$. This broader view of conflict is expressed by the measure of confusion C given by Eq. (6.42). Let us demonstrate this fact.

From Eq. (5.46) and the general property of basic assignments (satisfied for every $A \in \mathcal{P}(X)$),

$$\sum_{B \subseteq A} m(B) + \sum_{B \nsubseteq A} m(B) = 1 \tag{6.45}$$

we get

$$C(m) = -\sum_{A \in \mathcal{F}} m(A) \log_2 \left(1 - \sum_{B \nsubseteq A} m(B) \right). \tag{6.46}$$

The term

$$L(A) = \sum_{B \nsubseteq A} m(B)$$

in Eq. (6.46) expresses the sum of all evidential claims that conflict with the one focusing on the set A according to the following broader view of conflict: $m(B)$ conflicts with $m(A)$ whenever $B \nsubseteq A$. The reason for using the function

$$-\log_2[1 - L(A)]$$

instead of $L(A)$ in Eq. (6.46) is the same as already explained in the context of functional E. The conclusion is that $C(m)$ is the mean (expected) value of the conflict, viewed in the broader sense, among evidential claims within a given body of evidence $\langle \mathcal{F}, m \rangle$.

Functional C is also not fully satisfactory as a measure of conflicting evidential claims within a body of evidence, but for a different reason than functional E. Although it employs the broader, and more satisfactory, view of conflict, it does not properly scale each particular conflict of $m(B)$ with respect to $m(A)$ according to the degree of violation of the subsethood relation $B \subseteq A$. It is clear that the more this subsethood relation is violated, the greater the conflict. In addition, neither E nor C satisfy the essential axiomatic requirement of subadditivity.

To overcome the deficiencies of functionals E and C as adequate measures of conflict in evidence theory, a new functional, D, was proposed in the early 1990s:

$$D(m) = -\sum_{A \in \mathcal{F}} m(A) \log_2 \left(1 - \sum_{B \in \mathcal{F}} m(B) \frac{|B - A|}{|B|} \right). \tag{6.47}$$

Observe that the term

$$Con(A) = \sum_{B \in \mathcal{F}} m(B) \frac{|B - A|}{|B|} \tag{6.48}$$

in Eq. (6.47) expresses the sum of individual conflicts of evidential claims with respect to a particular set A, each of which is properly scaled by the degree to which the subsethood $B \subseteq A$ is violated. This conforms to the intuitive idea of conflict that emerged from the critical reexamination of functionals E and C. Clearly, $Con(A) \in [0, 1]$ and, furthermore,

$$K(A) \le Con(A) \le L(A). \tag{6.49}$$

The reason for using the function

$$-\log_2[1 - Con(A)]$$

instead of Con in Eq. (6.47) is exactly the same as previously explained in the context of functional E. This monotonic transformation extends the range of $Con(A)$ from $[0, 1]$ to $[0, \infty)$.

Functional D, which is called a measure of *discord*, is clearly a measure of the mean conflict (expressed by the logarithmic transformation of function Con) among evidential claims within each given body of evidence. It follows immediately from Eq. (6.49) that

$$E(m) \le D(m) \le C(m). \tag{6.50}$$

Observe that $|B - A| = |B| - |A \cap B|$ and, consequently, Eq. (6.47) can be rewritten as

$$D(m) = -\sum_{A \in \mathcal{F}} m(A) \log_2 \sum_{B \in \mathcal{F}} m(B) \frac{|A \cap B|}{|B|}. \tag{6.51}$$

It is obvious that

$$Bel(A) \le \sum_{B \in \mathcal{F}} m(B) \frac{|A \cap B|}{|B|} \le Pl(A). \tag{6.52}$$

Although functional D is intuitively more appealing than functionals E and C, further examination revealed that it has a conceptual defect. To explain the defect, let sets A and B in Eq. (6.47) be such that $A \subset B$. Then, according to function Con, the claim $m(B)$ is taken to be in conflict with the claim $m(A)$ to the degree $|B - A|/|B|$. This, however, should not be the case: the claim focusing on B is implied by the claim focusing on A (since $A \subset B$) and, hence, $m(B)$ should not be viewed in this case as contributing to the conflict with $m(A)$.

Consider, as an example, incomplete information regarding the age of a person, say Joe. Assume that the information is expressed by two evidential claims pertaining to the age of Joe: "Joe is between 15 and 17 years old" with degree $m(A)$, where $A = [15, 17]$, and "Joe is a teenager" with degree $m(B)$, where $B = [13, 19]$. Clearly, the weaker second claim does not conflict with the stronger first claim.

Now assume that $A \supset B$. In this case, the situation is reversed: the claim focusing on B is not implied by the claim focusing on A and, consequently, $m(B)$ does conflict with $m(A)$ to a degree proportional to number of elements in A that are not covered by B. This conflict is not captured by function Con, since $|B - A| = 0$ in this case.

It follows from these observations that the total conflict of evidential claims within a body of evidence $\langle \mathcal{F}, m \rangle$ with respect to a particular claim $m(A)$ should be expressed by functional

$$CON(A) = \sum_{B \in \mathcal{F}} m(B)\frac{|A - B|}{|A|} \qquad (6.53)$$

rather than functional Con given by Eq. (6.48). Replacing $Con(A)$ in Eq. (6.47) with $CON(A)$, we obtain a new function, which is better justified as a measure of conflict in DST than the functional D. This new functional, which is called *strife* and is denoted by ST, is defined by the formula

$$ST(m) = -\sum_{A \in \mathcal{F}} m(A)\log_2\left(1 - \sum_{B \in \mathcal{F}} m(B)\frac{|A - B|}{|A|}\right). \qquad (6.54)$$

It is trivial to convert this form into a simpler one,

$$ST(m) = -\sum_{A \in \mathcal{F}} m(A)\log_2 \sum_{B \in \mathcal{F}} m(B)\frac{|A \cap B|}{|A|}, \qquad (6.55)$$

where the term $|A \cap B|/|A|$ expresses the degree of subsethood of set A in set B. Equation (6.55) can also be rewritten as

$$ST(m) = GH(m) - \sum_{A \in \mathcal{F}} m(A)\log_2 \sum_{B \in \mathcal{F}} m(B)|A \cap B|, \qquad (6.56)$$

where GH is the generalized Hartley measure defined by Eq. (6.27). Furthermore, introducing

$$Z(m) = \sum_{A \in \mathcal{F}} m(A)\log_2 \sum_{B \in \mathcal{F}} m(B)|A \cap B|, \qquad (6.57)$$

we have

$$ST(m) = GH(m) - Z(m). \qquad (6.58)$$

It was later shown that the distinction between strife and discord reflects the distinction between disjunctive and conjunctive set–valued statements, respectively.

To describe this distinction, consider statements of the form "x is A," where A is a subset of a given universal set X and $x \in X$. We assume the framework of DST, in which the evidence supporting this proposition is expressed by the value $m(A)$ of the basic probability assignment. The statement may be interpreted either as a *disjunctive set–valued statement* or a *conjunctive set–valued statement*.

A statement "x is A" is disjunctive if it means that x conforms to one of the elements in A. For example "Mary is a teenager" is disjunctive because it means that the real age of Mary conforms to one of the values in the set {13, 14, 15, 16, 17, 18, 19} . Similarly "John arrived between 10:00 and 11:00 A.M." is disjunctive because it means that John's real arrival time was one value in the time interval between 10:00 and 11:00 A.M.

A statement "x is A" is conjunctive if it means that x conforms to all of the elements in A. For example, the statement "The compound consists of iron, copper, and aluminium" is conjunctive because it means that the compound in question conforms to all the elements in the set {iron, copper, aluminium}. Similarly, "John was in the doctor's office from 10:00 A.M. to 11:00 A.M." is conjunctive because it means that John was in the doctor's office not only at one time during the time interval, but all the time instances during the time interval between 10:00 and 11:00 A.M.

Let S_A and S_B denote, respectively, the statements "x is A" and "x is B." Assume that $A \subset B$ and the statements are disjunctive. Then, clearly, S_A implies S_B and, consequently, S_B does not conflict with S_A while S_A does conflict with S_B. For example, the statement S_B: "Mary is a teenager" does not conflict with the statement S_A: "Mary is fifteen or sixteen," while S_A conflicts with S_B.

Let S_A and S_B be conjunctive and assume again that $A \subset B$. Then, clearly, S_B implies S_A and, consequently, S_A does not conflict with S_B, while S_B does conflict with S_A. For example, the statement S_B: "Steel is a compound of iron, carbon, and nickel" does conflict with the statement S_A: "Steel is a compound of iron and carbon," while S_A does not conflict with S_B in this case.

This examination clearly shows that the measure of strife ST (Eq. (6.55)) expresses the conflict among disjunctive statements, while the measure of discord D (Eq. (6.51)) expresses the conflict among conjunctive statements. Since DST and possibility theory usually deal with disjunctive statements, the measure of strife is a better justified measure of conflict (or entropy-like measure) in DST and possibility theory.

It is reasonable to conclude that functional ST is well justified on intuitive grounds as a measure of conflict among evidential claims in DST when disjunctive statements are employed. Similarly, functional D is a well justified measure of conflict in evidence theory when conjunctive statements are employed. Unfortunately, neither of these functionals is subadditive.

After the functionals ST and D were rejected as generalizations of the Shannon entropy due to their violation of subadditivity, the next idea was to explore the sums of $ST + GH$ and $D + GH$. Unfortunately, these sums were found to violate the requirement of subadditivity as well. Specific counterexamples demonstrating these violations are examined in the following example.

EXAMPLE 6.8. Let $Z = X \times Y$, where $X = \{x_1, x_2\}$ and $Y = \{y_1, y_2\}$, and let $z_{ij} = \langle x_i, y_j \rangle$ $(i, j = 1, 2)$. The joint body of evidence specified visually in Figure 6.7a is an example for which the functional $ST + GH$ is not subadditive. All numbers in the figures are values of the basic probability assignment functions for the indicated focal elements. For this joint body of evidence and its associated marginal bodies of evidence (also shown in Figure 6.7a), we can immediately see that

$$GH(m_X) + GH(m_Y) - GH(m) = 0.$$

For ST, we obtain:

$$ST(m_X) = ST(m_Y) = -\frac{1}{3}\log_2\left(\frac{1}{3} + \frac{2}{3}\right) - \frac{2}{3}\log_2\left(\frac{1}{3} \cdot \frac{1}{2} + \frac{2}{3}\right)$$

$$= -\frac{2}{3}\log_2\frac{5}{6},$$

$$ST(m) = -2 \cdot \frac{1}{3}\log_2\left(\frac{1}{3} + \frac{1}{3} \cdot \frac{1}{2} + \frac{1}{3}\right) - \frac{1}{3}\log_2\left(2 \cdot \frac{1}{3} \cdot \frac{1}{2} + \frac{1}{3}\right)$$

$$= -\frac{2}{3}\log_2\frac{5}{6} - \frac{1}{3}\log_2\frac{2}{3}.$$

That is,

$$ST(m_X) + ST(m_Y) - ST(m) = \frac{1}{3}\log_2\left(\frac{24}{25}\right).$$

which is a negative number. This means that the functional $ST + GH$ violates the requirement of subadditivity. Similarly, the joint body of evidence in Figure 6.7b is an example of where the functional $D + GH$ violates subadditivity. We can easily see that

$$GH(m_X) + GH(m_Y) - GH(m) = 0$$

and

$$D(m_X) + D(m_Y) - D(m) = \frac{1}{3}\log_2 0.96,$$

which again is a negative number.

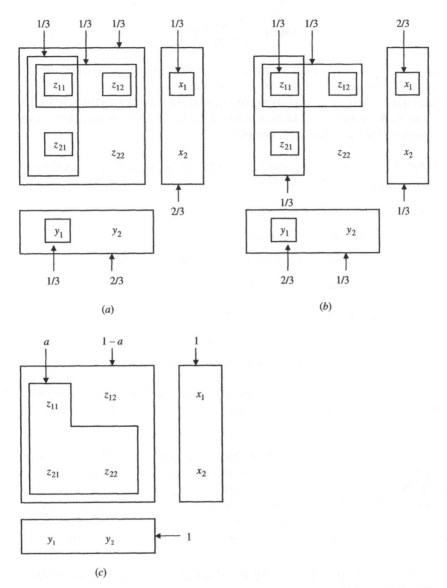

Figure 6.7. Counterexamples demonstrating the violation of subadditivity: (a) for $ST + GH$; (b) for $D + GH$; (c) for $AS + GH$ (Examples 6.8 and 6.9).

One additional prospective functional for generalized Shannon entropy in DST, quite different from the other ones, was also suggested in the literature. This functional, denoted here by AS, is defined as the average Shannon entropy for all probability distributions in the interaction representation (discussed in Section 4.3.3). That is,

$$AS(m) = \frac{1}{|X|!} \sum_{\mathbf{p} \in \mathcal{P}} S(\mathbf{p}), \tag{6.59}$$

where $\mathcal{P}$ denotes the set of all probability distributions in the interaction representation of m, and S denotes the Shannon entropy.

Although initially quite promising, functional AS was found deficient in the same way as the other candidates: neither AS nor $AS + GH$ is subadditive, as demonstrated by the following example.

EXAMPLE 6.9. Considering the family of bodies of evidence specified in Figure 6.7c, where $a \in [0, 1]$, we have:

$$AS(m_X) = AS(m_Y) = 0,$$

$$GH(m_X) = GH(m_Y) = 1,$$

$$AS(m) = [-a \log_2 a - (1 - a) \log_2 (1 - a)]/4,$$

$$GH(m) = a \log_2 3 + 2 - 2a.$$

For the subadditivity of $GH + AS$, the difference

$$\Delta = (GH_X + GH_Y + AS_X + AS_Y) - (GH + AS)$$
$$= [a \log_2 a + (1 - a) \log_2 (1 - a)]/4 + 2a - a \log_2 3$$

is required to be nonnegative for values $a \in [0, 1]$. However, Δ is negative in this case for any value $a \in [0, 0.58]$ and it reaches its minimum, $\Delta = -0.1$, at $a = 0.225$.

The long, unsuccessful, and often frustrating search for the generalized Shannon measure in DST was replaced in the early 1990s with the search for a justifiable aggregate measure of uncertainty, capturing both nonspecificity and conflict. An aggregated measure that possesses all the required mathematical properties was eventually found, but not as a composite of measures of uncertainty of the two types. This measure is examined in the next section.

6.6. AGGREGATE UNCERTAINTY IN DEMPSTER–SHAFER THEORY

Let AU denote a functional by which the *aggregate uncertainty* ingrained in any given body of evidence (expressed in terms of DST) can be measured. This functional is supposed to capture, in an aggregated fashion, both nonspecificity and conflict—the two types of uncertainty that coexist in DST. The functional AU may be expressed in terms of belief measures, plausibility measures, or basic probability assignments. Choosing, for example, belief measures, the functional has the following form

$$AU : \mathcal{B} \to [0, \infty),$$

where $\mathcal{B}$ is the set of all belief measures. Since there are one-to-one mappings between corresponding belief measures, plausibility measures, and basic probability assignments, the domain of this functional AU expressed in this form may be reinterpreted in terms of the corresponding plausibility measures or basic probability assignments.

To qualify as a meaningful measure of aggregate uncertainty in DST, the functional AU must satisfy the following requirements:

(AU1) Probability Consistency. Whenever Bel defines a probability measure (i.e., all focal subsets are singletons), AU assumes the form of the Shannon entropy

$$AU(Bel) = -\sum_{x \in X} Bel(\{x\}) \log_2 Bel(\{x\}).$$

(AU2) Set Consistency. Whenever Bel focuses on a single set (i.e., $m(A) = 1$ for some $A \subseteq X$), AU assumes the form of the Hartley measure

$$AU(Bel) = \log_2 |A|.$$

(AU3) Range. The range of AU is $[0, \log_2 |X|]$ when Bel is defined on $\mathcal{P}(X)$ and AU is measured in bits.

(AU4) Subadditivity. If Bel is an arbitrary joint belief function on $X \times Y$ and the associated marginal belief functions are Bel_X and Bel_Y, then

$$AU(Bel) \leq AU(Bel_X) + AU(Bel_Y).$$

(AU5) Additivity. If Bel is a joint belief function on $X \times Y$, and the marginal belief functions Bel_X and Bel_Y are noninteractive, then

$$AU(Bel) = AU(Bel_X) + AU(Bel_Y).$$

A measure of aggregate uncertainty in DST that satisfies all these requirements was conceived by several authors in the early 1990s (see Note 6.8). The measure is defined as follows.

Given a belief measure Bel on the power set of a finite set X, the aggregate uncertainty associated with Bel is measured by the functional

$$AU(Bel) = \max_{\mathcal{P}_{Bel}} \left\{ -\sum_{x \in X} p_x \log_2 p_x \right\}, \tag{6.60}$$

where the maximum is taken over all probability distributions that dominate the given belief measure. Thus, $\mathcal{P}_{Bel}$ in Eq. (6.60) consists of all probability distributions $\langle p_x | x \in X \rangle$ that satisfy the constraints

(a) $p_x \in [0, 1]$ for all $x \in X$ and $\sum_{x \in X} p_x = 1$.

(b) $Bel(A) \le \sum_{x \in A} p_x$ for all $A \subseteq X$.

The following five theorems establish that the functional AU defined by Eq. (6.60) satisfies the requirements (AU1)–(AU5). It is thus a well-justified measure of aggregate uncertainty in DST. Although its uniqueness is still an open problem, it is known that the function AU is the smallest measure of aggregate uncertainty in DST among all other measures (if they exist).

Theorem 6.5. The measure AU is probability consistent.

Proof. When *Bel* is a probability measure, all focal elements are singletons and this implies that $Bel(\{x\}) = p_x = Pl(\{x\})$ for all $x \in X$. Hence, the maximum is taken over a single probability distribution $\langle p_x | x \in X \rangle$ and $AU(Bel)$ is equal to the Shannon entropy of this distribution. ∎

Theorem 6.6. The measure AU is set consistent.

Proof. Let m denote the basic probability assignment corresponding to function *Bel* in Eq. (6.60). By assumption of set consistency, $m(A) = 1$ for some $A \subseteq X$, and this implies that $m(B) = 0$ for all $B \ne A$ (including $B \subset A$). This means that every probability distribution that sums to one for elements x in A and is zero for all x not in A is consistent with *Bel*. It is well known that the uniform probability distribution maximizes the entropy function and, hence, the uniform probability distribution on A will maximize AU. That is,

$$AU(Bel) = -\sum_{x \in X} \frac{1}{|A|} \log_2 \frac{1}{|A|} = \log_2 |A|.$$ ∎

Theorem 6.7. The measure AU has a range $[0, \log_2|X|]$.

Proof. Since $[0, \log_2|X|]$ is the range of the Shannon entropy for any probability distribution on X the measure AU cannot be outside these bounds. ∎

Theorem 6.8. The measure AU is *subadditive*.

Proof. Let *Bel* be a function on $X \times Y$ and let $\langle \hat{p}_{xy} | \langle x, y \rangle \in X \times Y \rangle$ denote a probability distribution for which

$$AU(Bel) = -\sum_{x \in X} \sum_{y \in Y} \hat{p}_{xy} \log_2 \hat{p}_{xy}$$

and

$$Bel(A) \le \sum_{(x,y) \in A} \hat{p}_{xy}$$

for all $A \subseteq X \times Y$ (this must be true for $\hat{p}_{xy}$ to dominate Bel). Furthermore, let

$$\hat{p}_x = \sum_{y \in Y} \hat{p}_{xy} \quad \text{and} \quad \hat{p}_y = \sum_{x \in X} \hat{p}_{xy}.$$

Using Gibbs' inequality in Eq. (3.28), we have

$$-\sum_{x \in X} \sum_{y \in Y} \hat{p}_{xy} \log_2 \hat{p}_{xy} \le \sum_{x \in X} \sum_{y \in Y} \hat{p}_{xy} \log_2 (\hat{p}_x \cdot \hat{p}_y)$$
$$= -\sum_{x \in X} \hat{p}_x \log_2 \hat{p}_x - \sum_{y \in Y} \hat{p}_y \log_2 \hat{p}_y.$$

Observe that, for all $A \subseteq X$

$$Bel_X(A) = Bel(A \times Y) \le \sum_{x \in X} \sum_{y \in Y} \hat{p}_{xy} = \sum_{x \in A} \hat{p}_x$$

and, analogously, for all $B \subseteq Y$,

$$Bel_Y(B) \le \sum_{y \in B} \hat{p}_y.$$

Considering all these facts, we conclude

$$AU(Bel) = -\sum_{x \in X} \sum_{y \in Y} \hat{p}_{xy} \log_2 \hat{p}_{xy}$$
$$\le -\sum_{x \in X} \hat{p}_x \log_2 \hat{p}_x - \sum_{y \in Y} \hat{p}_y \log_2 \hat{p}_y$$
$$\le AU(Bel_X) + AU(Bel_Y). \qquad \blacksquare$$

Theorem 6.9. The measure AU is *additive*

Proof. By subadditivity we know that $AU(Bel) \le AU(Bel_X) + AU(Bel_Y)$. When Bel is noninteractive we need to prove the reverse, $AU(Bel) \ge AU(Bel_X) + AU(Bel_Y)$, to conclude that the two quantities must be equal and that AU is additive.

Let Bel be a noninteractive belief function. Let $\langle \hat{p}_x | x \in X \rangle$ be the probability distribution for which

$$AU(Bel_X) = -\sum_{x \in X} \hat{p}_x \log_2 \hat{p}_x$$

and

$$Bel(A) \le \sum_{x \in A} \hat{p}_x$$

for all $A \subseteq X$; similarly, let $\langle \hat{p}_y | y \in Y \rangle$ be the probability distribution for which

$$AU(Bel_Y) = -\sum_{y \in Y} \hat{p}_y \log_2 \hat{p}_y$$

and

$$Bel_Y(B) \le \sum_{y \in B} \hat{p}_y,$$

for all $B \subseteq Y$. Define $\hat{p}_{xy} = \hat{p}_x \cdot \hat{p}_y$ for all $\langle x, y \rangle \in X \times Y$. Clearly, $\hat{p}_{xy}$ is a probability distribution on $X \times Y$. Moreover, for all $C \subseteq X \times Y$,

$$
\begin{aligned}
\sum_{\langle x,y \rangle \in C} \hat{p}_{xy} &= \sum_{\langle x,y \rangle \in C} \hat{p}_x \cdot \hat{p}_y \\
&= \sum_{x \in C_X} \hat{p}_x \sum_{\langle x,y \rangle \in C} \hat{p}_y \\
&\ge \sum_{x \in C_X} \hat{p}_x \sum_{B | \{x\} \times B \subseteq C} m_Y(B) \\
&= \sum_{x \in C_X} \sum_{B | \{x\} \times B \subseteq C} \hat{p}_x \cdot m_Y(B) \\
&= \sum_{B \subseteq A_Y} \sum_{x | \{x\} \times B \subseteq C} m_Y(B) \cdot \hat{p}_x \\
&\ge \sum_{B \subseteq A_Y} m_Y(B) \sum_{A \times B \subseteq C} m_X(A) \\
&= \sum_{A \times B \subseteq C} m_X(A) \cdot m_Y(B).
\end{aligned}
$$

This implies that

$$
\begin{aligned}
AU(Bel) &\ge -\sum_{x \in X} \sum_{y \in Y} \hat{p}_{xy} \log_2 \hat{p}_{xy} \\
&= -\sum_{x \in X} \hat{p}_x \log_2 \hat{p}_x - \sum_{y \in Y} \hat{p}_y \log_2 \hat{p}_y \\
&= AU(Bel_X) + AU(Bel_Y).
\end{aligned}
$$ ∎

6.6.1. General Algorithm for Computing the Aggregated Uncertainty

Since functional AU is defined in terms of the solution to a nonlinear optimization problem, its practical utility was initially questioned. Fortunately, a relatively simple and fully general algorithm for computing AU was developed. The algorithm is formulated as follows.

Algorithm 6.1. Calculating AU from a belief function.
Input. A frame of discernment X and a belief measure Bel on subsets of X.

Output. $AU(Bel)$, $\langle p_x | x \in X \rangle$ such that $AU(Bel) = -\sum_{x \in X} p_x \log_2 p_x$, $p_i \geq 0$, $\sum_{x \in X} p_x = 1$, and $Bel(A) \leq \sum_{x \in A} p_x$ for all $\emptyset \neq A \subseteq X$.

Step 1. Find a nonempty set $A \subseteq X$, such that $Bel(A)/|A|$ is maximal. If there are more such sets A than one, take the one with maximal cardinality.

Step 2. For $x \in A$, put $p_x = Bel(A)/|A|$.

Step 3. For each $B \subseteq X - A$, put $Bel(B) = Bel(B \cup A) - Bel(A)$.

Step 4. Put $X = X - A$.

Step 5. If $X \neq \emptyset$ and $Bel(X) > 0$, then go to Step 1.

Step 6. If $Bel(X) = 0$ and $X \neq \emptyset$, then put $p_x = 0$ for all $x \in X$.

Step 7. Calculate $AU(Bel) = -\sum_{x \in X} p_x \log_2 p_x$.

A proof that Algorithm 6.1 terminates after a finite number of steps, and that it produces the correct result (the maximum of the Shannon entropy within the set of probability distributions that dominate a given belief function) is covered in Appendix C.

EXAMPLE 6.10. Given the frame of discernment $X = \{a, b, c, d\}$, let a belief function Bel be defined by the associated basic probability assignment m:

$$m(\{a\}) = 0.26,$$
$$m(\{b\}) = 0.26,$$
$$m(\{c\}) = 0.26,$$
$$m(\{a, b\}) = 0.07,$$
$$m(\{a, c\}) = 0.01$$
$$m(\{a, d\}) = 0.01$$
$$m(\{b, c\}) = 0.01,$$
$$m(\{b, d\}) = 0.01,$$
$$m(\{c, d\}) = 0.01,$$
$$m(\{a, b, c, d\}) = 0.1$$

(only values for focal elements are listed). By Remark (c) in Appendix C, we do not need to consider the value of Bel for $\{d\}$ and $\emptyset$. The values of $Bel(A)$ and also the values of $Bel(A)/|A|$ for all other subsets A of X are listed in Table 6.3a. We can see from this table that the highest value of $Bel(A)/|A|$ is obtained for $A = \{a, b\}$. Therefore, we set $p_a = p_b = 0.295$. Now, we have to update our (now generalized) belief function Bel. For example,

$$Bel(\{c\}) = Bel(\{a, b, c\}) - Bel(\{a, b\})$$
$$= 0.87 - 0.59 = 0.28.$$

Table 6.3. Calculation of *AU* by Algorithm 6.1 in Example 6.10: (a) First Iteration; (b) Second Iteration

A	$Bel(A)$	$\dfrac{Bel(A)}{|A|}$		A	$Bel(A)$	$\dfrac{Bel(A)}{|A|}$
$\{a\}$	0.26	0.26		$\{c\}$	0.28	0.28
$\{b\}$	0.26	0.26		$\{d\}$	0.02	0.02
$\{c\}$	0.26	0.26		$\{c, d\}$	0.41	0.205
$\{a, b\}$	0.59	0.295				
$\{a, c\}$	0.53	0.65			(b)	
$\{a, d\}$	0.27	0.135				
$\{b, c\}$	0.53	0.265				
$\{b, d\}$	0.27	0.135				
$\{c, d\}$	0.27	0.135				
$\{a, b, c\}$	0.87	0.29				
$\{a, b, d\}$	0.61	0.20$\bar{3}$				
$\{a, c, d\}$	0.55	0.18$\bar{3}$				
$\{b, c, d\}$	0.55	0.18$\bar{3}$				
$\{a, b, c, d\}$	1	0.25				

(a)

All the values are listed in Table 6.3*b*. Our new frame of discernment X is now $\{c, d\}$. Since $X \neq \emptyset$ and $Bel(X) > 0$, we repeat the process. The maximum of $Bel(A)/|A|$ is now reached at $\{c\}$. We put $p_c = 0.28$, change $Bel(\{d\})$ to 0.13, and X to $\{d\}$. In the last pass through the loop we get $p_d = 0.13$. We can conclude that

$$AU(Bel) = - \sum_{i \in \{a,b,c,d\}} p_i \log_2 p_i$$
$$= -2 \times 0.295 \log_2 0.295 - 0.28 \log_2 0.28 - 0.13 \log_2 0.13$$
$$\doteq 1.9.$$

6.6.2. Computing the Aggregated Uncertainty in Possibility Theory

Due to the nested structure of possibilistic bodies of evidence, the computation of functional *AU* can be substantially simplified. It is thus useful to reformulate Algorithm 6.1 in terms of possibility profiles for applications in possibility theory. The following is the reformulated algorithm.

Algorithm 6.2. Calculating *AU* from a given possibility profile.
Input. $n \in \mathbb{N}, \mathbf{r} = \langle r_1, r_2, \dots, r_n \rangle$.
Output. $AU(Pos), \langle p_i | i \in \mathbb{N}_n \rangle$ such that $AU(Pos) = -\Sigma_{i=1}^n p_i \log_2 p_i$, with $p_i \geq 0$ for $i \in \mathbb{N}_n, \Sigma_{i=1}^n p_i = 1$, and $\Sigma_{x \in A} p_x \leq Pos(A)$ for all $\emptyset \neq A \subset X$.

Step 1. Let $j = 1$ and $r_{n+1} = 0$.
Step 2. Find maximal $i \in \{j, j + 1, \dots, n\}$ such that $(r_j - r_{i+1})/(i + 1 - j)$ is maximal.
Step 3. For $k \in \{j, j + 1, \dots, i\}$, put $p_k = (r_j - r_{i+1})/(i + 1 - j)$.

Step 4. Put $j = i + 1$.

Step 5. If $i < n$, then go to Step 2.

Step 6. Calculate $AU(Pos) = -\sum_{i=1}^{n} p_i \log_2 p_i$.

As already mentioned, it is sufficient to consider only all unions of focal elements of a given belief function, in this case a necessity measure. Since all focal elements of a necessity measure are nested, the union of a set of focal elements is the largest focal element (in the sense of inclusion) in the set. Therefore, we have to examine the values of $Nec(A)/|A|$ only for A being a focal element.

We show by induction on the number of passes through the loop of Steps 1–5 of Algorithm 6.1 that the following properties hold in a given pass:

(a) The "current" frame of discernment X is the set $\{x_j, x_{j+1}, \ldots, x_n\}$, where the value of j is taken in the corresponding pass through the loop of Steps 2–5 of Algorithm 6.2.

(b) All focal elements are of the form $A_i = \{x_j, x_{j+1}, \ldots, x_i\}$ for some $i \in \{j, j+1, \ldots, n\}$, where j has the same meaning as in (a).

(c) $Nec(A_i) = r_j - r_{i+1}$, where j is again as described in (a).

This implies that Algorithm 6.2 is a correct modification of Algorithm 6.1 for the case of possibility theory.

In the first pass, $j = 1$, $X = \{x_1, x_2, \ldots, x_n\}$; (b) holds due to our ordering convention regarding possibility profiles. Since $r_1 = 1$, we have

$$Nec(A_i) = 1 - Pos(\overline{A_i})$$
$$= 1 - \max_{k=i+1}^{n}\{r_j\}$$
$$= 1 - r_{i+1}.$$

So (c) is true.

Let us now assume that (a)–(c) were true in some fixed pass. We want to show that (a)–(c) hold in the next pass. Let l denote the value of i maximizing $Nec(A_i)/|A_i| = Nec(A_j)/(i + 1 - j)$. Now X becomes $X - A_l = \{x_{l+1}, x_{l+2}, \ldots, x_n\}$. Therefore (a) holds, since $j = l + 1$ and $Nec(A_i)$ becomes

$$Nec(A_i) = Nec(A_l \cup A_i) - Nec(A_l)$$
$$= \left[1 - Pos(\overline{A_l - A_i})\right] - \left[1 - Pos(\overline{A_l})\right]$$
$$= \left[1 - \max_{k=i+1}^{n}\{r_j\}\right] - \left[1 - \max_{k=l+1}^{n}\{r_j\}\right]$$
$$= r_{l+1} - r_{i+1}$$
$$= r_j - r_{i+1}$$

for $i \in \{j, j+1, \ldots, n\}$. This implies that (b) and (c) hold.

Table 6.4. Illustration of the Use of Algorithm 6.2 in Example 6.11

	$\dfrac{r_j - r_i + 1}{i + 1 - j}$	
Pass	1	2
i/j	1	4
1	0.1	
2	0.075	
3	0.1$\bar{6}$	
4	0.125	0
5	0.13	0.075
6	0.11$\bar{6}$	0.0$\bar{6}$
7	0.1286	0.1
8	0.125	0.1

EXAMPLE 6.11. Consider $X = \{1, 2, \ldots, 8\}$ and the possibility profile

$$\mathbf{p} = \langle 1, 0.9, 0.85, 0.5, 0.5, 0.35, 0.3, 0.1 \rangle.$$

The relevant values of $(r_j - r_{i+1})/(i + 1 - j)$ are listed in Table 6.4.

We can see there that in the first pass the maximum is reached at $i = 3$, and $p_1 = p_2 = p_3 = \frac{1}{3}$. In the second pass, the maximum is reached at both $i = 7$ and $i = 8$. We take the bigger one and put $p_4 = p_5 = p_6 = p_7 = p_8 = 0.1$. We finish by computing

$$AU(Pos) = -\sum_{i=1}^{8} p_i \log_2 p_i = \frac{3}{6} \log_2 6 + \frac{5}{10} \log_2 10$$
$$= 2.95.$$

6.7. AGGREGATE UNCERTAINTY FOR CONVEX SETS OF PROBABILITY DISTRIBUTIONS

Once the aggregated measure of uncertainty AU has been established in DST, it is fairly easy to generalize it for convex sets of probability distributions. Let this generalized version of AU be denoted by $\bar{S}$ to emphasize its definition in terms of the maximum Shannon entropy. Clearly, $\bar{S}$ is a functional defined on the family of convex sets of probability distribution functions. Given a convex set $\mathcal{D}$ of probability distribution functions p, $\bar{S}$ is defined by the formula

$$\bar{S}(\mathcal{D}) = \max_{p \in \mathcal{D}} \left\{ -\sum_{x \in X} p(x) \log_2 p(x) \right\}. \tag{6.61}$$

It is essential to show that $\bar{S}$ satisfies the following requirements, which are appropriate generalizations of their counterparts for AU:

($\bar{S}$1) **Probability Consistency.** When $\mathcal{D}$ contains only one probability distribution, $\bar{S}$ assumes the form of the Shannon entropy.

($\bar{S}$2) **Set Consistency.** When $\mathcal{D}$ consists of the set of all possible probability distributions on $A \subseteq X$, then

$$\bar{S}(\mathcal{D}) = \log_2 |A|.$$

($\bar{S}$3) **Range.** The range of $\bar{S}$ is $[0, \log_2|X|]$ provided that uncertainty is measured in bits.

($\bar{S}$4) **Subadditivity.** If $\mathcal{D}$ is an arbitrary convex set of probability distributions on $X \times Y$ and $\mathcal{D}_X$ and $\mathcal{D}_Y$ are the associated sets of marginal probability distributions on X and Y, respectively, then

$$\bar{S}(\mathcal{D}) \leq \bar{S}(\mathcal{D}_X) + \bar{S}(\mathcal{D}_Y).$$

($\bar{S}$5) **Additivity.** If $\mathcal{D}$ is the set of joint probability distributions on $X \times Y$ that is associated with independent marginal sets of probability distributions, $\mathcal{D}_X$ and $\mathcal{D}_Y$, which means that $\mathcal{D}$ is the convex hull of the set

$$\mathcal{D}_X \otimes \mathcal{D}_Y = \{p(x, y) = p_X(x) \cdot p_Y(y) \mid x \in X, y \in Y, p_X \in \mathcal{D}_X, P_Y \in \mathcal{D}_Y\},$$

then $\bar{S}(\mathcal{D}) = \bar{S}(\mathcal{D}_X) + \bar{S}(\mathcal{D}_Y)$.

It is obvious that the functional $\bar{S}$ defined by Eq. (6.61) satisfies requirements ($\bar{S}$1)–($\bar{S}$3). The remaining two properties, subadditivity and additivity, are addressed by the following two theorems.

Theorem 6.10. Functional $\bar{S}$ defined by Eq. (6.61) satisfies the requirement ($\bar{S}$4), that is, it is subadditive.

Proof. Given a convex set of joint probability distributions on $X \times Y$, let $\dot{\mathbf{p}}$ denote the joint probability distribution in $\mathcal{D}$ for which the maximum in Eq. (6.61) is obtained and let $\dot{\mathbf{p}}_X$ and $\dot{\mathbf{p}}_Y$ denote the marginal probability distributions of $\dot{\mathbf{p}}$. Then, by Gibbs' inequality (Eq. 3.28),

$$
\begin{aligned}
\bar{S}(\dot{\mathbf{p}}) &= -\sum_{y \in Y} \sum_{x \in X} \dot{p}(x, y) \log_2 \dot{p}(x, y) \\
&\leq -\sum_{y \in Y} \sum_{x \in X} \dot{p}(x, y) \log_2 [\dot{p}_X(x) \cdot \dot{p}_Y(y)] \\
&= -\sum_{y \in Y} \sum_{x \in X} \dot{p}(x, y) \log_2 \dot{p}_X(x) - \sum_{y \in Y} \sum_{x \in X} \dot{p}(x, y) \log_2 \dot{p}_Y(y) \\
&= -\sum_{x \in X} \dot{p}_X(x) \log_2 \dot{p}_X(x) - \sum_{y \in Y} \dot{p}_Y(y) \log_2 \dot{p}_Y(y) \\
&\leq \bar{S}(\dot{\mathbf{p}}_X) + \bar{S}(\dot{\mathbf{p}}_Y). \qquad \blacksquare
\end{aligned}
$$

Theorem 6.11. Functional $\bar{S}$ defined by Eq. (6.61) satisfies the requirement (S5), that is, it is additive.

Proof. From Theorem 6.10

$$\bar{S}(\mathcal{D}) \leq \bar{S}(\mathcal{D}_X) + \bar{S}(\mathcal{D}_Y).$$

It is thus sufficient to show that

$$\bar{S}(\mathcal{D}) \geq \bar{S}(\mathcal{D}_X) + \bar{S}(\mathcal{D}_Y)$$

under the assumption of independence. Let $\mathring{p}_X$ and $\mathring{p}_Y$ denote the marginal probability distributions on X and Y for which the maxima in Eq. (6.61) are obtained. Then

$$\bar{S}(\mathcal{D}_X) + \bar{S}(\mathcal{D}_Y) = -\sum_{x \in X} \mathring{p}_X(x) \log_2 \mathring{p}_X(x) - \sum_{y \in Y} \mathring{p}_Y(y) \log_2 \mathring{p}_Y(y)$$

$$= -\sum_{y \in Y} \sum_{x \in X} \mathring{p}_X(x) \cdot \mathring{p}_Y(y) \log_2 [\mathring{p}_X(x) \cdot \mathring{p}_y(y)]$$

$$\leq \bar{S}(\mathcal{D}). \qquad \blacksquare$$

EXAMPLE 6.12. Consider the following convex sets of marginal probability distributions on $X = \{x_1, x_2\}$ and $Y = \{y_1, y_2\}$:

$$\mathcal{D}_X = \{\langle p_X(x_1) = 0.2 + 0.2\lambda_X, \, p_X(x_2) = 0.8 - 0.2\lambda_X \rangle | \lambda_X \in [0, 1]\},$$

$$\mathcal{D}_Y = \{\langle p_Y(y_1) = 0.1 + 0.1\lambda_X, \, p_Y(y_2) = 0.9 - 0.1\lambda_Y \rangle | \lambda_Y \in [0, 1]\}.$$

The maximum entropy within $\mathcal{D}_X$ is obtained for $\lambda_X = 1$ (i.e., for $p_X(x_1) = 0.4$ and $p_X(x_2) = 0.6$). The maximum entropy within $\mathcal{D}_Y$ is obtained for $\lambda_Y = 1$ (i.e., for $p_Y(y_1) = 0.2$ and $p_Y(y_2) = 0.8$). Hence,

$$\bar{S}(\mathcal{D}_X) = S(0.4, 0.6) = 0.971,$$

$$\bar{S}(\mathcal{D}_Y) = S(0.2, 0.8) = 0.722,$$

$$\bar{S}(\mathcal{D}_X) + \bar{S}(\mathcal{D}_Y) = 1.693.$$

Assuming the independence of the marginal probabilities, the set $\mathcal{D}$ of the associated joint probabilities on $Z = X \times Y$ is the convex hull of the set

$$\mathcal{D}_X \otimes \mathcal{D}_Y = \{\langle p(z_{11}) = (0.2 + 0.2\lambda_X)(0.1 + 0.1\lambda_Y),$$
$$p(z_{12}) = (0.2 + 0.2\lambda_X)(0.9 - 0.1\lambda_Y),$$
$$p(z_{21}) = (0.8 - 0.2\lambda_X)(0.1 + 0.1\lambda_Y),$$
$$p(z_{22}) = (0.8 - 0.2\lambda_X)(0.9 - 0.1\lambda_Y)\rangle | \lambda_X, \lambda_Y \in [0, 1]\},$$

where $z_{i,j} = \langle x_i, y_j \rangle$ for all $i, j = 1, 2$. The maximum entropy within $\mathcal{D}_X \otimes \mathcal{D}_Y$ is obtained for $\lambda_X = \lambda_Y = 1$ and, hence,

$$\bar{S}(\mathcal{D}) = \bar{S}(\mathcal{D}_X) + \bar{S}(\mathcal{D}_Y) = 1.693.$$

Functional $\bar{S}$ is defined in this section for convex sets of probability distributions. However, it can be as well defined for the lower and upper probability functions or the Möbius representations obtained uniquely from convex sets of probability distributions. We can thus write not only $\bar{S}(\mathcal{D})$, but also $\bar{S}(\underline{\mu})$, $\bar{S}(\bar{\mu})$, or $\bar{S}(m)$.

Algorithm 6.1 (introduced in Section 6.6.1) is formulated and proved correct for belief functions. It is not necessarily applicable to lower probabilities that are more general or incomparable with belief functions. A lower probability for which the algorithm does not work is examined in the following example.

EXAMPLE 6.13. Consider the following convex set $\mathcal{D}$ of probability distributions on $Z = \{z_{11}, z_{12}, z_{21}, z_{22}\}$ defined in Example 6.7. The lower probability function, $\mu_{\mathcal{D}}$, associated with this set, which is derived by Eq. (4.11), is shown in Table 6.2. Recall that $\mu_{\mathcal{D}}$ is not 2-monotone in this example.

Applying Algorithm 6.1 results in probability distribution

$$p(z_{11}) = p(z_{12}) = 0.4, \qquad p(z_{21}) = p(z_{22}) = 0.1,$$

for which $S(0.4, 0.4, 0.1, 0.1) = 1.722$. However, this probability distribution is not a member of the given set of probability distributions. The plot of the Shannon entropy for $\lambda \in [0, 1]$ is shown in Figure 6.8. Its maximum is $= 1.69288$. It is obtained for $\mathbf{p} = \langle 0.48, 0.32, 0.12, 0.08 \rangle$. Algorithm 6.1 thus does not produce the correct value of $\bar{S}(\underline{\mu})$ in this example.

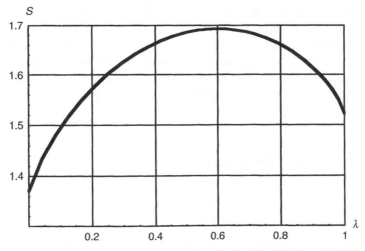

Figure 6.8. Values of the Shannon entropy for $\lambda \in [0, 1]$ in Example 6.13.

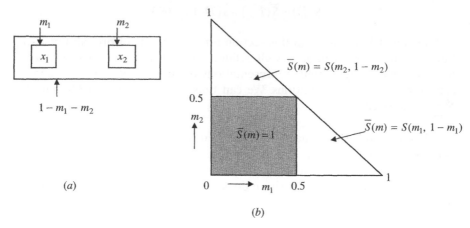

Figure 6.9. Illustration of the severe insensitivity of the aggregate uncertainty measure (Example 6.14).

6.8. DISAGGREGATED TOTAL UNCERTAINTY

Functional $\bar{S}$ defined by Eq. (6.60) is certainly well justified on mathematical grounds as an aggregate measure of uncertainty in the theory based on arbitrary convex sets of probability distributions and, consequently, to the various special theories of uncertainty as well. However, it has a severe practical shortcoming: it is highly insensitive to changes in evidence. To illustrate this undesirable feature of $\bar{S}$, let us examine the following example.

EXAMPLE 6.14. Consider the following class of simple bodies of evidence in DST: $X = \{x_1, x_2\}$, $m(\{x_1\}) = m_1$, $m(\{x_2\}) = m_1$, $m(X) = 1 - m_1 - m_2$, where m_1, $m_2 \in [0, 1]$ and $m_1 + m_2 \leq 1$ (Figure 6.9a). Clearly, $Bel(\{x_1\}) = m_1$, $Bel(\{x_2\}) = m_2$, and $Bel(X) = 1$. Then, $\bar{S}(Bel) = 1$ (or $AU(Bel) = 1$ when using the special notation in DST) for all $m_1 \in [0, 0.5]$ and $m_2 \in [0, 0.5]$. Moreover, when $m_1 > 0.5$, $\bar{S}(Bel) = S(m_1, 1 - m_1)$, where S denotes the Shannon entropy. Hence, $\bar{S}(Bel)$ is independent of m_2. Similarly, when $m_2 > 0.5$, $\bar{S}(Bel) = S(m_2, 1 - m_2)$ and, hence, $\bar{S}(Bel)$ is independent of m_1. These values of $\bar{S}(Bel)$ are shown in Figure 6.9b. The functional $\bar{S}$ in this example can also be expressed by the formula

$$\bar{S}(Bel) = S(\max\{m_1, m_2, 0.5\}, 1 - \max\{m_1, m_2, 0.5\}) \qquad (6.62)$$

for all $m_1, m_2 \in [0, 1]$ such that $m_1 + m_2 \leq 1$, where S denotes the Shannon entropy.

This example surely illustrates that the insensitivity of the functional $\bar{S}$ to changes in evidence is very severe. This feature makes the functional ill-suited for measuring uncertainty on conceptual and pragmatic grounds. However, recognizing that $\bar{S}$ is an aggregate of the two types of uncertainty—nonspecificity and conflict—it must be that

$$\bar{S}(\underline{\mu}) = GH(\underline{\mu}) + GS(\underline{\mu}), \tag{6.63}$$

where $\underline{\mu}$ is a lower probability function, which can be derived from any of the other representations of the same uncertainty (the associated convex set of probability distributions, Möbius representation, or upper probability function) or may be converted to any of these representations as needed. In Eq. (6.63) GH is the well-justified functional for measuring nonspecificity (defined at the most general level by Eq. (6.38) and referred to as the generalized Hartley measure) and GS is an unknown functional for measuring conflict (referred to as the generalized Shannon entropy).

Since functionals $\bar{S}$ and GH in Eq. (6.63) are well justified and GS is an unknown functional, it is suggestive to define GS from the equation as

$$GS(\underline{\mu}) = \bar{S}(\underline{\mu}) - GH(\underline{\mu}). \tag{6.64}$$

Here, GS is defined indirectly, in terms of two well-justified functionals, thus overcoming the unsuccessful attempts to define it directly (as discussed in Section 6.5).

Functional $\bar{S}$, which is well justified but practically useless due to its insensitivity, can now be disaggregated into two components, GH and GS, which measure two types of uncertainty that coexist in all uncertainty theories except the classical ones. A disaggregated total uncertainty, TU, is thus defined as the pair

$$TU = \langle GH, GS \rangle, \tag{6.65}$$

where GH and GS are defined by Eqs. (6.38) and (6.64), respectively. Since the sum of the two components of TU is $\bar{S}$, TU is as well justified as $\bar{S}$. One advantage of the disaggregated total uncertainty, TU, in comparison with its aggregated counterpart $\bar{S}$, is that it expresses amounts of both types of uncertainty (nonspecificity and conflict) explicitly, and consequently, it is highly sensitive to changes in evidence.

Another advantage of TU is that its components, GH and GS, need not satisfy all the mathematical requirements for measures of uncertainty. It only matters that their aggregate measure, $\bar{S}$, satisfies them. The lack of subadditivity of GH for arbitrary convex sets of probability distributions, established in Example 6.7, is thus of no consequence when GH is employed as one component in TU.

EXAMPLE 6.15. To appreciate the difference between $\bar{S}$ and TU, let us consider the following three bodies of evidence on X and let $|X| = n$ for convenience:

(a) In the case of total ignorance (when $m(X) = 1$), $\bar{S}(m) = \log_2 n$ and $TU(m)$ $= \langle \log_2 n, 0 \rangle$.

(b) When evidence is expressed by the uniform probability distribution on X (i.e., $m(\{x\}) = 1/n$ for all $x \in X$), then again $\bar{S}(m) = \log_2 n$, but $TU(m) = \langle 0, \log_2 n \rangle$.

(c) When evidence is expressed by $m(\{x\}) = a$ for all $x \in X$, where $a < 1/n$, and $m(X) = 1 - na$, then again $\bar{S}(m) = \log_2 n$, while $TU(m) = \langle (1 - na)\log_2 n, na\log_2 n \rangle$.

EXAMPLE 6.16. To illustrate the disaggregated total uncertainty $TU = \langle GH, GS \rangle$, let us consider again the class of simple bodies of evidence introduced in Example 6.14. Plots of the dependences of $\bar{S}$, GH, and GS on m_1 and m_2 are shown in the first column of Figure 6.10. We can see that both components of TU (GH and GS) are sensitive to changes of evidence. Plots of $\bar{S}$, GH, and GS for the extreme cases when either $m_2 = 0$ or $m_1 = 0$ (nested or possibilistic bodies of evidence) are shown in the second column of Figure 6.10. It is easy to determine that the maximum of GS is 0.585, and it is attained in these cases for $m_1 = 2/3$ and $m_2 = 0$ (or $m_1 = 0$ and $m_2 = 2/3$).

The plots in Figure 6.10 are based on the lower probability function, when $m_1 + m_2 \leq 1$. Similar plots can be made for the upper probability function, when $m_1 + m_2 \geq 1$. In Figure 6.11, values of $\bar{S}(m_1, m_2)$, $GH(m_1, m_2)$, $GS(m_1, m_2)$ are shown for all values of $m_1, m_2 \in [0, 1]$. Clearly,

$$\bar{S}(m_1, m_2) = \begin{cases} S(\max\{m_1, m_2, 0.5\}, \\ \quad 1 - \max\{m_1, m_2, 0.5\}) & \text{when } m_1 + m_2 \leq 1 \\ S(\max\{1 - m_1, 1 - m_2, 0.5\}, \\ \quad 1 - \max\{1 - m_1, 1 - m_2, 0.5\}) & \text{when } m_1 + m_2 \geq 1, \end{cases}$$

where S denotes the Shannon entropy, and

$$GH(m_1, m_2) = \begin{cases} 1 - m_1 - m_2 & \text{when } m_1 + m_2 \leq 1 \\ m_1 + m_2 - 1 & \text{when } m_1 + m_2 \geq 1 \end{cases}$$
$$GS(m_1, m_2) = \bar{S}(m_1, m_2) - GH(m_1, m_2),$$

To fully justify the disaggregated total uncertainty TU defined by Eq. (6.65), it remains to prove that its second component, the generalized Shannon entropy GS defined by Eq. (6.64) is always nonnegative. That is, we need to prove that the inequality

$$\bar{S}(\underline{\mu}) - GH(\underline{\mu}) \geq 0$$

holds for any lower probability function $\underline{\mu}$ (or any of its equivalent representations). This proof is presented for belief functions in Appendix D. A proof for lower probability functions outside DST is still needed.

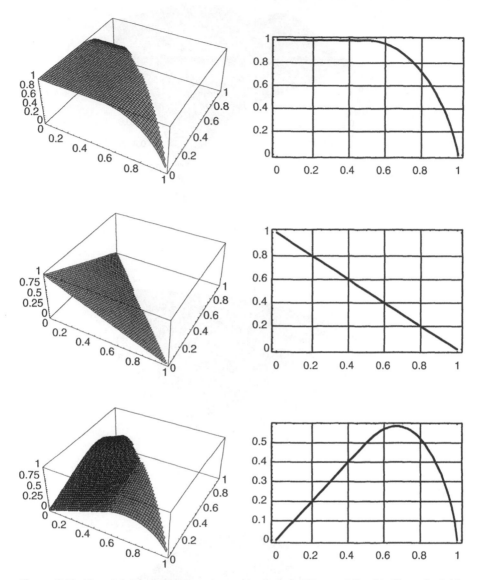

Figure 6.10. Uncertainties for the two-element bodies of evidence defined in Example 6.14. First column: $\bar{S}$, GH, and GS for all m_1 and m_2 such that $m_1 + m_2 \leq 1$; second column: $\bar{S}$, GH, and GS for either $m_1 = 0$ or $m_2 = 0$.

6.9. GENERALIZED SHANNON ENTROPY

The generalized Shannon entropy defined by Eq. (6.64) emerged fairly recently and has not been sufficiently investigated as yet. Nevertheless, some of its properties in Dempster–Shafer theory are derived in this section on the basis of Algorithm 6.1.

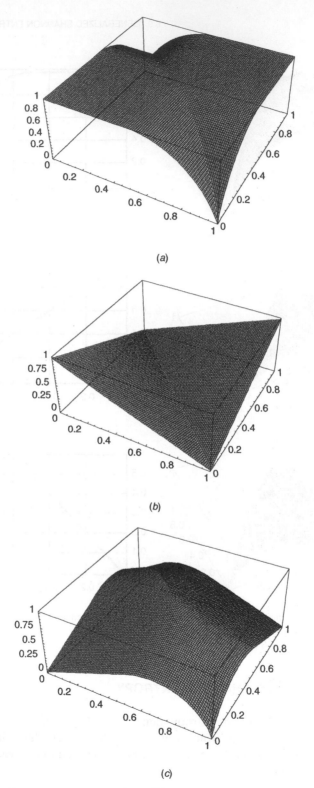

Figure 6.11. For all m_1, $m_2 \in [0, 1]$: (a) $\bar{S}(m_1, m_2)$; (b) $GH(m_1, m_2)$; (c) $GS(m_1, m_2)$.

To facilitate the following derivations, let

$$\mathcal{F} = \{A_i | A_i \in \mathcal{P}(X), i \in \mathbb{N}_q\}$$

denote the family of all focal sets of a given body of evidence in Dempster–Shafer theory. Furthermore, let

$$\mathcal{E} = \{B_k | B_k \in \mathcal{P}(X), k \in \mathbb{N}_r\}$$

denote the partition of $\cup_{k \in \mathbb{N}_q} A_i$ that is produced by Algorithm 6.1. Clearly, $r \leq q$ and

$$\left| \bigcup_{k \in \mathbb{N}_q} A_i \right| = \sum_{k \in \mathbb{N}_r} |B_k|.$$

For convenience, assume that block B_k is produced by the kth iteration of the algorithm, and let $Pro(B_k)$ denote the probability of B_k. Clearly,

$$\sum_{k \in \mathbb{N}_r} Pro(B_k) = 1. \tag{6.66}$$

According to Algorithm 6.1,

$$\overline{S}(Bel) = -\sum_{k \in \mathbb{N}_r} Pro(B_k) \log_2 [Pro(B_k)/|B_k|],$$

where Bel denote the belief measure associated with the given body of evidence. This equation can be rewritten as

$$\overline{S}(Bel) = -\sum_{k \in \mathbb{N}_r} Pro(B_k) \log_2 Pro(B_k) + \sum_{k \in \mathbb{N}_r} Pro(B_k) \log_2 |B_k|.$$

Due to Eq. (6.66), the first term on the right-hand side of this equation is the Shannon entropy of $\mathcal{E}$ and the second term is the generalized Hartley measure of $\mathcal{E}$. We can thus rewrite this equation as

$$\overline{S}(Bel) = S(Pro(B_k) | k \in \mathbb{N}_r) + GH(Pro(B_k) | k \in \mathbb{N}_r).$$

Using now Eq. (6.64), we obtain

$$GS(Bel) = S(Pro(B_k) | k \in \mathbb{N}_r) + GH(Pro(B_k) | k \in \mathbb{N}_r) - GH(m(A_i) | i \in \mathbb{N}_q). \tag{6.67}$$

This equation indicates that the generalized Shannon entropy is expressed by the Shannon entropy of $\mathcal{E}$ and the difference between the nonspecificity of $\mathcal{E}$ and the nonspecificity of the given body of evidence $\langle \mathcal{F}, m \rangle$.

In the rest of this section, some properties of the generalized Shannon entropy are derived for two special types of bodies of evidence, those in which focal elements are either disjoint or nested.

Theorem 6.12. Consider a given body of evidence $\langle \mathcal{F}, m \rangle$ in which $\mathcal{F}$ is a family of pairwise sets. Then,

$$GS(Bel) = -\sum_{i=1}^{q} m(A_i)\log_2 m(A_i),$$

where *Bel* denotes the belief measure based on *m*.

Proof. For convenience, let $m_i = m(A_i)$ and $a_i = |A_i|$. Assume (without any loss of generality) that

$$\frac{m_1}{a_1} \geq \frac{m_2}{a_2} \geq \cdots \geq \frac{m_q}{a_q}.$$

Then, $m_1 a_2 \geq m_2 a_1$. Now comparing m_1/a_1 with $(m_1 + m_2)/(a_1 + a_2)$, we obtain

$$\frac{m_1}{a_1} - \frac{m_1 + m_2}{a_1 + a_2} = \frac{m_1 a_2 - m_2 a_1}{a_1(a_1 + a_2)} \geq 0.$$

It is obvious that the same result is obtained when m_1/a_1 is compared with

$$\left(m_1 + \sum_{i \in I} m_i \right) \Big/ \left(a_1 + \sum_{i \in I} a_i \right)$$

for any $I \subseteq \{2, 3, \ldots, q\}$. Hence, Algorithm 6.1 assigns probabilities m_1/a_1 to all elements in A_i in the first iteration (observe that $Bel(A_i) = m(A_i)$ for all $i \in \mathbb{N}_q$ when focal elements are disjoint). By repeating the same argument for m_2/a_2, we can readily show that Algorithm 6.1 assigns probabilities m_2/a_2 to all elements of A_2 in the second iteration, and so forth. Hence,

$$\begin{aligned}
\bar{S}(Bel) &= -\sum_{i=1}^{q} a_i \left(\frac{m_i}{a_i} \right) \log_2 \left(\frac{m_i}{a_i} \right) \\
&= -\sum_{i=1}^{q} m_i \log_2 m_i + \sum_{i=1}^{q} m_i \log_2 a_i \\
&= -\sum_{i=1}^{q} m_i \log_2 m_i + GH(Bel).
\end{aligned}$$

Now,

$$GS(Bel) = \bar{S}(Bel) - GH(Bel)$$

$$= -\sum_{i=1}^{q} m_i \log_2 m_i.$$ ■

Now let us consider some properties of the generalized Shannon entropy for nested bodies of evidence. Let $X = \{x_i \mid i \in \mathbb{N}_n\}$ and assume that the elements of X are ordered in such a way that the family

$$\mathcal{A} = \{A_i = \{x_1, x_2, \ldots, x_i\} \mid i \in \mathbb{N}_n\}$$

contains all focal sets. According to this special notation (introduced in Section 5.2.1), $\mathcal{F} \subseteq \mathcal{A}$. For convenience, let $m_i = m(A_i)$ for all $i \in \mathbb{N}_n$.

Given a particular nested body of evidence $\langle \mathcal{F}, m \rangle$,

$$GS(m) = \bar{S}(m) - GH(m).$$

While $GH(m)$ in this equation is expressed by the simple formula

$$GH(m) = \sum_{i=2}^{n} m_i \log_2 i, \tag{6.68}$$

the expression of $\bar{S}(m)$ is not obvious at all. To investigate some properties of GS for nested bodies of evidence, let the following three cases be distinguished:

(a) $m_i \geq m_{i+1}$ for all $i \in \mathbb{N}_{n-1}$;
(b) $m_i \leq m_{i+1}$ for all $i \in \mathbb{N}_{n-1}$;
(c) Neither (a) nor (b).

Following the algorithm for computing $\bar{S}$, we obtain the formula

$$GS_a(m) = -\sum_{i=1}^{n} m_i \log_2 (m_i i) \tag{6.69}$$

for any function m that conforms to Case (a). By applying the method of Lagrange multipliers (Appendix E), we can readily find out that the maximum, $GS_a^*(n)$, of this functional for each $n \in \mathbb{N}$ is obtained for

$$m_i = (1/i) 2^{-(\alpha + 1/\ln 2)} (i \in \mathbb{N}_n),$$

where the value of α is determined by solving the equation

$$2^{-(\alpha + 1/\ln 2)} \sum_{i=1}^{n} (1/i) = 1.$$

Let $s_n = \sum\limits_{i=1}^{n}(1/i)$. Then,

$$\alpha = -\log_2(1/s_n) - (1/\ln 2)$$

and, hence,

$$m_i = (1/i)2^{\log_2(1/s_n)}$$
$$= \frac{1}{s_n i}.$$

Substituting this expression for m_i in Eq. (6.69), we obtain

$$GS_a^*(n) = \log_2 s_n. \tag{6.70}$$

The maximum, $GS_b^*(n)$, of this functional for some $n \in \mathbb{N}$, subject to the inequalities that are assumed in Case (b), is obtained for $m_i = 1/n$ for all $i \in \mathbb{N}_n$. Hence,

$$GS_b^*(n) = \log_2 \frac{n}{n!^{1/n}}. \tag{6.71}$$

For large n, $n!$ in this equation can be approximated by the Stirling formula

$$n! \approx (2\pi)^{1/2} n^{n+1/2} e^{-n}.$$

Then,

$$(n!)^{1/n} \approx (2\pi)^{1/2n} n^{1+(1/2n)} e^{-1}$$

and

$$\lim_{n\to\infty} \frac{n}{n!^{1/n}} = \lim_{n\to\infty} \frac{ne}{(2\pi)^{1/2n} n^{1+(1/2n)}} = e.$$

Hence, $GS_b^*(n) = \log_2 e \approx 1.442695$. That is, GS_b^* is bounded, contrary to $GS_a^*(n)$. Moreover $GS_b^*(n) < GS_a^*(n)$ for all $n \in \mathbb{N}$. Plots of $GS_a^*(n)$ and $GS_b^*(n)$ for $n \in \mathbb{N}_{1000}$ are shown in Figure 6.12.

Case (c) is more complicated for a general analytic treatment since it covers a greater variety of bodies of evidence with respect to the computation of GS. This follows from the algorithm for computing $\bar{S}$. For each given body of evidence, the algorithm partitions the universal set in some way, and distributes the value of the lower probability in each block of the partition uniformly. For

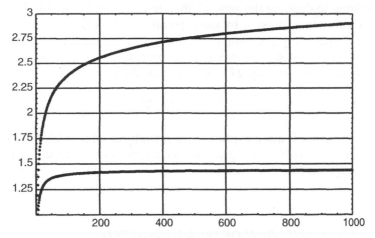

Figure 6.12. Plots of $GS_a^*(n)$ and $GS_b^*(n)$.

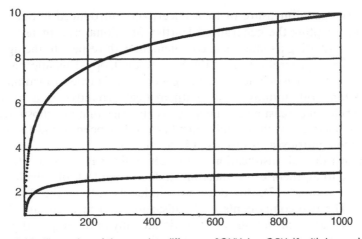

Figure 6.13. Illustration of the growing difference $[GH^*(n) - GS^*(n)]$ with increasing n.

nested bodies of evidence, the partitions preserve the induced order of elements of X. There are 2^{n-1}-order preserving partitions. The most refined partition and the least refined one are represented by Cases (a) and (b), respectively. All the remaining $2^{n-1} - 2$ partitions are represented by Case (c). A conjecture, based on a complete analysis for $n = 3$ and extensive simulation experiments for $n > 3$, is that the maxima of GS for all these partitions are for all $n \in \mathbb{N}$ smaller than the maximum $GS_a^*(n)$ for Case (a). According to this plausible conjecture, whose proof is an open problem, the difference between the maximum nonspecificity, $GH^*(n)$, and the maximum conflict, $GS_a^*(n)$, grows rapidly with n. This is illustrated by the plots of $GH^*(n)$ and $GS^*(n)$ for $n \in \mathbb{N}_{1000}$ in Figure 6.13.

In addition to defining the general Shannon entropy by Eq. (6.64), it seems also reasonable to express it by the interval $[\underline{S}(\mathcal{D}), \bar{S}(\mathcal{D})]$, where $\underline{S}(\mathcal{D})$ and $\bar{S}(\mathcal{D})$ are, respectively, the minimum and maximum values of the Shannon entropy within a given convex set $\mathcal{D}$ of probability distributions. Then, an alternative total uncertainty, TU', is defined as the pair

$$TU' = \langle GH, [\underline{S}, \bar{S}] \rangle. \tag{6.72}$$

While this measure is quite expressive, as it captures all possible values of the Shannon entropy within each given set $\mathcal{D}$, its properties and utility have yet to be investigated (see Note 6.10).

6.10. ALTERNATIVE VIEW OF DISAGGREGATED TOTAL UNCERTAINTY

The disaggregated measure of total uncertainty introduced in Section 6.8 is based on accepting the generalized Hartley functional as a measure of non-specificity and using its numerical complement with respect to the aggregated measure $\bar{S}$ as a generalization of the Shannon entropy. This approach to dis-aggregating $\bar{S}$ is reasonable since the generalized Hartley functional is well justified (on both intuitive and mathematical grounds) as a measure of nonspecificity in at least all the uncertainty theories that are subsumed under DST. No functional with a similar justification has been found to generalize the Shannon entropy, as is discussed in Section 6.5.

Although the full justification of the generalized Hartley functional does not extend to all theories of uncertainty, due to the lack of subadditivity (as shown in Example 6.7), this does not hinder its role in the disaggregated measure. The two components in the disaggregated measure are defined in such a way that whenever one of them violates any of the required properties, the other one compensates for these violations.

Recall that measure, $\bar{S}$, which is well justified in all uncertainty theories, aggregates two types of uncertainty: nonspecificity and conflict. In classical uncertainty theories, these types of uncertainty are measured by the Hartley functional and the Shannon functional, respectively. In the various general-izations of the classical theories, appropriate counterparts of these classical measures of nonspecificity and conflict are needed. This suggests looking for justifiable generalizations of the Hartley and Shannon functionals in the various nonclassical uncertainty theories. However, all attempts to generalize these functionals independently of each other have failed. This eventually led to the idea of disaggregating the aggregated measure $\bar{S}$. According to this idea, the generalization of the classical functionals should be constrained by the requirement that their sum always be equal to $\bar{S}$. One way of satisfying this requirement is to choose one of the components of the disaggregated measure and defined the other one as its numerical complement with respect to $\bar{S}$. Clearly, there are two ways to pursue this approach to the disaggregation of $\bar{S}$: the chosen component is either a measure of nonspecificity or a measure of conflict.

The disaggregated total uncertainty introduced in Section 6.8 is based on choosing the generalized Hartley functional as a measure of nonspecificity. The aim of this section is to explore the other possibility: to choose a particular measure of conflict, and to define the measure of nonspecificity as its numerical complement with respect to $\bar{S}$. Since, in this context, the chosen measure of conflict does not have to satisfy all the required properties, all the proposed generalizations of the Shannon entropy, which are discussed in Section 6.5, are in principle applicable. However, a measure that seems to capture best the generalized Shannon entropy is the functional $\underline{S}$ defined for any arbitrary closed and convex set $\mathcal{D}$ of probability distributions by the formula

$$\underline{S}(\mathcal{D}) = \min_{p \in \mathcal{D}}\{-\sum p(x)\log_2 p(x)\}. \tag{6.73}$$

This measure has been neglected due to its massive violations of the subadditivity requirement. However, this deficiency does not matter when $\underline{S}$ is used as one component of the disaggregated measure of total uncertainty.

Functional $\underline{S}$ is intuitively a good choice for a generalized Shannon entropy, since it measures (by the Shannon entropy itself) the essential (irreducible) amount of conflict embedded in any given credal set $\mathcal{D}$. When it is accepted as a measure of conflict, then the measure of nonspecificity, N, is defined for any given credal set $\mathcal{D}$ by the formula

$$N(\mathcal{D}) = \bar{S}(\mathcal{D}) - \underline{S}(\mathcal{D}). \tag{6.74}$$

This means that, according to this view of disaggregating $\bar{S}$, the measure of nonspecificity is not a generalized Hartley measure anymore.

Functionals $\underline{S}$ and N have been proved to possess the following properties (Note 6.10):

1. The range of both $\underline{S}$ and N is $[0, \log_2|X|]$ provided that the units of measurement are bits. The lower and upper bounds are reached under the following conditions:
   - $\underline{S}(\mathcal{D}) = 0$ iff $\sup_{p \in \mathcal{D}}\{p(x)\} = 1$ for some $x \in X$.
   - $\underline{S}(\mathcal{D}) = \log_2|X|$ iff $\mathcal{D}$ contains only the uniform probability distribution on X.
   - $N(\mathcal{D}) = 0$ iff $|\mathcal{D}| = 1$.
   - $N(\mathcal{D}) = \log_2|X|$ iff $\sup_{p \in \mathcal{D}}\{p(x)\} = 1$ for some $x \in X$ and $\mathcal{D}$ contains the uniform probability distribution on X.

2. $\underline{S}$ is monotone decreasing and N is monotone increasing with respect to subsethood relation between credal sets if $\mathcal{D} \subseteq \mathcal{D}'$, then $\underline{S}(\mathcal{D}) \geq \underline{S}(\mathcal{D}')$ and $N(\mathcal{D}) \leq N(\mathcal{D}')$.
3. Both $\underline{S}$ and N are additive.
4. Both $\underline{S}$ and N are continuous.

Taking into account all these properties, it makes sense to define for any given credal set $\mathcal{D}$ an alternative measure of disaggregated total uncertainty, ^aTU, by the formula

$$^aTU(\mathcal{D}) = \langle \overline{S}(\mathcal{D}) - \underline{S}(\mathcal{D}), \underline{S}(\mathcal{D}) \rangle. \tag{6.75}$$

In Figure 6.14, plots of the functionals involved in ^aTU (i.e., $\overline{S}$, $N = \overline{S} - \underline{S}$, and $\underline{S}$) are shown for the class of simple bodies of evidence introduced in Example 6.14. These are counterparts of the plots shown in Figure 6.11 of the functionals involved in TU. In this case,

$$\overline{S}(m) = \begin{cases} S(a, 1-a) & \text{when } m_1 + m_2 \leq 1 \\ S(b, 1-b) & \text{when } m_1 + m_2 \geq 1, \end{cases}$$

where $a = \max(m_1, m_2, 0.5)$ and $b = \max(1 - m_1, 1 - m_2, 0.5)$,

$$\underline{S}(m) = \begin{cases} S(c, 1-c) & \text{when } m_1 + m_2 \leq 1 \\ S(d, 1-d) & \text{when } m_1 + m_2 \geq 1, \end{cases}$$

where $c = \min(m_1, m_2)$ and $d = \min(1 - m_1, 1 - m_2)$, and $N(m) = \overline{S}(m) - \underline{S}(m)$; S denotes the Shannon entropy.

The practical utility of the alternative measure ^aTU of disaggregated total uncertainty is contingent on an efficient algorithm for computing $\underline{S}$. While it has been established that, due to the concavity of the Shannon entropy, $\underline{S}(\mathcal{D})$ is always obtained at some extreme point of $\mathcal{D}$, the computational complexity of searching through the extreme points is still very high. Since the upper count of the number of extreme point is $|X|!$, some ways of avoiding an exhaustive search is essential. Some results in this direction have been obtained already (Note 6.10).

EXAMPLE 6.17. Consider the credal set $\mathcal{D}$ defined in Example 6.7. Clearly, $\underline{S}(\mathcal{D})$ is obtained for the extreme point $\mathbf{p}_B$, as can be seen from the plot of $S(\lambda \mathbf{p}_A + (1 - \lambda)\mathbf{p}_B)$ in Figure 6.8, where $\mathbf{p}_A = \langle 0.4, 0.4, 0.2, 0 \rangle$ and $\mathbf{p}_B = \langle 0.6, 0.2, 0, 0.2 \rangle$. We have $\underline{S}(\mathcal{D}) = S(\mathbf{p}_B) = 1.371$. Since $\overline{S}(\mathcal{D}) = 1.693$ in this case (see Example 6.13), we have,

$$^aTU = \langle 0.21, 1.371 \rangle.$$

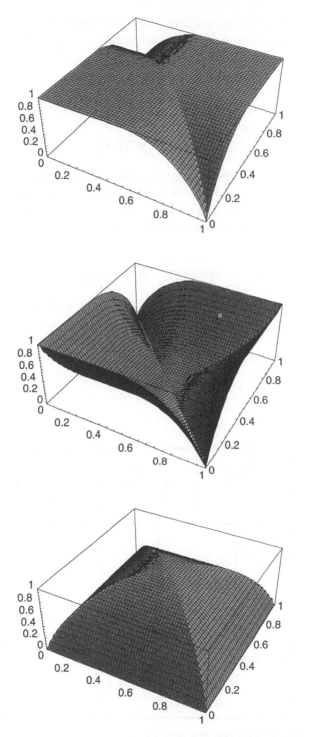

Figure 6.14. Counterparts of plots in Figure 6.11 for ^aTU defined by Eq. (6.75).

EXAMPLE 6.18. Consider the credal set $^{\mu_2}\mathcal{D}$ in Example 4.5, which is defined by convex combinations of four extreme points:

$$\mathbf{p}_1 = \langle 0.2, 0.3, 0.5 \rangle,$$
$$\mathbf{p}_2 = \langle 0.4, 0.1, 0.5 \rangle,$$
$$\mathbf{p}_3 = \langle 0.3, 0.3, 0.4 \rangle,$$
$$\mathbf{p}_4 = \langle 0.4, 0.2, 0.4 \rangle.$$

These points are obtained via the interaction representation of the given lower probability function μ_2, as is shown in Figure 4.5. Clearly, $S(\mathbf{p}_1) = 1.485$, $S(\mathbf{p}_2) = 1.361$, $S(\mathbf{p}_3) = 1.571$, and $S(\mathbf{p}_4) = 1.522$. Hence, $\underline{S}(^{\mu_2}\mathcal{D}) = S(\mathbf{p}_2) = 1.361$. Using Algorithm 6.1, it is easy to compute $\overline{S}(^{\mu_2}\mathcal{D}) = S(0.3, 0.3, 0.4) = S(\mathbf{p}_3) = 1.571$. Then,

$$^a TU(^{\mu_2}\mathcal{D}) = \langle 0.21, 1.361 \rangle.$$

A pictorial overview of justified measures of uncertainty in GIT is given in Figure 6.15. Although other disaggregations of $\overline{S}$ are possible, at least in principle, the two shown in the figure, TU and $^a TU$, seem to be the best choices on both intuitive and mathematical grounds. Needless to say that the issue of how to disaggregate $\overline{S}$ has not been sufficiently investigated as yet.

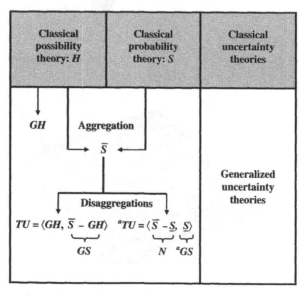

Figure 6.15. An overview of justified measures of uncertainty in GIT.

6.11. UNIFYING FEATURES OF UNCERTAINTY MEASURES

The formalization of uncertainty functions involves a considerable diversity, as is demonstrated in Chapter 5. However, it also involves some unifying features, which are examined in Chapter 4. Functionals that for each type of uncertainty function measure the amount of uncertainty are equally diverse while, at the same time, they also exhibit some unifying features. The following are some of these unifying features:

1. The aggregate uncertainty $\bar{S}$ is defined by one formula, expressed by Eq. (6.61), regardless of the uncertainty theory involved.

2. The definition of the generalized Hartley functional GH, when expressed by Eq. (6.38), does not change when we move from one uncertainty to another.

3. The measure of conflict, $\underline{S}$, is defined in the same way, expressed by Eq. (6.73), regardless of the uncertainty theory involved.

4. It follows from features 1 to 3 that the two versions of the disaggregated total uncertainty, TU, and ^aTU, are defined in the same way in all theories of uncertainty.

5. In all the generalized theories of uncertainty that have been adequately developed, we observe that the various equations established in classical uncertainty theories, which express how joint, marginal, and conditional uncertainty measures (and also the various information transmissions) are interrelated, hold as well.

These unifying features of uncertainty measures allow us to work with any set of recognized and well-developed theories of uncertainty as a whole. Some aspects of this issue are discussed further in Chapter 9.

NOTES

6.1. The U-uncertainty was proposed by Higashi and Klir [1983a]. They also formulated the axiomatic requirements for U-uncertainty given in Section 6.2.3 (except the branching axiom) and proved that the functional defined by Eq. (6.2) or, alternatively, Eq. (6.6), satisfies them. The possibilistic branching requirement was formulated by Klir and Mariano [1987], and they used it for proving the uniqueness of the U-uncertainty. It was shown by Ramer and Lander [1987] that the branching axiom is essential for obtaining a unique generalization of the Hartley measure for the theory of graded possibilities.

6.2. The generalization of the U-uncertainty to Dempster–Shafer theory was suggested by Dubois and Prade [1985c]. The uniqueness of this generalized Hartley measure, defined by Eq. (6.27), was proved by Ramer [1986, 1987]. The measure was also investigated and compared with other measures by Dubois and Prade [1987d].

6.3. An alternative measure of nonspecificity for graded possibilities, which is not a generalization of the Hartley measure, was proposed by Yager [1982c, 1983]. It is a functional a defined by the formula

$$a(m) = 1 - \sum_{A \in \mathcal{F}} \frac{m(A)}{|A|},$$

where $\langle \mathcal{F}, m \rangle$ is a given body of evidence in Dempster–Shafer theory. This functional does not satisfy some of the axiomatic requirements for a measure of uncertainty. In particular, it does not satisfy the requirements of subadditivity and additivity.

6.4. The ordering of bodies of evidence in DST was introduced by Dubois and Prade [1986a, 1987d]. A broader discussion of the concept of ordering bodies of evidence in DST is in [Dubois and Prade, 1986a].

6.5. The formulation of the symmetry and branching axioms and the proof of uniqueness of the generalized Hartley measure in DST, which is presented in Section 6.3 and in Appendix B, are due to Ramer [1986, 1987]. Ramer [1990a] also studied the relationship among the various axiomatic requirements for the generalized Hartley measure in possibility theory and DST.

6.6. The generalization of the Hartley measure for convex sets of probability distributions was proposed by Abellán and Moral [2000]. Relevant proofs regarding properties of this generalized measure are covered in the paper. Among them, the proof of monotonicity of the measure is the most complex one. Contrary to the claims in the paper, the generalized Hartley functional is not subadditive for arbitrary credal sets, as is demonstrated in Example 6.7. However, this is no deficiency in the context of the disaggregated measure of total uncertainty TU.

6.7. The following is a chronological list of main publications regarding the various unsuccessful attempts to generalize the Shannon entropy to DST that are surveyed in Section 6.5:

- Höhle [1982] proposed the measure of confusion C;
- Yager [1983] proposed the measure of dissonance E;
- Lamata and Moral [1988] suggested the sum $E + GH$;
- Klir and Ramer [1990] proposed the measure of discord D and further investigated its properties as well as the properties of $D + GH$ in Ramer and Klir [1993];
- Properties of discord in possibility theory were investigated by Geer and Klir [1991];
- Further examination of the measure of discord by Klir and Parviz [1992] revealed that it had a conceptual defect, and to correct it they suggested the measure of strife ST;
- Klir and Yuan [1993] showed that the distinction between strife and discord reflects the distinction between disjunctive and conjutive set–valued statements. The latter distinction is discussed, for example, in [Yager, 1987a];
- The violation of subadditivity was demonstrated for $D + GH$ by Vejnarová [1991], and for $ST + GH$ by Vejnarová and Klir [1993];

• Functional AS based on the interaction representation was suggested by Yager [2000] and Marichal and Roubens [2000] and its violation of subadditivity is shown in [Klir, 2003].

6.8. The idea of the aggregated measure of uncertainty in DST, which is discussed in Section 6.6, emerged in the early 1990s. It seems that it was conceived independently and almost simultaneously by several authors. The following are relevant publications: [Chokr and Kreinovich, 1991, 1994], [Maeda and Ichihashi, 1993], [Chau, et al., 1993], [Harmanec and Klir, 1994], and [Abellán and Moral, 1999]. Algorithms for computing the functional AU were investigated by Maeda et al. [1993], Meyerowitz et al. [1994], and Harmanec et al. [1996]; the algorithm and its proof of correctness that are presented in Sections 6.6.1 and 6.6.2 and Appendix C are adopted from the last reference. Harmanec [1995, 1996] made progress toward the proof of uniqueness of AU by proving that AU is the smallest functional among those that satisfy the requirements $(AU1)$–$(AU5)$, if any of them exist except AU. Abellán and Moral [2003a] generalized the aggregate measure of uncertainty to arbitrary credal sets and developed a more efficient algorithm for computing it within the theory based on reachable interval-valued probability distributions. They also illustrated a practical utility of the aggregated measure [Abellán and Moral, 2003b].

6.9. The disaggregated total uncertainty introduced in Section 6.8 was proposed and investigated by Smith [2000]. A part of his investigations is the proof of the proper range of the generalized Shannon entropy that is presented in Appendix D.

6.10. The importance of both $\underline{S}$ and $\bar{S}$ is discussed in Kapur et al. [1995] and in Kapur [1994, Chapter 23]. Also discussed in these references is the role of the difference $\bar{S} - \underline{S}$. More recently, Abellán and Moral [2004, 2005] investigated further properties of this difference and described an algorithm for calculating the value of $\underline{S}$ for any lower probability that is 2-monotone. They also suggested using the difference as a measure of nonspecificity. This suggestion led to the alternative view of disaggregated total uncertainty, which is discussed in Section 6.10.

EXERCISES

6.1. Derive Eq. (6.3) from Eq. (6.2).

6.2. For each of the possibility profiles on X in Table 6.5*a*, calculate:
   (a) The U-uncertainty by Eq. (6.2)
   (b) The U-uncertainty by Eq. (6.6)
   (c) The aggregate measure of uncertainty AU

6.3. For each pair of marginal possibility profiles on X and Y in Table 6.5*b*, calculate
   (a) The marginal U-uncertainties;
   (b) The joint U-uncertainty under the assumption of noninteraction of the marginal possibility profiles.

6.4. Express each of the marginal possibility profiles in Table 6.5*b* as a body of evidence in DST and calculate its nonspecificity. Then, determine by

Table 6.5. Supporting Information for Various Exercises

	X							
	x_1	x_2	x_3	x_4	x_5	x_6	x_7	x_8
${}^1r$	1.0	0.0	0.3	0.7	0.9	1.0	0.5	0.5
${}^2r$	0.4	0.5	0.5	1.0	0.8	0.8	0.6	0.5
${}^3r$	0.6	0.6	0.5	0.4	0.3	1.0	1.0	1.0
${}^4r$	0.7	0.9	1.0	1.0	1.0	0.9	0.7	0.5

(*a*)

	X		
	x_1	x_2	x_3
${}^1r_X$	0.6	1.0	0.8
${}^2r_X$	1.0	0.5	0.0
${}^3r_X$	1.0	0.3	0.2

${}^1r_Y$	${}^2r_Y$	${}^3r_Y$		
1.0	0.4	0.9	y_1	
1.0	0.7	1.0	y_2	Y
0.5	1.0	0.8	y_3	

(*b*)

		X			
	${}^1r$	x_1	x_2	x_3	x_4
	y_1	0.0	0.5	0.7	0.0
Y	y_2	0.8	1.0	0.9	0.7
	y_3	0.6	0.7	0.8	1.0
	y_4	0.4	0.6	0.8	1.0

		X			
	${}^2r$	x_1	x_2	x_3	x_4
	y_1	1.0	0.5	0.6	1.0
Y	y_2	0.7	0.0	0.0	0.8
	y_3	0.7	0.4	0.5	0.7
	y_4	1.0	0.8	0.9	1.0

		X			
	${}^3r$	x_1	x_2	x_3	x_4
	y_1	0.3	0.6	0.9	1.0
Y	y_2	0.5	0.7	0.8	0.9
	y_3	0.5	0.8	0.7	0.7
	y_4	0.3	0.9	0.6	0.4

(*c*)

the calculus of DST for each pair of these marginal bodies of evidence on X and Y the corresponding joint body of evidence under the assumption of their noninteraction and calculate:

(a) Its nonspecificity (compare these values with their counterparts in Exercise 6.3*b*);

(b) Its aggregate measure of uncertainty AU.

6.5. For each of the joint possibility profiles in Table 6.5c, calculate
  (a) The joint U-uncertainty and its associated marginal U-uncertainties;
  (b) Both conditional U-uncertainties;
  (c) The information transmission.

6.6. For each of the joint possibility profiles in Table 6.5c, calculate:
  (a) The aggregate measure of uncertainty AU and its marginal counterparts;
  (b) The total disaggregated uncertainty TU and its marginal counterparts.

6.7. Repeat Exercise 6.5 for the joint possibility profile derived from the respective marginal possibility profiles under the assumption of noninteraction.

6.8. For each of the possibility profiles in Table 6.5a and some $k \geq 3$, apply the branching axiom in Eq. (6.24) for calculating the U-uncertainty in two stages.

6.9. Let the range of real-valued variable x be $[0, 100]$. Assume that we are able to assess the value of the variable only approximately in terms of the possibility profile

$$r(x) = \begin{cases} 1 - \dfrac{|x-5|}{3} & \text{when } x \in [2, 8] \\ 0 & \text{otherwise.} \end{cases}$$

Plot the possibility profile and calculate:
  (a) The nonspecificity of this assessment;
  (b) The amount of information obtained by this assessment.

6.10. Repeat Exercise 6.9 for the following possibility profiles and ranges of the variable
  (a) $x \in [-10, 10]$ and $r(x) = \begin{cases} 1 - x^2 & \text{when } x \in [-1, 1] \\ 0 & \text{otherwise;} \end{cases}$

  (b) $x \in [0, 20]$ and $r(x) = \begin{cases} x^2 & \text{when } x \in [0, 1] \\ (2-x)^2 & \text{when } x \in [2, 2] \\ 0 & \text{otherwise;} \end{cases}$

  (c) $x \in [0, 10]$ and $r(x) = 1/(1 + 10(x - 2)^2)$.

6.11. For each of the Möbius representations of the joint lower probability functions on subsets $X \times Y = \{x_1, x_2\} \times \{y_1, y_2\}$ that are defined in Table 6.6, where $z_{ij} = \langle x_i, y_j \rangle$ for all $i, j = 1, 2$, calculate:
  (a) The generalized Hartley measure $GH(X \times Y)$ defined by Eq. (6.38);
  (b) The generalized Shannon measure $GS(X \times Y)$ defined by Eq. (6.64).

Table 6.6. Möbius Representations Employed in Exercises 6.11–6.13

	z_{11}	z_{12}	z_{21}	z_{22}	$m_1(C)$	$m_2(C)$	$m_3(C)$	$m_4(C)$	$m_5(C)$	$m_6(C)$
C:	0	0	0	0	0.0	0.0	0.0	0.0	0.0	0.00
	1	0	0	0	0.1	0.0	0.0	0.1	0.0	0.15
	0	1	0	0	0.1	0.1	0.1	0.1	0.0	0.03
	0	0	1	0	0.2	0.1	0.0	0.0	0.0	0.00
	0	0	0	1	0.4	0.2	0.2	0.3	0.0	0.10
	1	1	0	0	0.0	0.0	0.3	0.0	0.3	0.11
	1	0	1	0	0.0	0.2	0.0	0.0	0.0	0.00
	1	0	0	1	0.0	0.2	−0.2	0.0	0.0	0.30
	0	1	1	0	0.0	0.1	−0.1	0.2	0.0	0.00
	0	1	0	1	0.0	0.1	0.0	0.0	0.0	0.07
	0	0	1	1	0.0	0.3	0.3	0.1	0.5	0.00
	1	1	1	0	0.1	0.0	0.0	0.1	0.2	0.00
	1	1	0	1	0.1	0.0	0.2	0.1	0.2	0.22
	1	0	1	1	0.0	−0.2	0.1	0.1	0.1	0.00
	0	1	1	1	0.1	−0.1	0.1	0.1	0.1	0.00
	1	1	1	1	−0.1	0.0	0.0	−0.2	−0.4	0.02

6.12. Repeat Exercise 6.11 for:

  (a) The generalized marginal Hartley and Shannon measures;

  (b) The generalized conditional Hartley and Shannon measures;

  (c) The information transmissions based on the generalized Hartley and Shannon measures.

6.13. Repeat Exercise 6.11 by calculating the following:

  (a) The measure of dissonance defined by Eq. (6.41);

  (b) The measure of confusion defined by Eq. (6.42);

  (c) The measure of discord defined by Eq. (6.51);

  (d) The measure of strife defined by Eq. (6.55);

  (e) The measure of average Shannon entropy defined by Eq. (6.59).

6.14. Compare $\bar{S}(m)$, $TU(m)$, and $^aTU(m)$ for the simple body of evidence illustrated in Figure 6.9.

6.15. For each of the convex sets of probability distributions discussed in Example 6.6 (see also Table 6.1), compare $\bar{S}(^i D)$, $TU(^i D)$, and $^aTU(^i D)$.

6.16. Assume that a given body evidence $\langle \mathcal{F}, m \rangle$ consists of pairwise disjoint focal sets. Show that under this assumption

$$GS(m) = \sum_{A \in \mathcal{F}} m(A) \log m(A),$$

where GS is the generalized Shannon entropy defined by $\underline{S}$.

6.17. Consider the convex sets of probability distributions on $X = \{x_1, x_2, x_3\}$ defined by convex hulls of the following extreme points:

(a) $^1\mathbf{p} = \langle 0.4, 0.5, 0.1 \rangle$; $^2\mathbf{p} = \langle 0.6, 0.3, 0.1 \rangle$;
   $^3\mathbf{p} = \langle 0.6, 0.2, 0.2 \rangle$; $^4\mathbf{p} = \langle 0.4, 0.2, 0.4 \rangle$;
   $^5\mathbf{p} = \langle 0.1, 0.5, 0.4 \rangle$

(b) $^1\mathbf{p} = \langle 0.4, 0.5, 0.1 \rangle$; $^2\mathbf{p} = \langle 0.4, 0.2, 0.4 \rangle$;
   $^3\mathbf{p} = \langle 0.1, 0.5, 0.4 \rangle$

(c) $^1\mathbf{p} = \langle 0.4, 0.5, 0.1 \rangle$; $^2\mathbf{p} = \langle 0.6, 0.2, 0.2 \rangle$;
   $^3\mathbf{p} = \langle 0.4, 0.2, 0.4 \rangle$

(d) $^1\mathbf{p} = \langle 0.4, 0.5, 0.1 \rangle$; $^2\mathbf{p} = \langle 0.4, 0.2, 0.4 \rangle$

(e) $^1\mathbf{p} = \langle 0, 0.2, 0.8 \rangle$; $^2\mathbf{p} = \langle 0.2, 0.2, 0.8 \rangle$
   $^3\mathbf{p} = \langle 0.5, 0.1, 0.5 \rangle$; $^4\mathbf{p} = \langle 0, 0.5, 0.5 \rangle$

(f) $^1\mathbf{p} = \langle 0, 0.5, 0.5 \rangle$; $^2\mathbf{p} = \langle 0.5, 0, 0.5 \rangle$;
   $^3\mathbf{p} = \langle 0.5, 0.5, 0.5 \rangle$

(g) $^1\mathbf{p} = \langle 0, 1, 0 \rangle$; $^2\mathbf{p} = \langle 0, 0.5, 0.5 \rangle$;
   $^3\mathbf{p} = \langle 0.5, 0.5, 0 \rangle$

(h) $^1\mathbf{p} = \langle 0, 1, 0 \rangle$; $^2\mathbf{p} = \langle 1, 0, 0 \rangle$; $^3\mathbf{p} = \langle 0.4, 0.4, 0.4 \rangle$

For each of the convex sets of probability distribution functions, calculate TU and ^aTU.

6.18. Calculate TU and ^aTU for the interval-valued probability distribution $I = \langle [0, 0.2], [0, 0.6], [0.3, 0.5], [0.2, 0.4] \rangle$ on $X = \{x_1, x_2, x_3, x_4\}$.

7

FUZZY SET THEORY

Vagueness is no more to be done away with in the world of logic than friction in mechanics.

—Charles Sanders Peirce

7.1. AN OVERVIEW

The notion of the most common type of fuzzy sets, referred to as *standard fuzzy sets*, is introduced in Section 1.4. Recall that each standard fuzzy set is uniquely defined by a *membership function* of the form

$$A:X \to [0, 1],$$

where X is the universal set of concern. For each $x \in X$, the value $A(x)$ expresses the *degree* (or *grade*) *of membership* of the element x of X in standard fuzzy set A. For the sake of notational simplicity, the symbol of a given membership function, A, is also employed as a label of the standard fuzzy set defined by this function. It is obvious that no ambiguity is introduced by this double use of the same symbol.

Recall also from Chapter 1 that classical sets, when viewed as special fuzzy sets, are usually referred to as *crisp sets*. It is common in the literature to describe a standard fuzzy set A on a finite universal set X by the special form

$$A = a_1/x_1 + a_2/x_2 + \cdots + a_n/x_n,$$

Uncertainty and Information: Foundations of Generalized Information Theory, by George J. Klir
© 2006 by John Wiley & Sons, Inc.

where $x_i \in X$ for all $i \in \mathbb{N}_n$ and each a_i denotes the degree of membership of element x_i in A. Each slash is employed in this form to link an element of X with its degree of membership in A, and the plus signs indicate that the listed pairs collectively form the definition of the set A. The pairs with zero membership degrees are usually not listed.

The purpose of this chapter is to introduce the fundamentals of fuzzy set theory. This background is needed for understanding how the two classical uncertainty theories (surveyed in Chapters 2 and 3) and their generalizations based on the various types of monotone measures (discussed in Chapters 4–6) can be fuzzified. In other words, this background is needed to understand how these various uncertainty theories, all described in terms of the formalized language of classical set theory, can be further generalized via the more expressive formalized languages of fuzzy set theory. This generalization is manifested in the 2-dimensional array in Figure 1.3 by the horizontal expansion of column 1.

The common feature of all fuzzy sets is that the membership of any relevant object in any fuzzy set is a matter of degree. However, there are distinct ways of expressing membership degrees, which result in distinct categories of fuzzy sets. In standard fuzzy sets, membership degrees are expressed by real numbers in the unit interval. In other, nonstandard fuzzy sets, they may be expressed by intervals of real numbers, partially ordered qualitative descriptors of membership degrees, and in numerous other ways. Given any particular category of fuzzy sets, operations of set intersection, union, and complementation are not unique. Further distinctions within the category can thus be made by choosing various specific operations from the class of possible operations. Each choice induces an algebraic structure of some type on the given category of fuzzy sets. These algebraic structures are always weaker than Boolean algebra, that is, they are non-Boolean. The term "fuzzy set theory" thus stands for a collection of theories, each dealing with fuzzy sets in a particular category by specific operations and, consequently, based on a non-Boolean algebraic structure of some type. A fuzzified uncertainty theory is then obtained by formalizing a monotone measure of some type in terms of this algebraic structure, as illustrated in Figure 7.1.

Standard fuzzy sets have been predominant in the literature and, moreover, virtually all fuzzifications of the various uncertainty theories that are currently described in the literature are based on standard fuzzy sets. In this chapter it is thus natural to examine standard fuzzy sets in more detail than other types of fuzzy sets. However, the amount of research work on the various nonstandard categories of fuzzy sets has visibly increased during the last few years, primarily in response to emerging applications needs. Since the aim of generalized information theory (GIT) is to expand the development of uncertainty theories in both the dimensions depicted in Figure 1.3, it is essential to survey in this chapter those nonstandard types of fuzzy sets that have been proposed in the literature. However, this survey, which is the subject of Section 7.8, is only relevant to future research in GIT. It is not needed for understanding the

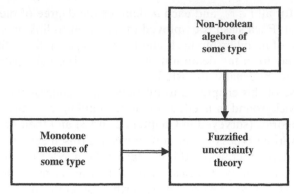

Figure 7.1. Components of a fuzzified uncertainty theory.

fuzzified theories of uncertainty examined in Chapter 8, which are, by and large, based on standard fuzzy sets.

7.2. BASIC CONCEPTS OF STANDARD FUZZY SETS

Since this section (and the subsequent Sections 7.3–7.7) deal only with standard fuzzy sets, the adjective "standard" is omitted for the sake of linguistic simplicity.

As in classical set theory, the concept of *subsethood* is one of the most fundamental concepts in fuzzy set theory. Contrary to classical set theory, however, this concept has two distinct meanings in fuzzy set theory. One of them is a *crisp subsethood*, the other one is a *fuzzy subsethood*. These two concepts of subsethood are introduced as follows.

Given two fuzzy sets A, B defined on the same universal set X, A is said to be a *subset* of B if and only if

$$A(x) \le B(x)$$

for all $x \in X$. The usual notation, $A \subseteq B$, is used to signify the subsethood relation. According to this definition, clearly, set A is *either* a subset of set B *or* it is not a subset of B (and, similarly, the other way around). Therefore, this definition of subsethood is called a crisp subsethood. The set of all fuzzy subsets of X (defined in this way) is called the *fuzzy power set* of X and is denoted by $\mathcal{F}(X)$. Observe that this set is crisp, even though its members are fuzzy sets. Moreover, this set is always infinite, even if X is finite.

When X is a finite universal set, a *fuzzy subsethood relation*, Sub, is defined by specifying for each pair of fuzzy sets on X, A and B, a degree of subsethood, $Sub(A, B)$, by the formula

$$Sub(A, B) = \frac{\sum_{x \in X} A(x) - \sum_{x \in X} \max\{0, A(x) - B(x)\}}{\sum_{x \in X} A(x)} \qquad (7.1)$$

The negative term in the numerator describes the sum of the degrees to which the subset inequality $A(x) \leq B(x)$ is violated, the positive term describes the largest possible violation of the inequality, the difference in the numerator describes the sum of the degrees to which the inequality is not violated, and the term in the denominator is a normalizing factor to obtain the range

$$0 \leq Sub(A, B) \leq 1.$$

When sets A and B are defined on a bounded subset of real numbers (i.e., X is a closed interval of real numbers), the three Σ terms in Eq. (7.1) are replaced with integrals over X.

For any fuzzy set A defined on a finite universal set X, its *scalar cardinality*, $|A|$, is defined by the formula

$$|A| = \sum_{x \in X} A(x). \qquad (7.2)$$

Scalar cardinality is sometimes referred to in the literature as a *sigma count*.

Among the most important concepts of standard fuzzy sets are the concepts of an *α-cut* and a *strong α-cut*. Given a fuzzy set A defined on X and a particular number α in the unit interval $[0, 1]$, the α-cut of A, denoted by $^{\alpha}A$, is a crisp set that consists of all elements of X whose membership degrees in A are greater than or equal to α. This can formally be written as

$$^{\alpha}A = \{x | A(x) \geq \alpha\}.$$

The strong α-cut, $^{\alpha+}A$, has a similar meaning, but the condition "greater than or equal to" is replaced with the stronger condition "greater than." Formally,

$$^{\alpha+}A = \{x | A(x) > \alpha\}.$$

The set ^{0+}A is called the *support* of A and the set ^{1}A is called the *core* of A. When the core A is not empty, A is called *normal*; otherwise, it is called *subnormal*. The largest value of A is called the *height* of A and it is denoted by h_A. The set of distinct values $A(x)$ for all $x \in X$ is called the *level set* of A and it is denoted by Λ_A.

All the introduced concepts are illustrated in Figure 7.2. We can see that

$$^{\alpha_1}A \subset {}^{\alpha_2}A \qquad \text{and} \qquad {}^{\alpha_1+}A \subset {}^{\alpha_2+}A$$

when $\alpha_1 \geq \alpha_2$. This implies that the set of all distinct α-cuts (as well as strong α-cuts) is always a *nested family* of crisp sets. When α is increased, the new α-

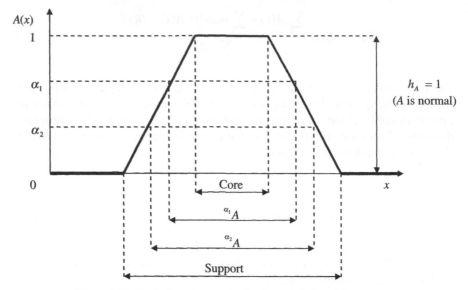

Figure 7.2. Illustration of some basic characteristics of fuzzy sets.

cut (strong α-cut) is always a subset of the previous one. Clearly, $^0A = X$ and $^{1+}A = \varnothing$.

It is well established that each fuzzy set is uniquely represented by the associated family of its α-cuts via the formula

$$A(x) = \sup\{\alpha \cdot {}^{\alpha}A(x) \mid \alpha \in [0,1]\}, \tag{7.3}$$

or by the associated family of its strong α-cuts via the formula

$$A(x) = \sup\{\alpha \cdot {}^{\alpha+}A(x) \mid \alpha \in [0,1]\}, \tag{7.4}$$

where sup denotes the supremum of the respective set and $^{\alpha}A$ (or $^{\alpha+}A$) denotes for each $\alpha \in [0, 1]$ the special membership function (characteristic function) of the α-cut (or strong α-cut, respectively).

EXAMPLE 7.1. To illustrate the meaning of Eq. (7.3) consider a fuzzy set A defined on $\mathbb{R}$ by the simple stepwise membership function shown in Figure 7.3a. Since $\Lambda(A) = \{0, 0.3, 0, 0.6, 1\}$ and $0 \cdot {}^0A(x) = 0$ for all $x \in \mathbb{R}$, A is fully represented by the three special fuzzy sets $\alpha \cdot {}^{\alpha}A(\alpha = 0.3, 0.6, 1)$, whose membership functions are shown in Figure 7.3b. The membership function A is uniquely reconstructed from these three membership functions by taking for each $x \in \mathbb{R}$ their supremum. The same can clearly be accomplished by the strong α-cuts and Eq. (7.4).

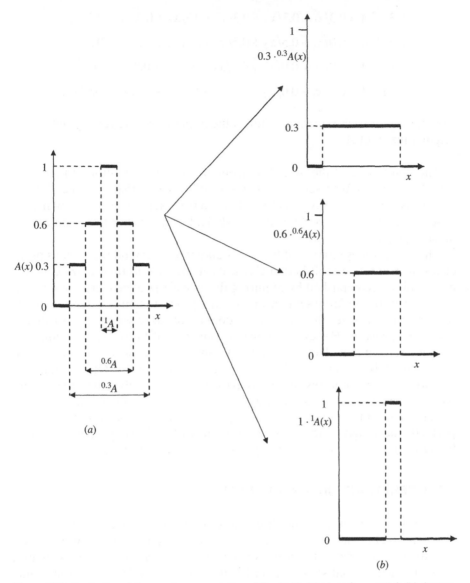

Figure 7.3. Illustration of the α-cut representation of Eq. (7.3): (a) given fuzzy set A; (b) decomposition of A into special fuzzy sets.

EXAMPLE 7.2. An alternative way to illustrate the meaning of Eq. (7.3) is to use a fuzzy set defined on a finite universal set. Let

$$A = 0.3/x_1 + 0.5/x_2 + 1.0/x_3 + 1.0/x_4 + 0.8/x_5 + 0.5/x_6 .$$

This fuzzy set is fully represented by the following four special fuzzy sets:

$$0.3 \cdot {}^{0.3}A = 0.3/x_1 + 0.3/x_2 + 0.3/x_3 + 0.3/x_4 + 0.3/x_5 + 0.3/x_6$$
$$0.5 \cdot {}^{0.5}A = 0.0/x_1 + 0.5/x_2 + 0.5/x_3 + 0.5/x_4 + 0.5/x_5 + 0.5/x_6$$
$$0.8 \cdot {}^{0.8}A = 0.0/x_1 + 0.0/x_2 + 0.8/x_3 + 0.8/x_4 + 0.8/x_5 + 0.0/x_6$$
$$1 \cdot {}^{1}A = 0.0/x_1 + 0.0/x_2 + 1.0/x_3 + 1.0/x_4 + 0.0/x_5 + 0.0/x_6.$$

Now taking for each x_i the maximum value in these sets, we readily obtain the original fuzzy set A.

The significance of the α-cut (or strong α-cut) representation of fuzzy sets is that it connects fuzzy sets with crisp sets. While each crisp set is a collection of objects that are conceived as a whole, each fuzzy set is a collection of nested crisp sets that are also conceived as a whole. Fuzzy sets are thus wholes of a higher category.

The α-cut representation of fuzzy sets allows us to extend the various properties of crisp sets, established in classical set theory, into their fuzzy counterparts. This is accomplished by requiring that the classical property be satisfied by all α-cuts of the fuzzy set concerned. Any property that is extended in this way from classical set theory into the domain of fuzzy set theory is called a *cutworthy property*. For example, when convexity of fuzzy sets is defined by the requirement that all α-cuts of a fuzzy convex set be convex sets in the classical sense, this conception of fuzzy convexity is cutworthy. Other important examples are the concepts of a fuzzy partition, fuzzy equivalence, fuzzy compatibility, and various kinds of fuzzy orderings that are cutworthy.

It is important to realize that many (perhaps most) properties of fuzzy sets, perfectly meaningful and useful, are not cutworthy. These properties cannot be derived from classical set theory via the α-cut representation.

7.3. OPERATIONS ON STANDARD FUZZY SETS

As is well known, each of the three basic operations on sets—*complementation*, *intersection*, and *union*—is unique in classical set theory. However, their counterparts in fuzzy set theory are not unique. Each of them consists of a class of functions that satisfy certain requirements. In this section, only some main features of these classes of functions are described.

7.3.1. Complementation Operations

Complementation operations on fuzzy sets are functions of the form

$$c : [0,1] \to [0,1]$$

that are order reversing and such that $c(0) = 1$ and $c(1) = 0$. Moreover, they are usually required to possess the property

$$c(c(a)) = a \tag{7.5}$$

for all $a \in [0, 1]$; this property is called an *involution*. Given a fuzzy set A and a particular complementation function c, the complement of A with respect to c, ^cA, is defined for all $x \in X$ by the formula

$$^cA(x) = c(A(x)). \tag{7.6}$$

An example of a practical class of involutive complementation functions, c_λ, is defined for each $a \in [0, 1]$ by the formula

$$c_\lambda(a) = (1 - a^\lambda)^{1/\lambda}, \qquad \lambda > 0, \tag{7.7}$$

where λ is a parameter whose values specify individual complements in this class. When $\lambda = 1$, the resulting complement is usually called a *standard complement*.

Which of the possible complements to choose is basically an experimental question. The choice is determined in the context of each particular application by eliciting the meaning of negating a given concept. This can be done by employing a suitable parametrized class of complementation functions, such as the one defined by Eq. (7.7), and determining a proper value of the parameter.

7.3.2. *Intersection* and *Union Operations*

Intersection and *union operations* on fuzzy sets are defined in terms of functions i and u, respectively, which assign to each pair of numbers in the unit interval $[0, 1]$ a single number in $[0, 1]$. These functions, i and u, are commutative, associative, monotone nondecreasing, and consistent with the characteristic functions of classical intersections and unions, respectively. They are often referred to in the literature as *triangular norms* (or *t-norms*) and *triangular conorms* (or *t-conorms*). It is well know that the inequalities

$$i_{\min}(a, b) \le i(a, b) \le \min(a, b), \tag{7.8}$$

$$\max(a, b) \le u(a, b) \le u_{\max}(a, b) \tag{7.9}$$

are satisfied for all $a, b \in [0, 1]$, where

$$i_{\min}(a, b) = \begin{cases} \min(a, b) & \text{when } \max(a, b) = 1 \\ 0 & \text{otherwise,} \end{cases} \tag{7.10}$$

$$u_{\max}(a, b) = \begin{cases} \max(a, b) & \text{when } \min(a, b) = 0 \\ 1 & \text{otherwise.} \end{cases} \tag{7.11}$$

Operations expressed by min and max are usually called *standard operations*; those expressed by $i_{\min}$ and $u_{\max}$ are called *drastic operations*.

Given fuzzy sets A and B, defined on the same universal set X, their intersections and unions with respect to i and u, $A \cap_i B$ and $A \cup_u B$, are defined for each $x \in X$ by the formulas

$$(A \cap_i B)(x) = i[A(x), B(x)], \tag{7.12}$$

$$(A \cup_u B)(x) = u[A(x), B(x)]. \tag{7.13}$$

It is useful to capture the full range of intersections and unions, as expressed by the inequalities (7.8) and (7.9), by suitable classes of functions. For example, all functions in the classes

$$i_\lambda(a, b) = 1 - \min\left\{1, \left[(1-a)^\lambda + (1-b)^\lambda\right]^{1/\lambda}\right\}, \tag{7.14}$$

$$u_\lambda(a, b) = \min\left\{1, (a^\lambda + b^\lambda)^{1/\lambda}\right\}, \tag{7.15}$$

where $\lambda > 0$, qualify as intersection and unions, respectively, and cover the whole range by varying the parameter λ. The standard operations are obtained in the limit for $\lambda \to \infty$, while the drastic operations are obtained in the limit for $\lambda \to 0$. In each particular application, a fitting operation is determined by selecting appropriate values of the parameter λ by knowledge acquisition techniques.

7.3.3. Combinations of Basic Operations

In classical set theory, the operations of complementation, intersection, and union possess the properties listed in Table 1.1. For each universal set X, these are properties of the Boolean algebra defined on the power set of X. The operations of intersection and union are dual with respect to the operation of complementation in the sense that they satisfy for any pair of subsets of X, A, and B, the De Morgan laws

$$\overline{A \cap B} = \overline{A} \cup \overline{B} \quad \text{and} \quad \overline{A \cup B} = \overline{A} \cap \overline{B}.$$

It is desirable that this duality be satisfied for fuzzy sets as well. It is obvious that only some combinations of t-norms, t-conorms, and complementation operations on fuzzy sets can satisfy this duality. We say that a t-norm i and a t-conorm u are dual with respect to a complementation operation c if and only if for all $a, b \in [0, 1]$

$$c[i(a, b)] = u[c(a), c(b)] \tag{7.16}$$

and

$$c[u(a, b)] = i[c(a), c(b)].$$ (7.17)

These equations describe the De Morgan laws for fuzzy sets. Triples $\langle i, u, c \rangle$ that satisfy these equations are called *De Morgan triples*.

None of the De Morgan triples satisfies all properties of the Boolean algebra on $\mathcal{P}(X)$. Depending on the operations used, various weaker algebraic structures are obtained. In some of them, for example, the laws of contradiction and excluded middle are violated. In others, these laws are preserved, but the laws of distributivity are violated.

7.3.4. Other Operations

Two additional types of operations are applicable to fuzzy sets, but they have no counterparts in classical set theory. They are called modifiers and averaging operations.

Modifiers are unary operations that are order preserving. Their purpose is to modify fuzzy sets that represent linguistic terms to account for *linguistic hedges* such as *very, fairly, extremely, more or less*, and the like. The most common modifiers either increase or decrease all values of a given membership function. A convenient class of functions, m_λ, that qualify as increasing or decreasing modifiers is defined for each $a \in [0, 1]$ by the simple formula

$$m_\lambda(a) = a^\lambda,$$ (7.18)

where $\lambda > 0$ is a parameter whose value determines which way and how strongly m_λ modifies a given membership function. Clearly, $m_\lambda(a) > a$ when $\lambda \in (0, 1)$, $m_\lambda(a) < a$ when $\lambda \in (1, \infty)$, and $m_\lambda(a) = a$ when $\lambda = 1$. The farther the value of λ from 1, the stronger the modifier m_λ.

Averaging operations are monotone nondecreasing and idempotent, but are not associative. Due to the lack of associativity, they must be defined as functions of n arguments for any $n \geq 2$. It is well known that any averaging operation, h, satisfies the inequalities

$$\min(a_1, a_2, \ldots, a_n) \leq h(a_1, a_2, \ldots, a_n) \leq \max(a_1, a_2, \ldots, a_n)$$ (7.19)

for any n-tuple $(a_1, a_2, \ldots, a_n) \in [0, 1]^n$. This means that the averaging operations fill the gap between intersections (t-norms) and unions (t-conorms).

One class of averaging operations, h_λ, which covers the entire interval between min and max operations, is defined for each n-tuple $(a_1, a_2, \ldots, a_n)$ in $[0, 1]^n$ by the formula

$$h_\lambda(a_1, a_2, \ldots, a_n) = \left(\frac{a_1^\lambda + a_2^\lambda + \cdots + a_n^\lambda}{n} \right)^{1/\lambda},$$ (7.20)

where λ is a parameter whose range is the set of all real numbers except 0. For $\lambda = 0$, function h_λ is defined by the limit

$$\lim_{\lambda \to 0} h_\lambda(a_1, a_2, \ldots, a_n) = (a_1, a_2, \ldots, a_n)^{1/n}, \tag{7.21}$$

which is the well-known geometric mean. Moreover

$$\lim_{\lambda \to -\infty} h_\lambda(a_1, a_2, \ldots, a_n) = \min(a_1, a_2, \ldots, a_n), \tag{7.22}$$

$$\lim_{\lambda \to -\infty} h_\lambda(a_1, a_2, \ldots, a_n) = \max(a_1, a_2, \ldots, a_n). \tag{7.23}$$

This indicates that the standard operations of intersection and union also may be viewed as extreme opposites in the range of averaging operations.

Other classes of averaging operations are now available, some of which use weighting factors to express the relative importance of the individual fuzzy sets involved. For example, the function

$$h(a_i, w_i \mid i = 1, 2, \ldots, n) = \sum_{i=1}^{n} w_i a_i,$$

where the weighting factors w_i usually take values in the unit interval $[0, 1]$ and

$$\sum_{i=1}^{n} w_i = 1,$$

expresses for each choice of values w_i the corresponding weighted average of values $a_i (i = 1, 2, \ldots, n)$. Again, the choice is an experimental issue.

7.4. FUZZY NUMBERS AND INTERVALS

Fuzzy sets that are defined on the set of real numbers, $\mathbb{R}$, have a special significance in fuzzy set theory. Among these sets, the most important are cut-worthy *fuzzy intervals*. They are defined by requiring that each α-cut be a single closed and bounded interval of real numbers for all $\alpha \in (0, 1]$. A fuzzy interval, A, may conveniently be represented for each $x \in \mathbb{R}$ by the canonical form

$$A(x) = \begin{cases} f_A(x) & \text{when } x \in [a, b] \\ 1 & \text{when } x \in [b, c) \\ g_A(x) & \text{when } x \in (c, d] \\ 0 & \text{otherwise,} \end{cases} \tag{7.24}$$

where a, b, c, d are specific real numbers such that $a \leq b \leq c \leq d, f_A$ is a real-valued function that is increasing, and g_A is a real-valued function that is decreasing. In most applications, functions f_A and g_A are continuous, but, in general, they may be only semicontinuous from the right and left, respectively. When $A(x) = 1$ for exactly one $x \in \mathbb{R}$ (i.e., $b = c$ in the canonical representation), A is called a *fuzzy number*.

Some common shapes of membership functions of fuzzy numbers or intervals are shown in Figure 7.4. Each of them represents, in a particular way, a fuzzy set of numbers described in natural language as "close to 1" (or "around 1"). Whether a particular membership function is appropriate for representing this linguistic description can be determined only in the context of each given application of the linguistic expression. Usually, however, the membership function that is supposed to represent a given linguistic expression in the context of a given application is constructed from the way in which the linguistic expression is interpreted in this application. The issue of constructing membership functions is addressed in Section 7.9.

In practical applications, the most common shapes of membership functions of fuzzy intervals are the trapezoidal ones, illustrated by the membership function in Figure 7.2 and also function B in Figure 7.4. They are easy to represent and manipulate. Each trapezoidal-shaped membership function is uniquely defined by the four real numbers a, b, c, d, in Eq. (7.24). Defining a trapezoidal fuzzy interval A by the quadruple

$$A = \langle a, b, c, d \rangle$$

means that

$$f_A(x) = \frac{x - a}{b - a} \quad \text{and} \quad g_A(x) = \frac{d - x}{d - c}$$

in Eq. (7.24). Clearly, triangular-shaped fuzzy numbers are special cases of the trapezoidal-shaped fuzzy intervals in which $b = c$.

For any fuzzy interval A expressed in the canonical form, the α-cuts of A are expressed for all $\alpha \in (0, 1]$ by the formula

$$^{\alpha}A = \begin{cases} [f_A^{-1}(\alpha), g_A^{-1}(\alpha)] & \text{when } \alpha \in (0,1), \\ [b, c] & \text{when } \alpha = 1, \end{cases} \tag{7.25}$$

where f_A^{-1} and g_A^{-1} are the inverse functions of f_A and g_A, respectively. When a membership function of fuzzy intervals has a trapezoidal shape, such as the function A in Figure 7.2, the α-cuts can readily be expressed in terms of the four real numbers a, b, c, d by the formula

$$^{\alpha}A = [a + (b - a)\alpha, d - (d - c)\alpha]. \tag{7.26}$$

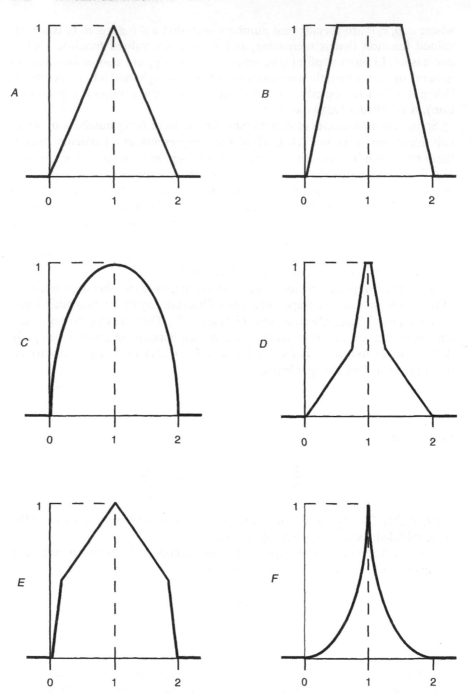

Figure 7.4. Examples of fuzzy sets of numbers that are "close to 1" (or "around 1").

EXAMPLE 7.3. Determine the α-cut representation of a fuzzy number F shown in Figure 7.4, whose membership function is defined for each $x \in \mathbb{R}$ by the formula

$$F(x) = \begin{cases} x^2 & \text{when } x \in [0,1) \\ (2-x)^2 & \text{when } x \in [1,2] \\ 0 & \text{otherwise.} \end{cases}$$

In this case, $f_F(x) = x^2$ and $g_F(x) = (2-x)^2$. Moreover, $f_F(1) = g_F(1) = 1$. For each $\alpha \in [0, 1]$, the left-end point, $\underline{x}$, in $^\alpha F$ is a function of α that is determined by the positive root of the equation $\alpha = \underline{x}^2$, and the right-end point, $\bar{x}$, in $^\alpha F$ is a function of α that is determined by the positive root of the equation $\alpha = (2-\bar{x})^2$. Hence,

$$^\alpha F = [\sqrt{\alpha}, 2-\sqrt{\alpha}]$$

for all $\alpha \in (0, 1]$

7.4.1. Standard Fuzzy Arithmetic

Consider fuzzy intervals A and B whose α-cut representations are

$$^\alpha A = {}^\alpha[\underline{a}, \bar{a}],$$
$$^\alpha B = {}^\alpha[\underline{b}, \bar{b}],$$

where $\underline{a}, \bar{a}, \underline{b}, \bar{b}$ are functions of α. Then, the individual arithmetic operations on these fuzzy intervals are defined for all $\alpha \in (0, 1]$ in terms of the well-established arithmetic operations on the closed intervals of real numbers by the formula

$$^\alpha(A * B) = {}^\alpha A * {}^\alpha B,$$

where $*$ denotes any of the four basic arithmetic operations (addition, subtraction, multiplication, and division). That is, arithmetic operations on fuzzy intervals are cutworthy. According to interval arithmetic,

$$^\alpha(A * B) = \{a * b \mid a \in {}^\alpha A,\ b \in {}^\alpha B\} \tag{7.27}$$

with the requirement that $0 \notin {}^\alpha B$ for all $\alpha \in (0, 1]$ when $*$ stands for division. The sets $^\alpha(A * B)$ (closed intervals of real numbers) defined by Eq. (7.27) can be obtained for the individual operations from the end points of $^\alpha A$ and $^\alpha B$ in the following ways:

$$^\alpha(A+B) = {}^\alpha[\underline{a}+\underline{b}, \overline{a}+\overline{b}], \tag{7.28}$$

$$^\alpha(A-B) = {}^\alpha[\underline{a}-\underline{b}, \overline{a}-\overline{b}]; \tag{7.29}$$

$$^\alpha(A \cdot B) = {}^\alpha\left[\min\left(\underline{a}\underline{b}, \underline{a}\overline{b}, \overline{a}\underline{b}, \overline{a}\overline{b}\right), \ \max\left(\underline{a}\underline{b}, \underline{a}\overline{b}, \overline{a}\underline{b}, \overline{a}\overline{b}\right)\right]; \tag{7.30}$$

$$^\alpha(A/B) = {}^\alpha[\min(\underline{a}/\underline{b}, \underline{a}/\overline{b}, \overline{a}/\underline{b}, \overline{a}/\overline{b}), \ \max(\underline{a}/\underline{b}, \underline{a}/\overline{b}, \overline{a}/\underline{b}, \overline{a}/\overline{b})]. \tag{7.31}$$

The operation of division, $^\alpha(A/B)$, requires that $0 \notin {}^\alpha[\underline{b}, \overline{b}]$ for all $\alpha \in (0, 1]$.

Fuzzy intervals together with these operations are usually referred to as *standard fuzzy arithmetic*. Algebraic expressions in which values of variables are fuzzy intervals are evaluated by standard fuzzy arithmetic in the same order as classical algebraic expressions are evaluated by classical arithmetic on real numbers.

EXAMPLE 7.4. Consider the following two triangular-shape fuzzy numbers, A and B, defined by the quadruples $\langle a, b, c, d \rangle$:

$$A = \langle 1, 2, 3, 4 \rangle \quad \text{and} \quad B = \langle 2, 4, 4, 5 \rangle.$$

To perform arithmetic operations with these fuzzy numbers, we need to determine their α-cuts. For triangular fuzzy numbers, this can be done via Eq. (7.26):

$$^\alpha A = [1 + \alpha, 3 - \alpha]$$
$$^\alpha B = [2 + 2\alpha, 5 - \alpha]$$

Then, by using Eqs. (7.28)–(7.31), we obtain, for example:

$$^\alpha(A+B) = [3 + 3\alpha, 8 - 2\alpha],$$
$$^\alpha(A-B) = [2\alpha - 4, 1 - 3\alpha],$$
$$^\alpha(B-A) = [3\alpha - 1, 4 - 2\alpha],$$
$$^\alpha(A \cdot B) = [(1 + \alpha)(2 + 2\alpha), (3 - \alpha)(5 - \alpha)],$$
$$^\alpha(A/B) = [(1 + \alpha)/(5 - \alpha), (3 - \alpha)/(2 + 2\alpha)],$$
$$^\alpha(B/A) = [(2 + 2\alpha)/(3 - \alpha), (5 - \alpha)/1 + \alpha].$$

7.4.2. Constrained Fuzzy Arithmetic

It turns out that standard fuzzy arithmetic does not take into account constraints among fuzzy intervals employed in algebraic expressions. For example, it does not distinguish between $^\alpha(A * B)$, where $^\alpha B$ happens to be equal to $^\alpha A$, and $^\alpha(A * A)$. In the latter case, both fuzzy intervals represent a value of the same variable, and, consequently, they are strongly constrained: whatever value we select in the first interval, the value in the second interval must be the same. Taking this *equality constraint* into account, Eq. (7.26) must be replaced with

$$^{\alpha}(A * A) = \{a * a \mid a \in {}^{\alpha}A\},$$

Thus, for example, $^{\alpha}A - {}^{\alpha}A = [0, 0] = 0$ under the equality constraint, while

$$^{\alpha}(A - A) = {}^{\alpha}[\underline{a} - \overline{a}, \overline{a} - \underline{a}]$$

under standard fuzzy arithmetic.

By ignoring constraints among fuzzy intervals involved, standard fuzzy arithmetic produces results that are, in general, deficient of information, even though they are principally correct. To avoid this information deficiency, all known constraints among fuzzy intervals in each application must be taken into account. This is a subject of *constrained fuzzy arithmetic.*

EXAMPLE 7.5. To compare standard fuzzy arithmetic with constrained fuzzy arithmetic, consider the simple equation $a = b/(b + 1)$, where a and b are real variables and $b \geq 0$. Assume that, due to information deficiency, the actual value of variable b in a given situation is known only imprecisely, by the triangular fuzzy number B in Example 7.4. Given this information, we want to determine as precisely as possible the value of variable a. Clearly, this value is expressed by another fuzzy number, A, such that $A = B/(B + 1)$. To determine A, we simply need to evaluate the expression on the right-hand side of this equation. Consider first the standard fuzzy arithmetic. Then,

$$^{\alpha}[B+1] = [2 + 2\alpha, 5 - \alpha] + [1,1] = [3 + 2\alpha, 6 - \alpha].$$

Now, we take this result and use it to calculate the whole expression:

$$^{\alpha}[B/(B+1)] = [(2 + 2\alpha)/(6 - \alpha), (5 - \alpha)/(3 + 2\alpha)] = {}^{\alpha}A.$$

As follows from Eq. (7.25), inverses of the left-end and right-end functions of $^{\alpha}A$ are functions f_A and g_A of the canonical form Eq. (7.24) of A. The inverses are obtained by solving the equations

$$\frac{2 + 2f_A(x)}{6 - f_A(x)} = x \quad \text{and} \quad \frac{5 - g_A(x)}{3 + 2g_A(x)} = x$$

for $f_A(x)$ and $g_A(x)$, respectively. The solutions are:

$$f_A(x) = \frac{6x - 2}{2 + x} \quad \text{and} \quad g_A(x) = \frac{5 - 3x}{2x + 1}.$$

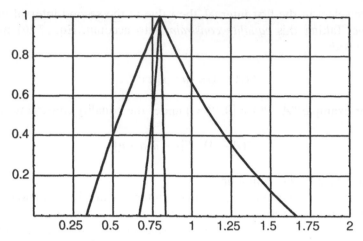

Figure 7.5. Illustration to Example 7.5.

We can easily determine from $^\alpha A$ that the support of A is the open interval $(1/3, 5/3)$ and its core $[0.8, 0.8]$. The wide graph in Figure 7.5 shows the whole membership function A obtained by standard fuzzy arithmetic. Its analytic form is

$$A(x) = \begin{cases} \dfrac{6x - 2}{2 + x} & \text{when } x \in [1/3, 0.8] \\ \dfrac{5 - 3x}{2x + 1} & \text{when } x \in [0.8, 5/3] \\ 0 & \text{otherwise.} \end{cases}$$

Now let us repeat the procedure by using constrained fuzzy arithmetic. This means to evaluate the expression $B/(B + 1)$ globally and to recognize that the two appearances of fuzzy number B in the expression represent the same variable b. This means, in turn, that whatever choice we make from any α-cut of B, we must make the same choice for both the B in the numerator and the one in the denominator. Since the function $b/(b + 1)$ is clearly increasing for all values of $b \geq 0$, the minimum and maximum values for each α-cut of the expression are obtained, respectively, for the left-end point and the right-end point of $^\alpha B$. Hence,

$$^\alpha[B/(B+1)] = [(2 + 2\alpha)/(3 + 2\alpha), (5 - \alpha)/(6 - \alpha)] = {}^\alpha A,$$

and the inverses of the left-hand and right-hand functions of the α-cut representation of A are

$$f_A(x) = \frac{2 - 3x}{2x - 2} \quad \text{and} \quad g_A(x) = \frac{6x - 5}{x - 1}.$$

In this case, the support of A is the open interval $(2/3, 5/6)$ and, again, the core is $[0.8, 0.8]$. The whole function A is depicted in Figure 7.5 by the narrow graph. Its analytic form is

$$A(x) = \begin{cases} \dfrac{2 - 3x}{2x - 2} & \text{when } x \in [2/3, 0.8] \\[2mm] \dfrac{6x - 5}{x - 1} & \text{when } x \in [0.8, 5/6] \\[2mm] 0 & \text{otherwise.} \end{cases}$$

We can clearly see that the result obtained by standard fuzzy arithmetic is considerably less precise than the one obtained by constrained fuzzy arithmetic. This is caused by the fact that standard fuzzy arithmetic ignores the equality constraint, which, in this case, contains a lot of information.

Observe that the given equation can be written as

$$a = 1 - \frac{1}{b + 1}.$$

In this form, the equality constraint is not applicable and, as can be easily verified, standard fuzzy arithmetic produces the same result we obtained by constrained fuzzy arithmetic. This demonstrates that results obtained by standard fuzzy arithmetic are dependent on the formula by which a given function is expressed.

For evaluating any arithmetic expression, $Exp(A_1, A_2, \ldots, A_n)$, where $A_i(i \in \mathbb{N})$ are symbols of fuzzy numbers or, more generally, fuzzy intervals (approximating the associated real numbers a_i), the equality constraint is utilized in the following way. First, all symbols that appear in the expression are grouped by the variables they represent. That is, all symbols that represent the same variable are placed in the same group. Each group thus consists of equal fuzzy numbers or intervals that represent the same variable. They are distinguished only by their distinct roles in the expression. Assume that the expression represents a function of m variables v_k whose values range over sets $V_k(k \in \mathbb{N}_m, m \leq n)$. Then, there are m groups of symbols appearing in the expression, each of which is associated with a particular variable v_k and the associated set of values V_k. The groups can be formally defined by a function

$$g : \mathbb{N}_n \to \mathbb{N}_m$$

such that

$$a_{g(i)} = v_k$$

where symbol a_i represents variable v_k. Using this function, let B_k be a common symbol for all the equal fuzzy numbers or intervals in group $k(k \in \mathbb{N}_m)$. Then,

$$
\begin{aligned}
{}^{\alpha}Exp(A_1, A_2, \ldots, A_n) &= \{Exp(a_{g(1)}, a_{g(2)}, \ldots, a_{g(n)}) | \\
\langle v_1, v_2, \ldots, v_m \rangle &\in \{{}^{\alpha}B_1 \times {}^{\alpha}B_2 \times \ldots \times {}^{\alpha}B_m\}.
\end{aligned}
\tag{7.32}
$$

This equation is a formal description of the following *equality rule*: for each group of fuzzy intervals denoted in an arithmetic expression by the same symbol and each $\alpha \in (0, 1]$, selections from these intervals are equal. If the function represented by the expression is monotone increasing or decreasing with respect to each variable $v_k(k \in \mathbb{N}_m)$, then the left-end and right-end points of ${}^{\alpha}Exp$ can be easily expressed in terms of the left-end and right-end points of the intervals ${}^{\alpha}B_1$, ${}^{\alpha}B_2$, ..., ${}^{\alpha}B_m$. Otherwise, the minimum and maximum in the set defined by Eq. (7.32) must be determined by some other means.

Although the equality constraints are perhaps the most common, any other constraints regarding fuzzy numbers or intervals in arithmetic expressions must be taken into account. Each constraint represents some information. Ignoring it inevitably results in the inflated imprecision of the resulting fuzzy number or interval.

A particularly important constraint, bearing upon the fuzzifications of uncertainty theories, is a *probabilistic constraint*. To illustrate this type of constraint, consider a sample space with two elementary events, A and B, whose probabilities are p_A and p_B. Assume that we can only estimate these probabilities in terms of fuzzy numbers P_A and P_B defined on $[0, 1]$. Since it is required that $p_A + p_B = 1$, arithmetic operations on P_A and P_B must take this constraint into account. That is, we must require that $P_A + P_B = 1$.

EXAMPLE 7.6. As a specific example, let $P_A = \langle 0.2, 0.3, 0.3, 0.4 \rangle$. By Eq. (7.26),

$$
{}^{\alpha}P_A = [0.2 + 0.1\alpha, 0.4 - 0.1\alpha]
$$

for all $\alpha \in (0, 1]$. Then,

$$
{}^{\alpha}P_B = 1 - {}^{\alpha}P_A = [0.6 + 0.1\alpha, 0.8 - 0.1\alpha].
$$

Graphs of P_A and P_B are shown in Figure 7.6a. Now, using standard fuzzy arithmetic, we obtain

$$
{}^{\alpha}(P_A + {}^{\alpha}P_B) = [0.8 + 0.2\alpha, 1.2 - 0.2\alpha]
$$

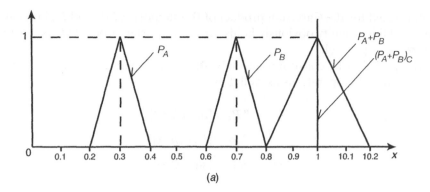

(a)

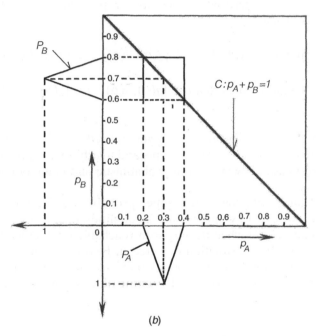

(b)

Figure 7.6. Fuzzy arithmetic with the probabilistic constraint.

by Eq. (7.28). The graph of $P_A + P_B$ is also shown in Figure 7.6a. This result is obtained by ignoring the probabilistic constraint,

$$C : p_A + p_B = 1,$$

which is illustrated in Figure 7.6b. Under this constraint,

$$(P_A + P_B)_C = 1,$$

as illustrated for the Cartesian product of the supports of P_A and P_B in Figure 7.6b, and this equality obviously holds for the Cartesian product ${}^\alpha P_A \times {}^\alpha P_B$ for any $\alpha \in (0, 1]$ as well.

The probabilistic constraint not only affects the operation of addition, but other arithmetic operations as well. Let

$$ {}^\alpha P_A = [\underline{a}(\alpha), \overline{a}(\alpha)] $$

$$ {}^\alpha P_B = [\underline{b}(\alpha), \overline{b}(\alpha)] $$

$$ {}^\alpha (P_A * P_B)_C = [\underline{s}(\alpha), \overline{s}(\alpha)], $$

for all $\alpha \in (0, 1]$, where $*$ denotes any of the four arithmetic operations and the subscript C indicates that the operation is performed under the probability constraint. Then

- $\underline{s}(\alpha) = \min\{a * b\}$ such that $a \in [\underline{a}(\alpha), \overline{a}(\alpha)]$, $b \in [\underline{b}(\alpha), \overline{b}(\alpha)]$, and $a + b = 1$;
- $\overline{s}(\alpha) = \max\{a * b\}$ such that $a \in [\underline{a}(\alpha), \overline{a}(\alpha)]$, $b \in [\underline{b}(\alpha), \overline{b}(\alpha)]$, and $a + b = 1$.

These optimization problems can be generalized in a fairly straightforward way to more then two elementary events and to arbitrary expressions. In these more general cases, however, both probabilistic and equality constraints are often involved.

When dealing with lower and upper probabilities, the probabilistic constraints are expressed by appropriate inequalities, such as the inequalities in Eqs. (5.61) and (5.62) when dealing with reachable interval-valued probability distributions. In the situation depicted in Figure 7.6b, for example, lower probabilities are constrained to points that are under the diagonal C, while upper probabilities are constrained to points that are over the diagonal.

7.5. FUZZY RELATIONS

When fuzzy sets are defined on universal sets that are Cartesian products of two or more sets, we refer to them as *fuzzy relations*. Individual sets in the Cartesian product of a fuzzy relation are called *dimensions* of the relation. When n-sets are involved in the Cartesian product, we call the relation n-dimensional ($n \geq 2$). Fuzzy sets may be viewed as degenerate, one-dimensional relations.

All concepts and operations applicable to fuzzy sets are applicable to fuzzy relations as well. However, fuzzy relations involve additional concepts and operations due to their multidimensionality. These additional operations include projections, cylindric extensions, compositions, joins, and inverses of fuzzy relations. Projections and cylindric extensions are applicable to any fuzzy relations, whereas compositions, joins, and inverses are applicable only to binary relations.

7.5.1. Projections and Cylindric Extensions

The operations of projection and cylindric extension are applicable to any n-dimensional fuzzy relation ($n \geq 2$). However, for the sake of simplicity, they are discussed here in terms of 3-dimensional relations. A generalization to higher dimensions is quite obvious.

Let R denote a 3-dimensional (ternary) fuzzy relation on $X \times Y \times Z$. A *projection* of R is an operation that converts R into a lower-dimensional fuzzy relation, which in this case is either a 2-dimensional or 1-dimensional (degenerate) relation. In each projection, some dimensions are suppressed (not recognized) and the remaining dimensions are consistent with R in the sense that each α-cut of the projection is a projection of the α-cut of R in the sense of classical set theory. Formally, the three 2-dimensional projections of R on $X \times Y$, $X \times Z$, and $Y \times Z$, R_{XY}, R_{XZ}, and R_{YZ}, are defined for all $x \in X$, $y \in Y$, $z \in Z$ by the following formulas:

$$R_{XY}(x, y) = \max_{z \in Z} R(x, y, z),$$

$$R_{XZ}(x, z) = \max_{y \in Y} R(x, y, z),$$

$$R_{YZ}(y, z) = \max_{x \in X} R(x, y, z).$$

Moreover, the three 1-dimensional projections of R on X, Y, and Z, R_X, R_Y, and R_Z, can then be obtained by similar formulas from the 2-dimensiional projections:

$$R_X(x) = \max_{y \in Y} R_{XY}(x, y)$$
$$= \max_{z \in Z} R_{XZ}(x, z)$$

$$R_Y(y) = \max_{x \in X} R_{XY}(x, y)$$
$$= \max_{z \in Z} R_{YZ}(y, z)$$

$$R_Z(z) = \max_{x \in X} R_{XZ}(x, z)$$
$$= \max_{y \in Y} R_{YZ}(y, z).$$

Any relation on $X \times Y \times Z$ that is consistent with a given projection is called an *extension* of the projection. The largest among the extensions is called a *cylindric extension*. Let R_{EXY} and R_{EX} denote the cylindric extensions of projections R_{XY} and R_X, respectively. Then, R_{EXY} and R_{EX} are defined for all triples $\langle x, y, z \rangle \in X \times Y \times Z$ by the formula

$$R_{EXY}(x, y, z) = R_{XY}(x, y),$$

$$R_{EX}(x, y, z) = R_X(x).$$

Cylindric extensions of the other 2-dimensional and 1-dimensional projections are defined in a similar way. This definition of cylindric extension for fuzzy relations is a cutworthy generalization of the classical concept of cylindric extension.

Given any set of projections of a given relation R, the standard fuzzy intersection of their cylindric extensions (expressed by the minimum operator) is called a *cylindric closure* of the projections. This is again a cutworthy concept. Regardless of the given projections, it is guaranteed that their cylindric closure contains the fuzzy relation R.

EXAMPLE 7.7. To illustrate the concepts introduced in this section, let us consider a binary relation R on $X \times Y$, where $X = \{x_1, x_2, \ldots, x_6\}$ and $Y = \{y_1, y_2, \ldots, y_{10}\}$, which is defined by the matrix

		y_1	y_2	y_3	y_4	y_5	y_6	y_7	y_8	y_9	y_{10}
	x_1	0.2	0.0	1.0	0.0	0.0	0.0	1.0	0.0	0.0	0.0
	x_2	1.0	0.0	0.0	0.3	0.0	0.4	0.0	0.0	1.0	0.0
$R =$	x_3	0.0	0.0	0.8	0.0	0.4	0.0	0.1	0.0	0.0	0.0
	x_4	0.0	0.1	0.0	0.0	0.0	0.0	0.0	0.9	0.7	0.5
	x_5	0.1	0.0	0.5	0.0	0.0	0.6	0.0	0.0	0.0	0.0
	x_6	0.0	1.0	0.0	0.0	0.2	0.0	1.0	0.0	0.0	0.5

The projections of R into X and Y are obtained for all $x_i \in X$ and for all $y_i \in Y$ by the equations

$$R_X(x_i) = \max_{y_j \in Y} R(x_i, y_j),$$

$$R_Y(y_j) = \max_{x_i \in X} R(x_i, y_j),$$

respectively. It is convenient to express them as vectors

	x_1	x_2	x_3	x_4	x_5	x_6
$R_X = [$	1.0	1.0	0.8	0.9	0.6	1.0],

	y_1	y_2	y_3	y_4	y_5	y_6	y_7	y_8	y_9	y_{10}
$R_Y = [$	1.0	1.0	1.0	0.3	0.4	0.6	1.0	0.9	1.0	0.5].

Now, the cylindric extensions R_{EX} and R_{EY} are obtained for all pairs $\langle x_i, y_j \rangle \in \langle X \times Y \rangle$ by the equations

$$R_{EX}(x_i, y_j) = R_X(x_i),$$

$$R_{EY}(x_i, y_j) = R_Y(y_j),$$

and their matrix representations are

$$
R_{EX} =
\begin{array}{c c}
 & \begin{array}{cccccccccc} y_1 & y_2 & y_3 & y_4 & y_5 & y_6 & y_7 & y_8 & y_9 & y_{10} \end{array} \\
\begin{array}{c} x_1 \\ x_2 \\ x_3 \\ x_4 \\ x_5 \\ x_6 \end{array} &
\left[\begin{array}{cccccccccc}
1.0 & 1.0 & 1.0 & 1.0 & 1.0 & 1.0 & 1.0 & 1.0 & 1.0 & 1.0 \\
1.0 & 1.0 & 1.0 & 1.0 & 1.0 & 1.0 & 1.0 & 1.0 & 1.0 & 1.0 \\
0.8 & 0.8 & 0.8 & 0.8 & 0.8 & 0.8 & 0.8 & 0.8 & 0.8 & 0.8 \\
0.9 & 0.9 & 0.9 & 0.9 & 0.9 & 0.9 & 0.9 & 0.9 & 0.9 & 0.9 \\
0.6 & 0.6 & 0.6 & 0.6 & 0.6 & 0.6 & 0.6 & 0.6 & 0.6 & 0.6 \\
1.0 & 1.0 & 1.0 & 1.0 & 1.0 & 1.0 & 1.0 & 1.0 & 1.0 & 1.0
\end{array} \right]
\end{array}
$$

$$
R_{EY} =
\begin{array}{c c}
 & \begin{array}{cccccccccc} y_1 & y_2 & y_3 & y_4 & y_5 & y_6 & y_7 & y_8 & y_9 & y_{10} \end{array} \\
\begin{array}{c} x_1 \\ x_2 \\ x_3 \\ x_4 \\ x_5 \\ x_6 \end{array} &
\left[\begin{array}{cccccccccc}
1.0 & 1.0 & 1.0 & 0.3 & 0.4 & 0.6 & 1.0 & 0.9 & 1.0 & 0.5 \\
1.0 & 1.0 & 1.0 & 0.3 & 0.4 & 0.6 & 1.0 & 0.9 & 1.0 & 0.5 \\
1.0 & 1.0 & 1.0 & 0.3 & 0.4 & 0.6 & 1.0 & 0.9 & 1.0 & 0.5 \\
1.0 & 1.0 & 1.0 & 0.3 & 0.4 & 0.6 & 1.0 & 0.9 & 1.0 & 0.5 \\
1.0 & 1.0 & 1.0 & 0.3 & 0.4 & 0.6 & 1.0 & 0.9 & 1.0 & 0.5 \\
1.0 & 1.0 & 1.0 & 0.3 & 0.4 & 0.6 & 1.0 & 0.9 & 1.0 & 0.5
\end{array} \right]
\end{array}
$$

The cylindric closure of projections R_X and R_Y, $Cyl\{R_X, R_Y\}$, is then obtained by taking the standard intersection of these cylindric extensions. That is,

$$Cyl\{R_x, R_Y\}(x_i, y_i) = \min\{R_{EX}(x_i, y_j), R_{EY}(x_i, y_j)\}$$

for all pairs $\langle x_i, y_j \rangle \in \langle X \times Y \rangle$. We obtain

$$
Cyl\{R_X, R_Y\} =
\begin{array}{c c}
 & \begin{array}{cccccccccc} y_1 & y_2 & y_3 & y_4 & y_5 & y_6 & y_7 & y_8 & y_9 & y_{10} \end{array} \\
\begin{array}{c} x_1 \\ x_2 \\ x_3 \\ x_4 \\ x_5 \\ x_6 \end{array} &
\left[\begin{array}{cccccccccc}
1.0 & 1.0 & 1.0 & 0.3 & 0.4 & 0.6 & 1.0 & 0.9 & 1.0 & 0.5 \\
1.0 & 1.0 & 1.0 & 0.3 & 0.4 & 0.6 & 1.0 & 0.9 & 1.0 & 0.5 \\
0.8 & 0.8 & 0.8 & 0.3 & 0.4 & 0.6 & 0.8 & 0.8 & 0.8 & 0.5 \\
0.9 & 0.9 & 0.9 & 0.3 & 0.4 & 0.6 & 0.9 & 0.9 & 0.9 & 0.5 \\
0.6 & 0.6 & 0.6 & 0.3 & 0.4 & 0.6 & 0.6 & 0.6 & 0.6 & 0.5 \\
1.0 & 1.0 & 1.0 & 0.3 & 0.4 & 0.6 & 1.0 & 0.9 & 1.0 & 0.5
\end{array} \right]
\end{array}
$$

We can see that $R \subseteq Cyl\{R_X, R_Y\}$, which is always the case, but this reconstructed relation from the projections R_X and R_Y is in this case much larger than the actual relation R. This means that information about R that is contained in the projection about R is in this case highly deficient.

7.5.2. Compositions, Joins, and Inverses

Consider two binary fuzzy relations P and Q that are defined on set $X \times Y$ and $Y \times Z$, respectively. Any such relations, which are connected via the common set Y, can be composed to yield a relation on $Y \times Z$. The standard composition of these relations, which is denoted by $P \circ Q$, produces a relation R on $X \times Z$ defined by the formula

$$R(x, z) = (P \circ Q)(x, z) = \max_{y \in Y} \min\{P(x, y), Q(y, z)\} \tag{7.33}$$

for all pairs $\langle x, z \rangle \in X \times Z$.

Other definitions of a composition of fuzzy relations, in which the min and max operations are replaced with other t-norms and t-conorms, respectively, are possible and useful in some applications. All compositions are associative:

$$(P \circ Q) \circ Q = P \circ (Q \circ Q).$$

However, the standard fuzzy composition is the only one that is cutworthy.

A similar operation on two connected binary relations, which differs from the composition in that it yields a 3-dimensional relation instead of a binary one, is known as the relational join. For the same fuzzy relations P and Q, the *standard relational join*, $P * Q$, is a 3-dimensional relation on $X \times Y \times Z$ defined by the formula

$$R(x, y, z) = (P * Q)(x, y, z) = \min\{P(x, y), Q(y, z)\} \tag{7.34}$$

for all triples $\langle x, y, z \rangle \in X \times Y \times Z$. Again, the min operation in this definition may be replaced with another t-norm. However, the relational join defined by Eq. (7.34) is the only one that is cutworthy.

The *inverse* of a binary fuzzy relation R on $X \times Y$, denoted by R^{-1}, is a relation on $Y \times X$ such that

$$R^{-1}(y, x) = R(x, y)$$

for all pairs $\langle y, x \rangle \in Y \times X$. When R is represented by a matrix, R^{-1} is represented by the transpose of this matrix. This means that rows are replaced with columns and vice versa. Clearly,

$$(R^{-1})^{-1} = R$$

holds for any binary relation.

It is convenient to perform compositions of binary fuzzy relations in terms of their matrix representations. Let

$$\mathbf{P} = [p_{ij}], \qquad \mathbf{Q} = [q_{jk}], \qquad \text{and} \qquad \mathbf{R} = [r_{ik}]$$

be matrix representations of fuzzy relations P, Q, R in Eq. (7.33). Then, using this matrix notation, we can write

$$[r_{ik}] = [p_{ij}] \circ [q_{jk}],$$

where

$$r_{ik} = \max_j \min\{p_{ij}, q_{jk}\}. \tag{7.35}$$

Observe that the same elements of matrices $\mathbf{P}$ and $\mathbf{Q}$ are used in the calculation of matrix $\mathbf{R}$ as would be used in the regular matrix multiplication, but the product and sum operations are replaced, respectively, with the min and max operations.

EXAMPLE 7.8. To illustrate the standard composition of fuzzy relations by their matrix representations, let symbols P, Q, R have the same meaning as in Eq. (7.33), where $X = \{x_1, x_2, x_3, x_4\}$, $Y = \{y_1, y_2, y_3\}$, and $Z = \{z_1, z_2, z_3, z_4, z_5\}$, and consider the following example of the matrix representation of equation $P \circ Q = R$:

$$
\begin{array}{c}
\begin{array}{ccc} y_1 & y_2 & y_3 \end{array} \\
\begin{array}{c} x_1 \\ x_2 \\ x_3 \\ x_4 \end{array}
\begin{bmatrix}
0.0 & 0.3 & 0.4 \\
0.2 & 0.5 & 0.3 \\
0.8 & 0.0 & 0.0 \\
0.7 & 0.7 & 1.0
\end{bmatrix}
\end{array}
\circ
\begin{array}{c}
\begin{array}{ccccc} z_1 & z_2 & z_3 & z_4 & z_5 \end{array} \\
\begin{array}{c} y_1 \\ y_2 \\ y_3 \end{array}
\begin{bmatrix}
0.7 & 0.0 & 0.0 & 0.3 & 0.6 \\
0.5 & 0.5 & 1.0 & 0.4 & 0.0 \\
0.0 & 0.7 & 0.2 & 0.9 & 0.0
\end{bmatrix}
\end{array}
$$

$$
=
\begin{array}{c}
\begin{array}{ccccc} z_1 & z_2 & z_3 & z_4 & z_5 \end{array} \\
\begin{array}{c} x_1 \\ x_2 \\ x_3 \\ x_4 \end{array}
\begin{bmatrix}
0.3 & 0.4 & 0.3 & 0.4 & 0.0 \\
0.5 & 0.5 & 0.5 & 0.4 & 0.2 \\
0.7 & 0.0 & 0.0 & 0.3 & 0.6 \\
0.7 & 0.7 & 0.7 & 0.9 & 0.6
\end{bmatrix}
\end{array}.
$$

Following Eq. (7.35), we have, for example,

$$0.3(= r_{11}) = \max\{\min\{p_{11}, q_{11}\}, \min\{p_{12}, q_{21}\}, \min\{p_{13}, q_{31}\}\}$$
$$= \min\{0, 0.7\}, \min\{0.3, 0.5\}, \min\{0.4, 0\}\}$$

$$0.4(= r_{24}) = \max\{\min\{p_{21}, q_{14}\}, \min\{p_{22}, q_{24}\}, \min\{p_{23}, q_{34}\}\}$$
$$= \min\{0.2, 0.3\}, \min\{0.5, 0.4\}, \min\{0.3, 0.9\}\}.$$

When the join operation defined by Eq. (7.34) is applied to the same fuzzy relations, P and Q, we obtain a 3-dimensional relation. Equation (7.34) can be expressed in terms of the associated arrays as

$$[r_{ijk}] = [p_{ij}] * [q_{jk}],$$

where

$$r_{ijk} = \min\{p_{ij}, q_{jk}\}.$$

The resulting 3-dimensional array **R** may conveniently be represented by the following four matrices, each associated with one element of set X:

	y_1	y_2	y_3	y_1	y_2	y_3	y_1	y_2	y_3	y_1	y_2	y_3
z_1	0.0	0.3	0.0	0.2	0.5	0.0	0.7	0.0	0.0	0.7	0.5	0.0
z_2	0.0	0.3	0.4	0.0	0.5	0.3	0.0	0.0	0.0	0.0	0.5	0.7
z_3	0.0	0.3	0.2	0.0	0.5	0.2	0.0	0.0	0.0	0.0	0.7	0.2
z_4	0.0	0.3	0.4	0.2	0.4	0.3	0.3	0.0	0.0	0.3	0.4	0.9
z_5	0.0	0.0	0.0	0.2	0.0	0.0	0.6	0.0	0.0	0.6	0.0	0.0
		x_1			x_2			x_3			x_4	

Equations (7.33), which describe $R = P \circ Q$ are called *fuzzy relation equations*. Normally, it is assumed that P and Q are given and R is determined by Eq. (7.33). However, two *inverse problems* play important roles in many applications. In one of them, R and P are given and Q is to be determined; in the other one, R and Q are given and P is to be determined. Various methods for solving these problems exactly as well as approximately have been developed.

It should also be mentioned that cutworthy fuzzy counterparts of the various classical binary relations on $X \times X$, such as equivalence relations, and the various ordering relations, have been extensively investigated. However, many types of fuzzy relations on $X \times X$ that are not cutworthy have been investigated as well and found useful in many applications.

7.6. FUZZY LOGIC

The term "fuzzy logic," as currently used in the literature, has two distinct meanings. It is viewed either in a *narrow sense* or in a *broad sense*. Fuzzy logic in the narrow sense is a generalization of the various many-valued logics, which have been investigated in the area of symbolic logic since the beginning of the 20th century. It is concerned with development of *syntactic aspects* (based on the notion of *proof*) and *semantic aspects* (based on the notion of *truth*) of a relevant logic calculus. In order to be acceptable, the calculus must be *sound* (provability implies truth) and *complete* (truth implies provability).

Fuzzy logic in the narrow sense is important since it provides theoretical foundations for fuzzy logic in the broad sense. The latter is viewed as a system of concepts, principles, and methods for reasoning that is approximate rather than exact. It utilizes the apparatus of fuzzy set theory for formulating various forms of sound approximate reasoning with fuzzy propositions. It is fuzzy logic in this broad sense that is primarily involved in dealing with fuzzified uncertainty theories.

To establish a connection between fuzzy set theory and fuzzy logic, it is essential to connect degrees of membership in fuzzy sets with degrees of truth of fuzzy propositions. This can only be done when the degrees of membership and the degrees of truth refer to the same objects. Let us consider first the simplest connection, in which only one fuzzy set is involved.

Given a fuzzy set A, its membership degree $A(x)$ for any x in the underlying universal set X can be interpreted as the degree of truth of the associated fuzzy proposition "x is a member of A." Conversely, given an arbitrary proposition of the simple form "x is A," where x is from X and A is a fuzzy set that represents an inherently vague linguistic term (such as *low, high, near, fast*), its degree of truth may be interpreted as the membership degree of x in A. That is, the degree of truth of the proposition is equal to the degree with which x belongs to A.

This simple correspondence between membership degrees and degrees of truth, which conforms well to our intuition, forms a basis for determining degrees of truth of more complex propositions. Moreover, negations, conjunctions, and disjunctions of fuzzy propositions are defined under this correspondence in exactly the same way as complement, intersections, and unions of fuzzy sets, respectively.

7.6.1. Fuzzy Propositions

Now let us examine basic propositional forms of fuzzy propositions. To do that, we need a convenient notation. Let X denote a variable whose states (values) are in set X, and let A denote a fuzzy set defined on X that represents an approximate description of the state of variable X by a linguistic term such as *low, medium, high, slow, fast*.

Using this notation, the simplest fuzzy propositions (unconditional and unqualified) are expressed in the *canonical propositional form*,

$$f_A(x): X \text{ is } A,$$

in which the fuzzy set A is called a *fuzzy predicate*. Given this propositional form, a fuzzy proposition, $f_A(x)$, is obtained when a particular state (value) x from X is substituted for variable X in the propositional form. That is,

$$f_A(x): X = x \text{ is } A,$$

where $x \in X$, is a particular *fuzzy proposition* of propositional form f_A. For simplicity, let $f_A(x)$ also denote the degree of truth of the proposition "x is A." This means that the symbol f_A denotes a propositional form as well as a function by which degrees of truth are assigned to fuzzy propositions based on the form. This double use of the symbol f_A does not create any ambiguity, since there is only one function for each propositional form that assigns degrees of truth to individual propositions subsumed under the form. In this case, the function is defined for all $x \in X$ by the simple equation

$$f_A(x) = A(x).$$

That is, $f_A(x)$ is true to the same degree to which x in a member of fuzzy set A.

The propositional form f_A can be modified by qualifying the claims for the degree of truth of the associated fuzzy propositions. Two types of qualified propositional forms are recognized: a truth-qualified propositional form and a probability-qualified propositional form. These two forms also can be combined.

The *truth-qualified propositional form*, $f_{T(A)}$, is obtained by adding a *truth qualifier*, T, to the basic form of f_A. That is,

$$f_{T(A)} : X \text{ is } A \text{ is } T,$$

where T is a fuzzy number or interval on $[0, 1]$ that represents a linguistic term by which the meaning of truth in fuzzy propositions associated with this propositional form is qualified. Linguistic terms such as *very true, fairly true, barely true, false, fairly false*, or *virtually false* are typical examples of linguistic truth qualifiers.

To obtain the degree of truth of a truth-qualified proposition $f_{T(A)}$, we need to compose the membership function A with the membership function of the truth qualifier T. That is,

$$f_{T(A)}(x) : T(A(x)) \tag{7.36}$$

for all $x \in X$.

EXAMPLE 7.9. Consider truth-qualified fuzzy propositions of the form

$$f_{T(A)} : \text{The rate of inflation is low is fairly true.}$$

In this example, X is the variable "the rate of inflation" (in %), A is a fuzzy set representing the predicate "low" (L), T is a fuzzy set representing the truth qualifier "fairly true" (FT). The relevant membership functions of L and FT are shown in Figure 7.7. Now assume that the actual rate of inflation is 2.5%. Then, $L(2.5) = 0.75$ and $FT(0.75) = 0.86$. Hence

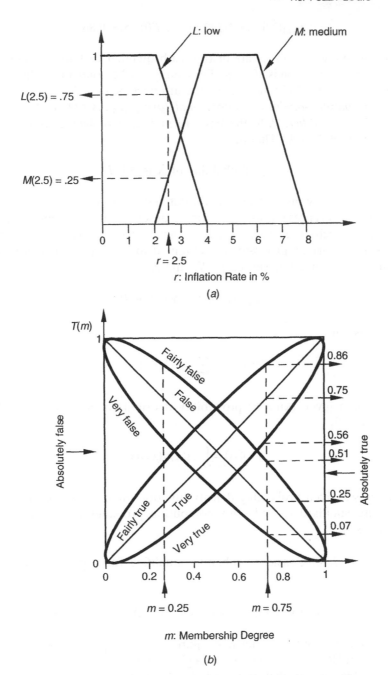

Figure 7.7. Illustration of truth-qualified unconditional fuzzy propositions.

$$f_{T(L)}(2.5) = FT(L(2.5)) = FT(0.75) = 0.86.$$

When we change the predicate or the truth qualifier or both, the degree of truth of the proposition is affected. For example, when only the truth qualifier is changed to "false" (F) the degree of truth becomes $F(0.75) = 0.25$.

The *probability-qualified propositional form*, $f_{P(A)}$, is obtained by adding a *probability qualifier*, P, to the basic form f_A and replacing "X is A" with "Probability of X is A." That is,

$$f_{P(A)} : \text{Probability of } X \text{ is } A \text{ is } P,$$

where P is a fuzzy number or interval defined on $[0, 1]$, called a *probability qualifier*, which represents an approximate description of probability by a linguistic term, such as *likely*, *very likely*, *extremely likely*, around 0.5. Given this propositional form, a proposition is obtained for each particular probability measure, *Pro*, defined on subsets of X. That is,

$$f_{P(A)} : (Pro(A)) : Pro(A) \text{ is } P.$$

When X is finite,

$$Pro(A) = \sum_{x \in X} A(x)p(x), \tag{7.37}$$

where p is a given (known) probability distribution function on X. When $X = [\underline{x}, \bar{x}] \subseteq \mathbb{R}$,

$$Pro(A) = \int_X A(x)q(x)\, dx, \tag{7.38}$$

where q is a given probability density function on X. Equations (7.37) and (7.38) are standard formulas for computing probabilities of fuzzy events (Section 8.4).

EXAMPLE 7.10. Consider probability-qualified fuzzy propositions of the form

$$f_{P(F)} : \text{Probability of temperature (at a given place and time) is } A \text{ is } P,$$

where A is a fuzzy number shown in Figure 7.8a that represents the fuzzy predicate "around 80°F," as understood in a given context, and P is a fuzzy number shown in Figure 7.8b that represents the linguistic term "unlikely." Assume now that the probabilities $p(t)$ for the individual temperatures, obtained from relevant statistical data over many years, are given in Table 7.1a. To determine the degree of truth of the proposition, we calculate first $Pro(A)$ by Eq. (7.37):

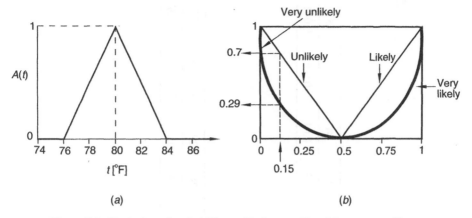

Figure 7.8. Illustration of probability-qualified unconditional fuzzy proposition.

$$Pro(A) = 0.25 \times 0.14 + 0.5 \times 0.11 + 0.75 \times 0.04 + 1 \times 0.02 + 0.75 \times 0.001$$
$$+ 0.5 \times 0.006 + 0.25 \times 0.004 = 0.15.$$

Now, applying this probability value to the probability qualifier "unlikely" in Figure 7.8b, we obtain 0.7. Hence,

$$f_{P(A)}(p) = P(Pro(A)) = P(0.15) = 0.7.$$

Fuzzy propositions that are probability-qualified as well as truth-qualified have the form

$$f_{T[P(A)]} : \text{Probability of } X \text{ is } A \text{ is } P \text{ is } T.$$

To obtain the degree of truth of any proposition of this form, we need to perform two compositions of functions:

$$f_{T[P(A)]}(p) = T\{P[Pro(A)]\},$$

where $Pro(A)$ is calculated either by Eq. (7.37) or by Eq. (7.38).

An important type of fuzzy proposition, which is essential for knowledge-based fuzzy systems, is the *conditional fuzzy proposition*. This type of proposition is based on the propositional form

$$f_{B|A} : \text{If } X \text{ is } A, \text{ then } \mathcal{Y} \text{ is } B,$$

where X and $\mathcal{Y}$ are variables whose states are in sets X and Y, respectively. These propositions may also be expressed in an alternative, but equivalent form

Table 7.1. Probabilities in (a) Example 7.10; (b) Exercise 7.29

t	71	72	73	74	75	76	77	78	79	80	81	82	83
$p(t)$	0.01	0.03	0.11	0.15	0.21	0.16	0.14	0.11	0.04	0.02	0.01	0.006	0.004

(a)

r	18	19	20	21	22	23	24	25	26	27	28	29	30	31	32
$p(r)$	0.005	0.005	0.01	0.02	0.05	0.05	0.04	0.05	0.05	0.05	0.1	0.3	0.2	0.04	0.02

(b)

$$f_{B|A} : (X, \mathcal{Y}) \text{ is } R,$$

where R is a fuzzy relation on $X \times Y$. It is assumed here that R is determined for each $x \in X$ and each $y \in Y$ by the formula

$$R(x, y) = J(A(x), B(x)),$$

where the symbol J stands for a binary operation on $[0, 1]$ that represents an appropriate *fuzzy implication* in the given application context. Clearly,

$$f_{B|A}(x, y) = R(x, y)$$

for all $\langle x, y \rangle \in X \times Y$. Moreover, if a truth qualification or a probability qualification is employed, R must be composed with the respective qualifiers to obtain for each $\langle x, y \rangle \in X \times Y$ the degree of truth of the conditional and qualified proposition.

As is well known, operations that qualify as fuzzy implications form a class of binary operations on $[0, 1]$, similarly as fuzzy intersections and fuzzy unions. An important class of fuzzy implications, referred to as *Lukasiewicz implications*, is defined for each $a \in [0, 1]$ and each $b \in [0, 1]$ by the formula

$$J(a, b) = \min[1, 1 - a^\lambda + b^\lambda]^{1/\lambda}, \tag{7.39}$$

where $\lambda > 0$ is a parameter by which individual implications are distinguished from one another.

Fuzzy propositions of any of the introduced types may also be quantified. In general, fuzzy quantifiers are fuzzy intervals. This subject is beyond the scope of this overview.

7.6.2. Approximate Reasoning

Reasoning based on fuzzy propositions of the various types is usually referred to as *approximate reasoning*. The most fundamental components of approximate reasoning are conditional fuzzy propositions, which may also be truth qualified, probability qualified, quantified, or any combination of these. Special procedures are needed for each of these types of fuzzy propositions. This large variety of fuzzy propositions makes approximate reasoning methodologically rather intricate. This reflects the richness of natural language and the many intricacies of common-sense reasoning, which approximate reasoning based upon fuzzy set theory attempts to emulate.

To illustrate the essence of approximate reasoning, let us characterize the fuzzy-logic generalization of one of the most common inference rules of classical logic: modus ponens. The *generalized modus ponens* is expressed by the following schema:

Fuzzy rule:	If X is A, then Y is B
Fuzzy fact:	X is F
Fuzzy conclusion:	Y is C.

Clearly, in this schema A and F are fuzzy sets defined on X, while B and C are fuzzy sets defined on Y. Assuming that the fuzzy rule is already converted to the alternative form

$$(X, Y) \text{ is } R,$$

where R represents the fuzzy implication employed, the fuzzy conclusion C is obtained by composing F with R. That is

$$B = F \circ R$$

or, more specifically,

$$B(y) = \max_{x \in X} \{\min[F(x), R(x, y)]\} \tag{7.40}$$

for all $y \in Y$. This way of obtaining the conclusion according to the generalized modus ponens schema is called a *compositional rule of inference*.

To use the compositional rule of inference, we need to choose a fitting fuzzy implication in each application context and express it in terms of a fuzzy relation R. There are several ways in which this can be done. One way is to derive from the application context (by observing or by expert's judgements) pairs F, C of fuzzy sets that are supposed to be inferentially connected (facts and conclusions). Relation R, which represents a fuzzy implication, is then determined by solving the inverse problem of fuzzy relation equations. This and other issues regarding fuzzy implications in approximate reasoning are discussed fairly thoroughly in the literature (see Note 7.9).

7.7. FUZZY SYSTEMS

In general, each classical system is ultimately a set of variables together with a relation among states (or values) of the variables. When the states of variables are fuzzy sets, the system is called a *fuzzy system*. In most typical fuzzy systems, the states are fuzzy intervals that represent linguistic terms such as *very small, small, medium, large, very large*, as interpreted in the context of each particular application. If they do, the variables are called *linguistic variables*.

Each linguistic variable is defined in terms of a base variable, whose values are usually real numbers within a specific range. A base variable is a variable in the classical sense, as exemplified by any physical variable (temperature, pressure, tidal range, grain size, etc.). Linguistic terms involved in a linguistic variable are used for approximating the actual values of the associated base variable. Their meanings are captured, in the context of each particular application, by appropriate fuzzy intervals. That is, each linguistic variable consists of:

- A *name*, which should reflect the meaning of the base variable involved.
- A *base variable* with its *range of values* (usually a closed interval of real numbers).
- A set of *linguistic terms* that refers to values of the base variable.
- A set of *semantic rules*, which assign to each linguistic term its meaning in terms of an appropriate fuzzy interval (or some other fuzzy set) defined on the range of the base variable.

An example of a linguistic variable is shown in Figure 7.9. Its name "interest rate" captures the meaning of the base variable—a real variable whose range is [0, 20]. Five linguistic states are distinguished by the linguistic terms *very small*, *small*, *medium large*, *large*, and *very large*. Each of these terms is represented by a trapezoidal-shaped fuzzy interval, as shown in the figure.

Consider a linguistic variable whose base variable has states (values) in set X. Fuzzy sets representing a finite set of linguistic states of the linguistic variable are often defined in such a way that the sum of membership degrees in these fuzzy sets is equal to 1 for each $x \in X$. Such a family of fuzzy subsets of X is called a *fuzzy partition* of X. Formally a finite family $\{A_i | A_i \in \mathcal{F}(X), i \in \mathbb{N}_n, n \geq 1\}$ of n fuzzy subsets of X is a fuzzy partition if and only if $A_i \neq \varnothing$ for all $i \in \mathbb{N}_n$ and

$$\sum_{i \in \mathbb{N}_n} A_i(x) = 1 \quad \text{for each } x \in X.$$

An example of a fuzzy partition of $X = [0, 20]$ is the family of fuzzy intervals in Figure 7.9.

7.7.1. Granulation

Representing states of variables by fuzzy sets (usually fuzzy numbers or intervals) is called a *granulation*. It is a fuzzy counterpart of classical *quantization*, which is any meaningful grouping of states of variables into quanta (or aggregates). An example of quantization of a real variable whose range is [0, 1] into eleven quanta (semiopen or closed intervals of real numbers) is shown in Figure 7.10*a*. One of many possible fuzzy counterparts of this quantization, a particular granulation of the variable by triangular fuzzy numbers, is shown in Figure 7.10*b*.

While viewing physical variables, such as temperature, pressure, and electric current, as real variables is mathematically convenient, this view introduces a fundamental inconsistency between the infinite precision

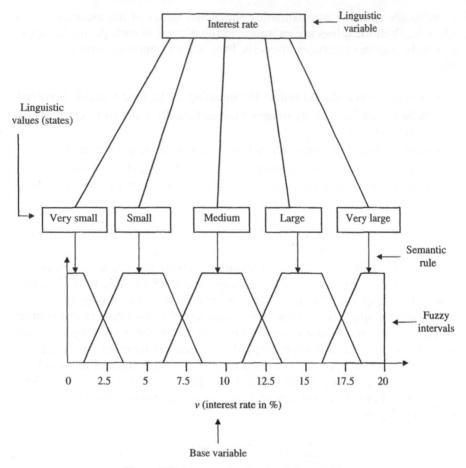

Figure 7.9. An example of a linguistic variable.

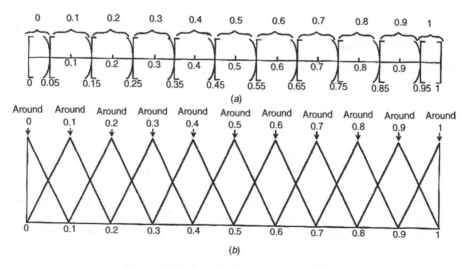

Figure 7.10. Quantization versus granulation.

required to distinguish real numbers and the finite precision of any measuring instrument. Appropriate quantization, whose coarseness reflects the precision of a given measuring instrument, is thus inevitable to resolve this inconsistency. Consider, for example, a real variable that represents electric current whose values range from 0 to 1 ampere. Assume, for the sake of simplicity, that measurements of the variable can be made to an accuracy of 0.1 ampere. Then, according to the usual quantization, the interval [0, 1] is partitioned into 10 semiopen intervals and one closed interval (quanta, aggregates), $[0, 0.5), [0.5, 1.5), [1.5, 2.5), \ldots, [0.85, 0.95), [0.95, 1]$, which are labeled by their ideal representatives $0, 0.1, 0.2, \ldots, 0.9, 1$, respectively. This is exactly the quantization shown in Figure 7.10a.

Although the usual quantization of real variables is capable of capturing the limited resolutions of measuring instruments employed, it completely ignores the issue of measurement errors. While representing the infinite number of values by a finite number of quanta (disjoint intervals of real numbers and their ideal representatives) the unavoidable measurement errors make the sharp boundaries between the quanta highly unrealistic. The representation can be made more realistic by a finite number of granules (fuzzy numbers or intervals), as illustrated for our example in Figure 7.10b. The fundamental difference is that transitions from each granule to its adjacent granules are smooth rather than abrupt. Moreover, any available knowledge regarding measurement errors in each particular application context can be utilized in molding the granules.

7.7.2. Types of Fuzzy Systems

In principle, fuzzy systems can be knowledge-based, model-based, or hybrid. In *knowledge-based fuzzy systems*, relationships between variables are described by collections of *fuzzy inference rules* (conditional fuzzy propositional forms). These rules attempt to capture the knowledge of a human expert, expressed often in natural language. *Model-based fuzzy systems* are based on traditional systems modeling, but they employ appropriate areas of fuzzy mathematics (fuzzy analysis, fuzzy differential equations, etc.). These mathematical areas, based on the notion of fuzzy numbers or intervals, allow us to approximate classical mathematical systems of various types via appropriate granulation to achieve tractability, robustness, and low computational cost. *Hybrid fuzzy systems* are combinations of knowledge-based and model-based fuzzy systems. At this time, knowledge-based fuzzy systems are more developed than model-based or hybrid fuzzy systems.

In knowledge-based fuzzy systems, the relation between input and output linguistic variables is expressed in terms of a set of fuzzy inference rules (conditional propositional forms). From these rules and any information describing actual states of input variables, the actual states of output variables are derived by an appropriate compositional rule of inference. Assuming that the input variables are $X_1, X_2, \ldots$, and the output variables are $Y_1, Y_2, \ldots$, we have the following general scheme of inference to represent the input–output relation of the system:

Rule 1:	If X_1 is A_{11} and X_2 is A_{21} and $\cdots$, then Y_1 is B_{11} and Y_2 is B_{21} and $\cdots$
Rule 2:	If X_1 is A_{12} and X_2 is A_{22} and $\cdots$, then Y_1 is B_{12} and Y_2 is B_{22} and $\cdots$

....

Rule n:	If X_1 is A_{1n} and X_2 is A_{2n} and $\cdots$ then Y_1 is B_{1n} and Y_2 is B_{2n} and $\cdots$
Fact:	X_1 is C_1 and X_2 is C_2 and $\cdots$
Conclusion:	Y_1 is D_1 and Y_2 is D_2 and $\cdots$

This general formulation of knowledge-based fuzzy systems involves many complex issues regarding the construction of the inference rules as well as procedures for using them to make inferences. These issues have been extensively addressed in the literature, but they are beyond the scope of this overview.

Although fuzzy systems that approximate relationships among numerical base variables have special significance, due to their extensive applicability, they are not the only fuzzy sets. Any classical (crisp) system whose variables are not numerical (e.g., ordinal-scale or nominal-scale variables) can be fuzzified as well.

7.7.3. Defuzzification

The result of each fuzzy inference based on a system of input and output linguistic variables that approximate numerical base variables is, in general, a set of fuzzy intervals, one for each output variable. Some applications (such as control or decision making) require that each of these fuzzy intervals be converted to a single real number that, in the context of a given application, best represent the fuzzy interval. This conversion of a given fuzzy interval A, to its real-number representation $d(A)$, is called a *defuzzification* of A.

The most common defuzzification method, which is called a *centroid method*, is defined by the formula

$$d(A) = \frac{\int_{\mathbb{R}} x \cdot A(x)\, dx}{\int_{\mathbb{R}} A(x)\, dx} \tag{7.41}$$

or, when A is defined on a finite universal set of numbers X, by the formula

$$d(A) = \frac{\sum_{x \in X} x \cdot A(x)}{\sum_{x \in X} A(x)}. \tag{7.42}$$

This method may be viewed as unbiased since it treats all values $A(x)$ equally. However, it has been observed that other methods of defuzzification are preferable in some applications. In these methods, differences in values $A(x)$ are either upgraded or downgraded to various degrees. Such methods are members of the general class of defuzzification methods defined either by the formula

$$d_\lambda(A) = \frac{\int_{\mathbb{R}} x \cdot A^\lambda(x)\, dx}{\int_{\mathbb{R}} A^\lambda(x)},$$

(7.43)

when A is defined on $\mathbb{R}$, or by the formula

$$d_\lambda(A) = \frac{\sum_{x \in X} x \cdot A^\lambda(x)}{\sum_{x \in X} A^\lambda(x)},$$

(7.44)

when A is defined on a finite universal set X. Individual methods in this class are distinguished by the value of parameter $\lambda \in (0, \infty)$. When $\lambda = 1$, we obtain the centroid method. When $\lambda > 1$, differences in values $A(x)$ are upgraded; when $\lambda < 1$, they are downgraded.

An alternative kind of defuzzification is to replace fuzzy intervals with representative crisp intervals rather than with individual real numbers. This kind of defuzzification can be formalized rationally in information–theoretic terms, as is shown in Chapter 9. Another problem, somewhat connected with defuzzification, is the problem of converting a given fuzzy set to a linguistic expression. This problem, which is usually referred to as linguistic approximation, can also be dealt with in information-theoretic terms.

7.8. NONSTANDARD FUZZY SETS

In addition to standard fuzzy sets, several other types of fuzzy sets have been introduced in the literature. Each of them is a nucleus of a particular formalized language, which may be viewed as a branch of the overall fuzzy set theory. Distinct types of fuzzy sets are distinguished from one another by the domain and range of their membership functions.

The following are definitions of the most visible of nonstandard fuzzy sets. In each of them, symbols X and A denote, respectively, the universal set of concern and the fuzzy set defined.

Interval-Valued Fuzzy Sets. Membership functions have the form

$$A : X \to CI([0, 1]),$$

where $CI([0, 1])$ denotes the set of all closed intervals contained in $[0, 1]$. For each $x \in X$, $A(x)$ is a closed interval of real numbers in $[0, 1]$. An alternative formulation:

$$A = \langle \underline{A}, \overline{A} \rangle,$$

where $\underline{A}$ and $\overline{A}$ are standard fuzzy sets such that $\underline{A}(x) \le \overline{A}(x)$ for all $x \in X$. Fuzzy sets defined in this way are sometimes called *gray fuzzy sets*. Clearly, for each $x \in X$

$$A(x) = \left[\underline{A}(x), \overline{A}(x) \right] \in CI([0, 1]).$$

An example of an interval-valued fuzzy set is shown in Figure 7.11.

Fuzzy Sets of Type 2. Membership functions have the form

$$A : X \rightarrow FI([0, 1]),$$

where $FI([0, 1])$ denotes the set of all fuzzy intervals defined on $[0, 1]$. For each $x \in X$, $A(x) = I_x$, where I_x is a fuzzy interval defined on $[0, 1]$. This fuzzy interval defines (imprecisely) the membership degree of x in A. An example is shown in Figure 7.12, where the fuzzy intervals are assumed to be of a trapezoidal shape. For each $x \in X$, the closed interval defined by the shaded area is the core of I_x, and the closed interval defined by the uppermost and lowermost curves is the support of I_x.

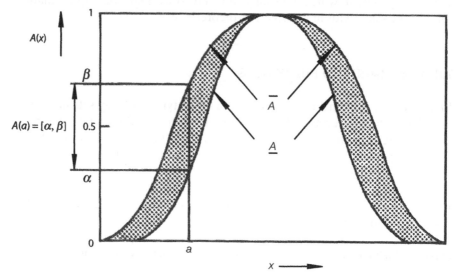

Figure 7.11. Example of an interval-valued fuzzy set (or a gray set).

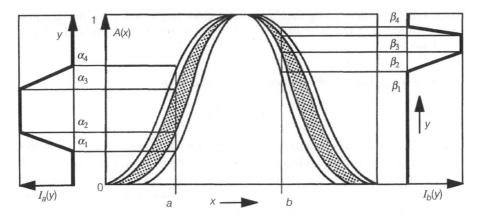

Figure 7.12. Example of a fuzzy set of type 2.

Fuzzy Sets of Type t (t > 2). For each $t > 2$, membership functions are defined recursively by the form

$$A : X \to FI^{t-1}([0, 1]),$$

where $FI^{t-1}([0, 1])$ denote the set of all fuzzy intervals of type $t - 1$. These types of fuzzy sets were introduced in the literature as theoretically possible generalizations of fuzzy sets of type 2. However, they have not been seriously investigated so far, and their practical utility remains to be seen.

Fuzzy Sets of Level 2. Membership functions have the form

$$A : \mathcal{F}(X) \to [0, 1],$$

where $\mathcal{F}(X)$ denotes a family of fuzzy sets defined on X. That is, a fuzzy set of level 2 is defined on a family of fuzzy sets, each of which is defined, in turn, on a given universal set X. This mathematical structure allows us to represent a higher-level concept by lower-level concepts, all expressed in imprecise linguistic terms of natural language.

Fuzzy Sets of Level l (l > 2). For each $l > 2$, membership functions are defined recursively by the form

$$A : \mathcal{F}^{(l-1)}(X) \to [0, 1],$$

where $\mathcal{F}^{(l-1)}(X)$ denotes a family of fuzzy sets of level $l - 1$. Sets of these types are natural generalizations of fuzzy sets of level 2. They are sufficiently expressive to facilitate representation of high-level concepts embedded in natural language. Notwithstanding their importance for linguistics, cognitive science, and knowledge representation, their theory has not been adequately developed as yet.

L-Fuzzy Sets. Membership functions have the form

$$A : X \to L,$$

where L denotes a recognized set of membership grades, which are not required to be numerical. However, the membership grades recognized in L are required to be at least partially ordered. Usually, L is assumed to be a complete lattice. These fuzzy sets are very general, and some of the other types of fuzzy sets may be viewed as special L-fuzzy sets.

Intuitionistic Fuzzy Sets. Fuzzy sets of this type are defined by pairs of standard fuzzy sets,

$$A = \langle AM, AN \rangle,$$

where AM and AN denote standard fuzzy sets on X such that

$$0 \le AM(x) + AN(x) \le 1$$

for each $x \in X$. The values $AM(x)$ and $AN(x)$ are interpreted for each $x \in X$ as, respectively, the degree of membership and the degree of nonmembership of x in A.

Rough Fuzzy Sets. These are fuzzy sets whose α-cuts are approximated by *rough sets*. That is, A_R is a rough approximation of a fuzzy set A based on an equivalence relation R on X. Symbols $\underline{A}_R$ and $\bar{A}_R$ denote, respectively, the lower and upper approximations of A in which the set of equivalence classes X/R is employed instead of the universal set X; for each $\alpha \in [0, 1]$, the α-cuts of $\underline{A}_R$ and $\bar{A}_R$ are defined by the formulas

$$^\alpha \underline{A}_R = \cup \{[x]_R | [x]_R \subseteq {}^\alpha A, x \in X\},$$
$$^\alpha \bar{A}_R = \cup \{[x]_R | [x]_R \cap {}^\alpha A \ne \varnothing, x \in X\},$$

where $[x]_R$ denotes the equivalence class in X/R that contains x. This combination of fuzzy sets with rough sets must be distinguished from another combination, in which a fuzzy equivalence relation is employed in the definition of a rough set. It is appropriate to refer to the sets that are based on the latter combination as *fuzzy rough sets*. These combinations, which have been discussed in the literature since the early 1990s, seem to have great utility in some application areas.

Observe that the introduced types of fuzzy sets are interrelated in numerous ways. For example, a fuzzy set of any type that employs the unit interval [0, 1] can be generalized by replacing [0, 1] with a complete lattice L; some of the types (e.g., standard, interval-valued, or type 2 fuzzy sets) can be viewed as special cases of L-fuzzy sets; or rough fuzzy sets can be viewed as special interval-valued sets. The overall fuzzy set theory is thus a broad formalized language based on an appreciable inventory of interrelated types of fuzzy sets, each associated with its own variety of concepts, operations, methods of computation, interpretations, and applications.

7.9. CONSTRUCTING FUZZY SETS AND OPERATIONS

Fuzzy set theory provides us with a broad spectrum of tools for representing and manipulating linguistic concepts, most of which are intrinsically vague and strongly dependent on the context in which they are used. Some linguistic concepts are represented by fuzzy sets, others are represented by operations on fuzzy sets. A prerequisite to each application of fuzzy set theory is to construct appropriate fuzzy sets and operations on fuzzy sets by which the intended meanings of relevant linguistic terms are adequately captured.

The problem of constructing fuzzy sets (i.e., their membership functions) or operations on fuzzy sets in the contexts of various applications is not a problem of fuzzy set theory per se. It is a problem of knowledge acquisition, which is a subject of a relatively new field referred to as knowledge engineering. The process of knowledge acquisition involves one or more experts in a specific domain of interest and a knowledge engineer. The role of the knowledge engineer is to elicit the knowledge of interest from experts, and express it in some operational form of a required type.

In applications of fuzzy set theory, knowledge acquisition involves basically two stages. In the first stage, the knowledge engineer attempts to elicit relevant knowledge in terms of propositions expressed in natural language. In the second stage, the knowledge engineer attempts to determine the meaning of each linguistic term employed in these propositions. It is during this second stage of knowledge acquisition that membership functions of fuzzy sets as well as appropriate operations on these fuzzy sets are constructed.

Many methods for constructing membership functions are described in the literature. It is useful to classify them into *direct methods* and *indirect methods*. In direct methods, the expert is expected to define a membership function either completely or exemplify it for some selected individuals in the universal set. To request a complete definition from the expert, usually in terms of a justifiable mathematical formula, is feasible only for a concept that is perfectly represented by some objects of the universal set, called *ideal prototypes* of the concept, and the compatibility of other objects in the universal set with these ideal prototypes can be expressed mathematically by a meaningful similarity function. For example, in pattern recognition of handwritten characters, any given straight line s can be defined as a member of a fuzzy set H of horizontal straight lines with the membership degree

$$H(s) = \begin{cases} 1 - |\theta|/45 & \text{when } |\theta| \in [0, 45] \\ 0 & \text{otherwise,} \end{cases}$$

where θ is the angle (measured in degrees) between s and an ideal horizontal straight line (an ideal prototype) that crosses it; it is assumed that $\theta \in [-90, 90]$, which means that the angle is required to be measured either in the first guadrant or the fourth guadrant, as relevant.

If it is not feasible to define the membership function in question completely, the expert should at least be able to exemplify it for some representative objects of the universal set. The exemplification can be facilitated by asking the expert questions regarding the compatibility of individual objects x with the linguistic term that is to be represented by fuzzy set A. These questions, regardless of their form, result in a set of pairs $\langle x, A(x) \rangle$ that exemplify the membership function under construction. This set is then used for constructing the full membership function. One way to do that is to select an appropriate class of functions (triangular, trapezoidal, S-shaped, bell-shaped, etc.) and employ some relevant curve-fitting method to determine the function that best fits the given samples. Another way is to use an appropriate neural network to construct the membership function by learning from the given samples. This approach has been so successful that neural networks are now viewed as a standard tool for constructing membership functions.

When a direct method is extended from one expert to multiple experts, the opinions of individual experts must be properly combined. Any averaging operation, including those introduced in Section 7.3.4, can be used for this purpose. The most common operation is the simple weighted average

$$A(x) = \sum_{i=1}^{n} c_i A_i(x),$$

where $A_i(x)$ denotes the valuation of the proposition "x belongs to A" by expert i, n denotes the number of experts involved, and c_i denote weights by which the relative significance of the individual experts can be expressed; it is assumed that

$$\sum_{i=1}^{n} c_i = 1.$$

Experts are instructed either to value each proposition by a number in $[0, 1]$ or to value it as true or false.

Direct methods based on exemplification have one fundamental disadvantage. They require the expert (or experts) to give answers that are overly precise and, hence, unrealistic as expressions of their qualitative subjective judgments. As a consequence, the answers are always somewhat arbitrary. Indirect methods attempt to reduce this arbitrariness by replacing the requested direct estimates of degrees of membership with simpler tasks.

In indirect methods, experts are usually asked to compare elements of the universal set in pairs according to their relative standing with respect to their membership in the fuzzy set to be constructed. The pairwise comparisons are often easier to estimate than the direct values, but they have to be somehow connected to the direct values. Numerous methods have been developed for dealing with this problem. They have to take into account possible inconsistencies in the pairwise estimates. Most of these methods deal with pairwise comparisons obtained from one expert, but a few methods are described in the literature that aggregate pairwise estimates from multiple experts. The latter methods are particularly powerful since they allow the knowledge engineer to determine the degrees of competence of the participating experts, which are then utilized, together with the expert's judgments for calculating the degrees of membership in question.

Methods for constructing membership functions of fuzzy sets and relevant operations on these fuzzy sets in the context of individual applications are essential for utilizing the enormous expressive power of fuzzy set theory for representing knowledge. Although many powerful methods are now available for this purpose, research in this area is still very active. To cover the various construction methods in detail and the ongoing research in this rather large problem area is far beyond the scope of this overview.

NOTES

7.1. Standard fuzzy sets and their basic properties were introduced in a seminal paper by Lotfi A. Zadeh [1965], even though the ideas of fuzzy sets and fuzzy logic were envisioned some 30 years earlier by the American philosopher Max Black in his penetrating discussion of the concept of vagueness [Black, 1937]. Zadeh is not only the founder of fuzzy set theory, but he has also been a key contributor to its development in various directions. His principal writings on fuzzy set theory, fuzzy logic, fuzzy systems, and other related areas during the period 1965–1995 are available in two books, edited by Yager et al. [1987] and Klir and Yuan [1996]. These books are perhaps the most important sources of information about the development of ideas in these areas. Since 1995, Zadeh has published several significant papers in which he examines the important role of fuzzy logic in computing with perceptions [Zadeh, 1996, 1997, 1999, 2002, 2005]. The idea of computing with perceptions introduces many new challenges, one of which is the translation from statements in natural language that approximate perceptions to their counterparts in the formalized language of fuzzy logic and the reverse translation (or retranslation) from statements in fuzzy logic (obtained by approximate reasoning) to the their counterparts in natural language. Some initial work in this area has been done by Dvořák [1999], Klir and Sentz [2005], and Yager [2004].

A historical overview of fuzzy set theory and fuzzy logic is presented in [Klir, 2001].

7.2. The literature on fuzzy set theory and related areas is abundant and rapidly growing. Two important handbooks, edited by Ruspini et al. [1998] and Dubois and Prade [2000], are recommended as convenient sources of information on virtually any aspect of fuzzy set theory and related areas; the latter one is the first volume in a multivolume series of handbooks on fuzzy set theory. From among the growing number of textbooks on fuzzy set theory, any of the following general textbooks is recommended for further study: Klir and Yuan [1995a], Lin and Lee [1996], Nguyen and Walker [1997], Pedrycz and Gomide [1998], Zimmermann [1996].

7.3. Research and education in fuzzy logic is now supported by numerous professional organizations. Many of them cooperate in a federation-like manner via the International Fuzzy Systems Association (IFSA), which publishes the prime journal in the field—*Fuzzy Sets and Systems*, and has organized the biennial World IFSA Congress since 1985. The oldest professional organization supporting fuzzy logic is the North American Fuzzy Information Processing Society (NAFIPS). Founded in 1981, NAFIPS publishes the *Journal of Approximate Reasoning* and organizes annual meetings.

7.4. Two valuable books were written in a popular genre by McNeil and Freiberger [1993] and Kosko [1993a]. Although both books characterize the relatively short but dramatic history of fuzzy set theory and discuss the significance of the theory, they have different foci. While the former book focuses on the impact of fuzzy set theory on high technology, the latter is concerned more with philosophical and cultural aspects; these issues are further explored in a more recent book by Kosko [1999]. Another book for the popular audience, which is worth reading, was written by De Bono [1991]. He argues that fuzzy logic (called in the book "water logic") is important in virtually all aspects of human affairs.

7.5. Operations called t-norms and t-cornoms were originally introduced by Menger [1942] for the study of statistical metric spaces [Schweizer and Sklar, 1983]. The most comprehensive treatment of these important operations for fuzzy set theory are two publications by Klement et al. [2000, 2004]. Procedures are now available by which various parameterized classes of operations of fuzzy intersections, unions, and complementations can be generated [Klir and Yuan, 1995a]. An excellent overview of the whole spectrum of aggregation operations on fuzzy sets was prepared by Dubois and Prade [1985b]. Linguistic modifiers are surveyed in [Kerre and DeCock, 1999].

7.6. The concept of a fuzzy number and the associated fuzzy arithmetic were introduced by Dubois and Prade [1978, 1979, 1987a]. They also developed basic ideas of fuzzy differential calculus [Dubois and Prade, 1982a]. Fuzzy differential equations were studied by Kaleva [1987]. Interval arithmetic, which is a basis for fuzzy arithmetic, is thoroughly covered in books by Alefeld and Herzeberger [1983], Hansen [1992], Moore [1966, 1979], and Neumaier [1990]. A specialized introductory book on fuzzy arithmetic was written by Kaufmann and Gupta [1985], and a more advanced book on using fuzzy arithmetic was written by Mareš [1994]. Basic ideas of constrained fuzzy arithmetic are developed in [Klir, 1997a, b] and [Klir and Pan, 1998]. It is worth of mentioning that the concept of a complex fuzzy numbers was also introduced and its utility explored in [Nguyen et al., 1998].

7.7. Basic ideas of fuzzy relations and the concepts of cutworthy fuzzy equivalence, compatibility, and partial ordering were introduced by Zadeh [1971] and were further investigated by many researchers. This subject is thoroughly covered in the book by Bělohlávek [2002], which also contains a large bibliography and extensive bibliographical comments. Another excellent book on fuzzy relations, focusing more on software tools and applications, was written by Peeva and Kyosev [2004]. The notion of fuzzy relation equations was first proposed by Sanchez [1976]. This important area of fuzzy set theory has been studied extensively, and many results emerging from these studies are covered in a dedicated monograph by Di Nola et al. [1989]. More recent results can be found in Chapter 6 (written by De Baets) in Dubois and Prade [2000], as well as in Gottwald [1993], De Baets and Kerre [1994], and Klir and Yuan [1995a].

7.8. An excellent and comprehensive survey of multivalued logics, which are the basis for fuzzy logic in the narrow sense, was prepared by Rescher [1969]. A survey of more recent developments was prepared by Wolf [1977], Bolc and Borowic [1992], and Malinowski [1993]. Fuzzy logic in the narrow sense is most comprehensively covered in [Hájek, 1998]. Other important publications in this area include a series of papers by Pavelka [1979], and books by Novák et al. [1999], Gottwald [2001], and Gerla [2001].

7.9. Literature dealing with approximate reasoning (or fuzzy logic in the broad sense) is very extensive. Two major references seem to capture the literature quite well. One of them is a pair of overview papers published together [Dubois and Prade, 1991], and the other one is a book edited by Bezdek et al. [1999], which is one of the handbooks in [Dubois and Prade, 1998–]. Two classical papers by Bandler and Kohout [1980a, b] on fuzzy implication operators are useful to read.

7.10. The concept of a linguistic variable was introduced and thoroughly investigated by Zadeh [1975–76]. It was also Zadeh who introduced the initial ideas of fuzzy systems in several articles that are included in Yager et al. [1987] and Klir and Yuan [1996]. The book by Negoita and Ralescu [1975] is an excellent early book on fuzzy systems. There are many books dealing with fuzzy modeling, often in the context of control, which is the most visible application of fuzzy systems. A very small sample consists of the excellent books by Babuška [1998], Piegat [2001], and Yager and Filev [1994]. Fuzzy systems are also established as universal approximators of a broad class of continuous functions, as is well discussed by Kreinovich et al. [2000]. Special fuzzy systems—fuzzy automata and languages—are thoroughly covered in a large book by Monderson and Malik [2002]. A good overview of defuzzification methods was prepared by Van Leekwijck and Kerre [1999].

7.11. Interval-valued fuzzy sets and type 2 fuzzy sets have been investigated since the 1970s. The book by Mendel [2001] is by far the most comprehensive coverage of these types of fuzzy sets, but a paper by John [1998] is a useful overview. Fuzzy sets of type k were introduced by Zadeh [1975–76] as a natural generalization of fuzzy sets of type 2, but their theory has not been developed as yet. Fuzzy sets of level 2 and higher levels were recognized in the late 1970s [Gottwald, 1979], but they have been rather neglected in the literature, in spite of their potential utility for representing complex concepts expressed in natural language. L-fuzzy sets were introduced by Goguen [1967] and they have been the source of various generalizations in fuzzy set theory. Intuitionistic fuzzy sets were introduced by Atanassov [1986] and are well developed in his more recent book [Atanassov, 2000]. Combinations of fuzzy sets with rough sets, which have already been proved useful in some applications, were first examined by Dubois and Prade [1987c, 1990b, 1992b]; rough sets were introduced by Pawlak [1982] and are further developed in his book [Pawlak, 1991].

7.12. The problem of constructing membership functions of fuzzy sets and operations on fuzzy sets has been addressed by many authors. Some representative publications dealing with this problem are [Bharathi-Devi and Sarma, 1985], [Chameau and Santamarina, 1987], and [Sancho-Royo and Verdegay, 1999]. Increasingly, the use of neural networks or genetic algorithms has begun to dominate this area, as is exemplified by the following references: [Lin and Lee, 1996], [Nauck et al., 1997] [Cordón et al. [2001], [Rutkowska, 2002], and [Rutkowski, 2004]

7.13. Since the late 1980s, many surprisingly successful applications of fuzzy set theory have been developed in virtually all areas of engineering as well as some other professions. An overview of these applications with relevant references can be found, for example, in Klir [2000] or Klir and Yuan [1995a]. A significant, more recent development is the use of fuzzy set theory and fuzzy logic in some areas of science, such as economics [Billot, 1992], chemistry [Rouvray, 1997], geology [Demicco and Klir, 2003; Bárdossy and Fodor, 2004], and social sciences [Ragin, 2000].

7.14. It has been argued in the literature (see, for example, [Klir, 2000]) that the emergence of fuzzy set theory initiated a scientific revolution (or a paradigm shift) according to the criteria introduced in the highly influential book by Thomas Kuhn [1962].

7.15. Fuzzy sets may be viewed as wholes, each capturing as a collection (potentially infinite) of classical sets—its α-cuts. The need for such wholes in mathematics is expressed very clearly in the classic book *Holism and Reductionism* by Smuts [1926].

EXERCISES

7.1. Show that the fuzzy subsethood *Sub* can be expressed in terms of the sigma count and standard operation of intersection as

$$Sub(A, B) = \frac{|A \cap B|}{|A|}. \tag{7.45}$$

7.2. Determine the strong α-cut representation of fuzzy set A in Figure 7.3a.

7.3. Derive Eq. (7.26) from Eq. (7.24) under the assumption that f_A and g_A are linear functions.

7.4. Membership function C in Figure 7.4 is defined for each $x \in \mathbb{R}$ by the formula

$$C(x) = \max\{0, 2x - x^2\}.$$

Determine the α-cut representation of C.

7.5. Derive general formulas for membership functions of fuzzy numbers whose shapes are exemplified in Figure 7.4 by functions D and E, and convert them to their respective α-cut representations.

7.6. Determine the membership functions for the α-cut representations in Example 7.4.

7.7. Show that for each $\lambda > 0$ the complementation function c_λ defined by Eq. (7.7) is involutive.

7.8. Show that for each $\lambda > -1$ the function

$$c_\lambda = \frac{1-a}{1+\lambda a}$$

is an involutive complementation function of fuzzy sets.

7.9. Show that the standard operations of intersection, union, and complementation of fuzzy sets possess the following properties:

(a) They satisfy the De Morgan laws expressed by Eqs. (7.16) and (7.17);

(b) They violate the law of excluded middle and the law of contradiction.

7.10. For the classes of intersection and union operations of fuzzy sets defined by Eqs. (7.14) and (7.15), respectively, show the following:

(a) The standard operations are obtained in the limit for $\lambda \to \infty$;

(b) The drastic operations are obtained in the limit for $\lambda \to 0$.

7.11. Show that the following combinations of fuzzy operations on fuzzy sets satisfy the law of excluded middle and the law of contradiction:

(a) Drastic intersection, drastic union, and standard complementation;

(b) $i(a, b) = \max\{0, a + b - 1\}$, $u(a, b) = \min\{1, a + b\}$, $c(a) = 1 - a$.

7.12. Show that for crisp sets the class of operations defined by Eq. (7.14) (or, alternatively, by Eq. (7.15)) collapse to a single operation that conforms to the classical set intersection (or, alternatively, the classical set union).

7.13. Show that averaging operations of any kind are not applicable within the domain of classical set theory.

7.14. Repeat Example 7.5 for the alternative form of the equation: $a = 1 - 1/(b + 1)$.

7.15. Consider a fuzzy number A defined by the formula

$$A(x) = \begin{cases} 1 - (1.25 - x)^2 & \text{when } x \in [0.25, 2.25] \\ 0 & \text{otherwise.} \end{cases}$$

and the fuzzy number C in Exercise 7.4. Determine each of the following fuzzy numbers by both standard and constrained fuzzy arithmetic:

(a) $A + C$

(b) $A - C$; $C - A$; and $C - C$

(c) $A \cdot C$

(d) C/A and A/A

(e) $A/C - C/A$

(f) $A \cdot C/(A + C)$

7.16. Repeat Exercise 7.15 for fuzzy number

$$A(x) = \begin{cases} 1 - |x - 5|/3 & \text{when } x \in [2, 8] \\ 0 & \text{otherwise,} \end{cases}$$

and a triangular fuzzy number $C = \langle -1, 0, 0, 1 \rangle$.

7.17. For P_A and $P_B = 1 - P_A$ in Example 7.6, determine the following operations with and without the probabilistic constraint:

(a) $P_A \cdot P_B$

(b) P_A/P_B and P_B/P_A

(c) $P_A - P_B$ and $P_B - P_A$

(d) $P_A + P_B/P_A$ and $P_A + P_B/P_B$

7.18. Using the standard and constrained fuzzy arithmetic, determine $A \cdot A$ for the following trapezoidal-shape fuzzy intervals:

(a) $A = \langle -1, 0, 1, 2 \rangle$

(b) $A = \langle 1, 2, 2, 4 \rangle$

(c) $A = \langle -5, -3, -2, -1 \rangle$

(d) $A = \langle -2, 0, 0, 2 \rangle$

7.19. Given a fuzzy interval whose α-cut representation is ${}^{\alpha}A = [\underline{a}(\alpha), \bar{a}(\alpha)]$, show that under the equality constraint

$$^\alpha(A \cdot A) = \begin{cases} \left[\overline{a}^2(\alpha), \underline{a}^2(\alpha)\right] & \text{when } \overline{a}(\alpha) < 0 \\ \left[\underline{a}^2(\alpha), \overline{a}^2(\alpha)\right] & \text{when } \underline{a}(\alpha) > 0 \\ \left[0, \max\{\overline{a}^2(\alpha), \underline{a}^2(\alpha)\}\right] & \text{when } 0 \in [\underline{a}(\alpha), \overline{a}(\alpha)] \end{cases}$$

for all $\alpha \in (0, 1]$.

7.20. Given arbitrary intervals A, B, C show that:
   (a) $A(B + C) \subseteq AB + AC$ when standard fuzzy arithmetic is used;
   (b) $A(B + C) = AB + AC$ when constrained fuzzy arithmetic is used.

7.21. Given the equation $A + X = B$, where A and B are given fuzzy intervals, show that $X = B - A$ under constrained fuzzy arithmetic, but not under standard fuzzy arithmetic

7.22. Given the equation $A \cdot X = B$, where A and B are given fuzzy intervals, show that $X = B/A$ (assuming that $0 \notin A$) under the constrained fuzzy arithmetic, but not under the standard fuzzy arithmetic.

7.23. Show that the standard operations of intersection and union on fuzzy sets are cutworthy.

7.24. Show that $(P \circ Q)^{-1} = Q^{-1} \circ P^{-1}$ for the standard composition and the inverse of any connected fuzzy binary relations P and Q.

7.25. For each of the three pairs of connected binary fuzzy relations defined in Table 7.2 by their matrix representations, determine:
   (a) The standard composition;
   (b) The standard join.

7.26. For some of the compositions performed in Exercise 7.25, verify the equation $(P \circ Q)^{-1} = Q^{-1} \circ P^{-1}$.

7.27. For some of the fuzzy relations defined in Table 7.2, determine the following:
   (a) Projections to each dimension;
   (b) Cylindric extensions from the projections;
   (c) Cylindric closure based on the projections.

7.28. Let X be the relative humidity (measured in %) at some particular place on the Earth, and let the property of *high humidity* be expressed by the trapezoidal-shape fuzzy interval $H = \langle 60, 80, 100, 100 \rangle$ defined on the interval $[0, 100]$. Using Figure 7.7b, determine the degree of truth of the truth-qualified fuzzy proposition

$$f_{T(H)}(x) : X = x \text{ is } H \text{ is } T$$

Table 7.2. Matrix Representations of Fuzzy Relations Employed in Exercises 7.25–7.27

$$P_1 = \begin{array}{c} \\ x_1 \\ x_2 \\ x_3 \end{array} \begin{array}{ccc} y_1 & y_2 & y_3 \\ \left[\begin{array}{ccc} 1.0 & 0.0 & 0.7 \\ 0.3 & 0.2 & 0.0 \\ 0.0 & 0.5 & 1.0 \end{array} \right] \end{array}$$

$$P_2 = \begin{array}{c} \\ x_1 \\ x_2 \\ x_3 \\ x_4 \end{array} \begin{array}{cccc} y_1 & y_2 & y_3 & y_4 \\ \left[\begin{array}{cccc} 1.0 & 0.9 & 0.8 & 0.7 \\ 0.6 & 1.0 & 0.5 & 0.4 \\ 0.3 & 0.2 & 1.0 & 0.1 \\ 0.2 & 0.3 & 0.4 & 1.0 \end{array} \right] \end{array}$$

$$P_3 = \begin{array}{c} \\ x_1 \\ x_2 \\ x_3 \\ x_4 \\ x_5 \\ x_6 \end{array} \begin{array}{ccccc} y_1 & y_2 & y_3 & y_4 & y_5 \\ \left[\begin{array}{ccccc} 0.9 & 1.0 & 1.0 & 1.0 & 0.9 \\ 0.7 & 0.8 & 0.9 & 0.7 & 0.6 \\ 0.6 & 0.5 & 0.8 & 0.6 & 0.5 \\ 0.3 & 0.4 & 0.7 & 0.3 & 0.4 \\ 0.2 & 0.5 & 0.6 & 0.5 & 0.2 \\ 0.1 & 0.3 & 0.5 & 0.3 & 0.1 \end{array} \right] \end{array}$$

$$Q_1 = \begin{array}{c} \\ y_1 \\ y_2 \\ y_3 \end{array} \begin{array}{cccc} z_1 & z_2 & z_3 & z_4 \\ \left[\begin{array}{cccc} 1.0 & 0.0 & 0.7 & 0.5 \\ 0.0 & 1.0 & 0.0 & 1.0 \\ 0.7 & 0.0 & 1.0 & 0.8 \end{array} \right] \end{array}$$

$$Q_2 = \begin{array}{c} \\ y_1 \\ y_2 \\ y_3 \\ y_4 \end{array} \begin{array}{cccccc} z_1 & z_2 & z_3 & z_4 & z_5 & z_6 \\ \left[\begin{array}{cccccc} 0.0 & 0.8 & 0.6 & 0.0 & 0.4 & 0.2 \\ 0.0 & 0.0 & 0.9 & 0.9 & 0.7 & 0.7 \\ 1.0 & 1.0 & 0.0 & 0.5 & 0.0 & 0.0 \\ 1.0 & 1.0 & 0.5 & 0.0 & 1.0 & 1.0 \end{array} \right] \end{array}$$

$$Q_3 = \begin{array}{c} \\ y_1 \\ y_2 \\ y_3 \\ y_4 \\ y_5 \\ y_6 \end{array} \begin{array}{ccccccc} z_1 & z_2 & z_3 & z_4 & z_5 & z_6 & z_7 \\ \left[\begin{array}{ccccccc} 1.0 & 0.9 & 0.9 & 0.8 & 0.6 & 0.4 & 0.2 \\ 1.0 & 0.9 & 1.0 & 0.8 & 0.7 & 0.7 & 0.5 \\ 1.0 & 1.0 & 0.8 & 0.9 & 0.0 & 0.0 & 0.0 \\ 0.0 & 0.0 & 0.0 & 0.7 & 0.8 & 0.0 & 0.0 \\ 0.0 & 0.0 & 0.0 & 0.0 & 0.6 & 0.7 & 0.0 \\ 0.0 & 0.0 & 0.0 & 0.5 & 0.5 & 0.5 & 0.6 \end{array} \right] \end{array}$$

under the following specifications:
(a) $x = 65\%$ and $T =$ True;
(b) $x = 65\%$ and $T =$ False;
(c) $x = 50\%$ and $T =$ True;
(d) $x = 50\%$ and $T =$ False;
(e) $x = 76\%$ and $T =$ Fairly true;
(f) $x = 76\%$ and $T =$ Very false.

7.29. Repeat Exercise 7.28 for the property of *medium humidity*, which is expressed by the trapezoidal-shaped fuzzy interval $M = \langle 20, 40, 60, 80 \rangle$.

7.30. The probabilities, $p(r)$, of daily receipts, r, of a shop (rounded to the nearest hundred dollars) that have been obtained from statistical data collected over many years are shown in Table 7.1*b*. Consider three fuzzy events: *low*, *medium*, and *high* receipts and assume that they are represented by trapezoidal-shaped fuzzy intervals $L = \langle 18, 18, 21, 23 \rangle$, $M = \langle 21, 23, 27, 29 \rangle$, and $H = \langle 27, 29, 32, 32 \rangle$, respectively. Determine the degrees of truth of the following propositions:

(a) The daily receipts are low;

(b) The daily receipts are medium;

(c) The daily receipts are high.

7.31. Modify the granulation defined in Figure 7.10*b* by replacing the triangular-shaped granules with:

(a) Trapezoidal-shaped granules, with small plateaus around the ideal values;

(b) Granules of the shape illustrated by membership function C in Figure 7.4 and defined in Exercise 7.4;

(c) Granules of the shape illustrated by membership function F in Figure 7.4 and defined in Example 7.3.

7.32. Show that the defuzzified value $d(A)$ obtained by the centroid defuzzification method may be interpreted as the expected value of x based on A.

7.33. Show that the defuzzified value $d(A)$ defined by Eq. (7.41) is the value of x for which the area under the graph of membership function A is divided into two equal areas.

7.34. Using the centroid defuzzification method defuzzify the following fuzzy sets:

(a) Fuzzy number $A(x) = \max\{0, 4x - 4 - (x-2)^2\}$;

(b) Fuzzy interval L in Figure 7.7*a*;

(c) The two fuzzy numbers obtained in Example 7.5 and shown in Figure 7.5;

(d) Fuzzy set A that is equal to the possibility profile r in Figure 6.1 (i.e., $A(x) = r(x)$ for all $x \in \mathbb{N}_{15}$);

(e) Fuzzy interval A defined by the following α-cut representation:

$$
{}^{\alpha}A = \begin{cases} [1.5(\alpha) + 1.75, 4 - 0.5\alpha] & \text{when } \alpha \in (0, 0.5] \\ [\alpha + 2.5, 4 - 0.5\alpha] & \text{when } \alpha \in (0.5, 1]. \end{cases}
$$

7.35. Repeat Exercise 7.34 for some other defuzzification methods defined by Eqs. (7.43) and (7.44). Choose at least one value of $\lambda < 1$ and one value of $\lambda > 1$.

7.36. Every rectangle may be considered to some degree a member of a fuzzy set of squares. Using common sense, define a possible membership function of such a fuzzy set.

8

FUZZIFICATION OF UNCERTAINTY THEORIES

The limit of language is the limit of the world.

—Stephen A. Tyler

8.1. ASPECTS OF FUZZIFICATION

Perhaps the best description of the nature of fuzzification is contained in the well-known classical paper by Goguen [1967]: "Fuzzification is a process of imparting a fuzzy structure to a definition (concept), a theorem, or even a whole theory." The strength of this deceivingly simple description is its sweeping generality. The single sentence captures the essence of a wide variety of issues that must be addressed when mathematical concepts, properties, or theories based on classical sets are generalized to their counterparts based on fuzzy sets of some type.

One method of fuzzifying various properties (concepts, operations, theorems) of classical set theory is to use the α-cut (or strong α-cut) representations of fuzzy sets. Since α-cuts (as well as strong α-cuts) are classical sets, every property of classical set theory applies to them. A property of classical sets is fuzzified (that is, it becomes a property of fuzzy sets) via the α-cut representation by requiring that it holds (in the classical sense) in all α-cuts of the fuzzy sets involved. Any property of fuzzy sets of some specific type that is derived from a property of classical sets in this way is called a *cutworthy property*.

For standard fuzzy sets, there are many properties that are cutworthy. One such property is convexity of fuzzy sets. A fuzzy set is said to be convex if and

Uncertainty and Information: Foundations of Generalized Information Theory, by George J. Klir
© 2006 by John Wiley & Sons, Inc.

only if all its α-cuts are convex sets in the classical sense. An important class of special convex fuzzy sets consists of fuzzy intervals. A fuzzy set on $\mathbb{R}$ is said to be a fuzzy interval if and only if all its α-cuts are closed intervals of real numbers (i.e., classical convex subsets of $\mathbb{R}$). Arithmetic operations on fuzzy intervals are also cutworthy. At each α-cut, they follow the rules of either standard or constrained arithmetic on closed intervals of real numbers.

It is significant that the standard operations of intersection and union of fuzzy sets (min and max operations) are cutworthy. As a consequence, other operations that are based solely on them are cutworthy as well. They include, for example, the max–min composition of binary fuzzy relations as well as the relational join defined by the min operation. The concept of cylindric closure of fuzzy relations is also cutworthy when the intersection of the cylindric extensions of the given fuzzy relations is defined by the min operation.

Important examples of cutworthy properties are some properties of binary fuzzy relations on X^2, such as equivalence, compatibility, or partial ordering. Classical equivalence relations, for example, are defined by three properties: (i) reflexivity; (ii) symmetry; and (iii) transitivity. Let these properties be defined for fuzzy relations, R, as follows:

(i) R is reflexive iff $R(x, x) = 1$ for all $x \in X$.
(ii) R is symmetric iff $R(x, y) = R(y, x)$ for all $x, y \in X$.
(iii) R is transitive (or, more specifically, max–min transitive) iff

$$R(x, z) \geq \max_{y \in Y} \min\{R(x, y), R(y, z)\} \tag{8.1}$$

for all pairs $\langle x, z \rangle \in X^2$.

Then, any binary fuzzy relation on X^2 that possesses these properties is a fuzzy equivalence relation in the cutworthy sense. That is, all α-cuts of any binary fuzzy relation that possesses these properties are classical equivalence relations. This follows from the fact that each of the three properties is cutworthy. In the case of reflexivity and symmetry, it is obvious. In the case of transitivity, it is cutworthy since it is defined in terms of the standard operations of intersection and union, which are cutworthy.

EXAMPLE 8.1. Consider the binary fuzzy relation R on X^2, where $X = \{x_i \mid i \in \mathbb{N}_7\}$, which is defined by the matrix

	x_1	x_2	x_3	x_4	x_5	x_6	x_7
x_1	1.0	0.8	0.0	0.4	0.0	0.0	0.0
x_2	0.8	1.0	0.0	0.4	0.0	0.0	0.0
x_3	0.0	0.0	1.0	0.0	1.0	0.9	0.5
$\mathbf{R} = x_4$	0.4	0.4	0.0	1.0	0.0	0.0	0.0
x_5	0.0	0.0	1.0	0.0	1.0	0.9	0.5
x_6	0.0	0.0	0.9	0.0	0.9	1.0	0.5
x_7	0.0	0.0	0.5	0.0	0.5	0.5	1.0

It is easy to see that this relation is reflexive and symmetric. It is also max–min transitive, but that is more difficult to verify. One convenient way to verify it is to calculate $(R \circ R) \cup R$ where $\circ$ denotes the max–min composition and $\cup$ denotes the standard union operation. Then, R is transitive if and only if

$$R = (R \circ R) \cup R.$$

The given relation satisfies this equation. Hence, it is max–min transitive and, due to its reflexivity and symmetry, it is a fuzzy equivalence relation. This can be verified by examining all its α-cuts.

The level set of the given relation R is $\Lambda_R = \{0, 0.4, 0.5, 0.8, 0.9, 1\}$. Therefore, R represents six classical equivalence relations, one for each $\alpha \in \Lambda_R$. Each of these equivalence relations, $^{\alpha}R$, partitions the set X in some particular way, $^{\alpha}\pi(X)$. Since $^{\alpha'}R \subseteq {}^{\alpha}R$ when $\alpha' \geq \alpha$, clearly

$$^{\alpha'}\pi(X) \leq {}^{\alpha}\pi(X) \qquad \text{when } \alpha' \geq \alpha.$$

The six partitions of the given relation are shown in the form of a partition tree in Figure 8.1. The partitions become increasingly more refined when values of α in Λ_R increase.

Other cutworthy types of binary fuzzy relations can be defined in a similar way. Examples are fuzzy compatibility relations (reflexive and symmetric) and fuzzy partial orderings (reflexive, antisymmetric, and transitive). For fuzzy partial orderings, the property of fuzzy antisymmetry is defined as follows: for all $x, y \in X$, if $R(x, y) > 0$ and $R(y, x) > 0$, then $x = y$. This, clearly, is a cutworthy property.

It is important to realize that there are many fuzzy-set generalizations of properties of classical sets that are not cutworthy. A fuzzy-set generalization of some classical property is required to reduce to its classical counterpart when membership grades are restricted to 0 and 1, but it is not required to be cutworthy. There often are multiple generalizations of a classical property, but only one or, in some cases, none of them is cutworthy. Examples of fuzzy-set generalizations that are not cutworthy are all operations of intersection and union of fuzzy sets (t-norms and t-conorms) except the standard ones (min and max). Even more interesting examples are operations of complementation of fuzzy sets, *none* of which is cutworthy, even though all of them are, by definition, generalizations of the classical complementation.

Another way of connecting classical set theory and fuzzy set theory is to fuzzify functions. Given a function

$$f : X \rightarrow Y,$$

where X and Y are crisp sets, we say that the function is *fuzzified* when it is extended to act on fuzzy sets defined on X and Y. That is, the fuzzified function maps, in general, fuzzy sets defined on X to fuzzy sets defined on Y. Formally, the fuzzified function, F, has a form

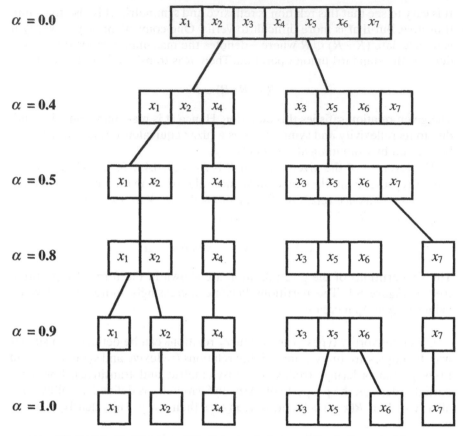

Figure 8.1. Partition tree of the fuzzy equivalence relation in Example 8.1.

$$F : \mathcal{F}(X) \to \mathcal{F}(Y),$$

where $\mathcal{F}(X)$ and $\mathcal{F}(Y)$ denote the fuzzy power sets (sets of all fuzzy subsets) of X and Y, respectively. To qualify as a fuzzified version of f, function F must conform to f within the extended domain $\mathcal{F}(X)$ and $\mathcal{F}(Y)$. This is guaranteed when a principle is employed that is called an *extension principle*. According to this principle

$$B = F(A)$$

is determined for any given fuzzy set $A \in \mathcal{F}(X)$ and all $y \in Y$ via the formula

$$B(y) = \begin{cases} \sup\{A(x) \mid x \in X, f(x) = y\} & \text{when } f^{-1}(y) \neq \varnothing \\ 0 & \text{otherwise.} \end{cases} \qquad (8.2)$$

The inverse function

$$F^{-1} : \mathcal{F}(Y) \to \mathcal{F}(X)$$

of F is defined, according to the extension principle, for any given $B \in \mathcal{F}(Y)$ and all $x \in X$, by the formula

$$[F^{-1}(B)](x) = B(y), \tag{8.3}$$

where $y = f(x)$. Clearly,

$$F^{-1}[F(A)] \supseteq A$$

for all $A \in \mathcal{F}(X)$, where the equality is obtained when f is a one-to-one function.

The use of the extension principle is illustrated in Figure 8.2, which shows how fuzzy set A is mapped to fuzzy set B via function F that is consistent with the given function f. That is, $B = F(A)$. For example, since

$$b = f(a_1) = f(a_2) = f(a_3),$$

we have

$$B(b) = \max\{A(a_1), A(a_2), A(a_3)\}$$

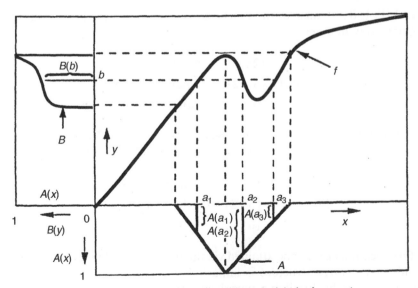

Figure 8.2. Illustration of the extension principle for fuzzy sets.

by Eq. (8.2). Conversely,

$$[F^{-1}(B)](a_1) = [F^{-1}(B)](a_2) = [F^{-1}(B)](a_3) = B(b)$$

by Eq. (8.3).

The introduced extension principle, by which functions are fuzzified, is basically described by Eqs. (8.2) and (8.3). These equations are direct generalizations of similar equations describing the extension principle of classical set theory. In the latter, symbols A and B denote characteristic functions of crisp sets.

To fuzzify a function of n variables of the form

$$f : X_1 \times X_2 \times \cdots \times X_n \to Y,$$

the formula in Eq. (8.2) has to be replaced with the more general formula

$$B(y) = \begin{cases} \sup\limits_{i \in \mathbb{N}_n} \min\{A_i(x_i) \mid x_i \in X_i, i \in \mathbb{N}_n, f(x_1, x_2, \ldots, x_n) = y\} & \text{when } f^{-1}(y) \neq \varnothing \\ 0 & \text{otherwise.} \end{cases} \tag{8.4}$$

Similarly, Eq. (8.3) has to be replaced with

$$[F^{-1}(B)](x_1, x_2, \ldots, x_n) = B(y). \tag{8.5}$$

Equation (8.4) can be further generalized by replacing the min operator with a t-norm.

EXAMPLE 8.2. Consider the function $y = \sqrt{x}$, where $x \in [0, 5]$, and the triangular fuzzy number

$$A(x) = \begin{cases} 2x - 1 & \text{when } x \in [0.5, 1) \\ 3 - 2x & \text{when } x \in [1, 1.5] \\ 0 & \text{otherwise,} \end{cases}$$

which represents an approximate assessment of value x. Using the extension principle, the value of y is approximately assessed by the fuzzy number

$$B(y = \sqrt{x}) = \begin{cases} 2x - 1 & \text{when } \sqrt{x} \in [\sqrt{0.5}, 1) \\ 3 - 2x & \text{when } \sqrt{x} \in [1, \sqrt{1.5}] \\ 0 & \text{otherwise.} \end{cases}$$

Fuzzification also can be attained via the mathematical theory of categories. In this way, a fuzzy structure is imparted into various categories of mathematical objects by fuzzifying morphisms through which the categories are defined. This is a powerful approach to fuzzification, which has played an important role in fuzzifying many areas of mathematics, such as topology, analysis, various algebraic theories, graphs, hypergraphs, geometry, and finite-state automata. Because the reader of this book is not expected to have sufficient background in category theory, this approach to fuzzification is not employed in this chapter.

There are usually multiple ways in which a given classical mathematical structure can be fuzzified. These ways of fuzzification are distinguished from one another by choosing which components of the structure are going to be fuzzified and how. These choices have to be made in each given application context on pragmatic grounds.

8.2. MEASURES OF FUZZINESS

The pragmatic value of fuzzy logic in the broad sense consists of its capability to represent and deal with vague linguistic expressions. Such expressions are typical of natural language. A linguistic expression is vague when its meaning is not fixed by sharp boundaries. Such a linguistic expression is always associated with uncertainty regarding its applicability. However, this type of uncertainty does not result from any information deficiency, but from the lack of linguistic precision. It is a *linguistic uncertainty* rather than *information-based uncertainty*. Consider, for example, a fuzzy set that was constructed in a given application context to represent the linguistic term "high temperature." Assume now that a particular measurement of the temperature was taken (say 92°F). This measurement belongs to the fuzzy set with a particular membership degree (say, 0.7). Clearly, this degree does not express any lack of information (the actual value of the temperature is known—it is 92°F), but rather the degree of compatibility of the known value with the imprecise (vague) linguistic term.

Since this linguistic uncertainty is represented by fuzzy sets, it is usually called *fuzziness*. Each fuzzy set is clearly associated with some amount of fuzziness, and it is desirable to be able to measure this amount in some meaningful way. Although the amount of fuzziness is not connected in any way to the quantification of information, it is an important trait of information representation. Clearly, our ability to measure it allows us to characterize information representation more completely, and thus enriches our methodological capabilities for dealing with information.

In general, a *measure of fuzziness* for some type of fuzzy sets is a functional

$$f : \mathcal{F}(X) \to \mathbb{R}^+,$$

where $\mathcal{F}(X)$ denotes the set of all fuzzy subsets of X of a given type. Thus far, however, the issue of measuring fuzziness has been addressed only for standard fuzzy sets. This restriction is thus followed in this section as well.

Several ideas of how to measure the fuzziness of fuzzy sets (i.e., standard fuzzy sets) have been pursued in the literature. One of them, which is currently predominant in the literature and is followed in this section, is to measure the fuzziness of a fuzzy set by the lack of distinction between the fuzzy set and its complement. This is a sound idea. It is precisely the lack of distinction between sets and their complements that distinguishes fuzzy sets from crisp sets. The less a set differs from its complement, the fuzzier it is.

Measuring fuzziness in terms of distinctions between sets and their complements is dependent on the definition of the complementation operation. The simplest way of expressing the local distinctions (one for each $x \in X$) of a given set A and its complement, $c(A)$, is to calculate the difference

$$|A(x) - c[A(x)]|.$$

The maximum of this difference, obtained for crisp sets, is 1. The lack of distinction between A and its complement is thus expressed for each $x \in X$ by the value

$$1 - |A(x) - c[A(x)]|.$$

When X is finite, a measure of fuzziness, $f(A)$, of the set A is then obtained by adding these values for all x in the support of A, $S(A)$. That is,

$$f(A) = \sum_{x \in S(A)} (1 - |A(x) - c[A(x)]|), \tag{8.6}$$

or, alternatively,

$$f(A) = |S(A)| - \sum_{x \in S(A)} |A(x) - c[A(x)]|. \tag{8.7}$$

Equations (8.6) and (8.7) are justified by the fact that $1 - |A(x) - c[A(x)]| = 0$ for all $x \notin S(A)$. Clearly,

$$0 \le f(A) \le |X|,$$

where $f(A) = 0$ if and only if A is a crisp set, and $f(A) = |X|$ if and only if

$$A(x) = c[A(x)]$$

for all $x \in X$, which means that the set is equal to its complement. When $f(A)$ is divided by $|X|$, we obtain a normalized value, $\hat{f}(A)$, of fuzziness. Clearly,

$$0 \le \hat{f}(A) \le 1.$$

When $X = [\underline{x}, \bar{x}]$ and $S(A) = [a, b] \subseteq X$, the summation in Eq. (8.6) must be replaced with integration. That is,

$$f(A) = \int_a^b (1 - |A(x) - c[A(x)]|) \, dx, \tag{8.8}$$

or alternatively,

$$f(A) = b - a - \int_a^b |A(x) - c[A(x)]| dx. \tag{8.9}$$

Clearly,

$$0 \leq f(A) \leq \bar{x} - \underline{x}$$

and for the associated normalized version, $\hat{f}(A) = f(A)/(\bar{x} - \underline{x})$,

$$0 \leq \hat{f}(A) \leq 1.$$

When c is the standard fuzzy complement, Eq. (8.7) becomes

$$f(A) = |S(A)| - \sum_{x \in S(A)} |2A(x) - 1|, \tag{8.10}$$

and Eq. (8.9) becomes

$$f(A) = b - a - \int_a^b |2A(x) - 1| \, dx. \tag{8.11}$$

EXAMPLE 8.3. To calculate the amount of fuzziness of the fuzzy relation R in Example 8.1 by Eq. (8.10) (i.e., assuming the use of standard operation of complementation), it is convenient to first calculate the following matrix of the differences $|2R(x) - 1|$ for all pairs $x = \langle x_i, x_j \rangle \in S(R)$:

	x_1	x_2	x_3	x_4	x_5	x_6	x_7
x_1	1.0	0.6	—	0.2	—	—	—
x_2	0.6	1.0	—	0.2	—	—	—
x_3	—	—	1.0	—	1.0	0.8	0.0
x_4	0.2	0.2	—	1.0	—	—	—
x_5	—	—	1.0	—	1.0	0.8	0.0
x_6	—	—	0.8	—	0.8	1.0	0.0
x_7	—	—	0.0	—	0.0	0.0	1.0

The sum of all these differences is 14.2, $|X| = 49$, and $|S(R)| = 25$. Hence, $f(R) = 25 - 14.2 = 10.8$ and $\hat{f}(R) = 10.8/49 = 0.22$.

EXAMPLE 8.4. To illustrate the use of Eq. (8.11), let

$$A(x) = \max\{0, 2x - x^2\}$$

for all $x \in X = [0, 5]$. It is obvious that the support of A is the interval $[0, 2]$. Hence,

$$f(A) = 2 - \int_0^2 |4x - 2x^2 - 1| \, dx.$$

The expression $4x - 2x^2 - 1$ is positive within the interval $[x_1, x_2]$ and negative within the intervals $[0, x_1]$ and $[x_2, 2]$, where $x_1 = (2 - \sqrt{2})/2$ and $x_2 = (2 + \sqrt{2})/2$. Hence,

$$f(A) = 2 - \int_0^{x_1} (1 + 2x^2 - 4x) \, dx - \int_{x_1}^{x_2} (4x - 2x^2 - 1) \, dx$$
$$- \int_{x_2}^2 (1 + 2x^2 - 4x) \, dx = 0.78,$$

and $f(A) = 0.78/5 = 0.156$.

Two alternative functionals for measuring the fuzziness of fuzzy sets are well known in the literature. Both of them are based on the idea that fuzziness is manifested by the lack of distinction between a fuzzy set and its complement. One of them, say functional f', expresses this lack of distinction by the standard intersection of the set and its complement. That is,

$$f'(A) = \sum_{x \in S(A)} \min\{A(x), c[A(x)]\} \tag{8.12}$$

when X is finite, or

$$f'(A) = \int_a^b \min\{A(x), c[A(x)]\} \, dx \tag{8.13}$$

when $S(A) = [a, b]$. The minimum, $f'(A) = 0$, is clearly obtained only if A is a crisp set. The maximum is obtained when

$$A(x) = c[A(x)]$$

for all $x \in X$. The value for which this equation is satisfied is called an *equilibrium* of complement c. For example, the equilibrium of the standard complement is 0.5. Denoting the equilibrium of complement c by $^e c$, it is obvious that

$$0 \le f'(A) \le |X| \cdot {}^e c$$

when X is finite, or

$$0 \le f'(A) \le (\bar{x} - \underline{x}) \cdot {}^e c$$

when $X = (\underline{x}, \bar{x})$. For the standard fuzzy complement, it is easy to show that $f(A) = 2f'(A)$ for any fuzzy set A, regardless whether it is defined on a finite set X or on $\mathbb{R}$. In this case, f and f' differ only in the measurement unit, which does not affect their normalized versions. Hence, $\hat{f}(A) = \hat{f}'(A)$ for all fuzzy sets.

Another functional, f'', extensively covered in the literature, is applicable only to the standard fuzzy complement. It is defined by the formula

$$f''(A) = \sum_{x \in X} [-A(x)\log_2 A(x) - (1 - A(x))\log_2(1 - A(x))] \qquad (8.14)$$

when X is finite, or by the formula

$$f''(A) = \int_{\mathbb{R}} [-A(x)\log_2 A(x) - (1 - A(x))\log_2(1 - A(x))] \, dx \qquad (8.15)$$

when $X = [\underline{x}, \bar{x}]$. This functional is based on the recognition that for each $x \in X$, the values of $A(x)$ and the standard fuzzy complement, $1 - A(x)$, add to 1. Hence, for each $x \in X$, the pair $\langle A(x), 1 - A(X) \rangle$ may be viewed as a probability distribution on two elements. The functional utilizes, for each $x \in X$, the Shannon entropy of this elementary distribution, $S(A(x), 1 - A(x))$, as a convenient measure of the lack of distinction between $A(x)$ and $1 - A(x)$.

EXAMPLE 8.5. To compare functionals f and f'', let us repeat Example 8.3 for f''. First, for each entry of the matrix we calculate the Shannon entropy $S(R(x), 1 - R(x))$:

	x_1	x_2	x_3	x_4	x_5	x_6	x_7
x_1	0.00	0.72	0.00	0.97	0.00	0.00	0.00
x_2	0.72	0.00	0.00	0.97	0.00	0.00	0.00
x_3	0.00	0.00	0.00	0.00	0.00	0.47	1.00
x_4	0.97	0.97	0.00	0.00	0.00	0.00	0.00
x_5	0.00	0.00	0.00	0.00	0.00	0.47	1.00
x_6	0.00	0.00	0.47	0.00	0.47	0.00	1.00
x_7	0.00	0.00	1.00	0.00	1.00	1.00	0.00

After adding all entries in this matrix, we obtain $f''(R) = 13.2$ and $\hat{f}''(R) = 13.2/49 = 0.27$.

EXAMPLE 8.6. To illustrate the use of Eq. (8.15), let us repeat Example 8.4 for f''. We have,

$$
\begin{aligned}
f''(A) &= -\int_0^2 [(2x-x^2)\log_2(2x-x^2)+(1-2x+x^2)\log_2(1-2x+x^2)]\,dx \\
&= \left[\frac{4x}{3}+\frac{4}{3}\log_2(x-2)+\frac{2}{3}\log_2(x-1)+\frac{1}{3}(x-3)x^2\log_2(2x-x^2)\right. \\
&\quad \left.-\frac{1}{3}x(3-3x-x^2)\log_2(1-2x+x^2)\right]_0^2 \\
&= 1.18.
\end{aligned}
$$

Moreover, $f''(A) = 1.18/5 = 0.236$.

8.3. FUZZY-SET INTERPRETATION OF POSSIBILITY THEORY

The term "possibility theory" is used in this section for the theory of graded possibilities introduced in Section 5.2. As a formal mathematical system, possibility theory has various interpretations, some of which are mentioned in Section 5.2.6. Perhaps the most visible and useful interpretation of possibility theory, which is the subject of this section, is its *fuzzy-set interpretation*. In this interpretation, possibility profiles are derived from information expressed in terms of fuzzy sets.

In order to explain this interpretation of possibility theory, let X denote a variable that takes values on a universal set X, and assume that information about the actual value of the variable is expressed by a proposition "X is F," where F is, in general, a fuzzy set on X. This clearly also covers the special case when F is a crisp set or even a singleton. To express information captured by this fuzzy proposition in measure–theoretic terms, it is natural, to interpret the membership degree $F(x)$ for each $x \in X$ as the degree of possibility that $X = x$. This interpretation induces a unique possibility profile r_F on X that is defined for all $x \in X$ by the equation

$$r_F(x) = F(x). \tag{8.16}$$

Given this possibility profile, the corresponding possibility measure, Pos_F, is then determined for all $A \in \mathcal{P}(X)$ by the formula

$$Pos_F(A) = \sup_{x \in X}\{\min\{\chi_A(x), r_F(x)\}\}, \tag{8.17}$$

where χ_A denotes the characteristic function of A. When A is a fuzzy set, the characteristic function in Eq. (8.17) is replaced with the membership function of A, which results in the more general formula

$$Pos_F(A) = \sup_{x \in X}\{\min\{A(x), r_F(x)\}\}. \tag{8.18}$$

The use of Eqs. (8.17) and (8.18) is illustrated in Figure 8.3a and 8.3b, respectively.

Equation (8.18) is also applicable when X is a multidimensional variable and F is then a fuzzy relation, generally n-dimensional ($n \geq 2$); the induced possibility distribution function r_F is also n-dimensional in this case. When the given fuzzy proposition is truth qualified, probability qualified, or modified in some other way, Eq. (8.18) is still applicable, provided that the fuzzy set F in Eq. (8.16) represents a relevant composition of functions involved in the modification, as is explained in Section 7.6.1.

The fuzzy-set interpretation of possibility theory emerges quite naturally from the similarity between the mathematical structures of possibility measures (or, alternatively, necessity measures) and fuzzy sets. In both cases, the

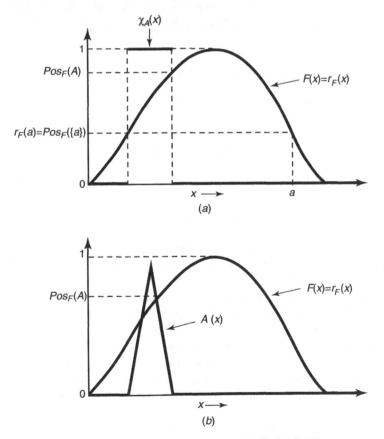

Figure 8.3. Illustration of (a) Eq. (8.17); (b) Eq. (8.18).

underlying mathematical structures are families of nested sets. In possibility theory, these families consist of focal elements; in fuzzy sets, they consist of α-cuts.

Equation (8.16) is usually referred to as the *standard fuzzy-set interpretation of possibility theory*. This interpretation is simple and natural, but it is applicable only to fuzzy sets that are normal. Recall that a fuzzy set F is normal when its height, h_F, is equal to 1. When a subnormal fuzzy set F is involved in Eq. (8.16), for which $h_F < 1$ by definition, then

$$\sup_{x \in X}\{r_F(x)\} = h_F < 1$$

and, by Eq. (8.18),

$$Pos(X) = h_F < 1.$$

This violates one of the axioms of possibility theory, and hence, the theory looses its coherence. To illustrate this problem, let S_F and $\bar{S}_F$ denote the support of F and its complement, respectively. When $h_F < 1$ for the fuzzy set F in Eq. (8.16), we obtain

$$Pos_F(S_F) = h_F \qquad \text{and} \qquad Pos_F(\bar{S}_F) = 0$$

by Eq. (8.18). Then, by the duality between *Pos* and *Nec*, expressed by Eq. (5.1), we have

$$Nec_F(S_F) = 1.$$

Hence,

$$Nec_F(S_F) > Pos_F(S_F),$$

which violates one of the key properties of possibility theory: whatever is necessary must be possible at least to the same degree, as expressed by the inequality in Eq. (5.12).

To assume that F in Eq. (8.16) is always normal is overly restrictive. For example, if the given fuzzy proposition "X is F" represents the conjunction of several fuzzy propositions that express information about X obtained from disparate sources, it is quite likely that F is subnormal. In this case, the value of $1 - h(F)$ indicates the degree of inconsistency among the individual information sources. It is thus important that a sound fuzzy-set interpretation of possibility theory be coherent for all fuzzy sets, regardless of whether they are normal or not.

In order to formulate a genuine fuzzy-set interpretation of possibility theory, we need to address the following question: How should we interpret information given in the form "X is F," where F is an arbitrary fuzzy subset of X, in terms of a possibility profile? An associated, but more specific, question is: How should we assign values of r_F to the values of $F(x)$ for all $x \in X$?

The position taken here is that the definition of the possibility profile r_F in terms of a given fuzzy set F should be such that it does not change the evidence conveyed by F and, at the same time, preserves the required possibilistic normalization, which is expressed in general by the equation $\sup_{x \in X}\{r_F(x)\} = 1$. Since the standard definition of r_F given by Eq. (8.16) does not satisfy the normalization for subnormal fuzzy sets, it must be appropriately modified.

To satisfy the possibilistic normalization, we must define $r_F(x) = 1$ for at least one $x \in X$. Since there is no reason to treat distinct elements of X differently, the only sensible way to achieve the required normalization is to increase the values of r_F equally for all $x \in X$ by the amount of $1 - h_F$. This means that the revised fuzzy-set interpretation of possibility theory is expressed for all $x \in X$ by the equation

$$r_F(x) = F(x) + 1 - h_F. \tag{8.19}$$

This is a generalized counterpart of the standard interpretation, Eq. (8.16); it is applicable to all fuzzy sets, regardless whether they are normal or not. For normal fuzzy sets, clearly, Eq. (8.19) collapses to Eq. (8.16).

The significance of the possibility profile r_F defined by Eq. (8.19) is that it is the only one that does not change the evidence conveyed by F. An easy way to show this uniqueness is to use the Möbius representation of this possibility profile (see Section 5.2.1). As in Section 5.2.1, assume that elements of the universal set $X = \{x_1, x_2, \ldots, x_n\}$ are ordered in such a way that

$$r_F(x_i) \geq r_F(x_{i+1})$$

for all $i \in \mathbb{N}_{n-1}$, and let $r_F(x_{i+1}) = 0$ by convention. Moreover, let $A_i = \{x_1, x_2, \ldots, x_i\}$. Then, the Möbius representation, m_F, of the possibility profile is given by the formula

$$m_F(A) = \begin{cases} r_F(x_i) - r_F(x_{i+1}) & \text{when } A = A_i \text{ for some } i \in \mathbb{N}_n \\ 0 & \text{when } A \neq A_i \text{ for all } i \in \mathbb{N}_n. \end{cases} \tag{8.20}$$

Now substituting for r_F from Eq. (8.19), we obtain

$$m_F(A) = \begin{cases} F(x_i) - F(x_{i+1}) & \text{when } A = A_i \text{ for some } i \in \mathbb{N}_{n-1} \\ \inf_{x_i \in X} F(x_i) + 1 - h_F & \text{when } A = X(= A_n) \\ 0 & \text{when } A \neq A_i \text{ for all } i \in \mathbb{N}_n. \end{cases}$$

We can see that the evidential support, $m_F(A)$, for the various sets $A \subset X$ is based on the original evidence expressed by the values $F(x_i)$ for all $i \in \mathbb{N}_i$. The only change is in the value of $m_F(X)$, which expresses the degree of our ignorance. Since inconsistency in evidence, which is expressed by the value $1 - h_F$, is a form of ignorance, it is natural to place it in $m_F(X)$.

The necessity function, Nec_F, which is based on the evidence expressed by F, is uniquely determined by m_F via the usual formula

$$Nec_F(A) = \begin{cases} \sum_{k=1}^{i} m(A_k) = 1 - r_F(x_{i+1}) & \text{when } A = A_i \text{ for some } i \in \mathbb{N}_n. \\ 1 - \max_{x_i \in A} r_F(x_i) & \text{otherwise.} \end{cases} \quad (8.21)$$

This function, again, does not modify in any way the evidence expressed by F. Moreover,

$$Pos_F(A_i) = \sum_{A_k \cap A_i \neq \emptyset} m(A_k) = 1 \quad \text{for all } i \in \mathbb{N}_n, \quad (8.22)$$

which is always the case for all possibilistic bodies of evidence.

It is easy to see that defining r_F in any way different from Eq. (8.19) would either violate the possibilistic normalization or the evidence conveyed by F. For example, it has often been suggested in the literature to obtain a normalized possibility profile, r_F', for a subnormal fuzzy set by the formula

$$r_F'(x) = \frac{F(x)}{h_F} \quad (8.23)$$

for all $x \in X$. However, this possibility profile does not preserve the evidence conveyed by F. Clearly,

$$m_F'(A_i) = r_F'(r_i) - r'(r_{i+1}) = \frac{1}{h_F}[F(x_i) - F(x_{i+1})],$$

for all $i \in \mathbb{N}_n$, where the symbols x_i, x_{i+1}, and A_i have the meaning introduced earlier in this section. We can see that the evidential support for sets A_i (focal elements), expressed by the values of $m_F'(A_i)$, is inflated by the factor of $1/h_F$.

Under the generalized fuzzy-set interpretation of possibility theory, Eq. (8.18) can be written more explicitly as

$$Pos_F(A) = \sup_{x \in X}\{\min\{A(x), F(x) + 1 - h_F\}\}, \quad (8.24)$$

where A and F are arbitrary fuzzy sets. Thus, for example,

$$\begin{aligned} Pos_F(S_F) &= \sup_{x \in X}\{\min\{S_F(x), F(x) + 1 - h_F\}\} \\ &= \sup_{x \in X}\{F(x) + 1 - h_F\} \\ &= 1 \end{aligned}$$

$$Pos_F(\bar{S}_F) = \sup_{x \in X}\{\min\{\bar{S}_F(x), F(x) + 1 - h_F\}\}$$

$$= \begin{cases} 1 - h_F & \text{when } \bar{S}_F \neq \emptyset \\ 0 & \text{when } \bar{S}_F = \emptyset. \end{cases}$$

$$Nec_F(S_F) = 1 - Pos_F(\bar{S}_F)$$

$$= \begin{cases} h_F & \text{when } S_F \neq X \\ 0 & \text{when } S_F = X \end{cases}$$

$$Nec_F(\bar{S}_F) = 1 - Pos_F(S_F)$$
$$= 0.$$

Now consider the two extreme cases when $F = X$ and $F = \emptyset$. In the former case, the proposition "X is X" does not carry any information; in the latter case, the proposition "X is $\emptyset$" represents evidence that is totally conflicting, and hence, it does not carry any information either. Applying Eq. (8.19), we obtain

$$r_X(x) = r_\emptyset(x) = 1$$

for all $x \in X$. This means that both propositions are represented by the same possibility and necessity measures:

$$Pos_X(A) = Pos_\emptyset(A) = \begin{cases} 1 & \text{when } A \neq \emptyset, \\ 0 & \text{when } A = \emptyset, \end{cases}$$

$$Nec_X(A) = Nec_\emptyset(A) = \begin{cases} 0 & \text{when } A \neq X, \\ 1 & \text{when } A = X, \end{cases}$$

These results are exactly the same as we would expect on intuitive grounds.

EXAMPLE 8.7. To illustrate the generalized fuzzy-set interpretation of possibility theory, let evidence regarding the relation between two discrete variables X and Y with states in sets $X = \{x_a | a \in \mathbb{N}_4\}$ and $Y = \{y_b | b \in \mathbb{N}_5\}$, respectively, be expressed in terms of a fuzzy relation R defined by the following matrix:

$$\mathbf{R} = \begin{array}{c} \\ x_1 \\ x_2 \\ x_3 \\ x_4 \end{array} \begin{array}{ccccc} y_1 & y_2 & y_3 & y_4 & y_5 \\ \left[\begin{array}{ccccc} 0.0 & 0.2 & 0.0 & 0.4 & 0.5 \\ 0.3 & 0.0 & 0.0 & 0.6 & 0.5 \\ 0.0 & 0.7 & 0.6 & 0.5 & 0.4 \\ 0.2 & 0.4 & 0.3 & 0.0 & 0.0 \end{array}\right] \end{array}.$$

The possibility profile, r_R, based on this evidence is determined by Eq. (8.19):

$$[r_R(x_i, y_j)] = \begin{array}{c} \\ x_1 \\ x_2 \\ x_3 \\ x_4 \end{array} \overset{\begin{array}{ccccc} y_1 & y_2 & y_3 & y_4 & y_5 \end{array}}{\begin{bmatrix} 0.3 & 0.5 & 0.3 & 0.7 & 0.8 \\ 0.6 & 0.3 & 0.3 & 0.9 & 0.8 \\ 0.3 & 1.0 & 0.9 & 0.8 & 0.7 \\ 0.5 & 0.7 & 0.6 & 0.3 & 0.3 \end{bmatrix}}.$$

The nested body of evidence associated with this 2-dimensional possibility profile is shown in Figure 8.4, where the pairs of integers denote the subscripts of pairs $\langle x_a, y_b \rangle$. Calculating the various components of uncertainty associated with this possibilistic body of evidence, we obtain: $GH(m_R) = 3.12$, $\bar{S}(Nec_R) = 4.29$, and $\underline{S}(Nec_R) = 0$. Viewing r_R as the normalized version, $\hat{R}$, of the given relation R, we readily obtain $f(\hat{R}) = 11.2$ and $\hat{f}(\hat{R}) = 0.56$, while for the given fuzzy relation R, we get $f(R) = 9.6$ and $\hat{f}(R) = 0.48$.

Consider now a real-valued variable X that takes values in $X = [\underline{x}, \bar{x}]$. Information about the actual value of the variable is expressed again by the proposition "X is F," where F is in this case a fuzzy interval defined on X. Differences in Eq. (8.20) are now differentials. Denoting the focal elements corresponding to the α-cuts $^\alpha F$ by $^\alpha r_F$, the necessity function for all focal elements is expressed by the simple formula

$$Nec_F\left({}^\alpha r_F \right) = \int\limits_1^\alpha -d\alpha$$
$$= 1 - \alpha. \tag{8.25}$$

For other crisp sets

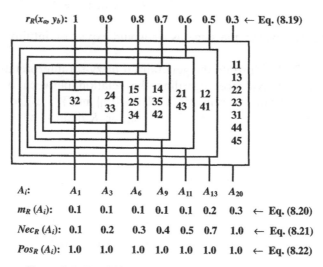

Figure 8.4. Possibilistic body of evidence in Example 8.7.

$$Nec_F(A) = 1 - \sup_{x \in \overline{A}} r_F(x) \qquad (8.26)$$

which can be easily generalized to fuzzy set A via Eq. (8.18).

Equations (8.25) and (8.26) are counterparts of Eq. (8.21) for any convex and bounded universal set $X \subset \mathbb{R}$. A generalization to $\mathbb{R}^n (n \geq 2)$ is straightforward, but it is not pursued here.

EXAMPLE 8.8. Let evidence regarding the value of a real-valued variable X, whose values are in the interval $X = [0, 4]$, be expressed by the proposition "X is F," where F is a triangular fuzzy number shown in Figure 8.5 and defined for each $x \in X$ by the formula

$$F(x) = \begin{cases} (x-1)/2 & \text{when } x \in [1,2) \\ (3-x)/2 & \text{when } x \in [2,3] \\ 0 & \text{otherwise.} \end{cases}$$

F is a subnormal fuzzy set; $h_F = 0.5$. According to Eq. (8.19), the associated possibility profile r_F is defined for each $x \in X$ by

$$r_F(x) = F(x) + 0.5,$$

and its graph is shown in Figure 8.5. For each $\alpha \in (0, 1]$, the focal elements of r_F are

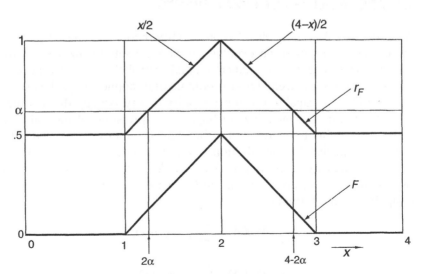

Figure 8.5. Illustration to Example 8.8.

$$
{}^{\alpha}r_F = \begin{cases} [0, 4] & \text{when } (0, 0.5] \\ [2\alpha, 4 - 2\alpha] & \text{when } (0.5, 1], \end{cases}
$$

and their Lebesgue measures are

$$
\mu({}^{\alpha}r_F) = \begin{cases} 4 & \text{when } \alpha \in [0, 0.5] \\ 4 - 4\alpha & \text{when } \alpha \in (0.5, 1]. \end{cases}
$$

The generalized Hartley-like measure of nonspecificity, *GHL*, is then calculated by the following integral:

$$
\begin{aligned}
GHL(r_F) &= \int_0^1 \log_2[1 + \mu({}^{\alpha}r_F)]\, d\alpha \\
&= \int_0^{0.5} \log_2 5\, d\alpha + \int_{0.5}^1 \log_2(5 - 4\alpha)\, d\alpha \\
&= [\alpha \log_2 5]_0^{0.5} - \left[\frac{1}{4}(5 - 4\alpha)\log_2(5 - 4\alpha) + \frac{\alpha}{\ln 2}\right]_{0.5}^1 \\
&= 2.077.
\end{aligned}
$$

Using Eq. (8.11), the degree of fuzziness of F and its normalized counterpart, $\hat{F} = r_F$, are $f(F) = 1$ and $f(\hat{F}) = 3$, and their normalized versions are $\hat{f}(F) = 0.25$ and $\hat{f}(\hat{F}) = 0.75$. The numbers can be obtained in this case by simple geometrical considerations in Fig. 8.5.

8.4. PROBABILITIES OF FUZZY EVENTS

A simple fuzzification of classical probability theory is obtained by extending the classical concept of an event from classical sets to fuzzy sets. Notwithstanding its simplicity, this fuzzification makes probability theory more expressive and, consequently, enlarges the domain of its applicability. The use of fuzzy events adds new capabilities to probability theory. Among them is the capability to capture the meanings of linguistic expressions of natural language, and the capability to express borderline uncertainties in measurement.

Given a probability distribution function p on X (when X is finite) or a probability density function q on X (when $X = \mathbb{R}^n$ for some $n \geq 1$), the classical probability measure, *Pro(A)*, can be expressed for each set A in a given σ-algebra by the formulas

$$
Pro(A) = \sum_{x \in X} \chi_A(x)p(x),
$$

$$
Pro(A) = \int_{\mathbb{R}^n} \chi_A(x)q(x)\, dx,
$$

respectively, where X_A denotes the characteristic function of set A. These formulas are readily generalized to fuzzy sets (events) A by replacing the characteristic function X_A with the membership function A:

$$Pro(A) = \sum_{x \in X} A(x)p(x), \tag{8.27}$$

$$Pro(A) = \int_{\mathbb{R}^n} A(x)q(x) \, dx. \tag{8.28}$$

Observe that the classical case of crisp events is captured by Eqs. (8.27) and (8.28) as well. Observe also that probabilities of fuzzy events preserve some properties of classical (additive) probabilities. For example, using the standard operations on fuzzy sets and assuming that X is finite, we obtain for any $A, B \in \mathcal{F}(X)$

$$\begin{aligned} Pro(A \cup B) &= \sum_{x \in X} \max\{A(x), B(x)\} \\ &= \sum_{x \in X} A(x) + \sum_{x \in X} B(x) - \sum_{x \in X} \min\{A(x), B(x)\} \\ &= Pro(A) + Pro(B) - Pro(A \cap B). \end{aligned}$$

This equation conforms to the calculus of classical probability theory, even though A and B are fuzzy events. However, some other properties of the calculus are not preserved for fuzzy events. For example, $Pro(A \cup \bar{A}) \le 1$ and $Pro(A \cap \bar{A}) \ge 0$ when A is a fuzzy event; the equalities are obtained only in the special case when A becomes a crisp event.

EXAMPLE 8.9. Consider a random variable whose values are positive real numbers and whose probability distribution function p (shown in Figure 8.6) is characterized by the probability density function

$$q(x) = \begin{cases} (x - 76)/40 & \text{when } x \in [76, 84) \\ (86 - x)/10 & \text{when } x \in [84, 86] \\ 0 & \text{otherwise.} \end{cases}$$

This function is also shown in Figure 8.6, together with the membership function

$$A(x) = \begin{cases} (x - 76)/4 & \text{when } x \in [76, 80) \\ (86 - x)/4 & \text{when } x \in [80, 84] \\ 0 & \text{otherwise.} \end{cases}$$

which represents (in a given application context) a fuzzy event "around 80." Using Eq. (8.28), the probability that x is around 80, $Pro(A)$, is calculated as follows:

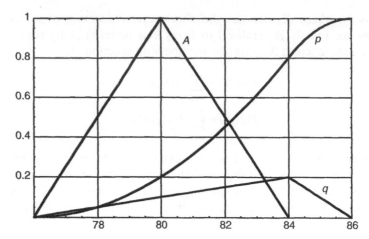

Figure 8.6. Probability distribution p, probability density function q, and fuzzy event A in Example 8.9.

$$Pro(A) = \int_{76}^{80} \frac{x-76}{4} \cdot \frac{x-76}{40} \, dx + \int_{80}^{84} \frac{84-x}{4} \cdot \frac{x-76}{40} \, dx = 0.4.$$

A random variable whose values are real numbers in an interval $X = [\underline{x}, \bar{x}]$ is often approximated by partitioning X into a finite set of disjoint intervals. This approximation is usually referred to as *quantization*. Assuming that the partition consists of n intervals (quanta) of equal length, $\Delta_n = (\underline{x} - \bar{x})/n$, the partition, $\pi_n(X)$, is defined in general terms as

$$\pi_n(X) = \{\{[\underline{x} + (i-1)\Delta_n, \underline{x} + i\Delta_n] \mid i \in \mathbb{N}_{n-1}\}, [\underline{x} + (n-1)\Delta_n, \underline{x} + n\Delta_n]\}.$$

Observe that all intervals except the last one in this set are semiclosed intervals; the last interval is closed to ensure that the union of all the intervals is equal to X (as required for a partition). For convenience, let

$$I_i = \begin{cases} [\underline{x} + (i-1)\Delta_n, \underline{x} + i\Delta_n] & \text{when } i \in \mathbb{N}_{n-1} \\ [\underline{x} + (i-1)\Delta_n, \underline{x} + i\Delta_n] & \text{when } i \in n. \end{cases}$$

Recall that these intervals may be represented by their characteristic functions χ_{I_i}.

An example of partition $\pi_n(X)$, where $X = [0, 100]$ and $n = 5$, is shown in Figure 8.7a. The five intervals in this partition may be characterized as classes of values of the variable that are *small*, *medium*, *large*, and so on, as is indicated in the figure. A more expressive representation of these classes is obtained when the intervals $I_i \in \pi_n(X)$ are replaced with appropriate fuzzy

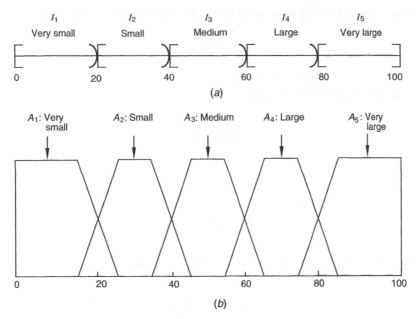

Figure 8.7. Examples of partitions of the interval [0, 100]: (*a*) crisp partition; (*b*) fuzzy partition.

intervals A_i, as shown in Figure 8.7*b*. The main advantage of fuzzy intervals is their capability to make transitions between adjacent classes gradual rather than abrupt.

A nonempty family of fuzzy intervals A_i, $i \in \mathbb{N}_n$, for some $n \geq 1$ (or, more generally, fuzzy sets) defined on X is called a *fuzzy partition* of X if and only if the family does not contain the empty set and

$$\sum_{i \in \mathbb{N}_n} A_i(x) = 1 \quad \text{for each } x \in X.$$

It is left to the reader to show that fuzzy partitions are not cutworthy. The five trapezoidal-shaped fuzzy intervals in Figure 8.7*b*, whose definitions are

$$A_1 = \langle 0, 0, 15, 25 \rangle$$

$$A_2 = \langle 15, 25, 35, 45 \rangle$$

$$A_3 = \langle 35, 45, 55, 65 \rangle$$

$$A_4 = \langle 55, 65, 75, 85 \rangle$$

$$A_5 = \langle 75, 85, 100, 100 \rangle,$$

form a fuzzy partition on [0, 100].

Now assume that we have a set of m observations of the variable

$$\{x_k \in X | k \in \mathbb{N}_m\}.$$

Using these data, probabilities of the individual intervals, $Pro(I_i)$, are calculated in the usual way. For each $i \in \mathbb{N}_n$,

$$Pro(I_i) = \frac{f_i}{m}, \tag{8.29}$$

where

$$f_i = \sum_{k=1}^{m} \chi_{I_i}(x_k). \tag{8.30}$$

Counterparts of Eqs. (8.29) and (8.30) for fuzzy intervals A_i are, respectively

$$Pro(A_i) = \frac{a_i}{m}, \tag{8.31}$$

where

$$a_i = \sum_{k=1}^{m} A_i(x_k).$$

If the family $\{A_i | i \in \mathbb{N}_n\}$ forms a fuzzy partition of X, then

$$\sum_{i \in \mathbb{N}_n} a_i = m \quad \text{and} \quad \sum_{i \in \mathbb{N}_n} Pro(A_i) = 1.$$

8.5. FUZZIFICATION OF REACHABLE INTERVAL-VALUED PROBABILITY DISTRIBUTIONS

A natural fuzzification of the theory of reachable interval-valued probability distributions (introduced in Section 5.5) is obtained by extending relevant probability intervals to fuzzy intervals via the α-cut representation. To discuss basic issues of this fuzzified uncertainty theory, let the following notation be used.

Assume, as in Section 5.5, that we deal with a finite set $X = \{x_i | i \in \mathbb{N}_n\}$ of considered alternatives. These alternatives may, in general, be viewed as states of variable X. Assume further that each alternative $x_i \in X$ is associated with an imprecise probability expressed by a fuzzy interval, F_i, defined on $[0, 1]$. It is assumed that $F_i(p)$ is defined for all $p \in [0, 1]$ in terms of the canonical form expressed by Eq. (7.25). As always, the fuzzy interval is uniquely represented by the family of its α-cuts,

$$^{\alpha}F_i = [l_i(\alpha), u_i(\alpha)]$$

for all $\alpha \in [0,1]$. In each particular application, evidence in this fuzzified uncertainty theory is expressed by an n-tuple, $\mathbf{F}$, of fuzzy intervals F_i. That is,

$$\mathbf{F} = \langle F_i \mid i \in \mathbb{N}_n \rangle. \tag{8.32}$$

Again, each $\mathbf{F}$ is uniquely represented by the family of its α-cuts,

$$^{\alpha}\mathbf{F} = \langle ^{\alpha}F_i \mid i \in \mathbb{N}_n \rangle$$

for all $\alpha \in [0,1]$. For convenience, let fuzzy intervals F_i in Eq. (8.32) be called *probability granules*, and let each tuple $\mathbf{F}$ defined by Eq. (8.32) be called a *tuple of probability granules*.

The various properties of tuples of probability intervals, which are introduced in Section 5.5, are extended to tuples of probability granules via the α-cut representations of the latter. Thus, for example, a tuple of probability granules is called *proper* iff all its α-cuts are proper in the classical sense; it is called *reachable* iff all its α-cuts are reachable in the classical sense; and so forth.

To make computation with probability granules as efficient as possible, it is desirable to represent them by trapezoidal membership functions. That is, it is desirable to deal only with special tuples of probability granules,

$$\mathbf{T} = \langle T_i = \langle a_i, b_i, c_i, d_i \rangle \mid i \in \mathbb{N}_n \rangle, \tag{8.33}$$

where each probability granule T_i is a trapezoidal-shaped fuzzy interval with support $[a_i, d_i]$ and core $[b_i, c_i]$. However, when the operations of multiplication or division of fuzzy arithmetic are applied to trapezoidal granules, the resulting granules are not trapezoidal. This is unfortunate since we often need to use the resulting granules as inputs for further processing. For efficient computation, it is thus desirable to approximate the membership functions of resulting probability granules at each stage of computation by appropriate trapezoidal granules.

A simple way of approximating an arbitrary granule, F, by a trapezoidal one, T, is to keep the values a, b, c, d of the canonical form of F (see Eq. (7.25)) unchanged and replace its nonlinear functions in the intervals $[a, b]$ and $[c, d]$ with their linear counterparts

$$\frac{x-a}{b-a} \quad \text{and} \quad \frac{d-x}{d-c},$$

respectively. The advantage of this approximation method is its simplicity. However, when multiplication or division operations are used repeatedly, the accumulated error can become sufficiently large to produce misleading results. Other approximation methods have been proposed to reduce the accumulated error. Although some recent results seem to indicate that this approximation problem can be adequately solved, further research is needed to obtain more conclusive results (see Note 8.11).

Assume, for the sake of simplicity, that we deal in this section only with tuples of trapezoidal probability granules and that we specify them either in the form of Eq. (8.33) or via the associated α-cut representation

$$^{\alpha}\mathbf{T} = \langle\, ^{\alpha}T_i = [a_i + (b_i - a_i)\alpha,\, d_i - (d_i - c_i)\alpha] \mid i \in \mathbb{N}_n \rangle. \qquad (8.34)$$

To be proper, a tuple of trapezoidal probability granules must satisfy the inequalities

$$\sum_{i \in \mathbb{N}_n} [a_i + (b_i - a_i)\alpha] \le 1,$$

$$\sum_{i \in \mathbb{N}_n} [d_i - (d_i - c_i)\alpha] \ge 1$$

for all $\alpha \in [0, 1]$. Clearly, if the inequalities are satisfied for $\alpha = 1$, then they are satisfied for all $\alpha \in [0, 1]$. It is thus sufficient to check the inequalities

$$\sum_{i \in \mathbb{N}_n} b_i \le 1 \qquad \text{and} \qquad \sum_{i \in \mathbb{N}_n} c_i \ge 1.$$

Moreover, this convenient feature holds for all tuples of probability granules, not only for the trapezoidal ones.

To be reachable, a tuple of trapezoidal probability granules must satisfy the inequalities

$$\sum_{j \ne i} [a_j + (b_j - a_j)\alpha] + d_i - (d_i - c_i)\alpha \le 1, \qquad (8.35)$$

$$\sum_{j \ne i} [d_j - (d_j - c_j)\alpha] + a_i + (b_i - a_i)\alpha \ge 1 \qquad (8.36)$$

for all $i \in \mathbb{N}_n$ and all $\alpha \in [0, 1]$. Since the left-hand sides of these inequalities are linear functions of α, it is sufficient to check the inequalities only for $\alpha = 0$ and $\alpha = 1$. That is, it is sufficient to check the inequalities

$$\sum_{j \ne i} a_j + d_i \le 1, \qquad (8.37)$$

$$\sum_{j \ne i} d_j + a_i \ge 1, \qquad (8.38)$$

$$\sum_{j\neq i} b_j + c_i \leq 1, \tag{8.39}$$

$$\sum_{j\neq i} c_j + b_i \geq 1, \tag{8.40}$$

for all $i \in \mathbb{N}_n$. Clearly, this convenient feature does not hold for tuples of probability granules that are not trapezoidal.

EXAMPLE 8.10. Consider a tuple with four trapezoidal probability granules, $\mathbf{T} = \langle T_i | i \in \mathbb{N}_4 \rangle$, where

$$T_1 = \langle 0, 0.1, 0.1, 0.2 \rangle,$$

$$T_2 = \langle 0.1, 0.2, 0.2, 0.25 \rangle,$$

$$T_3 = \langle 0.2, 0.3, 0.3, 0.4 \rangle,$$

$$T_4 = \langle 0.3, 0.4, 0.4, 0.6 \rangle.$$

This tuple, which is shown in Figure 8.8, is proper since

$$\sum_{i\in\mathbb{N}_4} b_i = \sum_{i\in\mathbb{N}_4} c_i = 1.$$

It is also reachable since the inequalities (8.37)–(8.40) are satisfied for all $i \in \mathbb{N}_4$, as can be easily verified.

When a given tuple of trapezoidal fuzzy granules,

$$\mathbf{T} = \langle T_i \mid i \in \mathbb{N}_n \rangle,$$

is proper but not reachable, it can be converted to its reachable counterpart,

$$\mathbf{F} = \langle F_i \mid i \in \mathbb{N}_n \rangle,$$

for all $\alpha \in [0, 1]$ and all $i \in \mathbb{N}_n$ via the formulas

$$l_i(\alpha) = \max\left\{ a_i + (b_i - a_i)\alpha, 1 - \sum_{j\neq i}[d_j - (d_j - c_j)\alpha] \right\} \tag{8.41}$$

$$u_i(\alpha) = \min\left\{ d_i - (d_i - c_i)\alpha, 1 - \sum_{j\neq i}[a_j + (b_j - a_j)\alpha] \right\} \tag{8.42}$$

where $[l_i(\alpha), u_i(\alpha)] = {}^{\alpha}F_i$. Observe that, due to the max and min operations in these formulas, granules F_i of the resulting tuple $\mathbf{F}$ may not have trapezoidal shapes.

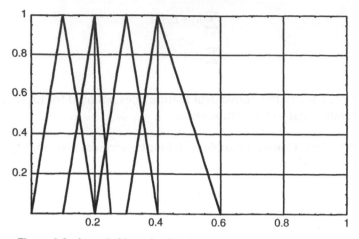

Figure 8.8. A reachable tuple of probability granules (Example 8.10).

EXAMPLE 8.11. Consider a tuple $\mathbf{T} = \langle T_1, T_2, T_3 \rangle$ of trapezoidal probability granules

$$T_1 = \langle 0.0, 0.0, 0.2, 0.4 \rangle$$

$$T_2 = \langle 0.1, 0.3, 0.4, 0.5 \rangle$$

$$T_3 = \langle 0.4, 0.6, 0.8, 0.9 \rangle$$

whose α-cuts are

$$^{\alpha}T_1 = [0, 0.4 - 0.2\alpha],$$

$$^{\alpha}T_2 = [0.1 + 0.2\alpha, 0.5 - 0.1\alpha],$$

$$^{\alpha}T_3 = [0.4 + 0.2\alpha, 0.9 - 0.1\alpha].$$

This tuple, which is shown in Figure 8.9a, is proper since $b_1 + b_2 + b_3 = 0.9 < 1$ and $c_1 + c_2 + c_3 = 1.4 > 1$. However, it is not reachable since, for example, the inequality (8.39) is violated for $i = 1$ ($b_2 + b_3 + c_1 = 1.1 \nleq 1$) and $i = 3$ ($b_1 + b_2 + c_3 = 1.1 \nleq 1$). To use the probability granules for further processing, for example, as prior probabilities in Bayesian inference, we need to convert $\mathbf{T}$ to its reachable counterpart $\mathbf{F}$. By using Eqs. (8.41) and (8.42) for each $i = 1, 2, 3$ and $\alpha \in [0, 1]$, we obtain the following α-cuts of the probability granules (not necessarily trapezoidal) in F:

$i = 1 : l_1(\alpha) = \max\{0, 0.2\alpha - 0.4\} = 0,$

$$u_1(\alpha) = \min\{0.4 - 0.2\alpha, 0.5 - 0.4\alpha\} = \begin{cases} 0.4 - 0.2\alpha & \text{when } \alpha \in [0, 0.5) \\ 0.5 - 0.4\alpha & \text{when } \alpha \in [0.5, 1], \end{cases}$$

$i = 2 : l_2(\alpha) = \max\{0.1 + 0.2\alpha, 0.3\alpha - 0.3\} = 0.1 + 0.2\alpha,$
$\quad u_2(\alpha) = \min\{0.5 - 0.1\alpha, 0.6 - 0.2\alpha\} = 0.5 - 0.1\alpha,$

$i = 3 : l_3(\alpha) = \max\{0.4 + 0.2\alpha, 0.1 + 0.3\alpha\} = 0.4 + 0.2\alpha,$
$\quad u_3(\alpha) = \min\{0.9 - 0.1\alpha, 0.9 - 0.2\alpha\} = 0.9 - 0.2\alpha.$

Graphs of F_1, F_2, F_3 are shown in Figure 8.9b. We can see that $F_1 \neq T_1$ and $F_3 \neq T_3$. Moreover, F_1 is not of the trapezoidal shape.

Once it is verified that a given tuple **F** of probability granules on set X is reachable, it can be used for calculating probability granules for any subset of X. These calculations are facilitated by applying Eqs. (5.71) and (5.72) to the α-cuts of **F**.

For each $A \in \mathcal{P}(X)$ and each $\alpha \in [0, 1]$, let

$$^{\alpha}F_A = [l_A(\alpha), u_A(\alpha)]$$

denote the α-cut of the probability granule, F_A, of set A. Then, using Eqs. (5.71) and (5.72), we have

$$l_A(\alpha) = \max\left\{ \sum_{x_i \in A} l_i(\alpha), 1 - \sum_{x_i \notin A} u_i(\alpha) \right\}, \tag{8.43}$$

$$u_A(\alpha) = \min\left\{ \sum_{x_i \in A} u_i(\alpha), 1 - \sum_{x_i \notin A} l_i(\alpha) \right\}. \tag{8.44}$$

EXAMPLE 8.12. Consider the reachable tuple of four probability granules on $X = \{x_1, x_2, x_3, x_4\}$ that is discussed in Example 8.10. Graphs of the granules are shown in Figure 8.8, and their α-cut representations are:

$$^{\alpha}T_1 = [0.1\alpha, 0.2 - 0.1\alpha],$$

$$^{\alpha}T_2 = [0.1 + 0.1\alpha, 0.25 - 0.05\alpha],$$

$$^{\alpha}T_3 = [0.2 + 0.1\alpha, 0.4 - 0.1\alpha],$$

$$^{\alpha}T_4 = [0.3 + 0.1\alpha, 0.6 - 0.2\alpha].$$

The α-cuts of probability granules $^{\alpha}F_A$ for all sets $A \in \mathcal{P}(X)$, calculated by Eqs. (8.43) and (8.44), are given in Table 8.1. For convenience, sets A are identified

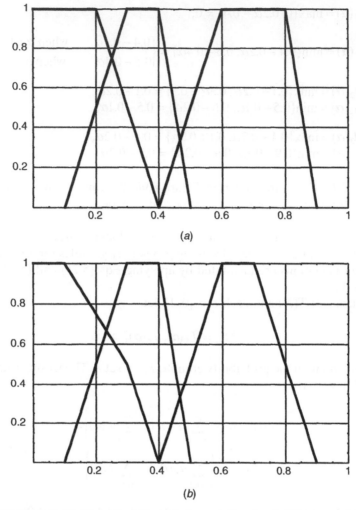

Figure 8.9. Illustration to Example 8.11: (*a*) A proper tuple of probability granules that is not reachable; (*b*) the reachable tuple of probability granules derived from the one in part (*a*).

by the index k defined in the table. For the singletons, clearly, $k = i$. For example, for set $A = \{x_1, x_2\}$, identified by $k = 5$, we have

$$
\begin{aligned}
l_5(\alpha) &= \max\{l_1(\alpha) + l_2(\alpha), 1 - u_3(\alpha) - u_4(\alpha)\} \\
&= \max\{0.1 + 0.2\alpha, 0.3\alpha\} \\
&= 0.1 + 0.2\alpha,
\end{aligned}
$$

$$
\begin{aligned}
u_5(\alpha) &= \min\{u_1(\alpha) + u_2(\alpha), 1 - l_3(\alpha) - l_4(\alpha)\} \\
&= \min\{0.45 - 0.15\alpha, 0.5 - 0.3\alpha\} \\
&= 0.45 - 0.15\alpha;
\end{aligned}
$$

for the set $A = \{x_1, x_3, x_4\}$, identified by $k = 13$, we have

$$l_{13}(\alpha) = \max\{l_1(\alpha) + l_3(\alpha) + l_4(\alpha), 1 - u_2(\alpha)\}$$
$$= \max\{0.5 + 0.3\alpha, 0.75 + 0.05\alpha\}$$
$$= 0.75 + 0.05\alpha,$$

$$u_{13}(\alpha) = \min\{u_1(\alpha) + u_3(\alpha) + u_4(\alpha), 1 - l_2(\alpha)\}$$
$$= \min\{1.2 - 0.4\alpha, 0.9 - 0.1\alpha\}$$
$$= 0.9 - 0.1\alpha;$$

and, similarly, all the remaining values of $l_A(\alpha)$ and $u_A(\alpha)$ in Table 8.1 are calculated. Observe the following:

(a) For all $A \in \mathcal{P}(X)$ and all $\alpha \in [0, 1]$,

$$u_A(\alpha) = 1 - l_{\bar{A}}(\alpha);$$

(b) For all $A \in \mathcal{P}(X)$, the probability granules F_A have triangular shapes;
(c) Due to the triangular shapes, the α-cuts $^\alpha F_A$ are additive for $\alpha = 1$ ($l_A(1) = u_A(1)$ for all $A \in \mathcal{P}(X)$).

Table 8.1. Lower and Upper Probability Granulas for the Reachable Tuple of Probability Granules Discussed in Example 8.10 and Shown in Figure 8.8 (illustration of Example 8.12)

	X			k	$^\alpha F_A$		$^\alpha m(\alpha)$
x_1	x_2	x_3	x_4		$l_A(\alpha)$	$u_A(\alpha)$	
A: 0	0	0	0	0	0	0	0
1	0	0	0	1	0.1α	$0.2 - 0.1\alpha$	0.1α
0	1	0	0	2	$0.1 + 0.1\alpha$	$0.25 - 0.05\alpha$	$0.1 + 0.1\alpha$
0	0	1	0	3	$0.2 + 0.1\alpha$	$0.4 - 0.1\alpha$	$0.2 + 0.1\alpha$
0	0	0	1	4	$0.3 + 0.1\alpha$	$0.6 - 0.2\alpha$	$0.3 + 0.1\alpha$
1	1	0	0	5	$0.1 + 0.2\alpha$	$0.45 - 0.15\alpha$	0
1	0	1	0	6	$0.2 + 0.2\alpha$	$0.6 - 0.2\alpha$	0
1	0	0	1	7	$0.35 + 0.15\alpha$	$0.7 - 0.2\alpha$	$0.05 - 0.05\alpha$
0	1	1	0	8	$0.3 + 0.2\alpha$	$0.65 - 1.5\alpha$	0
0	1	0	1	9	$0.4 + 0.2\alpha$	$0.8 - 0.2\alpha$	0
0	0	1	1	10	$0.55 + 0.15\alpha$	$0.9 - 0.2\alpha$	$0.05 - 0.05\alpha$
1	1	1	0	11	$0.4 + 0.2\alpha$	$0.7 - 0.1\alpha$	$0.1 - 0.1\alpha$
1	1	0	1	12	$0.6 + 0.1\alpha$	$0.8 - 0.1\alpha$	$0.15 - 0.15\alpha$
1	0	1	1	13	$0.75 + 0.05\alpha$	$0.9 - 0.1\alpha$	$0.15 - 0.15\alpha$
0	1	1	1	14	$0.8 + 0.1\alpha$	$1.0 - 0.1\alpha$	$0.15 - 0.15\alpha$
1	1	1	1	15	1	1	$-0.25 + 0.25\alpha$

Once the α-cuts of probability granules F_A are determined for all $A \in \mathcal{P}(X)$, the Möbius representation of these α-cuts, $^\alpha m$, can be calculated by the usual formula

$$^\alpha m(A) = \sum_{B|B \subseteq A} (-1)^{|A-B|} l_B(\alpha) \tag{8.45}$$

for all $A \in \mathcal{P}(X)$. For the lower probability in Example 8.12, the Möbius representation of is shown in Table 8.1. For example,

$$^\alpha m(\{x_1, x_2, x_3\}) = 0.4 + 0.2\alpha - (0.1 + 0.2\alpha) - (0.2 + 0.2\alpha) - (0.3 + 0.2\alpha)$$
$$+ 0.1\alpha + 0.1 + 0.1\alpha + 0.2 + 0.1\alpha = 0.1 - 0.1\alpha.$$

Observe that

$$\sum_{A \in \mathcal{P}(X)} {}^\alpha m(A) = 1$$

for each $\alpha \in [0, 1]$.

Using the Möbius representation, we can compute the average nonspecificity embedded in the associated tuple of probability granules. First, we calculate the generalized Hartley measure as a function of α,

$$GH(^\alpha m) = \sum_{A \in \mathcal{P}(X)} {}^\alpha m(A) \log_2 |A|, \tag{8.46}$$

which is followed by computing the average value, GH_a, via the integral

$$GH_a = \int_0^1 GH(^\alpha m) \, d\alpha. \tag{8.47}$$

Applying Eq. (8.46) to $^\alpha m$ in Example 8.12 (given in Table 8.1), we obtain

$$GH(^\alpha m) = 0.47 - 0.47\alpha.$$

Clearly, $GH(^0 m) = 0.47$ and $GH(^1 m) = 0$. Applying now Eq. (8.47), we obtain $GH_a = 0.235$.

Computing $\bar{S}$ is more difficult. Algorithm 6.1 can be employed, but it is now applicable to the ratios $l_A(\alpha)/|A|$, which are functions of α. First, we need to express the dependence of $\bar{S}$ on α, $\bar{S}(l_A(\alpha))$. Then, the average value, $\bar{S}_a$, is computed by the integral

$$\bar{S}_a = \int_0^1 \bar{S}(l_A(\alpha)) \, d\alpha. \tag{8.48}$$

Applying Algorithm 6.1 to the α-cuts $l_A(\alpha)$ in Table 8.1, we obtain the following probabilities of $\bar{S}(l_A(\alpha))$ in the given order, one at each iteration of the algorithm:

$$^{\alpha}p(x_4) = 0.3 + 0.1\alpha,$$

$$^{\alpha}p(x_3) = 0.25 + 0.05\alpha,$$

$$^{\alpha}p(x_2) = 0.25 - 0.05\alpha,$$

$$^{\alpha}p(x_1) = 0.2 - 0.1\alpha.$$

Now, we have

$$\bar{S}(l_A(\alpha)) = -(0.3 + 0.1\alpha)\log_2(0.3 + 0.1\alpha)$$
$$-(0.25 + 0.05\alpha)\log_2(0.25 + 0.5\alpha)$$
$$-(0.25 - 0.05\alpha)\log_2(0.25 - 0.05\alpha)$$
$$-(0.2 - 0.1\alpha)\log_2(0.2 - 0.1\alpha).$$

The plot of $\bar{S}(l_A(\alpha))$ for $\alpha \in [0, 1]$ is shown in Figure 8.10. The extreme values are $\bar{S}(l_A(0)) = 1.985$ and $\bar{S}(l_A(1)) = 1.846$. The average value, obtained by Eq. (8.48), is $\bar{S}_a = 1.93$.

When tuples of probability granules are employed in any computation involving arithmetic operations (such as computing expected values or posterior probabilities in Bayesian inference), it is essential that constrained fuzzy arithmetic be used. Computing with fuzzy probabilities requires that not only the requisite equality constraints be observed, but also the probabilistic constraints, as is explained in Section 7.4.2.

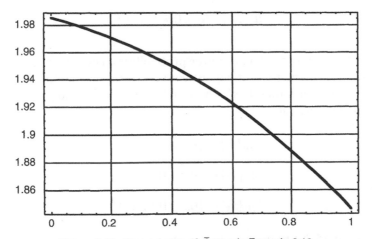

Figure 8.10. Dependence of $\bar{S}$ on α in Example 8.12.

8.6. OTHER FUZZIFICATION EFFORTS

Efforts to fuzzify uncertainty theories have not been restricted to the three fuzzifications described in Section 8.3–8.5. However, among the various efforts to fuzzify uncertainty, which have been discussed in the literature (see Note 8.12), the three fuzzifications described in this chapter have some visible advantages. They are well developed, conceptually and computationally sound, and they have features that are suitable for applications.

To illustrate the point that some uncertainty theories are less practical to fuzzify than others, let us examine the following very simple example of fuzzification within the theory of uncertainty based on the Sugeno λ-measures.

EXAMPLE 8.13. Consider the set $X = \{x_1, x_2, x_3\}$ of alternatives and assume that we want to deal with uncertainty regarding these alternatives in terms of λ-measures (Section 5.3). This requires that we are able to assess *either* the lower probabilities *or* the upper probabilities on singletons. From these assessments, we compute the value of parameter λ by Eq. (5.30), which makes it possible, in turn, to compute lower and upper probabilities for all subsets of X. Once uncertainty regarding the alternatives is formalized in this way, we can apply the rules of the calculus of λ-measures to manipulate the uncertainty in any desirable way within this theory.

Now assume that we are able to make only fuzzy assessments of lower (or upper) probabilities on singletons, and that these assessments are represented by appropriate fuzzy intervals or numbers. As a specific example, let the triangular-shape fuzzy numbers

$$\underline{\mu}(\{x_1\}) = \langle 0.1, 0.2, 0.2, 0.2 \rangle,$$

$$\underline{\mu}(\{x_2\}) = \langle 0.2, 0.3, 0.3, 0.3 \rangle,$$

$$\underline{\mu}(\{x_3\}) = \langle 0.3, 0.4, 0.4, 0.4 \rangle,$$

represent a given assessment of lower probabilities. These fuzzy numbers are reasonable representations of linguistic assessments such as "the lower probability of x_i is p_i or a little less," where $p_1 = 0.2$, $p_2 = 0.3$, and $p_2 = 0.4$. The α-cuts of the numbers are

$$^\alpha\mu(\{x_1\}) = [0.1 + 0.1\alpha, 0.2],$$

$$^\alpha\mu(\{x_2\}) = [0.2 + 0.1\alpha, 0.3],$$

$$^\alpha\mu(\{x_3\}) = [0.3 + 0.1\alpha, 0.4].$$

Applying Eq. (5.30) to these α-cuts results in the equation

$$(0.001\alpha^3 + 0.006\alpha^2 + 0.11\alpha + 0.006)\lambda^2 + (0.12\alpha + 0.11)\lambda + 0.3\alpha - 0.4 = 0,$$

whose positive root defines the value $\lambda(\alpha)$, for each $\alpha \in [0, 1]$. The plot of $\lambda(\alpha)$ is shown in Figure 8.11a. The analytic expression for $\lambda(\alpha)$ is too complex to be practical, even in this very simple example. It is more practical to calculate $\lambda(\alpha)$ for selected values of α, such as those shown in Figure 8.11b.

Once values of $\lambda(\alpha)$ are determined, the α-cuts of lower probabilities, $^\alpha\mu(A)$, can be determined via Eq. (5.28) for all subsets A of X, as is shown in Table 8.2. Considering the fact that $\lambda(\alpha)$ is determined by a rather complex expression, we can see that some expressions in this table would be quite formidable when completed. Clearly, expressions for the Möbius representation will be even more complex.

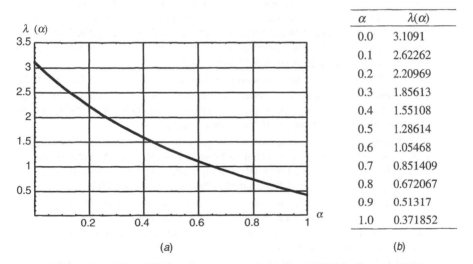

α	$\lambda(\alpha)$
0.0	3.1091
0.1	2.62262
0.2	2.20969
0.3	1.85613
0.4	1.55108
0.5	1.28614
0.6	1.05468
0.7	0.851409
0.8	0.672067
0.9	0.51317
1.0	0.371852

(a) (b)

Figure 8.11. Plot of $\lambda(\alpha)$ and some numerical values of $\lambda(\alpha)$ in Example 8.13.

Table 8.2. α-Cuts of Lower Probabilities, $^\alpha\underline{\mu}$ (A), in Example 8.13

	x_1	x_2	x_3	$^\alpha\underline{\mu}(A)$
A:	0	0	0	0
	1	0	0	$[0.1 + 0.1\alpha, 0.2]$
	0	1	0	$[0.2 + 0.1\alpha, 0.3]$
	0	0	1	$[0.3 + 0.1\alpha, 0.4]$
	1	1	0	$[0.3 + 0.2\alpha + \lambda(\alpha) (0.1 + 0.1\alpha) (0.2 + 0.1\alpha), 0.5 + \lambda(\alpha) \cdot 0.06]$
	1	0	1	$[0.4 + 0.2\alpha + \lambda(\alpha) (0.1 + 0.1\alpha) (0.3 + 0.1\alpha), 0.6 + \lambda(\alpha) \cdot 0.08]$
	0	1	1	$[0.5 + 0.2\alpha + \lambda(\alpha) (0.2 + 0.1\alpha) (0.3 + 0.1\alpha), 0.7 + \lambda(\alpha) \cdot 1.2]$
	1	1	1	1

It is sensible to conclude from this strikingly simple example that the uncertainty theory based on λ-measures is conceptually and computationally difficult to fuzzify. This contrasts with the theory based on reachable interval-valued probability distributions, whose fuzzification (discussed in Section 8.5) is much more transparent and computationally tractable. By comparing the fuzzifications of these two theories, it is reasonable to expect that only the latter will survive as useful for various applications. It is likely that among the many possible fuzzifications of uncertainty theories some will be found useful and some will be discarded.

NOTES

8.1. The concept of α-cuts of fuzzy sets was already introduced in the seminal paper by Zadeh [1965]. In another paper [Zadeh, 1971], he introduced the α-cut representation of fuzzy sets and showed how equivalence, compatibility, and ordering relations can be fuzzified via this representation. However, the term "cutworthy" was coined much later by Bandler and Kohout [1993]. A highly general and comprehensive investigation of cutworthy properties was made in terms fuzzy predicate logic by Bělohlávek [2003].

8.2. The first formulation of the extension principle was introduced by Zadeh [1965], even though it was described under the heading "fuzzy sets induced by mappings." The term "extension principle" was introduced in [Zadeh, 1975–76], where the principle and its utility are thoroughly examined.

8.3. Fuzzification via fuzzy morphisms and category theory was pioneered by Goguen [1967, 1968–69, 1974]. Among other notable references are [Arbib and Manes, 1975], [Rodabaugh et al., 1992], [Rodabaugh and Klement, 2003], [Höhle and Klement, 1995], [Höhle and Rodabaugh, 1999], and Walker [2003].

8.4. The idea that the degree of fuzziness of a fuzzy set can be most naturally expressed in terms of the lack of distinction between the set and its complement was proposed by Yager [1979, 1980b]. A general formulation based on this idea, which is applicable to all possible fuzzy complements, was developed by Higashi and Klir [1982]. They proved that every measure of fuzziness of this type can be expressed in terms of a metric distance that is based on an appropriate aggregate of the absolute values of the individual differences between membership grades of the given fuzzy set and its complement (of a type chosen in a given application) for all elements of the universal set.

8.5. Observe that the summation term in Eq. (8.7) and the integral in Eq. (8.9) express the Hamming distance between A and $c(A)$. It is of course possible to use other distance functions for this purpose. The whole range of measures of fuzziness based on the Minkowski class of distance functions is examined by Higashi and Klir [1982].

8.6. The measure of fuzziness based on local Shannon entropies (defined by Eq. (8.14) or Eq. (8.15)) was introduced by De Luca and Termini [1972, 1974]. This measure, which is usually called an *entropy of fuzzy sets*, has been investigated in the literature fairly extensively by numerous authors.

8.7. Broader classes of measures of fuzziness were characterized on axiomatic grounds by Knopfmacher [1975] and Loo [1977]. A good overview of the literature dealing with measures of fuzziness was prepared by Pal and Bezdek [1994].

8.8. The standard fuzzy set interpretation of possibility theory was introduced by Zadeh [1978a] as a natural tool for dealing with uncertainty associated with information expressed in natural language and represented by fuzzy propositions [Zadeh, 1978b, 1981]. The generalized fuzzy-set interpretation of possibility theory defined by Eq. (8.19) was introduced in a paper by Klir [1999]. The paper also surveys other, unsuccessful attempts to revise the standard fuzzy-set interpretation.

8.9. Probabilities of fuzzy events were introduced by Zadeh [1968], just three years after he introduced the concept of a fuzzy set [Zadeh, 1965].

8.10. Fuzzification of the uncertainty theory based on reachable interval-valued probability distributions was investigated by Pan [1997a] and Pan and Yuan [1997]. They also developed a fuzzified Bayesian inference within this uncertainty theory, which is based on methods of linear programming.

8.11. Some methods for approximating arbitrary probability granules by trapezoidal ones were examined by Giachetti and Young [1997] from the standpoint of the accumulated error for repeated use of multiplication or division. A method proposed in the paper is shown to lead to acceptable results in most practical cases. An interesting approximation method was proposed by Pan [1997a,b] in the context of Bayesian inference with probability granules. It is based on keeping the core unchanged and calculating new support in terms of the weighted least-square-error method in which the weights are monotone increasing with values of α according to some rule. This method seems promising, but no error analysis has been made for it as yet.

8.12. Perhaps the most fundamental approach to fuzzifying uncertainty theories, at least from the theoretical point of view, is to fuzzify monotone measures. This is explored in a paper by Qiao [1990], which is also reprinted in [Wang and Klir, 1992]. Interesting approaches to fuzzifying the Dempster–Shafer theory were proposed by Yen [1990] and Yang et al. [2003]. The concept of a fuzzy random variable is covered in the literature quite extensively. A broad introduction is in [Zwick and Wallsten, 1989] and [Negoita and Ralescu, 1987]; a Special Issue edited by Gil [2001] covers more recent developments, as well as valuable references to previous work. The following are some useful references that deal with fuzzy probabilities, even though they do not specifically address the issue of fuzzifying uncertainty theories: [Buckley, 2003; Cai, 1996; Fellin et al., 2005; Manton et al., 1994; Möller and Beer, 2004; Ross et al., 2002; Viertl, 1996]. The book by Mordeson and Nair [1998] is a good overview of other fuzzifications in mathematics.

EXERCISES

8.1. Show that the standard complement of a fuzzy set is not a cutworthy operation.

8.2. Show that the standard operations of intersection and union of fuzzy sets are cutworthy.

8.3. Verify that the binary fuzzy relation R defined by the matrix

$$
R = \begin{array}{c|cccccc}
 & x_1 & x_2 & x_3 & x_4 & x_5 & x_6 \\
\hline
x_1 & 1.0 & 0.2 & 1.0 & 0.6 & 0.2 & 0.6 \\
x_2 & 0.2 & 1.0 & 0.2 & 0.2 & 0.8 & 0.2 \\
x_3 & 1.0 & 0.2 & 1.0 & 0.6 & 0.2 & 0.6 \\
x_4 & 0.6 & 0.2 & 0.6 & 1.0 & 0.2 & 0.8 \\
x_5 & 0.2 & 0.8 & 0.2 & 0.2 & 1.0 & 0.2 \\
x_6 & 0.6 & 0.2 & 0.6 & 0.8 & 0.2 & 1.0
\end{array}
$$

is a fuzzy equivalence relation that is cutworthy and determine its partition tree.

8.4. Repeat Example 8.2 for function $y = 9 + 1/x$, where $x \in [1, 10]$, and two triangular fuzzy members by which the value x is approximately assessed: $A_1 = \langle 1.5, 2, 2, 2.5 \rangle$ and $A_2 = \langle 2.5, 3, 3, 3.5 \rangle$.

8.5. Assuming the standard complement and using the functional f defined by Eq. (8.10), calculate the amount of fuzziness and its normalized value for the following fuzzy sets:
   (a) $A = 0.5/x_1 + 0.7/x_2 + 1/x_3 + 0.9/x_4 + 0.6/x_5 + 0.4/x_6$;
   (b) Fuzzy relation R in Exercise 8.3.

8.6. Repeat Example 8.4 for the following fuzzy numbers:
   (a) Triangular fuzzy number $A = \langle 0, 1, 1, 2 \rangle$;
   (b) A fuzzy number B whose α-cut representation is ${}^{\alpha}B = [\sqrt{x}, 2 - \sqrt{x}]$;
   (c) $C(x) = e^{-|5x-10|}$ for all $x \in [1, 3]$ and $C(x) = 0$ for $x \notin [1, 3]$.

8.7. Repeat the calculations in Examples 8.3 and 8.4 and Exercises 8.5 and 8.6 for some nonstandard complements introduced in Section 7.3.1.

8.8. Let f and f' be defined by Eqs. (8.7) and (8.12), respectively, and let c be the standard complement of fuzzy sets. Show that for any fuzzy set A defined on a finite set X:
   (a) $f(A) = 2f'(A)$
   (b) $\hat{f}(A) = f'(A)$

8.9. Repeat Exercise 8.8 for f and f' defined by Eqs. (8.9) and (8.13), respectively, and $X = \mathbb{R}$.

8.10. Investigate the relationship between the functional f defined by Eq. (8.10) and the functional f'' defined by Eq. (8.14).

8.11. Using the functional f'' defined by Eq. (8.15), calculate the amount of fuzziness and its normalized value for the following fuzzy intervals:

   (a) A trapezoidal fuzzy interval $A = \langle 1, 3, 4, 7 \rangle$;

   (b) A trapezoidal fuzzy interval $B = \langle 0, 0, 1, 5 \rangle$;

   (c) $C(x) = e^{-|5x-10|}$ for all $x \in [1, 3]$ and $C(x) = 0$ for $x \notin [1, 3]$.

8.12. Verify the values of $GH(m_R)$, $\bar{S}(Nec_R)$, $\underline{S}(Nec_R)$, $f(R)$, and $f(\hat{R})$ in Example 8.7.

8.13. Verify the values of $f(R)$, and $f(\hat{R})$ in Example 8.8 by using Eq. (8.11), as well as by geometrical reasoning in Figure 8.5.

8.14. Consider a discrete variable X whose values are in the set $X = \mathbb{N}_{200}$. Given information that "X is F," where F is a fuzzy set defined by

$$F = 0.1/52 + 0.2/53 + 0.4/54 + 0.6/55 + 0.8/56 + 0.5/57 + 0.3/58 + 0.1/50,$$

calculate the following:

   (a) The nonspecificity of the possibilistic representation of the given information;

   (b) The degrees of possibility and necessity in the sets $A = \{55, 56, \ldots, 62\}$, $B = \{53, 54, 55\}$, and $C = \{54, 55, 56\}$.

8.15. Consider a variable X whose values are in the interval $X = [0, 100]$. Assume that the value of the variable is assessed in a given context in terms of a fuzzy propositions "X is F," in which F is a fuzzy interval defined for all $x \in X$ as follows:

$$F(x) = \begin{cases} 1.4(x - 6.5) & \text{when } x \in [6.5, 7], \\ 0.7 & \text{when } x \in (7, 8), \\ 3.5(10 - x) & \text{when } x \in [8, 10], \\ 0 & \text{otherwise.} \end{cases}$$

Calculate:

   (a) The amount of nonspecificity and its normalized version for the possibilistic representation of the given information;

   (b) The degrees of possibility and necessity in the fuzzy sets whose α-cut representations are

$$^{\alpha}A = [6 + \alpha, 9 - \alpha],$$

$$^{\alpha}B = [7 + 2\alpha, 10 - \alpha],$$

$$^{\alpha}C = [6.5 + \alpha, 8.5 - \alpha].$$

8.16. Show that fuzzy partitions are not cutworthy regardless of which operations are used for intersections and unions of fuzzy sets.

8.17. Calculate probabilities of crisp and fuzzy events defined in Figure 8.7*a* and 8.7*b*, respectively, for the following probability density functions q on $\mathbb{R}$:

(a) $q(x) = \begin{cases} 0.0016x - 0.032 & \text{when } x \in [20, 45) \\ 0.112 - 0.0016x & \text{when } x \in [45, 70] \\ 0 & \text{otherwise} \end{cases}$

(b) $q(x) = \begin{cases} 4.6875 \cdot 10^{-5} \cdot x^2 & \text{when } x \in [0, 40] \\ 0 & \text{otherwise} \end{cases}$

8.18. Suggest some reasonable ways of approximating the probability granule F_1 in Figure 8.9*b* by a trapezoidal granule (see also Note 8.11).

8.19. Determine for each of the following tuples $\mathbf{T} = \langle T_1, T_2, T_3 \rangle$ of trapezoidal probability granules on $X = \{x_1, x_2, x_3\}$ whether it is proper and reachable:

(a) $T_1 = \langle 0, 0, 0, 0.1 \rangle$, $T_2 = \langle 0.2, 0.3, 0.4, 0.5 \rangle$, $T_3 = \langle 0.4, 0.6, 0.8, 0.9 \rangle$

(b) $T_1 = \langle 0, 0, 0.3, 0.5 \rangle$, $T_2 = \langle 0, 0.2, 0.5, 0.6 \rangle$, $T_3 = \langle 0.2, 0.4, 0.4, 0.7 \rangle$

(c) $T_1 = \langle 0.2, 0.3, 0.3, 0.4 \rangle$, $T_2 = \langle 0, 0.2, 0.2, 0.5 \rangle$, $T_3 = \langle 0.3, 0.5, 0.5, 0.8 \rangle$

Convert each of the tuples that is proper but not reachable to its reachable counterpart.

8.20. In analogy with Example 8.12, determine for each of the reachable tuples (or their reachable counterparts) in Exercise 8.19 the following functions of α for all $A \in \mathcal{P}(X)$:

(a) $l_A(\alpha)$

(b) $u_A(\alpha)$

(c) $^\alpha m_A(A)$

8.21. Using the functions determined in Exercise 8.20, calculate:

(a) $GH(^\alpha m)$ and GH_a

(b) $\bar{S}(l_A(\alpha))$ and $\bar{S}_a$

8.22. For Example 8.13, determine the following:

(a) $\lambda(\alpha)$

(b)   The α-cuts of the lower probabilities;

(c)   The Möbius representation of the lower probabilities in Table 8.2.

9

METHODOLOGICAL ISSUES

There is nothing better than to know that you don't know.
Not knowing, yet thinking your know—
This is sickness.
Only when you are sick of being sick
Can you be cured.

—Lao Tsu

9.1. AN OVERVIEW

From the methodological point of view, two *complementary features of generalized information theory* (*GIT*) are significant. One of them is the great *diversity* of prospective uncertainty theories that are subsumed under GIT. This diversity increases whenever new types of formalized languages or monotone measures are recognized. At any time, of course, only some of the prospective uncertainty theories are properly developed.

Complementary to the diversity of uncertainty theories subsumed under GIT is their *unity*, which is manifested by common properties that the theories share. From the methodological point of view, particularly notable are the common forms of functionals that generalize the Hartley and Shannon measures of uncertainty in the various uncertainty theories, and the invariance of equations and inequalities that express the relationship among joint, marginal, and conditional measures of uncertainty.

The diversity of GIT offers an extensive inventory of distinct uncertainty theories, each characterized by specific assumptions embedded in its axioms.

Uncertainty and Information: Foundations of Generalized Information Theory, by George J. Klir
© 2006 by John Wiley & Sons, Inc.

This allows us to choose, in any given application context, a theory that is compatible with the application of concern. On the other hand, the unity of GIT allows us to work within GIT as a whole. That is, it allows us to move from one theory to another, as needed.

The primary aim of this chapter is to examine the following four methodological *principles of uncertainty*:

1. Principle of minimum uncertainty
2. Principle of maximum uncertainty
3. Principle of requisite generalization
4. Principle of uncertainty invariance

In general, these principles are epistemologically based prescriptive procedures that address methodological issues involving uncertainty that cannot be resolved solely by using calculi of the individual uncertainty theories. Due to the connection between uncertainty and uncertainty-based information, these principles also can be interpreted as *principles of information.*

The principle of minimum uncertainty is basically an arbitration principle. It facilitates the selection of meaningful alternatives from solution sets obtained by solving problems in which some amount of the initial information is inevitably lost. According to this principle, we should accept only those solutions for which the loss of the information is as small as possible. This means, in turn, that we should accept only solutions with minimum uncertainty.

The second principle, *the principle of maximum uncertainty*, is essential for any problem that involves *ampliative reasoning*. This is reasoning in which conclusions are not entailed in the given premises. Using common sense, the principle may be expressed as follows: in any ampliative inference, use all information supported by available evidence, but make sure that no additional information (unsupported by the given evidence) is unwittingly added. Employing the connection between information and uncertainty, this definition can be reformulated in terms of uncertainty: any conclusion resulting from ampliative inference should maximize the relevant uncertainty within constraints representing given premises. This principle guarantees that we fully recognize our ignorance when we attempt to make inferences that are beyond the information domain defined by the given premises and, at the same time, that we utilize all information contained in the premises. In other words, the principle guarantees that our inferences are maximally noncommittal with respect to information that is not contained in the premises.

The principles of minimum and maximum uncertainty are well developed within classical, probability-based information theory. They are referred to as *principles of minimum and maximum entropy*. These classical principles of uncertainty are extensively covered in the literature. A survey of this literature is presented in Notes 9.1–9.3. A few examples in Sections 9.2 and 9.3 illustrate the important role of these principles in classical information theory.

Optimization problems that emerge from the minimum and maximum uncertainty principles outside classical information theory have yet to be properly investigated and tested in praxis. One complication is that two types of uncertainty coexist in all the nonclassical uncertainty theories. Material presented in this chapter is thus by and large exploratory, oriented primarily to examining the issues involved and stimulating further research.

Principles of minimum and maximum uncertainty are fundamentally different from the remaining principles: the principle of requisite generalization and the principle of uncertainty invariance. While the former are applicable within each particular uncertainty theory, the latter facilitate transitions from one theory to another. Clearly, the latter principles have no counterparts in classical information theory.

The *principle of requisite generalization* is based on the assumption that we work within GIT as a whole. According to this principle, we should not a priori commit to any particular uncertainty theory. Our choice should be determined by the nature of the problem we deal with. The chosen theory should be sufficiently general to allow us to capture fully our ignorance. Moreover, when the chosen theory becomes incapable of expressing uncertainty resulting from deficient information at some problem-solving stage, we should move to a more general theory that has the capability of expressing the given uncertainty.

The last principle, the *principle of uncertainty invariance* (also called the *principle of information preservation*), was introduced in GIT to facilitate meaningful transformations between various uncertainty theories. According to this principle, the amount of uncertainty (and the associated uncertainty-based information) should be preserved in each transformation from one uncertainty theory to another. The primary use of this principle is to approximate in a meaningful way uncertainty formalized in a given theory by a formalization in a theory that is less general.

The roles of each of the four principles of uncertainty are illustrated in Figure 9.1, where uncertainty theory T_2 is assumed to be more general than uncertainty theory T_1. While the principles of minimum and maximum uncertainty (1 and 2) are applicable within a single theory (theory T_1 in the figure), the principles of requisite generalization (3) and uncertainty invariance (4) deal with transitions from one theory to another.

9.2. PRINCIPLE OF MINIMUM UNCERTAINTY

The principle of minimum uncertainty is applicable to all problems that are prone to lose information. The principle may be viewed as a safeguard against losing more information than necessary to solve a given problem. It is thus reasonable to also refer to it as the *principle of minimum information loss*. In this section, the principle is illustrated by two classes of problems: simplification problems and conflict-resolution problems.

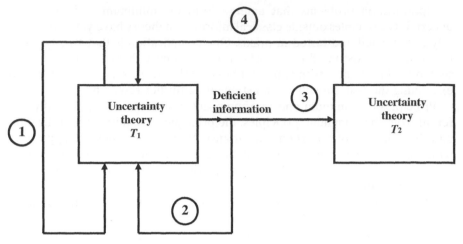

Figure 9.1. Principles of uncertainty: an overview. *Assumption:* Theory T_2 is more general than theory T_1. ① Principle of minimum uncertainty; ② principle of maximum uncertainty; ③ principle of requisite generalization; ④ principle of uncertainty invariance.

9.2.1. Simplification Problems

A major class of problems for which the principle of minimum uncertainty is applicable consists of *simplification problems*. When a system is simplified, it is usually unavoidable to lose some information contained in the system. The amount of information that is lost in this process results in the increase of an equal amount of relevant uncertainty. Examples of relevant uncertainties are predictive, retrodictive, or prescriptive uncertainty. A sound simplification of a given system should minimize the loss of relevant information (or the increase in relevant uncertainty) while achieving the required reduction of complexity. That is, we should accept only such simplifications of a given system at any desirable level of complexity for which the loss of relevant information (or the increase in relevant uncertainty) is minimal. When properly applied, the principle of minimum uncertainty guarantees that no information is wasted in the process of simplification.

There are many simplification strategies, all of which can perhaps be classified into three main classes:

- Simplifications made by eliminating some entities from the system (variables, subsystems, etc.).
- Simplifications made by aggregating some entities of the system (variables, states, etc.).
- Simplifications made by breaking overall systems into appropriate subsystems.

Regardless of the strategy employed, the principle of minimum uncertainty is utilized in the same way. It is an arbiter that decides which simplifications to choose at any given level of complexity or, alternatively, which simplifications to choose at any given level of acceptable uncertainty. Let us describe this important role of the principle of minimum uncertainty in simplification problems more formally.

Let Z denote a system of some type and let Q_Z denote the set of all simplifications of Z that are considered *admissible* in a given context. For example, Q_Z may be the set of simplifications of Z that are obtained by a particular simplification method. Let $\leq_c$ and $\leq_u$ denote *preference orderings* on Q_Z that are based upon complexity and uncertainty, respectively. In general, systems with smaller complexity and smaller uncertainty are preferred. The two preference orderings can be combined in a *joint preference ordering*, $\leq$, defined as follows: for any pair of systems $Z_i, Z_j \in Q_Z$, $Z_i \leq Z_j$ if and only if $Z_i \leq_c Z_j$ and $Z_i \leq_u Z_j$. The joint preference ordering is usually a partial ordering, even though it may be only a weak ordering in some simplification problems (reflexive and transitive relation on Q_Z). The use of the uncertainty preference ordering, which, in this case, exemplifies the principle of minimum uncertainty, enables us to reduce all admissible simplifications to a small set of *preferred simplifications*. The latter forms a *solution set*, SOL_Z, of the simplification problem, which consists of those admissible simplifications in Q_Z that are either equivalent or incomparable in terms of the joint preference ordering. Formally,

$$SOL_Z = \{Z_i \in Q_Z | \text{ for all } Z_j \in Q_Z, \ Z_j \leq Z_i \text{ implies } Z_i \leq Z_j\}.$$

Observe that the solution set SOL_Z in this formulation, which may be called an *unconstrained simplification problem*, contains simplifications at various levels of complexity and with various degrees of uncertainty. The problem can be constrained, for example, by considering simplifications admissible only when their complexities are at some designated level or when they are below a specified level, when their uncertainties do not exceed a certain maximum acceptable level, and the like. The formulation of these various *constrained simplification problems* differs from the preceding formulation only in the definition of the set of admissible simplifications Q_Z.

EXAMPLE 9.1. The aim of this example is to illustrate the role of the principle of minimum uncertainty in determining the solution set of preferable simplifications of a given system. The example deals with a simple diagnostic system, Z, with one input variable, x, and one output variable, y, whose states are in sets $X = \{0, 1, 2, 3\}$ and $Y = \{0, 1\}$, respectively. It is assumed that the states are ordered in the natural way: $0 \leq 1 \leq 2 \leq 3$. The relationship between the variables is expressed by the joint probability distribution function on $X \times Y$ that is given in Table 9.1a. The diagnostic uncertainty of the system is

$$\begin{aligned} S(Y \mid X) &= S(X, Y) - S(X) \\ &= 2.571 - 1.904 \\ &= 0.667. \end{aligned}$$

Table 9.1. Preferred Simplifications Determined by the Principle of Minimum Uncertainty (Example 9.1)

Z:	$p(x, y)$		Y		X	$p_X(x)$	
			0	1			
X	0	0.10		0.30	0	0.40	$S(X, Y) = 2.571$
	1	0.00		0.15	1	0.15	$S(X) = 1.904$
	2	0.15		0.05	2	0.20	$S(Y\|X) = 0.667$
	3	0.05		0.20	3	0.25	$I(Y\|X) = 0.333$

(a)

Z_1:	$p_1(x, y)$		Y		X_1	$p_{X_1}(x)$	
			0	1			
X_1	{0,1}	0.10		0.45	{0,1}	0.55	$S(X_1,Y) = 2.158$
	{2}	0.15		0.05	{2}	0.20	$S(X_1) = 1.439$
	{3}	0.05		0.20	{3}	0.25	$S(Y\|X_1) = 0.719$
							$I(Y\|X_1) = 0.281$

Z_2:	$p_2(x, y)$		Y		X_2	$p_{X_2}(x)$	
			0	1			
X_2	{0}	0.10		0.30	{0}	0.40	$S(X_2,Y) = 2.409$
	{1,2}	0.15		0.20	{1,2}	0.35	$S(X_2) = 1.559$
	{3}	0.05		0.20	{3}	0.25	$S(Y\|X_2) = 0.850$
							$I(Y\|X_2) = 0.150$

Z_3:	$p_3(x, y)$		Y		X_3	$p_{X_3}(x)$	
			0	1			
X_3	{0}	0.10		0.30	{0}	0.40	$S(X_3,Y) = 2.228$
	{1}	0.00		0.15	{1}	0.15	$S(X_3) = 1.458$
	{2,3}	0.20		0.25	{2,3}	0.45	$S(Y\|X_3) = 0.770$
							$I(Y\|X_3) = 0.230$

(b)

The amount of diagnostic information contained in the system, $I(Y \mid X)$, is the difference between the maximum uncertainty allowed within the experimental frame, $S_{\max}(Y \mid X) = \log_2 8 - \log_2 4 = 1$, and the actual uncertainty: $I(Y \mid X) = 1 - 0.667 = 0.333$. Now assume that we want to simplify the system by quantizing states of the input variable. Since the states of X are ordered, there are six meaningful quantizations (with at least two states), which are expressed by the following partitions on X:

Table 9.1. *Continued*

Z_4:	$p_4(x, y)$		Y		X_4	$p_{X4}(x)$	
			0	1			
X_4	$\{0,1\}$		0.10	0.45	$\{0,1\}$	0.55	$S(X_4,Y) = 1.815$
	$\{2,3\}$		0.20	0.25	$\{2,3\}$	0.45	$S(X_4) = 0.993$
							$\mathbf{S(Y\mid X_4) = 0.822}$
							$\mathbf{I(Y\mid X_4) = 0.178}$

Z_5:	$p_5(x, y)$		Y		X_5	$p_{X5}(x)$	
			0	1			
X_5	$\{0,1,2\}$		0.25	0.50	$\{0,1,2\}$	0.75	$S(X_5,Y) = 1.680$
	$\{3\}$		0.05	0.20	$\{3\}$	0.25	$S(X_5) = 0.811$
							$S(Y\mid X_5) = 0.869$
							$I(Y\mid X_5) = 0.131$

Z_6:	$p_6(x, y)$		Y		X_6	$p_{X6}(x)$	
			0	1			
X_6	$\{0\}$		0.10	0.30	$\{0\}$	0.40	$S(X_6,Y) = 1.846$
	$\{1,2,3\}$		0.20	0.40	$\{1,2,3\}$	0.60	$S(X_6) = 0.971$
							$S(Y\mid X_6) = 0.875$
							$I(Y\mid X_6) = 0.125$

(c)

$$X_1 = \{\{0,1\},\{2\},\{3\}\},$$
$$X_2 = \{\{0\},\{1,2\},\{3\}\},$$
$$X_3 = \{\{0\},\{1\},\{2,3\}\},$$
$$X_4 = \{\{0,1\},\{2,3\}\},$$
$$X_5 = \{\{0,1,2\},\{3\}\},$$
$$X_6 = \{\{0\},\{1,2,3\}\}.$$

Simplifications Z_1, Z_2, Z_3, which are based on partitions X_1, X_2, X_3, respectively, are shown in Table 9.1b. These simplifications have the same complexity, which is expressed in this example by the number of states of the input variable. Among them, simplification Z_1 has the smallest diagnostic uncertainty, $S(Y\mid X_1) = 0.719$, and the smallest loss of information with respect to the given system Z:

$$I(Y\mid X) - I(Y\mid X_1) = 0.052.$$

Simplifications Z_4, Z_5, Z_6, which are based on partitions X_4, X_5, X_6, respectively, are shown in Table 9.1c. They have the same complexity and, clearly, they are less complex than simplifications Z_1, Z_2, Z_3. Among them, simplification Z_4 has the smallest diagnostic uncertainty, $S(Y \mid X_4) = 0.822$. The loss of information with respect to Z is

$$I(Y \mid X) - I(Y \mid X_4) = 0.155,$$

and with respect to Z_1:

$$I(Y \mid X_1) - I(Y \mid X_4) = 0.103.$$

The solution set (including the given system Z) is:

$$SOL_Z = \{Z, Z_1, Z_4\}.$$

EXAMPLE 9.2. A relation R among variables x_1, x_2, x_3 is defined by the set of possible triples listed in Table 9.2a. Observe that the state sets of the variables are, respectively, $X_1 = \{0, 1\}$ and $X_2 = X_3 = \{0, 1, 2\}$. The relation may be utilized for providing us with useful information about the state of any one of the variables, given a joint state of the remaining two variables. Relevant amounts of uncertainty for this purpose, which are expressed in this case by the Hartley functional, are shown in Table 9.2b under the heading R. Also shown in the table are the associated amounts of information.

The aim of this example is to illustrate the role of the principle of minimum uncertainty in choosing simplifications based on the quantization of variables x_2 and x_3. Assuming that the states in X_2 and X_3 are ordered, each of the sets can be quantized by using one of the following functions:

$$
\begin{array}{ll}
q_1 : 0 \rightarrow 0 & q_2 : 0 \rightarrow 0 \\
\quad\;\; 1 \rightarrow 0 & \quad\;\; 1 \rightarrow 1 \\
\quad\;\; 2 \rightarrow 1 & \quad\;\; 2 \rightarrow 1.
\end{array}
$$

When we apply function q_1 to both X_2 and X_3 in relation R, we obtain the simplified relation R_{11} shown in Table 9.2a, where duplicate states are crossed out. By applying the other combinations of functions q_1 and q_2, we obtain the other simplified relations in Table 9.2a (R_{12}, R_{21}, and R_{22}), where each subscript indicates the combination employed. For each of the four simplified relations, relevant amounts of uncertainty and information are given in Table 9.2b. Observe that the simplifications with the minimum amount of uncertainty (indicated in the table by bold entries) are also *correctly* the ones with the maximum amount of information (or with minimum loss of information with respect to R). For variable x_1, any of relations R_{12}, R_{21}, and R_{22} qualifies as the best simplification; for variable x_2, R_{11} is the only minimum-uncertainty simplification; for x_3, both R_{21} and R_{22} are minimum-uncertainty simplifications. Observe that R_{12} does not preserve any information of R with respect to variable x_3.

Table 9.2. Preferred Simplification Determined by the Principle of Minimum Uncertainty (Example 9.2)

(a)

R			R11			R12			R21			R22		
x_1	x_2	x_3	x_1	x_2	x_3	x_1	x_2	x_3	x_1	x_2	x_3	x_1	x_2	x_3
0	0	1	0	0	0	0	0	1	0	0	0	0	0	1
0	0	2	0	0	1	0	0	0	0	0	1	0	0	1
0	1	0	0	1	0	0	0	0	0	1	0	0	1	0
0	1	1	0	1	1	0	1	0	0	1	1	0	1	1
0	2	0	0	1	0	0	1	1	0	1	0	1	0	0
0	2	1	1	0	0	1	0	0	1	0	0	1	1	0
1	0	0	1	1	0	1	0	1	1	1	0	1	1	1
1	0	1	1	1	1	1	1	1	1	1	1	1	1	1
1	1	0												
1	1	2												

(b)

	R	R11	R12	R21	R22
$H(X_1 \times X_2 \times X_3)$	$\log_2 10 = 3.322$	$\log_2 5 = 2.321$	$\log_2 6 = 2.585$	$\log_2 6 = 2.585$	$\log_2 6 = 2.585$
$H(X_1 \times X_2)$	$\log_2 5 = 2.321$	$\log_2 3 = 1.585$	$\log_2 3 = 1.585$	$\log_2 4 = 2.000$	$\log_2 4 = 2.000$
$H(X_1 \times X_3)$	$\log_2 6 = 2.585$	$\log_2 4 = 2.000$	$\log_2 4 = 2.000$	$\log_2 4 = 2.000$	$\log_2 4 = 2.000$
$H(X_2 \times X_3)$	$\log_2 8 = 3.000$	$\log_2 3 = 1.585$	$\log_2 4 = 2.000$	$\log_2 4 = 2.000$	$\log_2 4 = 2.000$
$H(X_1|X_2 \times X_3)$	0.322	0.736	**0.585**	**0.585**	**0.585**
$H(X_2|X_1 \times X_3)$	0.737	**0.321**	0.585	0.585	0.585
$H(X_3|X_1 \times X_2)$	1.000	0.736	1.000	**0.585**	**0.585**
$I(X_1|X_2 \times X_3)$	0.678	0.264	**0.415**	**0.415**	**0.415**
$I(X_2|X_1 \times X_3)$	0.848	**0.679**	0.415	0.415	0.415
$I(X_3|X_1 \times X_2)$	0.585	0.264	0.000	**0.415**	**0.415**

When departing from the two classical uncertainty theories, several options open for applying the principle of minimum uncertainty. Using Figure 6.15 as a guide, the following options can be identified:

1. To minimize the generalized Hartley functional GH.
2. To minimize the aggregated uncertainty $\bar{S}$.
3. To minimize either of the total uncertainty components in TU or ^aTU.
4. To consider both components in TU or in ^aTU and replace a single uncertainty preference ordering with two preference orderings, one for each component.

Among these alternatives, the first one seems conceptually the most fundamental. This alternative, which may be called a *principle of minimum nonspecificity*, guarantees that the imprecision in probabilities does not increase more than necessary when we simplify a system to some given level of complexity. This alternative is also computationally attractive since the functional to be minimized (the generalized Hartley measure) is a linear functional. The aim of the following example is to illustrate some of the other alternatives.

EXAMPLE 9.3. Consider a system Z with one input variable, x_1, and one output variable, x_2, in which the relationship between the variables is expressed in terms of the joint interval-valued probability distribution given in Table 9.3a. To illustrate the principle of minimum uncertainty, let us consider two simplifications of the systems by quantizing the input variable via either function q_1 or function q_2 introduced in Example 9.2. Interval-valued probability distributions of the two simplifications, Z_1 and Z_2, are shown in Table 9.3b. They are also shown with their marginals in Figure 9.2. Their complete formulations, including the Möbius representations, are in Table 9.4. Subsets A of the Cartesian product $\{0, 1\}^2$ are defined by their characteristic functions. Relevant conditional uncertainties (since x_1 is an input variable and x_2 is an output variable) are given in Table 9.3c. They are calculated by the differences between the uncertainties on $X_1 \times X_2$ (shown in Table 9.4) and the uncertainties on X_1 (shown in Figure 9.2). We can conclude that (a) Z_1 is preferred according to $\bar{S}$ and generalized Shannon (GS); (b) Z_2 is preferred according to GH; and (c) both Z_1 and Z_2 are accepted in terms of $TU = \langle GH, GS \rangle$, since they are not comparable in terms of the joint preference ordering.

9.2.2. Conflict-Resolution Problems

Another application of the principle of minimum uncertainty is the area of conflict-resolution problems. For example, when we integrate several overlapping subsystems into one overall system, the subsystems may be locally inconsistent in the following sense. An overall system composed of subsystems is said to be *locally inconsistent* if it contains at least one pair of subsystems that share some variables and whose uncertainty functions project to distinct mar-

Table 9.3. Illustration to Example 9.3

(a) Given System Z

x_1	x_2	$\underline{\mu}(x_1, x_2)$	$\bar{\mu}(x_1, x_2)$
0	1	0.1	0.3
0	1	0.3	0.4
1	0	0.0	0.0
1	1	0.2	0.4
2	0	0.0	0.2
2	1	0.2	0.4

(b) Simplifications Z_1 and Z_2 of System Z

		Z_1		Z_2	
x_1	x_2	$\underline{\mu}_1(s)$	$\bar{\mu}_1(s)$	$\underline{\mu}_2(s)$	$\bar{\mu}_2(s)$
0	1	0.1	0.3	0.1	0.3
0	1	0.5	0.7	0.3	0.4
1	0	0.0	0.2	0.0	0.2
1	1	0.2	0.4	0.4	0.6

(c) Uncertainties of Simplifications Z_1 and Z_2

Z_1	Z_2
$\bar{S}(X_2\|X_1) = 0.814$	$\bar{S}(X_2\|X_1) = 0.871$
$GH(X_2\|X_1) = 0.200$	$GH(X_2\|X_1) = 0.158$
$GS(X_2\|X_1) = 0.614$	$GS(X_2\|X_1) = 0.713$

ginal uncertainty functions based on the shared variables. An example of locally inconsistent probabilistic subsystems is shown in Figure 9.3a.

Local inconsistency among subsystems that form an overall system is a kind of conflict. If it is not resolved, the overall system is not meaningful. For example, the overall system in Figure 9.3a is not meaningful because no joint probability distribution function exists on $X \times Y \times Z$ that is consistent with both probability distribution functions ^1p, ^2p of the two given subsystems.

Two attitudes toward locally inconsistent collections of subsystems can be recognized. According to one of them, such collections should be rejected on the basis of the fact that they do not represent any overall systems. According to the other attitude, the local inconsistencies should be resolved by modifying the given uncertainty functions of the subsystems to achieve their consistency. However, this can usually be done in numerous ways. The right way to do that, on epistemological grounds, is to obtain the consistency with the smallest possible total loss of information contained in the given uncertainty functions. The total loss of information is expressed by the sum of information losses for the individual subsystems. Thus, resolving local inconsistency can be formulated, in generic terms, as the following optimization problem, where

Table 9.4. Complete Formulation of the Two Simplifications Discussed in Example 9.3

x_1x_2				Z_1			Z_2		
00	01	10	11	$\underline{\mu}_1(A)$	$\bar{\mu}_1(A)$	$m_1(A)$	$\underline{\mu}_2(A)$	$\bar{\mu}_2(A)$	$m_2(A)$
A: 0	0	0	0	0.0	0.0	0.0	0.0	0.0	0.0
1	0	0	0	0.1	0.3	0.1	0.1	0.3	0.1
0	1	0	0	0.5	0.7	0.5	0.3	0.4	0.3
0	0	1	0	0.0	0.2	0.0	0.0	0.2	0.0
0	0	0	1	0.2	0.4	0.2	0.4	0.6	0.4
1	1	0	0	0.6	0.8	0.0	0.4	0.6	0.0
1	0	1	0	0.1	0.3	0.0	0.1	0.3	0.0
1	0	0	1	0.3	0.5	0.0	0.5	0.7	0.0
0	1	1	0	0.5	0.7	0.0	0.3	0.5	0.0
0	1	0	1	0.7	0.9	0.0	0.7	0.9	0.0
0	0	1	1	0.2	0.4	0.0	0.4	0.6	0.0
1	1	1	0	0.6	0.8	0.0	0.4	0.6	0.0
1	1	0	1	0.8	1.0	0.0	0.8	1.0	0.0
1	0	1	1	0.3	0.5	0.0	0.6	0.7	0.1
0	1	1	1	0.7	0.9	0.0	0.7	0.9	0.0
1	1	1	1	1.0	1.0	0.2	1.0	1.0	0.1

$$GH(X_1 \times X_2) = 0.400 \qquad GH(X_1 \times X_2) = 0.358$$
$$\bar{S}(X_1 \times X_2) = 1.785 \qquad \bar{S}(X_1 \times X_2) = 1.871$$
$$GS(X_1 \times X_2) = 1.385 \qquad GS(X_1 \times X_2) = 1.513$$
$$TU_1 = \langle 0.4, 1.385 \rangle \qquad TU_2 = \langle 0.358, 1.513 \rangle$$

$I(^s u, \, ^s \hat{u})$ denotes the loss of information when uncertainty function $^s u$ is replaced with uncertainty function $^s \hat{u}$.

Given a family of subsystems $\{^s Z \mid s \in \mathbb{N}_n\}$ whose uncertainty functions, $^s u$ ($s \in \mathbb{N}_n$), formalized in uncertainty theory T, are locally inconsistent, determine locally consistent counterparts of these uncertainty functions, $^s \hat{u}$ ($s \in \mathbb{N}_n$), for which the functional

$$\sum_{s=1}^{n} I(^s u, \, ^s \hat{u})$$

reaches its minimum subject to the following constraints:

(a)   Axioms of theory T.

(b)   Conditions of local consistency of uncertainty functions $^s \hat{u}(s \in \mathbb{N}_n)$.

When for example, this optimization problem is formulated within probability theory, then $^s u$ and $^s \hat{u}$ are probability distribution functions: (a) are axioms of probability theory, (b) are linear algebraic equations, and

$$I(^s u, \, ^s \hat{u}) = \sum_{x \in X_s} {}^s u(x) \log_2 \frac{^s u(x)}{^s \hat{u}(x)},$$

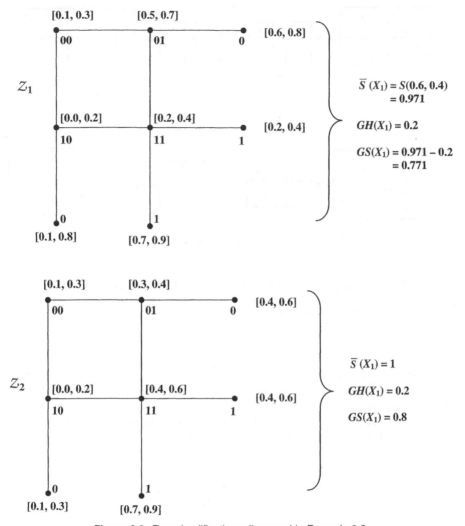

Figure 9.2. Two simplifications discussed in Example 9.3.

where X_s is the state set of subsystem Z_s, is the directed divergence introduced in Section 3.3 (see Eq. (3.56)).

The appearance of local inconsistencies among subsystems of an overall system is an indicator that the claims expressed by the uncertainty functions of the subsystems are not fully warranted under the given evidence. Thus, modifying them in a minimal way (with minimum loss of information) to achieve their consistency, as facilitated by the described optimization problems, is an epistemologically sound conflict-resolution strategy.

EXAMPLE 9.4. To illustrate the described optimization problem by a specific example, let us consider the two locally inconsistent subsystems described

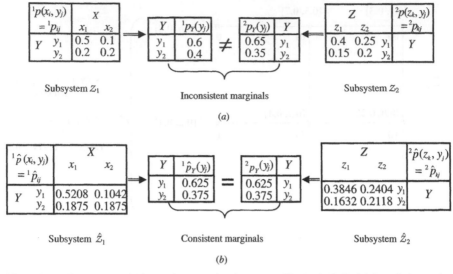

Figure 9.3. Resolving local inconsistency of subsystems (Example 9.4). (a) Locally inconsistent subsystems Z_1 and Z_2; (b) consistent subsystem $\hat{Z}_1$ and $\hat{Z}_2$ obtained by minimizing the loss of information in given subsystems Z_1 and Z_2.

in Figure 9.3a. Denoting, for convenience, $^1p(x_i, y_j) = {}^1p_{ij}$, $^2p(z_k, y_j) = {}^2p_{kj}$, $^1\hat{p}(x_i, y_j) = {}^1\hat{p}_{ij}$, and $^2\hat{p}(z_k, y_j) = {}^2\hat{p}_{kj}$ for all $i, j, k \in \{1, 2\}$, the optimization problem in this case has the following form.

Given probabilities $^1p_{ij}$ and $^2p_{kj}(i, j, k \in \{1, 2\})$, determine the values of $^1\hat{p}_{ij}$ and $^2\hat{p}_{kj}(i, j, k \in \{1, 2\})$ for which the functional

$$\sum_{i=1}^{2}\sum_{j=1}^{2} {}^1p_{ij} \log_2 \frac{^1p_{ij}}{^1\hat{p}_{ij}} + \sum_{k=1}^{2}\sum_{j=1}^{2} {}^2p_{kj} \log_2 \frac{^2p_{kj}}{^2\hat{p}_{kj}}$$

reaches its minimum under the following constraints:

($c1$) $^1\hat{p}_{11} + {}^1\hat{p}_{12} + {}^1\hat{p}_{21} + {}^1\hat{p}_{22} = 1$
($c2$) $^2\hat{p}_{11} + {}^2\hat{p}_{12} + {}^2\hat{p}_{21} + {}^2\hat{p}_{22} = 1$
($c3$) $^1\hat{p}_{ij} \geq 0$ for all $i, j \in \{1, 2\}$
($c4$) $^2\hat{p}_{kj} \geq 0$ for all $k, j \in \{1, 2\}$
($c5$) $^1\hat{p}_{11} + {}^1\hat{p}_{21} = {}^2\hat{p}_{11} + {}^2\hat{p}_{21}$
($c6$) $^1\hat{p}_{12} + {}^1\hat{p}_{22} = {}^2\hat{p}_{12} + {}^2\hat{p}_{22}$

Constraints ($c1$)–($c4$) capture axioms of probability theory; constraints ($c5$) and ($c6$) specify conditions for local consistency of the two subsystems.

The objective functional in the optimization problem, which measures the total loss of information for each modification of the probability distribution functions of subsystems Z_1 and Z_2, is within the given constraints positive (due to the Gibbs inequality) and convex. Hence, the optimization problem has a unique solution, which is given in Figure 9.3b. The loss of information in resolving the local inconsistency in this example is 0.0021.

9.3. PRINCIPLE OF MAXIMUM UNCERTAINTY

The *principle of maximum uncertainty* allows us to develop epistemologically sound procedures for dealing with a wide variety of problems that involve ampliative reasoning—any reasoning in which conclusions are not entailed in the given premises.

Ampliative reasoning is indispensable to science in a variety of ways. For example, whenever we utilize a given system for predictions, we employ ampliative reasoning. Similarly, when we want to estimate microstates from the knowledge of relevant macrostates and partial information regarding the microstates (as in image processing and many other problems), we must resort to ampliative reasoning. The problem of the identification of an overall system from some of its subsystems is another example that involves ampliative reasoning.

Ampliative reasoning is also common and important in our daily lives, where, unfortunately, the principle of maximum uncertainty is not always adhered to. Its violation leads almost invariably to conflicts in human communication, as was well expressed by Bertrand Russell in his *Unpopular Essays* [1950]:

> [W]henever you find yourself getting angry about a difference in opinion, be on your guard; you will probably find, on examination, that your belief is getting beyond what the evidence warrants.

9.3.1. Principle of Maximum Entropy

The principle of maximum uncertainty is well developed and broadly utilized within classical information theory, where it is called the *principle of maximum entropy*. It is formulated, in generic terms, as the following optimization problem: determine a probability distribution $\langle p(x) \mid x \in X \rangle$ that maximizes the Shannon entropy subject to given constraints $c_1, c_2, \ldots, c_n$, which express partial information about the unknown probability distribution, as well as the general constraints (axioms) of probability theory. The most typical constraints employed in practical applications of the maximum entropy principle are mean (expected) values of one or more random variables or various marginal probability distributions of an unknown joint distribution.

As an example, consider a random variable x with possible (given) non-negative real values $x_1, x_2, \ldots, x_n$. Assume that probabilities p_i of values x_i ($i \in \mathbb{N}_n$) are not known, although we do know the mean (expected) value $E(x)$ of the variable. Employing the maximum entropy principle, we estimate the unknown probabilities p_i, $i \in \mathbb{N}_n$, by solving the following optimization problem, in which values x_i ($i \in \mathbb{N}_n$) and their expected value are given and probabilities p_i ($i \in \mathbb{N}_n$) are to be determined.

Maximize the functional

$$S(p_1, p_2, \ldots, p_n) = -\sum_{i=1}^{n} p_i \ln p_i$$

subject to the constraints

$$E(x) = \sum_{i=1}^{n} p_i x_i \tag{9.1}$$

and

$$p_i \geq 0 \ (i \in \mathbb{N}_n), \qquad \sum_{i=1}^{n} p_i = 1. \tag{9.2}$$

For the sake of simplicity, in this formulation we use the natural logarithm in the definition of Shannon entropy. Clearly, the solution to this problem is not affected by changing the base of the logarithm in the objective functional. Equation (9.1) represents the available information; Eq. (9.2) represents the standard constraints imposed on p by probability theory.

First, we form the Lagrange function

$$L = -\sum_{i=1}^{n} p_i \ln p_i - \alpha \left(\sum_{i=1}^{n} p_i - 1 \right) - \beta \left(\sum_{i=1}^{n} p_i x_i - E(x) \right),$$

where α and β are the Lagrange multipliers that correspond to the two constraints. Second, we form the partial derivatives of L with respect to p_i ($i \in \mathbb{N}_n$), α, and β, and set them equal to zero; this results in the equations

$$\frac{\partial L}{\partial p_i} = -\ln p_i - 1 - \alpha - \beta x_i = 0 \quad \text{for each } i \in \mathbb{N}_n$$

$$\frac{\partial L}{\partial \alpha} = 1 - \sum_{i=1}^{n} p_i = 0$$

$$\frac{\partial L}{\partial \beta} = E(x) - \sum_{i=1}^{n} p_i x_i = 0.$$

The last two equations are exactly the same as the constraints of the optimization problem. The first n equations can be written as

$$
\begin{aligned}
p_1 &= e^{-1-\alpha-\beta x_1} = e^{-(1+\alpha)}e^{-\beta x_1} \\
p_2 &= e^{-1-\alpha-\beta x_2} = e^{-(1+\alpha)}e^{-\beta x_2} \\
&\vdots \qquad\qquad \vdots \\
p_n &= e^{-1-\alpha-\beta x_n} = e^{-(1+\alpha)}e^{-\beta x_n}.
\end{aligned}
$$

When we divide each of these equations by the sum of all of them (which must be one), we obtain

$$
p_i = \frac{e^{-\beta x_i}}{\sum_{k=1}^{n} e^{-\beta x_k}} \tag{9.3}
$$

for each $i \in \mathbb{N}_n$. In order to determine the value of β, we multiply the ith equation in Eq. (9.3) by x_i and add all of the resulting equations, thus obtaining

$$
E(x) = \frac{\sum_{i=1}^{n} x_i e^{-\beta x_i}}{\sum_{i=1}^{n} e^{-\beta x_i}}
$$

and

$$
\sum_{i=1}^{n} x_i e^{-\beta x_i} - E(x)\sum_{i=1}^{n} e^{-\beta x_i} = 0.
$$

Multiplying this equation by $e^{\beta E(x)}$ results in

$$
\sum_{i=1}^{n} [x_i - E(x)]e^{-\beta[x_i-E(x)]} = 0. \tag{9.4}
$$

This equation must now be solved (numerically) for β and the solution substituted for β in Eq. (9.3), which results in the estimated probabilities p_i ($i \in \mathbb{N}_n$).

EXAMPLE 9.5. To illustrate this application of the maximum entropy principle, first consider an "honest" (unbiased) die. Here $x_i = i$ for $i \in \mathbb{N}_6$ and $E(x) = 3.5$. Equation (9.4) has the form

$$
-2.5e^{2.5\beta} - 1.5e^{1.5\beta} - 0.5e^{0.5\beta} + 0.5e^{-0.5\beta} + 1.5e^{-1.5\beta} + 2.5e^{-2.5\beta} = 0.
$$

The solution is clearly $\beta = 0$; when this value is substituted for β into Eq. (9.3), we obtain the uniform probability $p_i = 1/6$ for all $i \in \mathbb{N}_6$.

Now consider a biased die for which it is known that $E(x) = 4.5$. Equation (9.4) assumes a different form

$$-3.5e^{3.5\beta} - 2.5e^{2.5\beta} - 1.5e^{1.5\beta} - 0.5e^{0.5\beta} + 0.5e^{-0.5\beta} + 1.5e^{-1.5\beta} = 0.$$

When solving this equation (by a suitable numerical method), we obtain $\beta = -0.37105$. Substitution of this value for β into Eq. (9.3) yields the maximum entropy probability distribution:

$$p_1 = \frac{1.45}{26.66} = 0.05$$

$$p_2 = \frac{2.10}{26.66} = 0.08$$

$$p_3 = \frac{3.04}{26.66} = 0.11$$

$$p_4 = \frac{4.41}{26.66} = 0.17$$

$$p_5 = \frac{6.39}{26.66} = 0.24$$

$$p_6 = \frac{9.27}{26.66} = 0.35.$$

Our only knowledge about the random variable x in the examples discussed is the knowledge of its expected value $E(x)$. It is expressed by Eq. (9.1) as a constraint on the set of relevant probability distributions. If $E(x)$ were not known, we would be totally ignorant about x, and the maximum entropy principle would yield the uniform probability distribution (the only distribution for which the entropy reaches its absolute maximum). The entropy of the probability distribution given by Eq. (9.3) is usually smaller than the entropy of the uniform distribution, but it is the largest entropy from among all the entropies of the probability distributions that conform to the given expected value $E(x)$.

A generalization of the principle of maximum entropy is the *principle of minimum cross-entropy*. It can be formulated as follows: given a prior probability distribution function p' on a finite set X and some relevant new evidence, determine a new probability distribution function p that minimizes the cross-entropy $\hat{S}$ given by Eq. (3.56) subject to constraints $c_1, c_2, \ldots, c_n$, which represent the new evidence, as well as to the standard constraints of probability theory.

New evidence reduces uncertainty. Hence, uncertainty expressed by p is, in general, smaller than uncertainty expressed by p'. The principle of minimum cross-entropy helps us to determine how much smaller it should be. It allows us to reduce the uncertainty of p' by the smallest amount necessary to satisfy the new evidence. That is, the posterior probability distribution function p estimated by the principle has the largest uncertainty among all other probability distribution functions that conform to the evidence.

9.3.2. Principle of Maximum Nonspecificity

When the principle of maximum uncertainty is applied within the classical possibility theory, where the only recognized type of uncertainty is nonspecificity, it is reasonable to describe this restricted application of the principle by a more descriptive name—a *principle of maximum nonspecificity*. This specialized principle is formulated as an optimization problem in which the objective functional is based on the Hartley measure (basic or conditional, Hartley-based information transmission, etc.). Constraints in this optimization problem consist of the axioms of classical possibility theory and any information pertaining to possibilities of the considered alternatives.

According to the principle of maximum nonspecificity in classical possibility theory, any of the considered alternatives that do not contradict given evidence should be considered possible. An important problem area, in which this principle is crucial, is the identification of n-dimensional relations from the knowledge of some of their projections. It turns out that the solution obtained by the principle of maximum nonspecificity in each of these identification problems is the cylindric closure of the given projections. Indeed, the cylindric closure is the largest and, hence, the most nonspecific n-dimensional relation that is consistent with the given projections. The significance of this solution is that it always contains the true but unknown overall relation.

A particular method for computing cylindric closure is described and illustrated in Examples 2.3 and 2.4. A more efficient method is to simply join all the given projections by the operation of relational join (introduced in Section 1.4) and, if relevant, eliminate inconsistent outcomes. This method is illustrated in the following example.

EXAMPLE 9.6. Consider a possibilistic system with three 2-valued variables, x_1, x_2, x_3, that is discussed in Example 2.4. The aim is to identify the unknown ternary relation among the variables (a subset of the set of overall states listed in Figure 9.4a) solely from the knowledge of some of its projections. It is assumed that we know two of the binary projections specified in Figure 9.4b or all of them. As is shown in Example 2.4, the identification is not unique in any of these cases. When applying the principle of maximum nonspecificity, we obtain in each case the least specific ternary relation, which is the cylindric closure of the given projections. The aim of this example is to show that an efficient way of determining the cylindric closure is to apply the operation of relational join.

Assume that projections R_{12} and R_{23} are given. Taking their relational join $R_{12} * R_{23}$, as illustrated in Figure 9.4c, we readily obtain their cylindric closure (compare with the same result in Example 2.4). In a similar way, we obtain the

States	x_1	x_2	x_3
s_0	0	0	0
s_1	0	0	1
s_2	0	1	0
s_3	0	1	1
s_4	1	0	0
s_5	1	0	1
s_6	1	1	0
s_7	1	1	1

(a)

R_{12}:

x_1	x_2
0	0
0	1
1	0

R_{23}:

x_2	x_3
0	0
0	1
1	1

R_{13}:

x_1	x_3
0	0
0	1
1	0

(b)

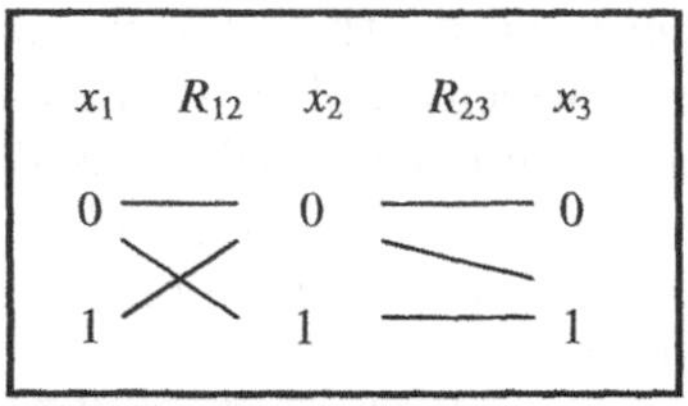

States	x_1	x_2	x_3
s_0	0	0	0
s_1	0	0	1
s_3	0	1	1
s_4	1	0	0
s_5	1	0	1

(c)

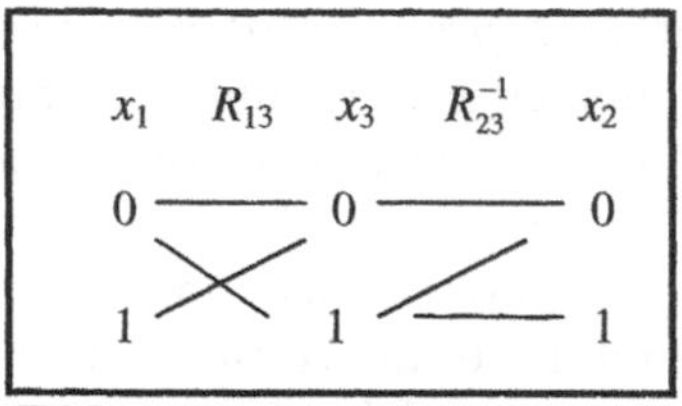

States	x_1	x_2	x_3
s_0	0	0	0
s_1	0	0	1
s_3	0	1	1
s_4	1	0	0

(d)

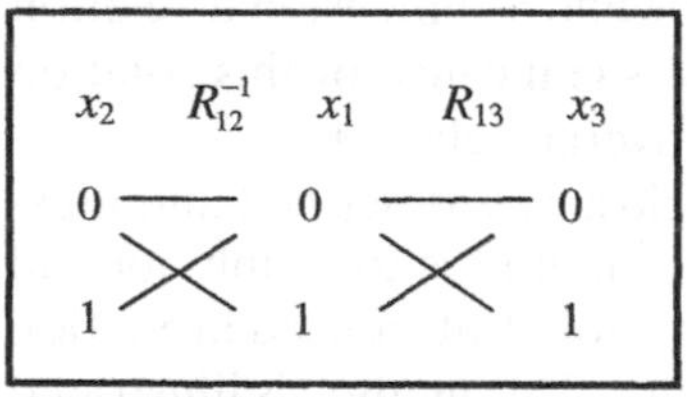

States	$R_{12}^{-1} * R_{13}$		
	x_1	x_2	x_3
s_0	0	0	0
s_1	0	0	1
s_2	0	1	0
s_3	0	1	1
s_4	1	0	0

(e)

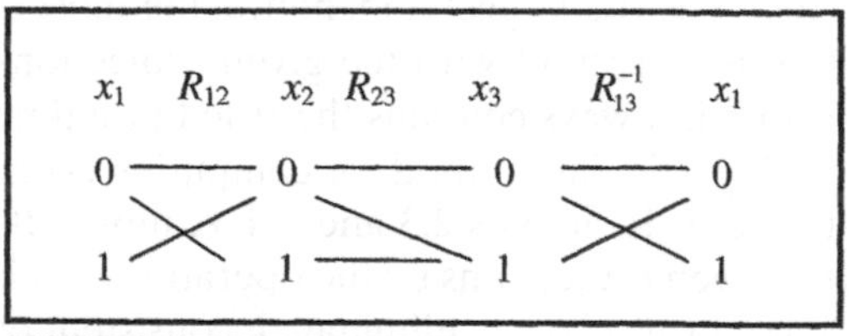

States	$(R_{12} * R_{23}) * R_{13}^{-1}$			
	x_1	x_2	x_3	x_1
s_0	0	0	0	0
—	0	0	0	1
s_1	0	0	1	0
s_3	0	1	1	0
—	1	0	0	0
s_4	1	0	0	1
—	1	0	1	0

(f)

Figure 9.4. Computation of cylindric closures (Example 9.6).

cylindric closures for the other pairs of projections, as shown in Figure 9.4*d* and 9.4*e*. Observe, however, that we need to use the inverse of one of the projections in these cases to be able to apply the relational join. The order in which the join is performed is not significant. When the relational join is applied to all three binary projections, as shown in Figure 9.4*f*, the outcomes are quadruples, in which one variable appears twice (variable x_1 in our case). Any quadruple in which the two appearances of this variable have distinct values must be excluded (they are shaded in the figure); such a quadruple indicates that the triple obtained by the join $R_{12} * R_{23}$ is inconsistent with projection R_{13}.

The idea of cylindric closure as the least specific identification of an n-dimensional relation from some of its projections is applicable to convex subsets of n-dimensional Euclidean space as well. The main difference is that there is a continuum of distinct projections in the latter case.

9.3.3. Principle of Maximum Uncertainty in GIT

In all uncertainty theories, except the two classical ones, two types of uncertainty coexist, which are measured by the appropriately generalized Hartley and Shannon functionals. To apply the principle of maximum uncertainty, we can use one of these types of uncertainty or both of them. In addition, we can also use the well-established aggregate of both types of uncertainty. This means that we can formulate the principle of maximum uncertainty in terms of four distinct optimization problems, which are distinguished from one another by the following objective functionals:

(a) Generalized Hartley measure GH: Eq. (6.38).

(b) Generalized Shannon measure GS: Eq. (6.64).

(c) Aggregated measure of total uncertainty $\bar{S}$: Eq. (6.61).

(d) Disaggregated measure of total uncertainty $TU = \langle GH, GS \rangle$: Eq. (6.65).

(e) Alternative disaggregated measure of total uncertainty $^aTU = \langle \bar{S} - \underline{S}, \underline{S} \rangle$: Eq. (6.75).

Which of the five optimization problems to choose depends a great deal on the context of each application. However, a few general remarks regarding the four options readily can be made. First, options (d) and (e) are clearly the most expressive ones, but their relationship is not properly understood yet. The utility of option (c) is somewhat questionable since functional $\bar{S}$ is known to be highly insensitive to changes in evidence. Of the two remaining options, (a) seems to be conceptually more fundamental than (b); it is a generalization of the principle of maximum nonspecificity discussed in Section 9.3.2. Moreover, option (a) is computationally attractive due to the linearity of the generalized Hartley measure.

None of the five optimization problems has been properly developed so far in any of the nonstandard theories of uncertainty. The rest of this section is thus devoted to illustrating the optimization problems by simple examples.

EXAMPLE 9.7. To illustrate the principle of maximum nonspecificity in evidence theory, let us consider a finite universal set X and three of its nonempty subsets that are of interest to us: A, B, and $A \cap B$. Assume that the only evidence on hand is expressed in terms of two numbers, a and b, that represent the total beliefs focusing on A and B, respectively, (a, $b \in [0, 1]$). Our aim is to estimate the degree of support for $A \cap B$ based on this evidence.

As a possible interpretation of this problem, let X be a set of diseases considered in an expert system designed for medical diagnosis in a special area of medicine, and let A and B be sets of diseases that are supported for a particular patient by some diagnostic tests to degrees a and b, respectively. Using this evidence, it is reasonable to estimate the degree of support for diseases in $A \cap B$ by using the principle of maximum nonspecificity. This principle is a safeguard that does not allow us to produce an answer (diagnosis) that is more specific than warranted by the evidence.

The use of the principle of maximum nonspecificity in our example leads to the following optimization problem: Determine values $m(X)$, $m(A)$, $m(B)$, and $m(A \cap B)$ for which the functional

$$GH(m) = m(X)\log_2|X| + m(A)\log_2|A| + m(B)\log_2|B|$$
$$m(A \cap B)\log_2|A \cap B|$$

reaches its maximum subject to the constraints

$$m(A) + m(A \cap B) = a$$
$$m(B) + m(A \cap B) = b$$
$$m(X) + m(A) + m(B) + m(A \cap B) = 1$$
$$m(X), m(A), m(B), m(A \cap B) \geq 0,$$

where a, $b \in [0, 1]$ are given numbers.

The constraints are represented in this case by three linear algebraic equations of four unknowns and, in addition, by the requirement that the unknowns be nonnegative real numbers. The first two equations represent our evidence, the third equation and the inequalities represent general constraints of evidence theory. The equations are consistent and independent. Hence, they involve one degree of freedom. Selecting, for example, $m(A \cap B)$ as the free variable, we readily obtain

$$m(A) = a - m(A \cap B)$$
$$m(B) = b - m(A \cap B) \qquad (9.5)$$
$$m(X) = 1 - a - b + m(A \cap B).$$

Since all the unknowns must be nonnegative, the first two equations set the upper bound of $m(A \cap B)$, whereas the third equation specifies its lower bound; the bounds are

$$\max\{0, a+b-1\} \le m(A \cap B) \le \min\{a, b\}. \tag{9.6}$$

Using Eqs. (9.5), the objective function now can be expressed solely in terms of the free variable $m(A \cap B)$. After a simple rearrangement of terms, we obtain

$$GH(m) = m(A \cap B)[\log_2|X| - \log_2|A| - \log_2|B| + \log_2|A \cap B|] + (1 - a - b)\log_2|X| + a\log_2|A| + b\log_2|B|.$$

Clearly, only the first term in this expression can influence its value, so that we can rewrite the expression as

$$GH(m) = m(A \cap B)\log_2 K_1 + K_2, \tag{9.7}$$

where

$$K_1 = \frac{|X| \cdot |A \cap B|}{|A| \cdot |B|}$$

and

$$K_2 = (1 - a - b)\log_2|X| + a\log_2|A| + b\log_2|B|,$$

are constant coefficients. The solution to the optimization problem depends only on the value of K_1. Since A, B, and $A \cap B$ are assumed to be nonempty subsets of X, $K_1 > 0$. If $K_1 < 1$, then $\log_2 K_1 < 0$ and we must minimize $m(A \cap B)$ to obtain the maximum of $GH(m)$; hence, $m(A \cap B) = \max\{0, a + b - 1\}$ due to Eq. (9.6). If $K_1 > 1$, then $\log_2 K_1 > 0$, and we must maximize $m(A \cap B)$; hence, $m(A \cap B) = \min\{a, b\}$ as given by Eq. (9.6). When $K = 1$, $\log_2 K_1 = 0$, and $GH(m)$ is independent of $m(A \cap B)$; this implies that the solution is not unique or, more precisely, that any value of $m(A \cap B)$ in the range of Eq. (9.6) is a solution to the optimization problem. The complete solution thus can be expressed by the following equations:

$$m(A \cap B) = \begin{cases} \max\{0, a+b-1\} & \text{when } K_1 < 1 \\ [\max\{0, a+b-1\}, \min\{a, b\}] & \text{when } K_1 = 1 \\ \min\{a, b\} & \text{when } K_1 > 1. \end{cases}$$

The three types of solutions are illustrated visually in Figure 9.5 and given for specific numerical values in Table 9.5.

EXAMPLE 9.8. The aim of this example is to illustrate an application of the principle of maximum nonspecificity within the uncertainty theory based on λ-measures. Assume that marginal λ-measures, ${}^\lambda\mu_X$ and ${}^\lambda\mu_Y$, are given, where $X = Y = \{0, 1\}$, and we want to determine the unknown joint λ-measure, ${}^\lambda\mu$, by

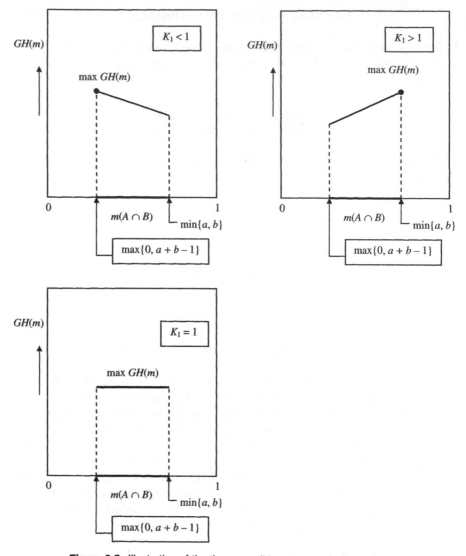

Figure 9.5. Illustration of the three possible outcomes in Example 9.7.

using the principle of maximum nonspecificity. A simple notation to be used in this example is introduced in Figure 9.6: values x_1, x_2, y_1, y_2 are given; values a, b, c, d are to be determined. Observe that the given values must satisfy the equations:

$$\frac{1-x_1-x_2}{x_1 x_2} = \lambda,$$

$$\frac{1-y_1-y_2}{y_1 y_2} = \lambda,$$

Table 9.5. Examples of the Three Types of Possible Solutions Discussed in Example 9.7

				(a) Three Particular Examples			
Example	$|X|$	$|A|$	$|B|$	$|A \cap B|$	a	b	$GH[m(A \cap B)]$
(i)	10	5	5	2	0.7	0.5	$-0.322m(A \cap B) + 2.122$
(ii)	10	5	4	2	0.8	0.6	1.497
(iii)	20	10	12	7	0.4	0.5	$0.222m(A \cap B) + 3.553$

	(b) Solutions for Three Particular Examples Obtained by the Principle of Maximum Nonspecificity				
Example	Type	$m(A \cap B)$	$m(A)$	$m(B)$	$m(X)$
(i)	$K_1 < 1$	0.2	0.5	0.3	0.0
(ii)	$K_1 = 1$	[0.4, 0.6]	[0.2, 0.4]	[0.0, 0.2]	[0.0, 0.2]
(iii)	$K_1 > 1$	0.4	0.0	0.1	0.5

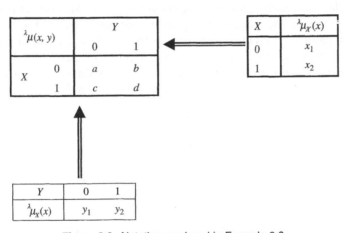

Figure 9.6. Notation employed in Example 9.8.

The relationship between the joint and marginal measures is expressed by the following equations:

$$a + b + \lambda ab = x_1$$
$$c + d + \lambda cd = x_2$$
$$a + c + \lambda cd = y_1$$
$$b + d + \lambda bd = y_2.$$

To determine the solution set of nonnegative values for a, b, c, d, we need one free variable. For example, choosing c as the free variable, we readily obtain the following dependencies of the remaining variables on c:

$$a = \frac{y_1 - c}{1 + \lambda c} \qquad \text{(from the third equation)},$$

$$d = \frac{x_2 - c}{1 + \lambda c} \qquad \text{(from the second equation)},$$

$$b = \frac{x_1(1 + \lambda c) + c - y_1}{1 + \lambda y_1} \qquad \text{(from the first equation)}.$$

Since it is required that $a, b, c, d \geq 0$, we obtain the following range of acceptable values of c:

$$\max\left\{0, \frac{y_1 - x_1}{\lambda x_1 + 1}\right\} \leq c \leq \min\{x_2, y_1\}. \qquad (9.8)$$

Each value of c within this range defines a particular joint λ-measure, ${}^{\lambda}\mu_c$, that is consistent with the given marginals. One way of determining the least specific of these measures is to search through the range of c by small increments and calculate for each value of c the measure ${}^{\lambda}\mu_c$, its Möbius representation, and the value of the generalized Hartley measure. The measure with the maximum value of the Hartley measure is then chosen. In a similar way, joint λ-measures that maximize functionals $\bar{S}$ or GS can be determined.

To illustrate the procedure for some numerical values, let $x_1 = 0.6$, $x_2 = 0.1$, $y_1 = 0.4$, and $y_2 = 0.2$. Then, clearly, $\lambda = 5$, $c \in [0, 0.1]$, and

$$a = \frac{0.4 - c}{1 + 5c},$$

$$b = 0.067 + 1.333c,$$

$$d = \frac{0.1 - c}{1 + 5c}.$$

The maximum of the generalized Hartley measure is 0.573 and it is obtained for $c = 0.069$. The least specific λ-measure (the one for $c = 0.069$) and its Möbius representation are shown in Table 9.6. Also shown in the table are measures that maximize functionals $\bar{S}$ (obtained for $c = 0.039$) and GS (obtained for $c = 0.036$). All these measures represent lower probabilities (since λ is positive). The corresponding upper probabilities can readily be calculated via the duality relation.

The problem of identifying an unknown n-dimensional relation from some of its projections by using the principle of maximum nonspecificity is illustrated in Example 9.6 for relations formalized in classical possibility theory. Two common methods for dealing with the problem are: (1) constructing the intersection of cylindric extensions of the given projections; and (2) combining the given projection by the operation of relational join

Table 9.6. Resulting λ-measures in Example 9.8

	$X \times Y$				max GH		max $\bar{S}$		max GS	
	00	01	01	11	$^5\mu(A)$	$m(A)$	$^5\mu(A)$	$m(A)$	$^5\mu(A)$	$m(A)$
A:	0	0	0	0	0.000	0.000	0.000	0.000	0.000	0.000
	1	0	0	0	0.246	0.246	0.302	0.302	0.309	0.309
	0	1	0	0	1.159	1.159	0.119	0.119	0.115	0.115
	0	0	1	0	0.069	0.069	0.039	0.039	0.036	0.036
	0	0	0	1	0.023	0.023	0.051	0.051	0.054	0.054
	1	1	0	0	0.600	0.195	0.600	0.179	0.600	0.177
	1	0	1	0	0.400	0.085	0.400	0.059	0.400	0.056
	1	0	0	1	0.298	0.028	0.430	0.077	0.446	0.084
	0	1	1	0	0.282	0.055	0.181	0.023	0.171	0.021
	0	1	0	1	0.200	0.018	0.200	0.030	0.200	0.031
	0	0	1	1	0.100	0.008	0.100	0.010	0.100	0.010
	1	1	1	0	0.876	0.067	0.756	0.035	0.744	0.032
	1	1	0	1	0.692	0.023	0.804	0.046	0.817	0.048
	1	0	1	1	0.469	0.010	0.553	0.015	0.563	0.015
	0	1	1	1	0.338	0.006	0.278	0.006	0.272	0.006
	1	1	1	1	1.000	0.008	1.000	0.009	1.000	0.009

and, if relevant, excluding inconsistencies. Both methods can be easily generalized to the theory of graded possibilities, as is illustrated in the following example.

EXAMPLE 9.9. Consider a system with three variables, x_1, x_2, x_3, that is similar to the one in Example 9.6. The set of all considered overall states of the system, specified in Table 9.7a, is the same as in Example 9.6. Again, we want to identify the unknown ternary relation from its projections by using the principle of maximum nonspecificity. The difference in this example is that the projections are fuzzy relations, which are defined by their membership functions in Table 9.7b. Since these projections provide us with partial information about the overall relation, we can interpret them as basic possibility functions. As in Example 9.6, the least specific overall relation that is consistent with the given projections is the cylindric closure of the projections. The construction of cylindric closures for fuzzy relations is discussed in Section 7.6.1. Applying this construction to the projections in this example, we first construct cylindric extensions of the projections, ER_{12}, ER_{23}, ER_{13} (as shown in Table 9.7c). Then, we determine the cylindric closure by taking the standard intersection (based on the minimum operator) of the cylindric extensions (shown in Table 9.7d). Recall that the standard operation of intersection of fuzzy sets is the only cutworthy one. This means that each α-cut of the cylindric closure of some fuzzy projections is a cylindric closure in the classical sense. An alternative, more efficient way of constructing cylindric closures of fuzzy projections is to use the relational join introduced in Section 7.5.2 (Eq. (7.34)).

Table 9.7. Illustration to Example 9.9

States	x_1	x_2	x_3
s_0	0	0	0
s_1	0	0	1
s_2	0	1	0
s_3	0	1	1
s_4	1	0	0
s_5	1	0	1
s_6	1	1	0
s_7	1	1	1

(a)

x_1	x_2	$R_{12}(x_1, x_2)$	x_2	x_3	$R_{23}(x_2, x_3)$	x_1	x_3	$R_{13}(x_1, x_3)$
0	0	1.0	0	0	0.5	0	0	0.9
0	1	0.8	0	1	1.0	0	1	1.0
1	0	0.6	1	0	0.7	1	0	1.0
1	1	0.0	1	1	0.4	1	1	0.3

(b)

x_1	x_2	x_3	ER_{12}	x_1	x_2	x_3	ER_{23}	x_1	x_2	x_3	ER_{13}
0	0	0	1.0	0	0	0	0.5	0	0	0	0.9
0	0	1	1.0	1	0	0	0.5	0	1	0	0.9
0	1	0	0.8	0	0	1	1.0	0	0	1	1.0
0	1	1	0.8	1	0	1	1.0	0	1	1	1.0
1	0	0	0.6	0	1	0	0.7	1	0	0	1.0
1	0	1	0.6	1	1	0	0.7	1	1	0	1.0
1	1	0	0.0	0	1	1	0.4	1	0	1	0.3
1	1	1	0.0	1	1	1	0.4	1	1	1	0.3

(c)

x_1	x_2	x_3	Cyl
0	0	0	0.5
0	0	1	1.0
0	1	0	0.7
0	1	1	0.4
1	0	0	0.5
1	0	1	0.3
1	1	0	0.0
1	1	1	0.0

(d)

9.4. PRINCIPLE OF REQUISITE GENERALIZATION

GIT, as an ongoing research program, offers us a steadily growing inventory of diverse uncertainty theories. Each of the theories is a formal mathematical system, which means that it is subject to some specific assumptions that are inherent in its axioms. If these assumptions are violated in the context of a given application, the theory is ill-suited for the application.

The growing diversity of uncertainty theories within GIT makes it increasingly more realistic to find an uncertainty theory whose assumptions are in harmony with each given application. However, other criteria for choosing a suitable theory are often important as well, such as low computational complexity or high conceptual transparency of the theory.

Due to the common properties of uncertainty theories recognized within GIT, emphasized especially in Chapters 4, 6, and 8, it is also feasible to work within GIT as a whole. In this case, we would move from one theory to another as needed when dealing with a given application. There are basically two reasons for moving from one uncertainty theory to another:

1. The theory we use is not sufficiently general to capture uncertainty that emerges at some stage of the given application. A more general theory is needed.
2. The theory we use becomes inconvenient at some stage of the given application (e.g., its computational complexity becomes excessive) and it is desirable to replace it with a more convenient theory.

These two distinct reasons for replacing one uncertainty theory with another lead to two distinct principles that facilitate these replacements: a principle of requisite generalization, which is introduced in this section, and a principle of uncertainty invariance, which is introduced in Section 9.5.

The following is one way of formulating the *principle of requisite generalization*: Whenever, at some stage of a problem-solving process involving uncertainty, a given uncertainty theory becomes incapable of representing emerging uncertainty of some type, it should be replaced with another theory, sufficiently more general, that is capable of representing this type of uncertainty. As suggested by the name of this principle, the extent of generalization is not optional, but requisite, determined by the nature of the emerging uncertainty.

It seems pertinent to compare the principle of requisite generalization with the principle of maximum uncertainty. Both these principles clearly aim at an epistemologically honest representation of uncertainty. The difference between them, a fundamental one, is that the former principle applies to GIT as a whole, while the latter applies to each individual uncertainty theory within GIT.

The principle of requisite generalization is introduced here for the first time. There is virtually no experience with its practical applicability. At this point, the best way to illustrate it seems to be to describe some relevant examples.

Table 9.8. Three Ways of Handling Incomplete Data (Example 9.10)

v_1	v_2	i
0	0	0
0	1	1
1	0	2
1	1	3

(a)

i:	0	1	2	3	$N(A)$	$Pro_1(A)$	$Pro_2(A)$	$m(A)$	$Bel(A)$	$Pl(A)$
A:	1	0	0	0	106	0.546	0.369	0.106	0.106	0.632
	0	1	0	0	55	0.284	0.237	0.055	0.055	0.419
	0	0	1	0	25	0.129	0.246	0.025	0.025	0.467
	0	0	0	1	8	0.041	0.148	0.008	0.008	0.288
	1	1	0	0	121	0.830	0.606	0.121	0.373	0.839
	1	0	1	0	314	0.675	0.615	0.314	0.445	0.785
	0	1	0	1	152	0.325	0.385	0.152	0.215	0.555
	0	0	1	1	128	0.170	0.394	0.128	0.161	0.627

(b)

EXAMPLE 9.10. Consider two variables, v_1, v_2, each of which has two possible states, 0 and 1. For convenience, let the joint states of the variables be labeled by an index i, as defined in Table 9.8a. Assume that we work within classical probability theory. Assume further that we have a record of an appreciable number of observations of the variables. Some of the observations contain a value of both variables, some of them contain a value of only one variable due to some measurement or communication constraints (not essential for our discussion). When only one variable is observed, it is known that this observation represents one of two joint states of the variable, but it is not known which one it actually is. For example, observing that $v_1 = 1$ (and not knowing the state of v_2) means that the joint state may be $i = 2$ or $i = 3$, but we do not know which of them is the actual joint state. There are basically three ways to handle these incomplete data:

(i) We can just ignore all the incomplete observations. This means, however, that we ignore information that may be useful, or even crucial, in some applications.

(ii) We can apply the principle of maximum entropy to fill any gaps in the data. This results in a particular probability measure, which is obtained by uniformly distributing the uncertain observations in each pair of relevant states. Using this principle allows us to stay in probability theory (and it is epistemologically the best way to deal with incom-

plete data within probability theory), but the use of the uniform distribution is totally arbitrary (imposed only by the axioms of probability theory) and does not represent the actual uncertainty in this case.

(iii) We can apply the principle of requisite generalization and move from
probability theory to Dempster–Shafer theory (DST), which is sufficiently general to represent the actual uncertainty in this case. Observing a state of one variable only is easily described in DST as observing
a set of two joint states. For example, observing that $v_1 = 1$ (and not
knowing the state of v_2) can be viewed as observing the subset $\{2, 3\}$
of the four joint states. In this way, uncertainty is represented fully and
accurately.

A numerical example illustrating the three options is shown in Table 9.8*b*.
It is assumed that we have a record of 1000 observations of the variables, some
containing values of both variables and some containing a value of only one
of the variables. Listed in the table are only subsets A of joint states that are
supported by observations (these are also all focal elements of the representation in DST). Symbols in the table have the following meanings:

$N(A)$: The number of observations of the relevant subsets A of the joint
sets.

$Pro_1(A)$: Values of the probability measure based on ignoring incomplete
observations.

$Pro_2(A)$: Values of the probability measure derived by the maximum
entropy principle.

$m(A)$: Values of the basic probability assignment function in DST, based on
the frequency interpretation.

$Bel(A)$ and $Pl(A)$: Values of the associated belief and plausibility measures,
respectively.

We can see that DST is the right (or requisite) generalization in this
example. It is capable of fully representing both types of uncertainty that are
inherent in the given evidence. No further generalization is needed. The probabilistic representation derived by the principle of maximum entropy is not
capable of capturing nonspecificity, and hence, it is not a full representation of
uncertainty in this case.

EXAMPLE 9.11. Consider the simplest possible marginal problem in probability theory, which is depicted in Figure 9.7: given marginal probabilities on
the two-element sets, what are the associated joint probabilities? This question can be answered in two very different ways, depending on whether it is
required or not required to remain in probability theory. If it is required, then
we need to determine one particular joint probability distribution function.
The proper way to do that (as is explained in Section 9.3) is to use the prin-

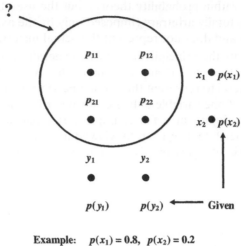

Example: $p(x_1) = 0.8$, $p(x_2) = 0.2$
$p(y_1) = 0.6$, $p(y_2) = 0.4$

Figure 9.7. The marginal problem in probability theory (Example 9.11).

ciple of maximum entropy. From subadditivity and additivity of the Shannon entropy, it follows immediately that the maximum-entropy joint probability distribution function p is the product of the given marginal probability distribution functions. That is, using the notation introduced in Figure 9.7,

$$p_{ij} = p(x_i) \cdot p(y_j) \quad \text{for } i, j = 1, 2.$$

For the numerical values given in Figure 9.7, we get $p_{11} = 0.48$, $p_{12} = 0.32$, $p_{21} = 0.12$, and $p_{22} = 0.08$.

If it is not required that we remain in probability theory, we should apply the principle of requisite generalization and move to another theory of uncertainty that is sufficiently general to capture the full uncertainty in this problem. The full uncertainty is expressed in terms of the set of all joint probability distributions that are consistent with the given marginal probability distributions. Here we can utilize Example 4.3, in which this set is determined. The following is its formulation for the numerical values in Figure 9.7:

$$p_{11} \in [0.4, 0.6]$$
$$p_{12} = 0.8 - p_{11},$$
$$p_{21} = 0.6 - p_{11},$$
$$p_{22} = p_{11} - 0.4.$$

From this set of joint probability distributions, we can derive the associated lower and upper probabilities, μ and $\bar{\mu}$, and the Möbius representation, which are shown in Table 9.9. The monotone measure representing lower

Table 9.9. Illustration of the Principle of Requisite Generalization (Example 9.11)

	x_1 y_1	x_1 y_2	x_2 y_1	x_2 y_2	$\underline{\mu}(A)$	$\bar{\mu}(A)$	$m(A)$	$Pro(A)$
A:	0	0	0	0	0.0	0.0	0.0	0.0
	1	0	0	0	0.4	0.6	0.4	0.48
	0	1	0	0	0.2	0.4	0.2	0.32
	0	0	1	0	0.0	0.2	0.0	0.12
	0	0	0	1	0.0	0.2	0.0	0.08
	1	1	0	0	0.8	0.8	0.2	0.80
	1	0	1	0	0.6	0.6	0.2	0.60
	1	0	0	1	0.4	0.8	0.0	0.56
	0	1	1	0	0.2	0.6	0.0	0.44
	0	1	0	1	0.4	0.4	0.2	0.40
	0	0	1	1	0.2	0.2	0.2	0.20
	1	1	1	0	0.8	1.0	−0.2	0.92
	1	1	0	1	0.8	1.0	−0.2	0.88
	1	0	1	1	0.6	0.8	−0.2	0.68
	0	1	1	1	0.4	0.6	−0.2	0.52
	1	1	1	1	1.0	1.0	0.4	1.00

probabilities is superadditive, but it is not 2-monotone (there are eight viola-tions of 2-monotonicity pertaining to subsets with these elements). Hence, to fully represent uncertainty in this example, we need to move to the most general uncertainty theory. This is requisite, not a matter of choice.

9.5. PRINCIPLE OF UNCERTAINTY INVARIANCE

The multiplicity of uncertainty theories within GIT has opened a new area of inquiry—the study of transformations between the various uncertainty theo-ries. The primary motivation for studying these transformations stems from questions regarding uncertainty approximations. How to approximate, in a meaningful way, uncertainty represented in one theory by a representation in another, less general theory? From the practical point of view, uncertainty approximations are sought to reduce computational complexity or to make the representation of uncertainty more transparent to the user. From the the-oretical point of view, studying uncertainty approximations enhances our understanding of the relationships between the various uncertainty theories.

In this section, all uncertainty approximation problems are expressed in terms of one common principle, which is referred to as the *principle of uncer-tainty invariance*. To formulate this principle, assume that T_1 and T_2 denote two uncertainty theories such that T_2 is less general than or incomparable to T_1. Assume further that uncertainty pertaining to a given problem-solving situa-tion is expressed by a particular uncertainty function u_1, and that this function is to be approximated by some uncertainty function u_2 of theory T_2. Then, according to the principle of uncertainty invariance, it is required that the amounts of uncertainty contained in u_1 and u_2 be the same. That is, the transformation from T_1 to T_2 is required to be invariant with respect to the amount of uncertainty. This explains the name of the principle. Due to the unique connection between uncertainty and uncertainty-based information, the principle may also be viewed as a *principle of information invariance* or *information preservation*. It seems reasonable to compare this principle, in a metaphoric way, with the principle of energy preservation in physics.

The following are examples of some obvious types of uncertainty approximations, some of which are examined in this section:

- Approximating belief measures by probability measures, measures of graded possibilities, λ-measures, or k-additive measures.
- Approximating probability measures by measures of graded possibilities and vice versa.
- Approximating measures of graded possibilities by crisp possibility measures.
- Approximating k-monotone measures (k finite) by belief measures.
- Approximating 2-monotone measures by measures based on reachable interval-valued distributions.
- Approximating fuzzified measures of some type by their crisp counterparts.

Observe that in virtually all these approximations, the principle of uncertainty invariance can be formulated in terms of the total aggregated uncertainty, $\bar{S}$, or in terms of one or both on its components, GH and GS. These variants of the principle lead, of course, to distinct approximations. Their choice depends on what type of uncertainty we desire to preserve in each application.

Preserving the amount of uncertainty of a certain type need not be the only requirement employed in formulating the various problems of uncertainty approximation. Other requirements may be added to the individual formulations in the context of each application. A common requirement is that uncertainty function u_1 can be converted to its counterpart u_2 by an appropriate scale, at least ordinal.

In the rest of this section, only some of the approximation problems are examined and illustrated by specific examples. This area is still not sufficiently developed to warrant a more comprehensive coverage.

9.5.1. Computationally Simple Approximations

Assume that the measure of uncertainty employed in the principle of uncertainty invariance is the functional $\bar{S}$. Then, all uncertainty approximations in which a given uncertainty function u_1 is to be approximated by a probability

Table 9.10. Probabilistic Approximation of a Given λ-Measure (Example 9.12)

		X			Given		Approximated	
a	b	c	d	$\underline{\mu}(A)$	$\bar{\mu}(A)$	$m(A)$	$Pro(A)$	$m(A)$
A: 0	0	0	0	0.000	0.000	0.000	0.000	0.000
1	0	0	0	0.302	0.722	0.302	0.302	0.302
0	1	0	0	0.119	0.447	0.119	0.298	0.298
0	0	1	0	0.039	0.196	0.039	0.196	0.196
0	0	0	1	0.051	0.244	0.051	0.204	0.204
1	1	0	0	0.600	0.900	0.179	0.600	0.000
1	0	1	0	0.400	0.800	0.059	0.498	0.000
1	0	0	1	0.430	0.819	0.077	0.506	0.000
0	1	1	0	0.181	0.570	0.023	0.494	0.000
0	1	0	1	0.200	0.600	0.030	0.502	0.000
0	0	1	1	0.100	0.400	0.010	0.400	0.000
1	1	1	0	0.756	0.949	0.035	0.796	0.000
1	1	0	1	0.804	0.961	0.046	0.804	0.000
1	0	1	1	0.553	0.881	0.015	0.702	0.000
0	1	1	1	0.278	0.698	0.006	0.698	0.000
1	1	1	1	1.000	1.000	0.009	1.000	0.000

measure u_2 are conceptually trivial and computationally simple. Just recall that when we calculate $\bar{S}(u_1)$ by Algorithm 6.1 (or in any other way), we obtain, as a by-product, a probability distribution function for which the Shannon entropy is equal to $\bar{S}(u_1)$. This probability distribution function is thus the probabilistic approximation of the uncertainty function u_1 based on the principle of uncertainty invariance and the use of functional $\bar{S}$.

EXAMPLE 9.12. Consider the lower and upper probabilities $\underline{\mu}$ and $\bar{\mu}$, defined in Table 9.10. It can be easily checked that $\underline{\mu}$ is a λ-measure for $\lambda = 5$. Applying Algorithm 6.1 to $\underline{\mu}$, we obtain $\bar{S}(\underline{\mu}) = 1.\overline{9}71$ and, as a by-product, the probability distribution

$$\mathbf{p} = \langle 0.302, 0.298, 0.196, 0.204 \rangle.$$

Probability measure Pro based on this distribution (also shown in Table 9.10) thus can be viewed as a probabilistic approximation of $\underline{\mu}$ for which

$$\bar{S}(\underline{\mu}) = \bar{S}(Pro)(= S(\mathbf{p})).$$

Although probabilistic approximations based on invariance of the aggregated total uncertainty $\bar{S}$ are computationally simple, their utility is questionable due to the notorious insensitivity of this uncertainty measure. When either GH or GS is employed instead of $\bar{S}$, then, clearly, these approximations must be handled differently, as is illustrated in the next section.

9.5.2. Probability–Possibility Transformations

Let the n-tuples $\mathbf{p} = \langle p_1, p_2, \ldots, p_n \rangle$ and $\mathbf{r} = \langle r_1, r_2, \ldots, r_n \rangle$ denote, respectively, a probability distribution and a possibility profile on a finite set X with n or more elements. Assume that these tuples are ordered in the same way and do not contain zero components. Hence,

(a) $p_i \in (0, 1]$ and $r_i \in (0, 1]$ for all $i \in \mathbb{N}_n$.
(b) $p_i \geq p_{i+1}$ and $r_i \geq r_{i+1}$ for all $i \in \mathbb{N}_{n-1}$.
(c) $\sum_{i=1}^{n} p = 1$ (probabilistic normalization).
(d) $r_1 = 1$ (possibilistic normalization).
(e) If $n < |X|$, then $p_i = r_i = 0$ for all $i = n+1, n+2, \ldots, |X|$.

Assume further that values r_i correspond to values p_i for all $i \in \mathbb{N}_n$ by some scale, at least ordinal.

Assuming that $\mathbf{p}$ and $\mathbf{r}$ are connected by a scale of some type means that certain properties (such as ordering or proportionality of values) are preserved when $\mathbf{p}$ is transformed to $\mathbf{r}$ or vice versa. Transformations between probabilities and possibilities are thus very different under different types of scales. Let us examine these fundamental differences for the five most common scale types: ratio, difference, interval, log-interval, and ordinal scales.

1. *Ratio Scales.* Ratio-scale transformations $\mathbf{p} \rightarrow \mathbf{r}$ have the form $r_i = \alpha p_i$ for all $i \in \mathbb{N}_n$, where α is a positive constant. From the possibilistic normalization (d), we obtain $1 = \alpha p_1$ and, consequently,

$$r_i = \frac{p_i}{p_1}$$

for all $i \in \mathbb{N}_n$. For the inverse transformation, $\mathbf{r} \rightarrow \mathbf{p}$, we have $p_i = r_i/\alpha$. From the probabilistic normalization (c), we get $1 = (r_1 + r_2 + \cdots + r_n)/\alpha$ and, consequently,

$$p_i = \frac{r_i}{r_1 + r_2 + \cdots + r_n}$$

for all $i \in \mathbb{N}_n$.

These transformations are clearly too rigid to preserve any other property than the property of ratio scales, $r_i/p_i = \alpha$ or $p_i/r_i = 1/\alpha$, for all $i \in \mathbb{N}_n$. Hence, the principle of uncertainty invariance cannot for formulated in terms of the ratio scales. Since there is no obvious reason why the ratios should be preserved, the utility of ratio-scale transformations between probabilities and possibilities is questionable.

2. *Difference Scales.* Difference-scale transformations use the form $r_i = p_i + \beta$ for all $i \in \mathbb{N}_n$, where β is a positive constant. From the normalization requirements (d) and (c), we readily obtain

$$r_i = 1 - (p_1 - p_i),$$

$$p_i = r_i - \frac{r_2 + r_3 + \cdots + r_n}{n}$$

for all $i \in \mathbb{N}_n$, respectively. These transformations are as rigid as the ratio-scale transformations and, consequently, their utility is severely limited for the same reasons.

3. *Interval Scales.* Interval-scale transformations are of the form $r_i = \alpha p_i + \beta$ for all $i \in \mathbb{N}_n$, where α and β are constant ($\alpha > 0$). Determining the value of β from the possibilistic normalization (d), we obtain

$$r_i = 1 - \alpha(p_1 - p_i) \tag{9.9}$$

for all $i \in \mathbb{N}_n$, and determining its value from the probabilistic normalization, we obtain

$$p_i = \frac{r_i - a_r}{\alpha} + \frac{1}{n} \tag{9.10}$$

for all $i \in \mathbb{N}_n$, where

$$a_r = \frac{1}{n} \sum_{i=1}^{n} r_i$$

is the average value of possibility profile **r**. To determine α, we can now apply the principle of uncertainty invariance by requiring that the amounts of uncertainty associated with **p** and **r** be equal. While the uncertainty in **p** is uniquely measured by the Shannon entropy $S(\mathbf{p})$, there are, at least in principle, three options for **r**: $GH(r)$, $GS(r)$, $\bar{S}(r)$. The most sensible choice seems to be GH, which is supported at least by the following three arguments: (1) GH is a natural generalization of the Hartley measure, which is the unique uncertainty measure in classical possibility theory; (2) in possibilistic bodies of evidence, nonspecificity (measured by GH) is more significant than conflict (measured by GS) and it dominates the overall uncertainty, especially for large sets of alternatives; and (3) the remaining option, $\bar{S}$, is overly insensitive. It is thus reasonable to express the requirement of uncertainty invariance by the equation

$$S(\mathbf{p}) = GH(\mathbf{r}). \tag{9.11}$$

This equation may help, in principle, to determine the value of α in Eqs. (9.9) and (9.10) for which the probability–possibility transformation, $\mathbf{p} \leftrightarrow \mathbf{r}$, is uncertainty-invariant in the given sense. However, it is known (see Note 9.5) that no value of α exists for the some possibility profiles that satisfy Eq. (9.11). It is also known that we do not encounter this limitation under a different, but closely connected type of scales—the log-interval scales. It is thus reasonable to abandon interval scales and formulate probability–possibility transformations in terms of the more satisfactory log-interval scales.

4. *Log-Interval Scales.* Log-interval scale transformations have the form $r_i = \beta p_i^{\alpha}$ for all $i \in \mathbb{N}_n$, where α and β are positive constants. Determining the value of β from the possibilistic normalization (d), we obtain

$$r_i = \left(\frac{p_i}{p_1}\right)^{\alpha} \tag{9.12}$$

for all $i \in \mathbb{N}_n$. The value of α is then determined by applying Eq. (9.11), which expresses the requirement of uncertainty invariance. This equation assumes the form

$$S(\mathbf{p}) = \sum_{i=2}^{n} \left(\frac{p_i}{p_1}\right)^{\alpha} \log_2 \frac{i}{i-1}, \tag{9.13}$$

where $S(\mathbf{p})$ is the Shannon entropy of the given probability distribution $\mathbf{p}$. After solving the equation (numerically) and applying the resulting value of α to Eq. (9.12), we obtain the desired possibility profile $\mathbf{r}$, one that is connected to $\mathbf{p}$ via a log-interval scale and contains the same amount of uncertainty as $\mathbf{p}$ in the sense of Eq. (9.11).

For the inverse transformation, from $\mathbf{r}$ to $\mathbf{p}$, we determine the value of β via the probabilistic normalization (c), and obtain

$$p_i = \frac{r_i^{1/\alpha}}{s} \tag{9.14}$$

for all $i \in \mathbb{N}_n$, where

$$s = \sum_{k=1}^{n} r_k^{1/\alpha}.$$

Equation (9.11) now assumes the form

$$GH(\mathbf{r}) = -\sum_{i=1}^{n} \frac{r_i^{1/\alpha}}{s} \log_2 \frac{r_i^{1/\alpha}}{s}, \tag{9.15}$$

where $GH(\mathbf{r})$ is the generalized Hartley measure of the given possibility profile $\mathbf{r}$. Solving this equation for α and applying the solution to Eq. (9.14) results in the sought probability distribution $\mathbf{p}$.

Figure 9.8 is a convenient overview of the procedures involved in probability–possibility transformation that are uncertainty-invariant (in the sense of Eq. (9.11)) and are based on log-interval scales.

EXAMPLE 9.13. Let $\mathbf{p} = \langle 0.7, 0.2, 0.075, 0.025 \rangle$. The uncertainty invariant log-interval scale transformation of $\mathbf{p}$ into $\mathbf{r}$ is determined as follows. First, we calculate $S(\mathbf{p}) = 1.2379$. Then, we solve Eq. (9.13), which assumes the form

$$1.2379 = 0.2857^{\alpha} + 0.585 \times 0.1071^{\alpha} + 0.415 \times 0.0357^{\alpha}.$$

We obtain (by a numerical method) $\alpha = 0.2537$. Using Eq. (9.12), we calculate components of the desired possibility profile $\mathbf{r} = \langle 1, 0.728, 0.567, 0.429 \rangle$.

EXAMPLE 9.14. Let $\mathbf{r} = \langle 1, 0.7, 0.5 \rangle$. To determine $\mathbf{p}$ connected to $\mathbf{r}$ via a log-interval scale and the principle of uncertainty invariance expressed by Eq. (9.15), we need to calculate $GH(\mathbf{r}) = 0.99281$ first. The equation then assumes the form

$$0.99281 = -\frac{1}{s}\log_2 \frac{1}{s} - \frac{0.7^{1/\alpha}}{s}\log_2 \frac{0.7^{1/\alpha}}{s} - \frac{0.5^{1/\alpha}}{s}\log_2 \frac{0.5^{1/\alpha}}{s},$$

where

$$s = 1 + 0.7^{1/\alpha} + 0.5^{1/\alpha}.$$

Solving this equation results in $\alpha = 0.262215$. Applying this value to Eq. (9.14), we readily obtain

$$\mathbf{p} = \langle 0.753173, 0.193264, 0.0535633 \rangle.$$

Observe that $S(\mathbf{p}) = 0.992482$, which is virtually the same as $GH(\mathbf{r})$, except for a tiny difference at the sixth decimal place due to rounding.

Let us now take the resulting probability distribution $\mathbf{p}$ as input and convert it to the associated possibility profile by the same variation of the principle of uncertainty invariance and log-interval scales. We now apply Eq. (9.13), which has the form

$$0.992482 = \left(\frac{p_2}{p_1}\right)^{\alpha} + \left(\frac{p_3}{p_1}\right)^{\alpha}\log_2 1.5.$$

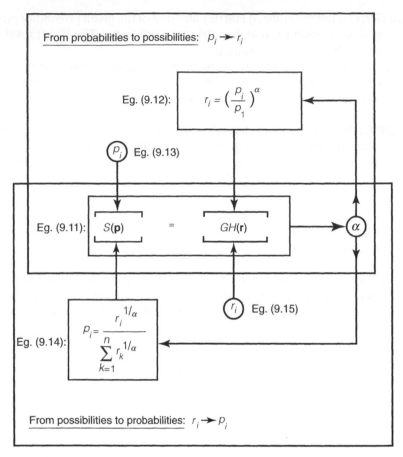

Figure 9.8. Uncertainty-invariant probability–possibility transformations based on log-interval scales.

Its solution is $\alpha = 0.262215$, which, as expected, is the same value as the one obtained for the inverse transformation $\mathbf{r} \to \mathbf{p}$. Applying this value to Eq. (9.12), we readily obtain

$$\mathbf{r} = \langle 1, 0.7, 0.5 \rangle,$$

which is the same as the initial possibility profile in this example. This again conforms to our expectation.

5. *Ordinal Scales.* Scales of one additional type are applicable to uncertainty-invariant transformations between probabilities and possibilities. These are known as *ordinal scales*. As the name suggests, ordinal scales are only required to preserve the ordering of components in the *n*-tuples $\mathbf{p}$ and $\mathbf{r}$. Their

form is $r_i = f(p_i)$ for all $i \in \mathbb{N}_n$, where f is a strictly increasing function. Contrary to uncertainty-invariant transformations $\mathbf{p} \leftrightarrow \mathbf{r}$ based on log-interval scales, those based on ordinal scales are in general not unique. This is not necessarily a disadvantage, since the additional degrees of freedom offered by ordinal scales allow us to satisfy various additional requirements.

EXAMPLE 9.15. As a simple example to illustrate probability–possibility transformations under ordinal scales, let the probability distribution

$$\mathbf{p} = \langle 0.4, 0.15, 0.15, 0.1, 0.1, 0.1 \rangle$$

be given and let the aim be to determine the corresponding possibility profile $\mathbf{r}$ by using the principle of uncertainty invariance under ordinal scales. Then,

$$\mathbf{r} = \langle r_1 = 1, a, a, b, b, b \rangle,$$

where $a = f(0.15)$ and $b = f(0.1)$. Now applying the principle of uncertainty invariance, we obtain the equation

$$\begin{aligned} S(\mathbf{p}) &= (a - b)\log_2 3 + b\log_2 6 \\ &= a\log_2 3 + b. \end{aligned}$$

This means that we still have one remaining degree of freedom. Choosing, for example, a as the free variable, we have

$$b = S(\mathbf{p}) - a \log_2 3. \tag{9.16}$$

Since it is required that $b \le a$, we obtain from the last equation the inequality

$$S(\mathbf{p}) - a\log_2 3 \le a,$$

which can be rewritten as

$$a \ge \frac{S(\mathbf{p})}{1 + \log_2 3}.$$

Since it is also required that $a \le 1$, the range of a is defined by the inequalities

$$\frac{S(\mathbf{p})}{1 + \log_2 3} \le a \le 1. \tag{9.17}$$

The set of all possibility profiles $\mathbf{r}$ for which

$$GH(\mathbf{r}) = S(\mathbf{p})$$

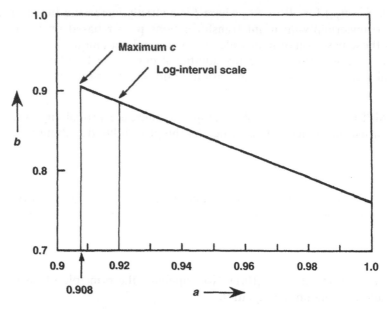

Figure 9.9. The sets of possibility profiles derived in Example 9.15.

is thus characterized by the set of all pairs that satisfy Eq. (9.16) within the restricted range of a defined by inequality (9.17). For the given probability distribution **p**, the set of possibility profiles **r** for which $GH(\mathbf{r}) = S(\mathbf{p}) = 2.346$ is characterized by equation

$$b = 2.346 - 1.585a,$$

where $a \in [0.908, 1]$. This set is shown in Figure 9.9. Also shown in the figure is the possibility profile based on log-interval scales and the one for which the probability–possibility consistency index c, defined for any **p** and **r** by the formula

$$c = \sum_{i=1}^{n} p_i r_i, \tag{9.18}$$

reaches its maximum. Observe that each value of a in the given range defines an ordinal scale characterized by a function f_a. For example, when $a = 0.95$, then $f_a(0.4) = 1, f_a(0.15) = 0.95$, and $f_a(0.1) = 0.84$; moreover $GH(f_a(\mathbf{p})) = 2.346$.

Now consider a generalization of Example 9.15, in which the given probability distribution contains d distinct values $(d \le n)$. Then, clearly, there are $d - 2$ free variables, since we have two equations for d distinct values in **r**: $r_1 = 1$ (possibilistic normalization) and $GH(\mathbf{r}) = S(\mathbf{p})$. In Example 9.15, $d = 3$, and hence, there is only one free variable. The next example illustrates a generalization to $d = 4$.

EXAMPLE 9.16. Given the probability distribution

$$\mathbf{p} = \langle 0.5, 0.1, 0.1, 0.1, 0.05, 0.05, 0.02, 0.02, 0.02, 0.02, 0.02 \rangle,$$

the corresponding possibility profiles,

$$\mathbf{r} = \langle r_i \mid i \in \mathbb{N}_{11} \rangle,$$

based on ordinal scales can be expressed in the form

$$\mathbf{r} = \langle b_1 = 1, b_2, b_2, b_2, b_3, b_3, b_4, b_4, b_4, b_4, b_4 \rangle$$

where b_1, b_2, b_3, b_4 represent distinct values of possibilities. Now, the principle of uncertainty invariance is expressed by the equation

$$S(\mathbf{p}) = (b_2 - b_3)\log_4 4 + (b_3 - b_4)\log_2 6 + b_4 \log_2 11.$$

One of the three variable (say b_4) can be expressed in terms of the other two variables, chosen as free variables. After calculating $S(\mathbf{p}) = 2.493$ and inserting it into the equation, we obtain

$$b_4 = 2.852 - 2.288 b_2 - 0.669 b_3.$$

The ordinal-scale transformation is characterized by the equalities

$$1 - b_2 \geq 0$$
$$b_2 - b_3 \geq 0$$
$$b_3 - b_4 \geq 0$$
$$b_4 \geq 0.$$

After expressing b_4 in the last two inequalities in terms of b_2 and b_3, we obtain the following set of four inequalities regarding the two free variables b_2 and b_3, which characterize the set of all possibility profiles whose nonspecificity is equal to $S(\mathbf{p})$:

$$1 - b_2 \geq 0 \qquad \text{(a)}$$
$$b_2 - b_3 \geq 0 \qquad \text{(b)}$$
$$2.288 b_2 + 1.669 b_3 - 2.852 \geq 0 \qquad \text{(c)}$$
$$-2.288 b_2 - 0.669 b_3 + 2.852 \geq 0. \qquad \text{(d)}$$

This set is illustrated visually by the shaded area in Figure 9.10. The probability–possibility index c defined by Eq. (9.18) reaches its maximum at $b_2 = b_3 =$

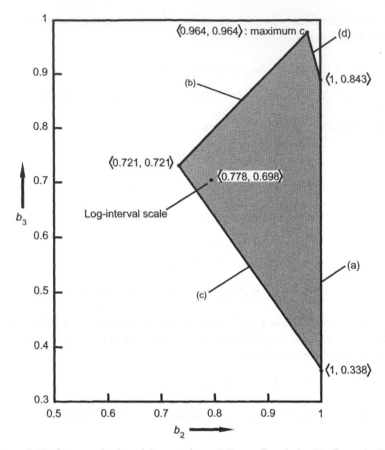

Figure 9.10. Characterization of the set of possibility profiles derived in Example 9.16.

0.964. The log-interval scaling transformation is obtained for $b_2 = 0.778$ and $b_3 = 0.698$ ($\alpha = 0.156122$).

The examined examples clearly demonstrate that uncertainty–invariant transformations from probabilities to possibilities under ordinal scales can be computed without any insurmountable difficulties. However, the complexity of these computations grows extremely rapidly with the number of distinct values in the given probability distributions. The opposite transformations from possibilities to probabilities are more difficult. The main obstacle is Eq. (9.11). It is, in these transformations, a nonlinear equation, that makes any symbolic treatment virtually impossible. Hence, we need to determine numerically all combinations of values of $d - 1$ unknowns (one of the unknowns is eliminated by the probabilistic normalization) that satisfy the equation and all the inequalities characterizing ordinal scales. More research is needed to deal with this particular problem.

9.5.3. Approximations of Belief Functions by Necessity Functions

In this section, the uncertainty–invariance principle is illustrated by the problem of approximating given belief functions of DST, by necessity functions of the theory of graded possibilities. The primary reason for pursuing these approximations is to reduce computational complexity. It is assumed that the uncertainty to be preserved is the total aggregated uncertainty $\bar{S}$. That is, the principle is expressed by the equation

$$\bar{S}(Bel) = \bar{S}(Nec), \tag{9.19}$$

where *Bel* and *Nec* denote, respectively, a given belief function and a necessity function that is supposed to approximate *Bel*. It is assumed that *Bel* and *Nec* are defined on $\mathcal{P}(X)$.

In addition to Eq. (9.19), it is also required that *Bel* and *Nec* satisfy the inequality

$$Nec(A) \le Bel(A) \tag{9.20}$$

for all $A \in \mathcal{P}(X)$. The motivation for using this requirement stems from the fact that belief functions are more general than necessity functions. Therefore, they are capable of expressing given evidence more precisely. This in turn, means that the set containment

$$[Bel(A), Pl(A)] \subseteq [Nec(A), Pos(A)]$$

should hold for all $A \in \mathcal{P}(X)$ when *Nec* approximates *Bel*. Functions *Bel* and *Nec* that are defined on the same universal set and satisfy inequality (9.20) are said to be *consistent*.

The approximation problem addressed in this section is thus formulated as follows: Given a belief function, *Bel*, determine a necessity function, *Nec*, that satisfies the requirements (9.19) and (9.20). The class of necessity functions that satisfy these requirements is characterized by the following theorem.

Theorem 9.1. Let *Bel* denote a given belief function on X (with $|X| = n$), and let $\{A_i\}_{i=1}^{s}$ (with $s \in \mathbb{N}_n$) denote the partition of X consisting of the sets A_i chosen by Step 1 of Algorithm 6.1 in the ith pass through the loop of Steps 1 to 5 during the computation of $\bar{S}(Bel)$. A necessity function *Nec* satisfies the requirements (9.19) and (9.20) if and only if

$$Nec(\textstyle\bigcup_{j=1}^{i} A_j) = Bel(\textstyle\bigcup_{j=1}^{i} A_j) \tag{9.21}$$

for each $i \in \mathbb{N}_s$.

Proof. [Harmanec and Klir, 1997]. ∎

We see from the theorem that, unless $s = n$ and $|A_i| = 1$ for all $i \in \mathbb{N}_s$, the solution to the approximation problem is not unique. The question is what criteria should be used to choose one particular approximation. Two criteria seem to be most natural. According to one of them, we choose the necessity function with the maximum nonspecificity; according to the other one, we choose the necessity function that is in some sense closest to the given belief function. Choices based on the first criterion are addressed by the following theorem.

Theorem 9.2. Among the necessity functions that satisfy Eq. (9.21) for given belief function *Bel*, the one that maximizes nonspecificity is given by the formula

$$Nec(B) = Bel\left(\bigcup_{j=1}^{i} A_j\right) \tag{9.22}$$

for all $B \in \mathcal{P}(X)$, where i is the largest integer such that $\bigcup_{j=1}^{i} A_j \subseteq B$, and it is assumed, by convention, that $\bigcup_{j=1}^{0} A_j = \varnothing$.

Proof. [Harmanec and Klir, 1997]. ∎

EXAMPLE 9.17. To illustrate the meaning of Eq. (9.22), consider the belief function *Bel* defined in Table 9.11. When applying Algorithm 6.1 to this belief function, we obtain

$$Bel(A_1) = Bel(\{a\}) = 0.3,$$
$$Bel(A_2) = Bel(\{b,c\}) = 0.5,$$
$$Bel(A_3) = Bel(\{d\}) = 0.2.$$

Hence, the *unique* necessity function, Nec_1, that approximates *Bel* (i.e., that satisfies requirements (9.19) and (9.20)) and maximizes nonspecificity is the one specified in Table 9.11, together with the associated functions Pos_1 and m_1. This approximation may conveniently be represented by the possibility profile

$$\mathbf{r}_1 = \langle 1, 0.7, 0.7, 0.2 \rangle.$$

In order to apply the second criterion operationally, we need a meaningful way of measuring the closeness of a necessity function to a given belief function. For this purpose, it is reasonable to use functional D_{Bel} defined by the formula

$$D_{Bel}(Nec) = \sum_{A \in \mathcal{P}(X)} [Bel(A) - Nec(A)] \tag{9.23}$$

for each necessity function *Nec* that is consistent with a given belief function *Bel*. We want to minimize $D_{Bel}(Nec)$ for all necessity functions that are consistent with *Bel* and satisfy Eq. (9.19) or, according to Theorem 9.1, satisfy Eq. (9.21).

Table 9.11. Necessity Approximations of a Belief Function (Examples 9.17 and 9.18)

	X			Given			Approximation 1			Approximation 2		
a	b	c	d	Bel(A)	Pl(A)	m(A)	Nec₁(A)	Pos₁(A)	m₁(A)	Nec₂(A)	Pos₂(A)	m₂(A)
A: 0	0	0	0	0.0	0.0	0.0	0.0	0.0	0.0	0.0	0.0	0.0
1	0	0	0	0.3	0.6	0.3	0.3	1.0	0.3	0.3	1.0	0.3
0	1	0	0	0.0	0.4	0.0	0.0	0.7	0.0	0.0	0.7	0.0
0	0	1	0	0.1	0.4	0.1	0.0	0.7	0.0	0.0	0.6	0.0
0	0	0	1	0.0	0.2	0.0	0.0	0.2	0.0	0.4	0.2	0.0
1	1	0	0	0.4	0.9	0.1	0.3	1.0	0.0	0.3	1.0	0.1
1	0	1	0	0.4	1.0	0.0	0.3	1.0	0.0	0.3	1.0	0.0
1	0	0	1	0.5	0.6	0.2	0.3	1.0	0.0	0.0	0.7	0.0
0	1	1	0	0.4	0.5	0.3	0.0	0.7	0.0	0.0	0.7	0.0
0	1	0	1	0.0	0.6	0.0	0.0	0.7	0.0	0.0	0.6	0.0
0	0	1	1	0.1	0.6	0.0	0.0	0.7	0.0	0.0	1.0	0.4
1	1	1	0	0.8	1.0	0.0	0.8	1.0	0.5	0.8	1.0	0.0
1	1	0	1	0.6	0.9	0.0	0.3	1.0	0.0	0.4	1.0	0.0
1	0	1	1	0.6	1.0	0.0	0.3	1.0	0.0	0.3	1.0	0.0
0	1	1	1	0.4	0.7	0.0	0.0	0.7	0.0	0.0	0.7	0.0
1	1	1	1	1.0	1.0	0.0	1.0	1.0	0.2	1.0	1.0	0.2

Observe that Eq. (9.23) may be rewritten as

$$D_{Bel}(Nec) = \sum_{A \in \mathcal{P}(X)} Bel(A) - \sum_{A \in \mathcal{P}(X)} Nec(A).$$

Since $\sum_{A \in \mathcal{P}(X)} Bel(A)$, $Nec(\emptyset)$, and $Nec(X)$ are constants, minimizing D_{Bel} is equivalent to maximizing the expression

$$\sum_{A \in \mathcal{P}(X) - \{\emptyset, X\}} Nec(A) \tag{9.24}$$

over all necessity functions Nec that satisfy Eq. (9.21).

EXAMPLE 9.18. Considering again the belief function defined in Table 9.11, let Nec_2 denote a necessity function that satisfies Eq. (9.21) and maximizes the expression (9.24). To satisfy Eq. (9.21) in this case means that $Nec_2(\{a\}) = 0.3$, $Nec_2(\{a, b, c\}) = 0.8$, and $Nec_2(X) = 1$. Due to the nested structure of focal subsets in possibility theory, $Nec_2(\{b, c\}) = 0$ and either $Nec_2(\{a, b\}) = 0.3$ and $Nec_2(\{a, c\}) \in [0.3, 0.4]$ or $Nec_2(\{a, b\}) \in [0.3, 0.4]$ and $Nec_2(\{a, c\}) = 0.3$. The maximum of expression (9.24) is clearly obtained for either $Nec_2(\{a, c\}) = 0.4$ and $Nec_2(\{a, b\}) = 0.3$ or $Nec_2(\{a, b\}) = 0.4$ and $Nec_2(\{a, c\}) = 0.3$. There are thus two minima of D_{Bel} in this case; the second one is shown in Table 9.11.

9.5.4. Transformations Between λ-Measures and Possibility Measures

A common property of λ-measures and possibility measures is that they are fully represented by their values on singletons. It is thus reasonable to consider uncertainty–invariant transformations between λ-measures and possibility measures by connecting the associated values on singletons via scales of some type.

Considering a universal set X with n elements, let

$$\mathbf{r} = \langle r_i \mid i \in \mathbb{N}_n \rangle$$

denote an ordered possibility profile on X with $r_i \geq r_{i+1}$ for all $i \in \mathbb{N}_{n-1}$, and let

$$^{\lambda}\boldsymbol{\mu} = \langle {}^{\lambda}\mu_i \mid i \in \mathbb{N}_n \rangle$$

denote the associated profile of a λ-measure on X with $\lambda < 0$ (i.e., the λ-measure represents an upper probability function) and $^{\lambda}\mu_i \geq {}^{\lambda}\mu_{i+1}$ for all $i \in \mathbb{N}_{n-1}$. Then, components of $\mathbf{r}$ and $^{\lambda}\boldsymbol{\mu}$ can be connected by scales of some type.

As an illustration of these transformations, let us examine transformations from $^{\lambda}\boldsymbol{\mu}$ to $\mathbf{r}$ under log-interval scales and under the requirement

$$GS(^{\lambda}\boldsymbol{\mu}) = GH(\mathbf{r}). \tag{9.25}$$

In this framework,

$$r_i = \left(\frac{^\lambda \mu_i}{^\lambda \mu_1} \right)^o \tag{9.26}$$

for all $i \in \mathbb{N}_n$. These equations are derived in a way similar to Eq. (9.12) for transformations from probabilities to possibilities. Equation (9.25) assumes the form

$$GS(^\lambda \mu) = \sum_{i=2}^{n} \left(\frac{^\lambda \mu_i}{^\lambda \mu_1} \right)^\alpha \log_2 \frac{i}{i-1}, \tag{9.27}$$

where $GS(^\lambda \mu)$ and $^\lambda \mu_i$ for all $i \in \mathbb{N}_n$ are given and the scaling parameter α is to be determined. Once the value of α is determined, components of $\mathbf{r}$ are calculated by Eq. (9.26).

EXAMPLE 9.19. Let $^\lambda \mu = \langle 0.576, 0.35, 0.32, 0.1 \rangle$. Then, $\lambda = -0.625$ and the λ-measure, which represents an upper probability function, is fully determined by the λ-rule. For this λ-measure, we obtain $\bar{S}(^\lambda \mu) = 1.890$, $GH(^\lambda \mu) = 0.329$, and $GS(^\lambda \mu) = 1.561$ (Exercise 9.19). Equation (9.27) has the form

$$1.561 = \left(\frac{0.35}{0.576} \right)^\alpha + \left(\frac{0.32}{0.576} \right)^\alpha + \left(\frac{0.1}{0.576} \right)^\alpha.$$

The solution is $\alpha = 0.333$. Substituting this value into Eq. (9.26), we readily obtain

$$\mathbf{r} = \langle 1, 0.847, 0.822, 0.558 \rangle.$$

Clearly, $\bar{S}(\mathbf{r}) = 2$, $GH(\mathbf{r}) = 1.560$ (the small difference from $GS(^\lambda \mu)$ at the third decimal place is due to rounding errors), and $GS(\mathbf{r}) = 0.440$.

The associated pairs of λ-measures and possibility measures, whose profiles are connected via log-interval scales and satisfy Eq. (9.25), may be viewed as complementary representations of uncertainty. While the λ-measure in each pair represents uncertainty primarily in terms of GS, the associated possibility measure represents it primarily in terms of GH.

The inverse transformations, from possibility measures to λ-measures, are more difficult, primarily due to the difficulty of formulating Eq. (9.25). They require further research.

9.5.5. Approximations of Graded Possibilities by Crisp Possibilities

Interesting applications of the principle of uncertainty invariance are approximations of graded possibilities by crisp possibilities. These approximations are especially useful when graded possibilities are interpreted in terms of fuzzy sets. Approximating fuzzy sets that result from approximate reasoning by crisp sets is often desirable, since the latter are easier to comprehend. A crisp approximation of a fuzzy set also can be utilized as an intermediary step in its defuzzification.

Recall that each basic function r of graded possibilities on X is associated with a family of basic functions of crisp possibilities, $\{^\alpha r \mid \alpha \in (0, 1]\}$, where

$$^\alpha r(x) = \begin{cases} 1 & \text{when } r(x) \geq \alpha \\ 0 & \text{otherwise} \end{cases}$$

for all $x \in X$. In general, any function in this family may be taken as a crisp approximation of r. However, according to the principle of uncertainty invariance, we should take the one for which the amount of nonspecificity is the same as for r. When X is finite, this requirement is expressed by the equation

$$\int_0^1 \log_2 |^\alpha r| d\alpha = \log_2 |^\alpha r|, \tag{9.28}$$

where $|^\alpha r|$ denotes the number of possible alternatives according to function $^\alpha r$. When X is infinite, the requirement is expressed by the alternative equation

$$\int_0^1 UL(^\alpha r) \, d\alpha = UL(^\alpha r), \tag{9.29}$$

where UL is defined by Eq. (6.26).

EXAMPLE 9.20. To illustrate the application of Eq. (9.28), consider function r defined in Figure 6.1, whose nonspecificity is 2.99 (according to the calculation in Example 6.1). Hence, Eq. (9.28) has the form

$$2.99 = \log_2 |^\alpha r|$$

and we need to solve it for $^\alpha r$. Values of $\log_2|^\alpha r|$ for all values α that are distinguished in this example are shown in Table 9.12. We can see that none of the values is exactly equal to 2.99 (due to discreteness of function r) and there are two of them that are equally close to 2.99 (for $\alpha = 0.6$ and 0.7) which are indicated in Table 9.12 in boldface.

EXAMPLE 9.21. To illustrate the application of Eq. (9.29), first consider the case of $X = \mathbb{R}$. Let

Table 9.12. Illustration to Example 9.21

α	$|^{\alpha}r|$	$\log_2|^{\alpha}r|$
0.1	15	3.91
0.3	12	3.59
0.4	11	3.46
0.6	**9**	**3.17**
0.7	**7**	**2.81**
0.9	5	2.32
1.0	3	1.58

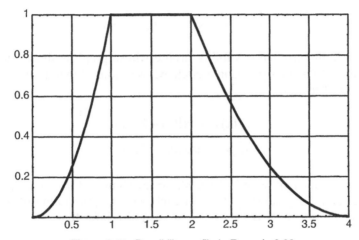

Figure 9.11. Possibility profile in Example 9.22.

$$r(x)\begin{cases} x^2 & \text{when } x \in [0, 1) \\ 1 & \text{when } x \in [1, 2) \\ [2 - (x/2)]^2 & \text{when } x \in (2, 4) \\ 0 & \text{otherwise} \end{cases}$$

be a given possibility profile whose graph is shown in Figure 9.11. Then,

$$^{\alpha}r = [\sqrt{\alpha}, 4 - 2\sqrt{\alpha}]$$

and

$$\int_0^1 UL(^{\alpha}r)\, d\alpha = \int_0^1 \log_2(5 - 3\sqrt{\alpha})\, d\alpha$$
$$= 1.546.$$

Hence, Eq. (9.29) has the form

$$1.546 = \log_2(5 - 3\sqrt{\alpha}),$$

and its solution is $\alpha = 0.481$. The crisp approximation of r is thus the closed interval

$$[\sqrt{0.481}, 4 - 2\sqrt{0.481}] = [0.694, 2.613].$$

EXAMPLE 9.22. This example illustrates the application of Eq. (9.29) when r is defined on $\mathbb{R}^2$. Let

$$r(x, y) = \max\{0, 1 - 2(x^2 + y^2)\}$$

be a given 2-dimensional possibility profile whose graph is shown in Figure 9.12a. Clearly $^\alpha r$ defines points in a circle whose radius depends on α. Projections of r are functions

$$r_X(x) = \max\{0, 1 - 2x^2\}$$
$$r_Y(y) = \max\{0, 1 - 2y^2\}.$$

Their graphs, which are identical, are shown in Figure 9.12b. From r_X (as well as r_Y), we can readily conclude that for each $\alpha \in (0, 1]$ the radius of the circle representing points of $^\alpha r$ is $\sqrt{(1-\alpha)/2}$. We can also infer that

$$^\alpha r_X = {}^\alpha r_Y = \left[-\sqrt{\frac{1-\alpha}{2}}, \sqrt{\frac{1-\alpha}{2}} \right].$$

Using all these facts, we have

$$UL(^\alpha r) = \log_2\left[1 + 4\sqrt{\frac{1-\alpha}{2}} + \pi\left(\frac{1-\alpha}{2}\right) \right]$$

and

$$\int_0^1 UL(^\alpha r)\, d\alpha = 1.796.$$

Hence, Eq. (9.29) assumes the form

$$1.796 = \log_2\left[1 + 4\sqrt{\frac{1-\alpha}{2}} + \pi\left(\frac{1-\alpha}{2}\right) \right],$$

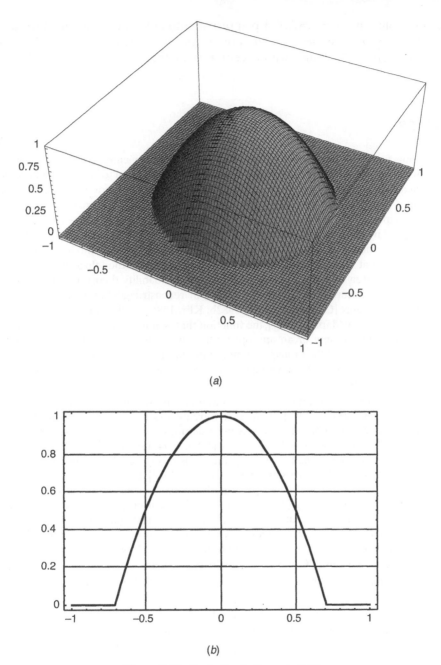

(a)

(b)

Figure 9.12. Illustration to Example 9.22.

and its solution is $\alpha = 0.585$. Crisp approximation of r is thus obtained for $\alpha = 0.585$. This is the set of all points in the circle whose radius is $\sqrt{(1 - 0.585)/2} = 0.456$ and whose center is in the origin of the coordinate system.

NOTES

9.1. Thus far, the principle of minimum uncertainty has been employed predominantly within the domain of probability theory, where the function to be minimized is the Shannon entropy (usually some of its conditional forms) or another function based on it (information transmission, directed divergence, etc.). The great utility of this principle in dealing with a broad spectrum of problems is perhaps best demonstrated by the work of Christensen [1980–81, 1985, 1986]. Another important user of the principle of minimum entropy is Watanabe [1981, 1985], who has repeatedly argued that entropy minimization is a fundamental methodological tool in the problem area of pattern recognition. Outside probability theory, the principle of minimum uncertainty has been explored in reconstructability analysis of possibilistic systems [Cavallo and Klir, 1982b; Klir, 1985, 1990b; Klir et al., 1986, Klir and Way, 1985; Mariano, 1997]; the function that is minimized in these explorations is the U-uncertainty or an appropriate function based on it [Higashi and Klir, 1983b]. The use of the principle of minimum uncertainty for resolving local inconsistencies in systems was investigated for probabilistic and possibilistic systems by Mariano [1985, 1987].

9.2. The principle of maximum entropy was founded, presumably, by Jaynes in the early 1950s [Rosenkrantz, 1983]. Perhaps the greatest skill in using this principle in a broad spectrum of applications has been demonstrated by Christensen [1980–81, 1985, 1986], Jaynes [1979, 2003], Kapur [1989, 1994, 1994/1996], and Tribus [1969]. The literature concerned with this principle is extensive. The following are a few relevant books of special significance: Batten [1983], Buck and Macaulay [1991], Kapur and Kesavan [1987, 1992], Karmeshu [2003], Levine and Tribus [1979], Theil [1967], Theil and Fiebig [1984], Webber [1979], Wilson [1970]. A rich literature resource regarding research on the principle of maximum entropy, both basic and applied, are edited volumes *Annual Workshops on Maximum Entropy and Bayesian Methods* that have been published since the 1980s by several publishers, among them Kluwer, Cambridge University Press, and Reidel.

9.3. The principle of maximum entropy has been justified by at least three distinct arguments:

1. The maximum entropy probability distribution is the only *unbiased distribution*, that is, the distribution that takes into account all available information but no additional (unsupported) information (bias). This follows directly from the facts that

   a. All available information (but nothing else) is required to form the constraints of the optimization problem, and

   b. The chosen probability distribution is required to be the one that represents the maximum uncertainty (entropy) within the constrained set of probability distributions. Indeed, any reduction of uncertainty is an equal gain of information. Hence, a reduction of uncertainty from its maximum value, which would occur when any distribution other than the one with maximum entropy were chosen would mean that some information from outside the available evidence was unwittingly added.

This argument of justifying the maximum entropy principle is covered in the literature quite extensively. Perhaps its best and most thorough presentation is given in a paper by Jaynes [1979], which also contains an excellent historical survey of related developments in probability theory, and in a book by Christensen [1980–81, Vol. 1].

2. It was shown by Jaynes [1968], strictly on combinatorial grounds, that the maximum probability distribution is the *most likely distribution* in any given situation.

3. It was demonstrated by Shore and Johnson [1980] that the principle of maximum entropy can be deductively derived from the following *consistency axioms* for inductive (or ampliative) reasoning:

Axiom (ME1) Uniqueness. The result should be unique.

Axiom (ME2) Invariance. The choice of coordinate system (permutation of variables) should not matter.

Axiom (ME3) System Independence. It should not matter whether one accounts for independent systems separately in terms of marginal probabilities or together in terms of joint probabilities.

Axiom (ME4) Subset Independence. It should not matter whether one treats an independent subset of system states in terms of separate conditional probabilities or together in terms of joint probabilities.

The rationale for choosing these axioms is expressed by Shore and Johnson [1980] as follows: Any acceptable method of inference must be such that different ways of using it to take the same information into account lead to consistent results. Using the axioms, they derive the following proposition: Given some information in terms of constraints regarding the probabilities to be estimated, there is only one probability distribution satisfying the constraints that can be chosen by a method that satisfies the consistency axioms; this unique distribution can be attained by maximizing the Shannon entropy (or any other function that has exactly the same maxima as the entropy) subject to the given constraints. Alternative derivations of the principle of maximum entropy were demonstrated by Smith [1974], Avgers [1983], and Paris and Vencovská [1990].

The principle of minimum cross-entropy can be justified by similar arguments. In fact, Shore and Johnson [1981] derived both principles and showed that the principle of maximum entropy is a special case of the principle of minimum cross-entropy. The latter principle is further examined by Williams [1980], who shows that it generalizes the well-known Bayesian rule of conditionalization. A broader range of applications of the principle of minimum cross-entropy was developed by Rubinstein and Kroese [2004].

In general, the principles of maximum entropy and minimum cross-entropy as well as the various principles of uncertainty that extend beyond probability theory are tools for dealing with a broad class of problems referred to as *inverse problems* or *ill-posed problems* [McLaughlin, 1984; Tarantola, 1987]. A common characteristic of these problems is that they are *underdetermined* and, consequently, do not have unique solutions. The various maximum uncertainty principles allow us to obtain unique solutions to underdetermined problems by injecting uncertainty of some type to each solution to reflect the lack of information in the formulation of the problem.

As argued by Bordley [1983], the assumption that "the behavior of nature can be described in such a way as to be consistent" (so-called *consistency premise*) is essential for science to be possible. This assumption, when formulated in a particular context in terms of appropriate consistency axioms, leads to an optimization principle. According to Bordley, the various optimization principles, which are exemplified by the principles of maximum entropy and minimum cross-entropy, are central to science.

9.4. The principle of maximum nonspecificity was first conceived by Dubois and Prade [1987b], who also demonstrated its significance in approximate reasoning based on possibility theory [Dubois and Prade, 1991]. Applications of the principle were developed in the area of image processing by Wierman [1994] and in the area of fuzzy control by Padet [1996].

9.5. The principle of uncertainty invariance was introduced in the context of probability–possibility transformations by Klir [1989b, 1990a]. These transformations were mathematically investigated by Geer and Klir [1992] and Harmanec and Klir [1997] for the discrete case, and by Wonneberger [1994] for the continuous case of n dimensions. It follows from the work of Geer and Klir [1992] that the existence of transformations based on the principle of uncertainty invariance expressed by Eq. (9.11) is not guaranteed under interval scales, but it is guaranteed under log-interval scales. Probability–possibility transformations were also investigated experimentally (by a series of simulation experiments) and compared with other probability–possibility transformations in terms of several criteria [Klir and Parviz, 1992]. Approximations of belief measures by necessity measures by preserving $\bar{S}$ were studied by Harmanec and Klir [1997].

9.6. The material presented in Section 9.5.3 is based on a paper by Harmanec and Klir [1997]. Many details presented in the paper are omitted in Section 9.5.3, in particular proofs of theorems and an efficient algorithm to facilitate relevant computations.

9.7. One methodological area that exemplifies the utility of the principles of uncertainty discussed in this chapter is known in the literature as *reconstructability analysis* (RA). The purpose of RA is to deal in information–theoretic terms with two broad classes of problems that are associated with the relationship between overall systems and their various subsystems: *identification problems* and *reconstruction problems*. In identification problems, a set of subsystems is given and the aim is to make meaningful inference about the associated overall system from information in the subsystems and, possibly, some additional background information. This may involve the use of the principle of minimum uncertainty to resolve local inconsistencies among the given subsystems, the use of the principle of maximum uncertainty to identify one particular overall system that is formalized within the same theory as the subsystems, or the use of the principle of requisite generalization by identifying the family of all overall systems that are consistent with the given information, and moving thus to a more general theory that is capable of representing this family. In reconstruction problems, an overall system is given and the aim is to investigate, in a systematic way, which sets of subsystems at each complexity level preserve most of the information contained in the overall systems. This requires determining how accurately the overall system can be reconstructed from each considered sets of subsystems.

Reconstruction problems and identification problems of RA were first recognized in terms of n-dimensional relations (i.e., within the classical possibility theory) by Ashby [1964] and Madden and Ashby [1972], respectively. They were further investigated within probability theory by Klir [1976, 1985, 1986a,b, 1990b], Broekstra [1976–77], Cavallo and Klir [1979, 1981, 1982a], Jones [1982, 1985, 1986], and Conant [1988], and within the theory of graded possibilities by Cavallo and Klir [1982b], Higashi et al. [1984], and Klir et al. [1986].

The few references on RA given in the previous paragraph cover only some of the early, historically significant contributions to RA. The literature on RA is now too extensive to be covered here more completely. However, RA is a broad and important application area of GIT and, therefore, it is appropriate to refer at least to key sources of further information on RA:

- **Overview articles of RA:** [Klir and Way, 1985; Pittarelli, 1990; Zwick, 2004].
- **Special Issues on RA:**
  —*International Journal of General Systems,* **7**(1), 1981, pp. 1–107;
  —*International Journal of General Systems,* **29**(3), 2000, pp. 357–495;
  —*Kybernetes,* **33**(5–6), 2004, pp. 873–1062.
- **Bibliographies of RA:** *International Journal of General Systems,* **7**(1), 1981; **24**(1–2), 1996.

EXERCISES

9.1. Consider the relation defined in Table 2.2 and assume that variables x_1, x_2, x_3 are input variables and variable x_4 is an output variable. Using the principle of minimum uncertainty, determine which of the three input variables can be excluded to lose the least amount of information.

9.2. Consider the probabilistic systems in Table 3.3 and assume in each case that x and y are input variables and z is an output variable. Using the principle of minimum uncertainty, determine which of the two input variables is preferable to exclude in each case.

9.3. Consider the joint possibility profile in Figure 5.5a and assume that X and Y are states of an input variable and an output variable, respectively. Using the principle of minimum uncertainty, determine the best quantization of the input variable with:

(a) Four quantized states;

(b) Three quantized states.

9.4. Let a finite set of states contains n states, where $n \geq 2$. How many meaningful quantizations exist for each $n = 2, 3, \ldots, 10$, provided that:

(a) The states are totally ordered;

(b) The states are not ordered.

9.5. Let sets X and Y in Table 9.13 denote, respectively, state sets of an input variable x and an output variable y. The dependence of the output variable on the input variable is expressed in Table 9.13a, b, c in terms of crisp possibilities, graded possibilities, and probabilities, respectively. Using the principle of minimum uncertainty, determine in each case the set of all admissible simplifications obtained by quantizing states of the input variable, provided that

(a) The states are totally ordered;

(b) The states are not ordered.

9.6. Using the principle of minimum uncertainty, resolve the local inconsistency of the two probabilistic subsystems defined in Table 9.14 that form an overall system with variables x, y, z.

9.7. Consider a finite random variable x that takes values in the set X. It is known that the mean (expected) value of the variable is $E(x)$. Estimate, using the maximum entropy principle, the probability distribution on X provided that:

(a) $X = \mathbb{N}_2$ and $E(x) = 1.7$ (a biased coin if 1 and 2 represent, for example, the head and the tail, respectively);

(b) $X = \mathbb{N}_6$ and $E(x) = 3$ (a biased die);

(c) $X = \mathbb{N}_{10}$ and $E(x) = 4$;

(d) $X = \mathbb{N}_{10}$ and $E(x) = 6$.

Table 9.13. Illustration to Exercise 9.5

$r(x,y)$		Y			$r(x,y)$		Y			$p(x,y)$		Y	
		0	1	2			0	1	2			0	1	2
X	0	1	0	1	X	0	0.8	0.2	1.0	X	0	0.02	0.05	0.10
	1	1	0	0		1	1.0	0.4	0.2		1	0.30	0.10	0.20
	2	1	1	0		2	1.0	0.8	0.3		2	0.01	0.01	0.01
	3	0	0	1		3	0.0	0.5	1.0		3	0.00	0.00	0.20

(a)	(b)	(c)

Table 9.14. Illustration to Exercise 9.6

x	y	$^1p(x,y)$	y	z	$^2p(y,z)$
0	0	0.4	0	0	0.3
0	1	0.3	0	1	0.2
1	0	0.2	1	0	0.4
1	1	0.1	1	1	0.1

9.8. Use the maximum entropy principle to derive a formula for the joint probability distribution function, p, under the assumption that only marginal probability distributions, p_X and p_Y, on finite sets X and Y are known.

9.9. Consider a universal set X, four subsets of which are of interest to us: $A \cap B$, $A \cap C$, $B \cap C$, and $A \cap B \cap C$. The only evidence we have is expressed in terms of DST by the equations

$$m(A \cap B) + m(A \cap B \cap C) = 0.2$$
$$m(A \cap C) + m(A \cap B \cap C) = 0.5$$
$$m(B \cap C) + m(A \cap B \cap C) = 0.1$$

Using the maximum nonspecificity principle, estimate the values of $m(A \cap B)$, $m(A \cap C)$, $m(B \cap C)$, and $m(A \cap B \cap C)$, and $m(X)$.

9.10. Repeat Example 9.8 with the following numerical values:
   (a) $x_1 = 0.5$; $x_2 = 0.2$; $y_1 = 0.6$; $y_2 = 0.14$
   (b) $x_1 = 0.6$; $x_2 = 0.7$; $y_1 = 0.4$; $y_2 = 0.84$
   (c) $x_1 = 0.9$; $x_2 = 0.8$; $y_1 = 0.6$; $y_2 = 0.96$

9.11. Construct the cylindric closure in Example 9.9 by using the operation of relational join defined by Eq. (7.34).

9.12. Show that the function μ defined in Table 9.9 is a monotone and superadditive measure, but it is not 2-monotone. Identify all its violations of 2-monotonicity.

9.13. For each of the following probability distributions, $\mathbf{p}$, determine the corresponding possibility profile, $\mathbf{r}$, by the transformations summarized in Figure 9.8:
   (a) $\mathbf{p} = \langle 0.5, 0.3, 0.2 \rangle$
   (b) $\mathbf{p} = \langle 0.3, 0.2, 0.2, 0.1, 0.1, 0.1 \rangle$
   (c) $\mathbf{p} = \langle 0.30, 0.20, 0.15, 0.10, 0.10, 0.05, 0.03, 0.03, 0.02, 0.02 \rangle$
   (d) $\mathbf{p} = \langle 0.2, 0.2, 0.2, 0.2, 0.2 \rangle$

9.14. For each of the following possibility profiles, $\mathbf{r}$, determine the corresponding probability distribution, $\mathbf{p}$, by the transformations summarized in Figure 9.8:
   (a) $\mathbf{r} = \langle 1, 0.8, 0.4 \rangle$
   (b) $\mathbf{r} = \langle 1, 1, 0.6, 0.6, 0.2, 0.2, 0.2 \rangle$
   (c) $\mathbf{r} = \langle 1, 0.9, 0.7, 0.65, 0.6, 0.5, 0.35, 0.2 \rangle$
   (d) $\mathbf{r} = \langle 1, 1, 1, 1, 1 \rangle$

9.15. For some cases in Exercises 9.13 and 9.14, show the inverse transformation, which should result in the same tuple from which you started.

9.16. Show that the maximum of the probability–possibility consistency index c defined by Eq. (9.18) is obtained for:

(a) $a = 0.908$ in Example 9.15

(b) $b_2 = b_3 = 0.964$ in Example 9.16

9.17. Determine which of the possibility profiles in Examples 9.15 and 9.16 are based on log-interval scale.

9.18. Apply the principle of uncertainty invariance expressed by Eq. (9.11) to transform each of the following probability distributions to the corresponding possibility profile via ordinal scales:

(a) $\mathbf{p} = \langle 0.5, 0.3, 0.2 \rangle$

(b) $\mathbf{p} = \langle 0.3, 0.2, 0.2, 0.1, 0.1, 0.1 \rangle$

(c) $\mathbf{p} = \langle 0.2, 0.2, 0.1, 0.1, 0.1, 0.1, 0.1, 0.05, 0.05 \rangle$

(d) $\mathbf{p} = \langle 0.4, 0.3, 0.2, 0.1 \rangle$

9.19. Determine the λ-measure $^\lambda\mu$ in Example 9.19 and calculate the values of $\bar{S}(^\lambda\mu)$, $GH(^\lambda\mu)$, and $GS(^\lambda\mu)$.

9.20. Using the principle of uncertainty invariance, determine the crisp approximation of the following basic possibility function r defined on $\mathbb{R}$:

(a) $r(x) = 1.25 \max\{0, 2^{-|3(x-3)|} - 0.2\}$

(b) $r(x) = \max\{0, \min\{1, 2.3x - x^2\}\}$

(c) $r(x) = \max\{0, 0.2(x - 50) - 0.1(x - 50)^2\}$

(d) $r(x) = 2x - x^2$

$$(\mathbf{e})\ r(x) = \begin{cases} x/2 & \text{when } x \in [0, 2) \\ 3 - x & \text{when } x \in [2, 2.5) \\ 0.5 & \text{when } x \in [2.5, 4) \\ (5 - x)/2 & \text{when } x \in [4, 5) \\ 0 & \text{otherwise} \end{cases}$$

9.21. Using the principle of uncertainty invariance, determine the crisp approximation of the following basic possibility functions r defined on $\mathbb{R}^2$:

(a) $r(x) = e^{-x^2 - y^2}$

(b) $r(x) = \max\{0, \min\{1, 1.5 - x^2 - y^2\}\}$

10

CONCLUSIONS

To be uncertain is uncomfortable,
but to be certain is ridiculous.

—Chinese Proverb

10.1. SUMMARY AND ASSESSMENT OF RESULTS IN GENERALIZED INFORMATION THEORY

A turning point in our understanding of the concept of uncertainty was reached when it became clear that there are several types of uncertainty. This new insight was obtained by examining uncertainty emerging from mathematical theories more general than classical set theory and classical measure theory. Recognizing more than one type of uncertainty raised numerous questions regarding the nature and scope of the very concept of uncertainty and its connection with the concept of information. Studying these questions evolved eventually into a new area of research, which became known in the early 1990s as *generalized information theory* (GIT).

The principal aims of GIT have been threefold: (1) to liberate the notions of uncertainty and uncertainty-based information from the narrow confines of classical set theory and classical measure theory; (2) to conceptualize a broad framework within which any conceivable type of uncertainty can be characterized; and (3) to develop theories for the various types of uncertainty that emerge from the framework at each of the four levels: formalization, calculus, measurement, and methodology. Undoubtedly, these are long-term aims, which

Uncertainty and Information: Foundations of Generalized Information Theory, by George J. Klir
© 2006 by John Wiley & Sons, Inc.

may perhaps never be fully achieved. Nevertheless, they serve well as a blueprint for a challenging, large-scale research program. Significant results emerging from this research program can readily be identified by scanning material covered in this book, especially in Chapters 4–6, 8, and 9. They include:

(a) Calculi of several nonclassical theories of uncertainty are now well developed. These are theories based on generalized (graded) possibility measures, Sugeno λ-measures, Choquet capacities of order ∞, and reachable interval-valued probability distributions.

(b) The nonclassical theories listed in part (a) have also been viewed and studied as special theories subsumed under a more general theory of imprecise probabilities. From this point of view, some common properties of the theories are recognized and utilized. They all are based on a pair of dual measures—lower and upper probabilities—but may also be represented in terms of closed and convex sets of probability distributions or functions obtained by the Möbius transform of lower or upper probabilities. Numerous global results, not necessarily restricted to the special theories listed in part (a), have already been obtained for imprecise probabilities.

(c) The Hartley and Shannon functionals for measuring the amount of uncertainty in classical theories of uncertainty have been adequately generalized not only to the special theories listed in part (a) but also to other theories dealing with imprecise probabilities.

(d) Only some limited efforts have been made thus far to fuzzify the various uncertainty theories. They include a fuzzification of classical probabilities to fuzzy events, a fuzzification of the theory based on reachable interval-valued probability distributions, several distinct fuzzifications of the Dempster–Shafer theory, and the fuzzy-set interpretation of the theory of graded possibilities.

(e) Some limited results have been obtained for formulating and using the principles of minimum and maximum uncertainty within the various nonclassical uncertainty theories. Two new principles emerged from GIT: the principle of requisite generalization and the principle of uncertainty invariance. Some of their applications are examined in Sections 9.4 and 9.5, respectively.

Among the results summarized in the previous paragraphs, those listed under parts (a)–(c) are the most significant and satisfactory. That is, we have now several well-developed nonclassical uncertainty theories, which, in addition, are integrated into a more general theory of imprecise probabilities. Furthermore, we have well-justified generalizations of the classical Hartley and Shannon functionals for measuring amounts of uncertainty of the two types that coexist in each of the nonclassical theories: nonspecificity and conflict.

On the other hand, the results listed under parts (d) and (e) are considerably less satisfactory. Thus far, fuzzifications of uncertainty theories have been very limited. It was proposed in an ad hoc fashion in most cases, and involved only standard fuzzy sets. Similarly underdeveloped are the various principles of uncertainty that are examined in Chapter 9. These principles are dependent on the capability of measuring uncertainty. Unfortunately, this capability became available only recently, when the well-justified generalizations of the Hartley and Shannon functionals were established for the various theories of imprecise probabilities. (Recall the difficulties in generalizing the Shannon functional, which are discussed in Chapter 6.)

10.2. MAIN ISSUES OF CURRENT INTEREST

It seems reasonable to expect that research on the various principles of uncertainty, which are briefly examined in Chapter 9, will dominate the area of GIT in the near future. This research will have to address a broad spectrum of questions. Some of them will be conceptual. For example, which of the measures, from among GH, GS, $\bar{S}$, $TU = \langle GH, GS \rangle$, or $^aTU = \langle \bar{S} - \underline{S}, \underline{S} \rangle$—should be used when one of the principles is applied to problems of a given type? Other questions will undoubtedly involve computational issues. For example, can an existing computational method be adapted for applications of a given principle to problems of a given type? In the case that the answer is negative, a substantial research effort will be needed to develop a fitting algorithm. Since it is not likely that many algorithms will be found that could be easily adapted for problems in this area, it is reasonable to expect that work on designing new algorithms will dominate research on uncertainty principles in the near future.

Since there is growing interest in utilizing linguistic information, we can expect some efforts in the near future to fuzzify the existing uncertainty theories in a more systematic way. It is not likely, however, that these efforts will involve nonstandard fuzzy sets.

There are also some theoretical issues of current interest. One of them is the issue of uniqueness of functional $\bar{S}$ as an aggregate measure of the respective types of uncertainty in all uncertainty theories. Although the uniqueness of $\bar{S}$ is still an open question, some progress has been made toward establishing it, at least in the Dempster–Shafer theory (DST). It was proved that the functional $\bar{S}$ is the smallest one in the DST to measure aggregate uncertainty. Its uniqueness can thus be proved by showing that it is also the largest functional to measure aggregate uncertainty in the DST.

An important area of theoretical research in GIT in the near future will undoubtedly be a comprehensive study of possible disaggregations of $\bar{S}$. One particular issue in this area is clarifying the relationship of the two versions of disaggregated total uncertainty introduced in Chapter 6, TU and ^aTU, and to identify their application domains.

It also remains to study the range of applicability of Algorithm 6.1 for computing $\bar{S}$. Thus far, the algorithm has been proved correct within DST. Its applicability outside DST, while plausible, has yet to be formally established. We also need to derive more algebraic properties of $\bar{S}$ from the algorithm, in addition to those employed in proving that $\bar{S} \geq GH$ in Appendix D.

Of course, there are many other issues regarding the development of efficient algorithms for GIT. We need, for example, efficient algorithms for computing $\underline{S}$, for converting credal sets to lower or upper probabilities and vice versa, and for computing Möbius representations of given lower and upper probabilities.

10.3. LONG-TERM RESEARCH AREAS

A respectable number of nonclassical uncertainty theories have already been developed, as surveyed in this book. However, this is only a tiny fraction of prospective theories that are of interest in GIT. Each of these theories is based on some generalization of classical measures, or some generalization of classical sets, or some generalization of both. It is clear that most of these theories are yet to be developed, and this undoubtedly will be a long-term area of research in GIT.

Further explorations of the 2-dimensional GIT territory can be pursued by focusing either on some previously unexplored types of generalized measures or on some previously unexplored types of generalized sets. Examples of the former are Choquet capacities of various finite orders, decomposable measures, and k-additive measures. The coherent family of Choquet capacities seems to be especially important for facilitating the new principle of requisite generalization introduced in Section 9.4. Decomposable measures and k-additive measures, on the other hand, are eminently suited to be approximators of more complex measures.

Focusing on unexplored types of generalized sets is a huge undertaking. It involves all the nonstandard types of fuzzy sets. This direction is perhaps the prime area of long-term research in GIT. Among the many challenges of this research area is the use of fuzzified measures (i.e., measures defined on fuzzy sets) for fuzzifying existing uncertainty theories or for developing new ones. Research in this area will undoubtedly be crucial for developing a perception-based theory of probabilistic reasoning.

The route of exploring the area of GIT will undoubtedly be guided, at least to some extent, by application needs. Some of the developed theories will survive and some, with questionable utility, will likely sink into obscurity.

Finally, it should be emphasized that the broad framework introduced in this book for formalizing uncertainty (and uncertainty-based information) still may not be sufficiently general to capture all conceivable types of uncertainty. It is not clear, for example, whether the *information-gap* conception of uncertainty, which has been developed by Ben-Haim [2001], can be formalized within this framework. This is an open research question. If the answer turns out to be negative, then, clearly, the framework will have to be appropriately extended to conform to the goal of the research program of GIT.

10.4. SIGNIFICANCE OF GIT

GIT is an outcome of two generalizations in mathematics. In one of them, classical measures are generalized by abandoning the requirement of additivity; in the other one, classical sets are generalized by abandoning the requirement of sharp boundaries between sets. Generalizing mathematical theories has been a visible trend in mathematics since about the middle of the 20th century, and the two generalizations of interest in this book embody this trend well. Other examples include generalizations from ordinary geometry (Euclidean as well as non-Euclidean) to fractal geometry; from ordinary automata to cellular automata; from regular languages to developmental languages; from precise analysis to interval analysis; from graphs to hypergraphs; and many others.

Each generalization of a mathematical theory usually results in a conceptually simpler theory. This is a consequence of the fact that some properties of the former theory are not required in the latter. At the same time, the more general theory always has a greater expressive power, which, however, is achieved only at the cost of greater computational demands. This explains why these generalizations are closely connected with the emergence of computer technology and steady increases in computing power. By generalizing mathematical theories, we not only enrich our insights but, together with computer technology, also extend our capabilities for modeling the intricacies of the real world.

Generalizing classical measures by abandoning the requirement of additivity broadens their applicability. Contrary to classical measures, generalized measures are capable of formalizing, for example, synergetic or inhibitory effects manifested by some properties measured on sets, data gathering in the face of unavoidable measurement errors, or evidence expressed in terms of a set of probability distributions.

Generalizing classical sets by abandoning sharp boundaries between sets is an extremely radical idea, at least from the standpoint of contemporary science. When accepted, one has to give up classical bivalent logic, generally presumed to be the principal pillar of science. Instead, we obtain a logic in which propositions are not required to be either true or false, but may be true or false to different degrees. As a consequence, some laws of bivalent logic do not hold any more, such as the law of excluded middle or the law of contradiction. At first sight, this seems to be at odds with the very purpose of science. However, this is not the case. There are at least four reasons why allowing membership degrees in sets and degrees of truth in propositions enhance scientific methodology considerably.

1. Fuzzy sets and fuzzy logic possess far greater capabilities than their classical counterparts to capture irreducible measurement uncertainties in their various manifestations. As a consequence, their use considerably improves the *bridge between mathematical models and the associated physical reality*. It is paradoxical that, in the face of the inevitability of measurement errors, fuzzy

data are always more accurate than their crisp counterparts. This greater accuracy is gained by replacing quantization of variables involved with granulation, as is explained in Section 7.7.1.

2. Fuzzy sets and fuzzy logic are powerful tools for *managing complexity and controlling computational cost.* This is primarily due to granulation of systems variables, which is a natural way of making the imprecision of systems models compatible with tasks for which they are constructed. Not only are complexity and computational cost substantially reduced by appropriate granulation, but the resulting solutions tend also to be more useful.

3. An important feature of fuzzy set theory is its capability of capturing the *vagueness* of linguistic terms in statements expressed in *natural languages.* Vagueness of a symbol (a linguistic term) in a given language results from the existence of objects for which it is intrinsically impossible to decide whether the symbol does or does not apply to them according to linguistic habits of some speech community using the language. That is, vagueness is a kind of uncertainty that does not result from information deficiency, but rather from imprecise meanings of linguistic terms, which particularly abound in natural languages. Classical set theory and classical bivalent logic are not capable of expressing the imprecision in meanings of vague terms. Hence, propositions in natural language that contain vague terms were traditionally viewed as unscientific. This view is extremely restrictive. As we increasingly recognize, natural language is often the only way in which meaningful knowledge can be expressed. The applicability of science that shies away from natural language is thus severely limited. This traditional limitation of science is now overcome by fuzzy set theory, which has the capability of dealing in mathematical terms with problems that require the use of natural language. Even though the vagueness inherent in natural language can be expressed via fuzzy sets only in a crude way, this capability is invaluable to modern science.

4. The apparatus of fuzzy set theory and fuzzy logic also enhances our capabilities of modeling human common-sense reasoning, decision making, and other aspects of human cognition. These capabilities are essential for knowledge acquisition from human experts, for knowledge representation and manipulation in expert systems in a human-like manner, and, generally, for designing and building human-friendly machines with high intelligence.

The basic claim of GIT, that uncertainty is a broader concept than the concept of classical probability, has been debated in the literature since the late 1980s (for an overview, see Note 10.4). As a result of this ongoing debate, as well as convincing advances in GIT, limitations of classical probability theory are now better understood. GIT is a continuation of classical, probability-based information theory, but without the limitations of probability theory.

The role of information in human affairs has become so predominant that it is now quite common to refer to our society as an "information society." It is thus increasingly important for us to develop a good understanding of the broad concept of information. In the generalized information theory, the concept of uncertainty is conceived in the broadest possible terms, and uncertainty-based information is viewed as a commodity whose value is its *potential* to reduce uncertainty pertaining to relevant situations. The theory does not deal with the issues of how much uncertainty of relevant users (cognitive agents) is *actually* reduced in the context of each given situation, and how valuable this uncertainty reduction is to them. However, the theory, when adequately developed (see Note 10.6.), will be a solid base for developing a conceptual structure to capture semantic and pragmatic aspects relevant to information users under various situations of information flow. Only when this is adequately accomplished, will a genuine science of information be created.

NOTES

10.1. It was proved by Harmanec [1995] that the functional $\bar{S}$ is the smallest one that satisfies axioms for aggregate uncertainty in DST (stated in Section 6.6). This result is, in some sense, relevant to the proof of the uniqueness of $\bar{S}$. The uniqueness of $\bar{S}$ can now be proved by showing that it is also the largest functional that satisfies the axioms.

10.2. The prospective role of functional $\underline{S}$, complementary to $\bar{S}$, was raised by Kapur et al. [1995]. In particular, it was suggested that the difference $\bar{S} - \underline{S}$, referred to as *uncertainty gap*, may be viewed as a measure of uncertainty about the probability distribution. Alternatively, it may be viewed as a measure of information contained in constraints involved in applying the principle of maximum entropy.

10.3. Ben-Haim [2001] introduced an unorthodox theory of uncertainty, oriented primarily to decision making. The essence of the theory is well captured in a review of Ben-Haim's book written by Jim Hal [*International Journal of General Systems*, **32**(2), 2003, pp. 204–203], from which the following brief characterization of the theory is reproduced:

> An info-gap analysis has three components: a system model, an info-gap uncertainty model and performance requirements. The system model describes the structure and behaviour of the system in question, using as much information as is reasonably available. The system model may, for example, be in the form of a set of partial differential equations, a network model of a project or process, or indeed a probabilistic model such as a Poisson process. The uncertainty in the system model is parameterized with an uncertainty parameter α (a positive real number), which defines a family of nested sets that bound regions or clusters of system behaviour. When $\alpha = 0$ the prediction from the system model converges to a point, which is the anticipated system behaviour, given current available information. However, it is recognised that the system model is incomplete so there will be a range of variation around the nominal behaviour. Uncertainty, as defined by the parameter α, is therefore a range of variation of the actual around the nominal. No further commitment is made to the structure of

uncertainty. The α is not normalized and has no betting interpretation, so is clearly distinct from a probability.

Uncertainty in Ben-Haim's theory is not explicitly defined in measure–theoretic terms. How to describe it within the GIT framework, if possible at all, is an open question.

10.4. The following is a chronology of those major debates regarding uncertainty and probability that are well documented in the literature:

- Probability theory versus evidence theory in artificial intelligence (AI)

  —Shafer, Lindley, Spiegelhalter, Watson, Dempster, Kong;

  —*Statistical Science*, 1987, **2**(1), 1–44.

- Bayesian (probabilistic) approach to managing uncertainty in AI versus other approaches;

  —Peter Cheeseman and 23 correspondents;

  —*Computational Intelligence* (Canadian), 1988, **4**(1), 57–142.

- "An AI View of the treatment of uncertainty" by A. Saffioti;

  —11 correspondents;

  —*The Knowledge Engineering Review*, 1988, **2**(1), 59–91.

- Cambridge debate: Bayesian approach to dealing with uncertainty versus other approaches;

  —Peter Cheeseman $\leftrightarrow$ George Klir, Cambridge University, August 1988;

  —*International Journal of General Systems*, 1989, **15**(4), 347–378.

- Fuzziness versus probability;

  —Michael Laviolette and John Seaman and 8 correspondents;

  —*IEEE Transactions on Fuzzy Systems*, February 1994, **2**(1), 1–42.

- "The Paradoxical Success of Fuzzy Logic";

  —Charles Elkan and 22 correspondents;

  —*IEEE Expert*, August 1994, **9**(4), 2–49.

- "Probablistic and Statistical View of Fuzzy Methods" by M. Laviolette, J.W. Seaman, J.D. Barrett, and W.H. Woodall;

  —6 responses;

  —*Technometrics*, 1995, **37**(3), 249–292.

In many of these debates, the focus is on the relationship between probability theory and the various novel uncertainty theories. The issues discussed and the claims presented by defenders of probability theory vary from debate to debate. The most extreme claims are expressed by Lindley in the first debate [italics added]:

> *The only* satisfactory description of uncertainty is probability. By this I mean that *every* uncertainty statement *must be* in the form of a probability; that several uncertainties *must be* combined using the rules of probability; and that the calculus of probabilities is adequate to handle *all situations* involving uncertainty. Probability is *the only* sensible description of uncertainty and is adequate for *all problems* involving uncertainty. *All* other methods are inadequate. . . . *Anything* that can be done with fuzzy

logic, belief functions, upper and lower probabilities, or *any* other alternative to probability, can *better* be done with probability.

These extreme claims are often justified by referring to a theorem by Cox [1946, 1961], which supposedly established that the only coherent way of representing and dealing with uncertainty is to use rules of probability calculus. However, it has been shown that either the theorem does not hold under assumptions stated explicitly by Cox [Halpern, 1999], or it holds under additional, hidden assumptions that are too strong to capture all aspects of uncertainty [Colyvan, 2004] (see Colyvan's paper for additional references on this topic).

10.5. The distinction between the broad concept of uncertainty and the narrower concept of probability has been obscured in the literature on classical, probability-based information theory. In this extensive literature, the Hartley measure is routinely portrayed as a special case of the Shannon entropy that emerges from the uniform probability distribution. This view is ill-conceived since the Hartley measure is totally independent of any probabilistic assumptions, as correctly recognized by Kolmogorov [1965] and Rényi [1970b]. Strictly speaking, the Hartley measure is based on one concept only: a finite set of possible alternatives, which can be interpreted as experimental outcomes, states of a system, events, messages, and the like, or as sequences of these. In order to use this measure, possible alternatives must be distinguished, within a given universal set, from those that are not possible. It is thus the *possibility* of each relevant alternative that matters in the Hartley measure. Hence, the Hartley measure can be meaningfully generalized only through broadening the notion of possibility. This avenue is now available in terms of the theory of graded possibilities and other nonclassical uncertainty theories that are the subject of GIT.

10.6. As shown by Dretske [1981, 1983], a study of semantic aspects of information requires a well-founded underlying theory of uncertainty-based information. While in his study Dretske relies only on information expressed in terms of the Shannon entropy, the broader view of GIT allows us now to approach the same study in a more flexible and, consequently, more meaningful fashion. Studies regarding pragmatic aspects of information, such as those by Whittemore and Yovits [1973, 1974] and Yovits et al. [1981], should be affected likewise. Hence, the use of the various novel uncertainty theories, uncertainty measures, and uncertainty principles in the study of semantic and pragmatic aspects of information will likely be another main, long-term direction of research.

10.7. The aims of GIT are very similar to those of *generalized theory of uncertainty* (GTU), which have recently been proposed by Lotfi Zadeh [2005]. However, the two theories are formulated quite differently. In GTU, information is viewed in terms of generalized constraints on values of given variables. These constraints are expressed, in general, in terms of the granular structure of linguistic variables. In the absence of any constraint regarding the variables of concern, we are totally ignorant about their actual state. In this situation, our uncertainty about the actual state of the variables is maximal. Any known constraint regarding the variables reduces this uncertainty, and may thus be viewed as a source of information. The concept of a generalized constraint, which is central to GTU, has many distinct modalities. Choosing any of them reduces the level of generality. Making further choices will eventually result in one of the classical theories of uncertainty. This approach to dealing with uncertainty is clearly complementary to the one employed in GIT. In the latter, we start with the two classical theories of uncertainty and generalize each of them by relaxing its axiomatic requirements. The GTU approach to uncertainty and information is thus *top-down*—from general to specific. On the contrary, the GIT approach to uncertainty and information is *bottom-up*—from specific to general. Both approaches are certainly meaningful and should be pursued. In the long run, results obtained by the two complementary approaches will eventually meet. When this happens, our understanding of information-based uncertainty and uncertainty-based information will be quite satisfactory.

APPENDIX A

UNIQUENESS OF THE *U*-UNCERTAINTY

In the following lemmas, functional U is assumed to satisfy requirements (U2), (U3), (U5), (U8), and (U9) as axioms: additivity, monotonicity, expansibility, branching, and normalization, respectively (p. 205). The notation $\mathbf{1}_m = 1, 1, 1, \ldots, 1$ and $\mathbf{0}_m = 0, 0, 0, \ldots, 0$ is carried over from the formulation of
$\underbrace{}_{m}$ $\underbrace{}_{m}$
the branching requirement.

Lemma A.1. For all $q, k \in \mathbb{N}, 2 \leq q \leq k \leq n$, with $\mathbf{r} = \{r_1, r_2, \ldots, r_n\}$

$$
U(\mathbf{r}) = U\left(r_1, r_2, \ldots, r_{k-q}, \underbrace{r_k, r_k, \ldots, r_k}_{q}, r_{k+1}, \ldots, r_n \right)
$$
$$
+ (r_{k-q} - r_k)U\left(\mathbf{1}_{k-q}, \frac{r_{k-q+1} - r_k}{r_{k-q} - r_k}, \frac{r_{k-q+2} - r_k}{r_{k-q} - r_k}, \ldots, \frac{r_{k-1} - r_k}{r_{k-q} - r_k}, \mathbf{0}_{n-k+1} \right)
$$
$$
- (r_{r-q} - r_k)U(\mathbf{1}_{k-q}, \mathbf{0}_{n-k+q}). \tag{A.1}
$$

Proof. By the branching axiom we have

$$
U(\mathbf{r}) = U(r_1, r_2, \ldots, r_{k-2}, r_k, r_k, r_{k+1}, \ldots, r_n)
$$
$$
+ (r_{k-2} - r_k)U\left(\mathbf{1}_{k-2}, \frac{r_{k-1} - r_k}{r_{k-2} - r_k}, \mathbf{0}_{n-k+1} \right)
$$
$$
- (r_{r-2} - r_k)U(\mathbf{1}_{k-2}, \mathbf{0}_{n-k+2}). \tag{A.2}
$$

This also shows that the lemma is true for $q = 2$. We now proceed with a proof by induction on q. Applying the induction hypothesis to the first term on the right-hand side of Eq. (A.2), we get

---

Uncertainty and Information: Foundations of Generalized Information Theory, by George J. Klir
© 2006 by John Wiley & Sons, Inc.

425

$$U(r_1, r_2, \ldots, r_{k-2}, r_k, r_k, r_{k+1}, \ldots, r_n)$$

$$= U\left(r_1, r_2, \ldots, r_{k-q}, \underbrace{r_k, r_k, \ldots, r_k}_{q-1}, r_k, r_{k+1}, \ldots, r_n\right)$$

$$+ (r_{k-q} - r_k)U\left(1_{k-q}, \frac{r_{k-q+1} - r_k}{r_{k-q} - r_k}, \frac{r_{k-q+2} - r_k}{r_{k-q} - r_k}, \ldots, \frac{r_{k-2} - r_k}{r_{k-q} - r_k}, 0_{n-k+2}\right)$$

$$- (r_{k-q} - r_k)U(1_{k-q}, 0_{n-k+q}). \tag{A.3}$$

If we substitute this back into Eq. (A.2), we deduce that

$$U(\mathbf{r}) = U\left(r_1, r_2, \ldots, r_{k-q}, \underbrace{r_k, r_k, \ldots, r_k}_{q}, r_{k+1}, \ldots, r_n\right)$$

$$+ (r_{k-q} - r_k)U\left(1_{k-q}, \frac{r_{k-q+1} - r_k}{r_{k-q} - r_k}, \frac{r_{k-q+2} - r_k}{r_{k-q} - r_k}, \ldots, \frac{r_{k-2} - r_k}{r_{k-q} - r_k}, 0_{n-k+2}\right)$$

$$- (r_{k-q} - r_k)U(1_{k-q}, 0_{n-k+q})$$

$$+ (r_{k-2} - r_k)U\left(1_{k-2}, \frac{r_{k-1} - r_k}{r_{k-2} - r_k}, 0_{n-k+1}\right)$$

$$- (r_{k-2} - r_k)U(1_{k-2}, 0_{n-k+2}). \tag{A.4}$$

Now apply the branching axiom to the quantity

$$(r_{k-q} - r_k)U\left(1_{k-q}, \frac{r_{k-q+1} - r_k}{r_{k-q} - r_k}, \frac{r_{k-q+2} - r_k}{r_{k-q} - r_k}, \ldots, \frac{r_{k-1} - r_k}{r_{k-q} - r_k}, 0_{n-k+1}\right) \tag{A.5}$$

to produce

$$(r_{k-q} - r_k)U\left(1_{k-q}, \frac{r_{k-q+1} - r_k}{r_{k-q} - r_k}, \frac{r_{k-q+2} - r_k}{r_{k-q} - r_k}, \ldots, \frac{r_{k-1} - r_k}{r_{k-q} - r_k}, 0_{n-k+1}\right)$$

$$= (r_{k-q} - r_k)U\left(1_{k-q}, \frac{r_{k-q+1} - r_k}{r_{k-q} - r_k}, \frac{r_{k-q+1} - r_k}{r_{k-q} - r_k}, \ldots, \frac{r_{k-2} - r_k}{r_{k-q} - r_k}, 0_{n-k+2}\right)$$

$$+ (r_{k-2} - r_k)U\left(1_{k-2}, \frac{r_{k-1} - r_k}{r_{k-2} - r_k}, 0_{n-k+1}\right)$$

$$- (r_{k-2} - r_k)U(1_{k-2}, 0_{n-k+2}). \tag{A.6}$$

All three terms on the right-hand side of the equality in Eq. (A.6) are present in the right-hand side of Eq. (A.3), so that by a simple substitution we conclude that which was to be proved:

$$U(\mathbf{r}) = U\left(r_1, r_2, \ldots, r_{k-q}, \underbrace{r_k, r_k, \ldots, r_k}_{q}, r_{k+1}, \ldots, r_n \right)$$

$$+ (r_{k-q} - r_k)U\left(\mathbf{1}_{k-q}, \frac{r_{k-q+1} - r_k}{r_{k-q} - r_k}, \frac{r_{k-q+2} - r_k}{r_{k-q} - r_k}, \ldots, \frac{r_{k-1} - r_k}{r_{k-q} - r_k}, \mathbf{0}_{n-k+1} \right)$$

$$- (r_{k-q} - r_k)U(\mathbf{1}_{k-q}, \mathbf{0}_{n-k+q}). \qquad\qquad \blacksquare(\text{A.7})$$

Lemma A.2. For all $k \in \mathbb{N}$, $k < n$, define $g(k) = U(\mathbf{1}_k, \mathbf{0}_{n-k})$, then $g(k) = \log_2 k$.

Proof. By the expansibility axiom we know that $U(\mathbf{1}_k, \mathbf{0}_{n-k}) = U(\mathbf{1}_k)$. By the additivity axiom $g(k + l) = g(k) + g(l)$, so by a proof identical to that of Theorem 2.1 we can conclude that $g(k) = \log_2 k$. $\qquad\qquad \blacksquare$

Lemma A.3. For $k \in \mathbb{N}$, $k \geq 2$, and $\rho \in \mathbb{R}$, $0 \leq \rho \leq 1$, we have

$$U(\mathbf{1}_k \boldsymbol{\rho}_{n-k}) = (1 - \rho)\log_2 k + \rho \log_2 n, \qquad\qquad (\text{A.8})$$

where $\boldsymbol{\rho}_{n-k}$ represents $\underbrace{\rho, \rho, \rho, \ldots, \rho}_{n-k}$.

Proof. Let $\mathbf{r}$ be the possibility profile $\mathbf{1}_k, \boldsymbol{\rho}_{n-k}$. Form the joint possibility profile $\mathbf{r}^2$, where both marginal possibility profiles are equivalent to $\mathbf{r}$. It is simple to check that $\mathbf{r}^2 = \mathbf{1}_{k^2}, \boldsymbol{\rho}_{n^2-k^2}$ has this property. Now $U(\mathbf{r}^2) = U(\mathbf{1}_{k^2}, \boldsymbol{\rho}_{n^2-k^2})$. First, we can deduce from the additivity axiom that

$$U(\mathbf{r}^2) = U(\mathbf{r}) + U(\mathbf{r}) = 2U(\mathbf{r}). \qquad\qquad (\text{A.9})$$

Second, by applying Lemma A.1, we get

$$
\begin{aligned}
U(\mathbf{r}^2) &= U(\mathbf{1}_{k^2}, \boldsymbol{\rho}_{n^2-k^2}) \\
&= U(\mathbf{1}_k, \boldsymbol{\rho}_{n^2-k}) \\
&\quad + (1-\rho)U(\mathbf{1}_{k^2}, \mathbf{0}_{n^2-k^2}) \\
&\quad - (1-\rho)U(\mathbf{1}_k, \mathbf{0}_{n^2-k})
\end{aligned}
\qquad\qquad (\text{A.10})
$$

We also know by the expansibility axiom that $U(\mathbf{r}) = U(\mathbf{r}, 0)$ for any possibility profile $\mathbf{r}$, and we can apply Lemma A.1 to any possibility profile and this will make the final q possibility values zero. Let us perform this operation upon the first term on the right-hand side of the equality in Eq. (A.10),

$$
\begin{aligned}
U(\mathbf{1}_k, \boldsymbol{\rho}_{n^2-k}) &= U(\mathbf{1}_{k,n-k}, \mathbf{0}_{n^2-n}) \\
&\quad + \rho U(\mathbf{1}_{n^2}) \\
&\quad - \rho U(\mathbf{1}_n, \mathbf{0}_{n^2-n}).
\end{aligned}
$$

Substituting Eq. (A.7) into Eq. (A.10), we calculate that

$$U(\mathbf{1}_{k^2}, \boldsymbol{\rho}_{n^2-k^2}) = U(\mathbf{1}_k, \boldsymbol{\rho}_{n^2-k}) + (1-\rho)U(\mathbf{1}_{k^2}, \mathbf{0}_{n^2-k^2}) - (1-\rho)U(\mathbf{1}_k, \mathbf{0}_{n^2-k})$$
$$= U(\mathbf{1}_{k,n-k}, \mathbf{0}_{n^2-n}) + \rho U(\mathbf{1}_{n^2}) - \rho U(\mathbf{1}_n, \mathbf{0}_{n^2-n}). \tag{A.11}$$

Now using the expansibility axiom to drop all final zeros and Lemma A.2 to replace $U(\mathbf{1}_k)$ with $\log_2 k$,

$$U(\mathbf{1}_{k^2}, \boldsymbol{\rho}_{n^2-k^2}) = U(\mathbf{1}_k, \boldsymbol{\rho}_{n-k}) + \rho \log_2 n^2 - \rho \log_2 n$$
$$+ (1-\rho)\log_2 k^2 - (1-\rho)\log_2 n. \tag{A.12}$$

We now finish the proof by remembering that $U(\mathbf{1}_{k^2}, \boldsymbol{\rho}_{n^2-k^2}) = U(\mathbf{r}^2) = 2U(\mathbf{r})$, that $U(\mathbf{1}_k, \boldsymbol{\rho}_n) = U(\mathbf{r})$, and that $\log_2 x^2 = 2\log_2 x$:

$$2U(\mathbf{r}) = U(\mathbf{r}) + 2\rho \log_2 n - \rho \log_2 n$$
$$+ 2(1-\rho)\log_2 k - (1-\rho)\log_2 n$$

$$U(\mathbf{r}) = \rho \log_2 n + (1-\rho)\log_2 k. \qquad \blacksquare \text{(A.13)}$$

Theorem A.1. The *U*-uncertainty is the only functional that satisfies the axioms of expansibility, monotonicity, additivity, branching, and normalization.

Proof. The proof is by induction on n, the length of the possibility profile. If $n = 2$, then we can apply Lemma A.3 to get the desired result immediately. Assume now that the theorem is true for $n - 1$; then we can use the expansibility and branching axioms in the same way as was used in the proof of Lemma A.3 to replace r_n with zero:

$$U(\mathbf{r}) = U(r_1, r_2, \dots, r_n)$$
$$= U(r_1, r_2, \dots, r_n, 0)$$
$$= U(r_1, r_2, \dots, r_{n-1}, 0, 0)$$
$$+ r_{n-1}U\left(\mathbf{1}_{n-1}, \frac{r_n}{r_{n-1}}, 0\right)$$
$$- r_{n-1}U(\mathbf{1}_{n-1}, 0, 0). \tag{A.14}$$

We can now apply Lemma A.3 to the term $U(\mathbf{1}_{n-1}, (r_n/r_{n-1}), 0)$ and drop terminal zeros,

$$U(\mathbf{r}) = U(r_1, r_2, \dots, r_{n-1})$$
$$+ r_{n-1}\left[\left(1 - \frac{r_n}{r_{n-1}}\right)\log_2(n-1) + \frac{r_n}{r_{n-1}}\log_2 n\right]$$
$$- r_{n-1}\log_2(n-1)$$
$$= U(r_1, r_2, \dots, r_{n-1})$$
$$+ (r_{n-1} - r_n)\log_2(n-1) + r_n \log_2 n$$
$$- r_{n-1}\log_2(n-1)$$
$$= U(r_1, r_2, \dots, r_{n-1}) - r_n \log_2(n-1) + r_n \log_2 n$$
$$= U(r_1, r_2, \dots, r_{n-1}) + r_n \log_2 \frac{n}{n-1}. \tag{A.15}$$

Applying the induction hypothesis, we conclude

$$
\begin{aligned}
U(\mathbf{r}) &= U(r_1, r_2, \ldots, r_{n-1}) + r_n \log_2 \frac{n}{n-1} \\
&= \sum_{i=2}^{n-1} r_i \log_2 \frac{i}{i-1} + r_n \log_2 \frac{n}{n-1} \\
&= \sum_{i=2}^{n} r_i \log_2 \frac{i}{i-1}.
\end{aligned}
\qquad \blacksquare \text{(A.16)}
$$

APPENDIX B

UNIQUENESS OF THE GENERALIZED HARTLEY MEASURE IN THE DEMPSTER–SHAFER THEORY

A proof of uniqueness of the generalized Hartley measure in the Dempster–Shafer theory (DST) that is presented here is based on the following six axiomatic requirements:

Axiom (GH1) Subadditivity. For any given joint body of evidence, $\langle \mathcal{F}, m \rangle$ on $X \times Y$, and the associated marginal bodies of evidence, $\langle \mathcal{F}_x, m_x \rangle$ and $\langle \mathcal{F}_Y, m_Y \rangle$

$$GH(m) \leq GH(m_x) + GH(m_Y).$$

Axiom (GH2) Additivity. For any noninteractive bodies of evidence $\langle \mathcal{F}_x, m_x \rangle$ and $\langle \mathcal{F}_Y, m_Y \rangle$, on X and Y, respectively, and the associated joint body of evidence, $\langle \mathcal{F}, m \rangle$, where

$$\mathcal{F} = \mathcal{F}_X \times \mathcal{F}_Y \text{ and } m(A \times B) = m_X(A) \cdot m_Y(B)$$

for all $A \in \mathcal{F}_x$ and all $B \in \mathcal{F}_Y$,

$$GH(m) = GH(m_X) + GH(m_Y).$$

Axiom (GH3) Symmetry. GH is invariant with respect to permutations of values of the basic probability assignment function within each group of subsets of X with equal cardinalities.

Uncertainty and Information: Foundations of Generalized Information Theory, by George J. Klir © 2006 by John Wiley & Sons, Inc.

Axiom (GH4) Branching. $GH(m) = GH(m_1) + GH(m_2)$ for any three bodies of evidence on X, $\langle \mathcal{F}, m \rangle$, $\langle \mathcal{F}_1, m_1 \rangle$, and $\langle \mathcal{F}_2, m_2 \rangle$, such that

$$\mathcal{F} = \{A, B, C, \ldots\},$$
$$\mathcal{F}_1 = \{A_1, B, C, \ldots\},$$
$$\mathcal{F}_2 = \{A, B_1, C_1, \ldots\},$$

where

$$A_1 \subseteq A, \qquad B_1 \subseteq B, \qquad C_1 \subseteq C, \ldots,$$
$$|A_1| = |B_1| = |C_1| = \cdots = 1,$$

and

$$m(A) = m_1(A) = m_2(A)$$
$$m(B) = m_1(B) = m_2(B_1)$$
$$m(C) = m_1(C) = m_2(C_1)$$

Axiom (GH5) Continuity. GH is a continuous functional.

Axiom (GH6) Normalization. $GH(m) = 1$ when $m(A) = 1$ and $|A| = 2$.

The requirements of subadditivity, additivity, continuity, and normalization are obvious generalizations of their counterparts for the U-uncertainty. The requirement of symmetry states that the functional GH (measuring nonspecificity) should depend only on values of the basic probability assignment functions and the cardinalities of the sets to which these values are allocated and not on the sets themselves.

The branching requirement states the following: Given a basic probability assignment function on X, if we maintain its values, but one of its focal subsets (e.g., subset A) is replaced with a singleton (thus obtaining basic probability assignment function m_1) and, separately, replace all remaining focal subsets of m (i.e., all except A) with singletons (thus obtaining function m_2), then the sum of the nonspecificities of m_1 and m_2 should be the same as the nonspecificity of the original basic probability assignment function m.

It can be now established, via the following theorem, that the generalized Hartley measure defined by Eq. (6.27) is uniquely characterized by the axiomatic requirements (GH1)–(GH6).

Theorem B.1. Let $\mathcal{M}$ denote the set of all basic probability assignments functions in DST and let GH denote a functional of the form

$$GH: \mathcal{M} \to [0, \infty).$$

If the functional GH satisfies the axiomatic requirements of subadditivity, additivity, symmetry, branching, and normalization (i.e., the axiomatic requirements (GH1)–(GH6)), then for any $m \in \mathcal{M}$,

$$GH(m) = \sum_{A \in \mathcal{F}} m(A) \log_2 |A|,$$

where $\mathcal{F}$ denotes the set of focal elements associated with m.

Proof. To facilitate the proof, let us introduce a convenient notation. Let $(^{\alpha}A, ^{\beta}B, ^{\gamma}C, \ldots)$ denote a basic probability assignment function such that $m(A) = \alpha$, $m(B) = \beta$, $m(C) = \gamma \ldots$, where $A, B, C, \ldots$ are the focal elements of m. When we are concerned only with the cardinalities of the focal elements, the basic probability assignment function can be written in the form $(^{\alpha}a, ^{\beta}b, ^{\gamma}c, \ldots)$ where $a, b, c, \ldots$ are positive integers that stand for the cardinalities of the focal elements; occasionally, when $\alpha = 1$, we write (a) instead of (^{1}a). For the sake of simplicity, let us refer to the value $m(A)$ of a basic assignment m as the *weight* of A.

(i) First, we prove that $GH(^{1}a) = \log_2 a$. For convenience, let $W(a) = GH(^{1}a)$ (that is, W is a nonnegative real-valued function on $\mathbb{N}$). From additivity $(GH2)$, we have

$$W(ab) = W(a) + W(b). \tag{B.1}$$

When $a = b = 1$, $W(1) = W(1) + W(1)$ and, consequently, $W(1) = 0$. To show that $W(a) \leq W(a = 1)$, let us take set A of $a(a + 1)$ elements, which is a subset of a Cartesian product $X \times Y$. By symmetry $(GH3)$, $GH(A)$ does not depend on how the elements of A are arranged. Let us consider two possible arrangements. In the first arrangement, A is viewed as a rectangle $a \times (a + 1)$; here $\times$ denotes a Cartesian product of sets with cardinalities a and $a + 1$. Then

$$GH(A) = W(a) + W(a + 1) \tag{B.2}$$

by additivity. In the second arrangement, we view A as a subset of a square $(a + 1) \times (a + 1)$ such that at least one diagonal of the square is fully covered. In this case, the projections onto both X and Y have the same cardinality $a + 1$. Hence, by subadditivity $(GH1)$, we have

$$GH(A) \leq W(a + 1) + W(a + 1). \tag{B.3}$$

It follows immediately from Eqs. (B.1) and (B.2) that

$$W(a) \leq W(a + 1).$$

Function W is thus monotonic nondecreasing. Since it is also additive in the sense of Eq. (B.1) and normalized by Axiom $(GH6)$, it is equivalent to the Hartley information H. Hence, by Theorem 2.1

$$W(a) = GH(^1a) = \log_2 a. \tag{B.4}$$

(ii) We prove that $GH(^\alpha 1, {}^\beta 1, {}^\gamma 1, \ldots) = 0$. First, we show that the equality holds only for two focal elements, that is, we show that $GH(^\alpha 1, {}^\beta 1) = 0$. Consider a set A of a^2 elements for some sufficiently large, temporarily fixed a. Let $B = A \times (^\alpha 1, {}^\beta 1)$. Then $GH(A) = 2\log_2 a$ and

$$GH(B) = 2\log_2 a + GH(^\alpha 1, {}^\beta 1) \tag{B.5}$$

by Eq. (B.4) and additivity. Set B can be considered as consisting of two squares $a \times a$ with weights α and β. Let us now place both of these squares into a larger square $(a+1) \times (a+1)$ in such a way that at least one of the diagonals of the larger square is totally covered by elements of the smaller squares and, at the same time, the $a \times a$ squares do not completely overlap. Clearly, both projections of this arrangement of set B have the same cardinality $a + 1$. Hence,

$$GH(B) \leq GH(a+1) + GH(a+1)$$

by subadditivity. Substituting for $GH(B)$ from Eq. (B.5) and applying Eq. (B.4), we obtain

$$GH(^\alpha 1, {}^\beta 1) + 2\log_2 a \leq 2\log_2 (a+1)$$

or, alternatively,

$$GH(^\alpha 1, {}^\beta 1) \leq 2[\log_2 (a+1) - \log_2 a].$$

This inequality must be satisfied for all $a \in \mathbb{N}$. When a goes to infinity, the left-hand side of the inequality remains constant, whereas its right-hand side converges to zero. Hence,

$$GH(^\alpha 1, {}^\beta 1) = 0.$$

By repeating the same argument for $(^\alpha 1, {}^\beta 1, {}^\gamma 1, \ldots)$ with n focal elements, we readily obtain

$$GH(^\alpha 1, {}^\beta 1, {}^\gamma 1, \ldots) \leq n[2[\log_2 (a+1) - \log_2 a]],$$

and, again, by allowing a to go to infinity, we obtain

$$GH(^{\alpha}1,^{\beta}1,^{\gamma}1,\ldots)=0. \tag{B.6}$$

Applying this result and the additivity of GH, we also have for an arbitrary basic assignment m the following:

$$GH(m\cdot(^{\alpha}1,^{\beta}1,^{\gamma}1,\ldots))=GH(m)+GH(^{\alpha}1,^{\beta}1,^{\gamma}1,\ldots)=GH(m). \tag{B.7}$$

This means that we can replicate all focal elements of m the same number of times, splitting their weights in a fixed proportion $\alpha, \beta, \gamma, \ldots$, and the value of GH does not change.

(iii) Next, we prove that $GH(m_1) = GH(m_2)$, where $m_1 = (^{\alpha}a, ^{\beta+\gamma}b)$ and $m_2 = (^{\alpha}a, ^{\beta}b, ^{\gamma}b)$. Since the actual weights are not essential in this proof, we omit them for the sake of simplicity; if desirable, they can be easily reconstructed. The proof is accomplished by showing that $GH(m_1) \leq GH(m_2)$ and, at the same time, $GH(m_2) \leq GH(m_1)$. To demonstrate the first inequality, let us view the focal elements m_1 and m_2 as collections of intervals of lengths a, b, and a, b, b, respectively. Furthermore, let us place both intervals of m_1 side by side, the first and second interval of m_2 side by side, and the third interval of m_2 above the second interval of m_2. According to this arrangement, the two projections of m_2 consist of m_1 and a pair of singletons with appropriate weights, say ($^{x}1$, $^{y}1$). It then follows from the subadditivity that

$$GH(m_2)\leq GH(m_1)+GH(^{x}1,^{y}1).$$

The last term in this inequality is 0 by Eq. (B.6) and, consequently, $GH(m_2) \leq GH(m_1)$.

To prove the opposite inequality, let $m_3 = (^{\beta}1, ^{\gamma}1)$ so that $m_1 \cdot m_3$ assigns the same weights to the b's as does m_2. We select a sufficiently large integer n, temporarily fixed. Let s and s' denote squares $n \times n$ and $(n + 1) \times (n + 1)$, respectively. We can view $m_1 \cdot m_3 \cdot s$ as a collection of four parallelepipeds, two with edges a, n, n, and the other two with edges b, n, n. Similarly, $m_2 \cdot s'$ can be viewed as a collection of three parallelepipeds, one with edges $a, n + 1, n + 1$, and two with edges $b, n + 1, n + 1$. We now place two blocks with edges a, n, n of $m_1 \cdot m_3 \cdot s$ inside one block with edges $a, n + 1, n + I$ of m_2 so that they cover the main diagonal. Furthermore, we place two blocks with edges b, n, n of $m_1 \cdot m_3 \cdot s$ inside the separate blocks with edges $b, n + 1, n + 2$ of $m_2 \cdot s'$ so that they again cover the diagonals. Using additivity and Eq. (B.4), the construction results in the equation

$$GH(m_1 \cdot m_3 \cdot s) = GH(m_1)+GH(^{\beta}1,^{\gamma}1)+2\log_2 n$$
$$= GH(m_1)+2\log_2 n.$$

Using subadditivity and the projections of the construction, we obtain

$$GH(m_1 \cdot m_3 \cdot s) \le GH(m_2) + GH(n+1) + GH(n+1).$$

Hence,

$$GH(m_1) + 2\log_2 n \le GH(m_2) + 2\log_2 n$$

and

$$GH(m_2) - GH(m_1) \ge 2[\log_2 n - \log_2(n+1)].$$

For n going to infinity, the right-hand side of this inequality converges to 0 and, consequently,

$$GH(m_2) - GH(m_1) \ge 0$$

or

$$GH(m_2) \ge GH(m_1).$$

The proof can easily be extended to the case of a general basic probability assignment function in which any given cardinality may repeat more than twice and the number of different cardinalities is arbitrary.

(iv) Repeatedly applying the branching property, we obtain

$$GH(^\alpha a, ^\beta b, ^\gamma c, \ldots) = GH(^\alpha a, ^{x_1}1, ^{x_2}1, \ldots) + GH(^\beta b, ^{y_1}1, ^{y_2}1, \ldots) + \ldots.$$

According to property (iii), we can combine the singletons so that the proof of the theorem reduces to the determination of $GH(^\alpha a, {}^{1-\alpha}1)$ for arbitrary a and α. Moreover,

$$GH(^1 a) = GH(^{1/2}a, ^{1/2}a)$$

and, by the branching axiom,

$$GH(^{1/2}a, ^{1/2}a) = 2GH(^{1/2}a, ^{1/2}1).$$

Hence,

$$GH(^1 a) = 2GH(^{1/2}a, ^{1/2}1).$$

Similarly,

$$GH(^1 a) = 3GH(^{1/3}a, ^{2/3}1) = 3GH(^{1/4}a, ^{3/4}1) \ldots.$$

Using Eq. (B.4) we obtain for each $t = 1/n$ the equation

$$GH({}^{t}a, {}^{1-t}1) = t \log_2 a.$$

This formula can easily be shown to hold for any rational t. Since an arbitrary real number can be approximated by a monotonic sequence of rational numbers, property (iii) implies that the formula holds also for any $t \in [0, 1]$. This concludes the proof. ∎

APPENDIX C

CORRECTNESS OF ALGORITHM 6.1

The correctness of the algorithm is stated in Theorem C.1. The proof employs the following two lemmas.

Lemma C.1. Let $x > 0$ and $c - x > 0$. Denote

$$L(x) = -[(c-x)\log_2(c-x) + x\log_2 x].$$

Then $L(x)$ is strictly increasing in x when $(c - x) > x$.

Proof. $L'(x) = \log_2(c - x) - \log_2 x$ so that $L'(x) > 0$ whenever $(c - x) > x$. ∎

Lemma C.2. [Dempster, 1967b] Let X be a frame of discernment, *Bel* a generalized belief function* on X, and m the corresponding generalized basic probability assignment; then a tuple $\langle p_x | x \in X \rangle$ satisfies the constraints

$$0 \le p_x \le 1, \qquad \forall x \in X,$$

$$\sum_{x \in X} p_x = Bel(X),$$

and

$$Bel(A) \le \sum_{x \in X} p_x, \qquad \forall A \subseteq X,$$

---

* A generalized belief function, *Bel*, is a function that satisfies all requirements of a belief measure except the requirement that $Bel(X) = 1$. Similarly, values of a generalized basic probability assignment are not required to add to one. *Bel* in Algorithm 6.1 is clearly a generalized belief function after the first iteration of the algorithm.

Uncertainty and Information: Foundations of Generalized Information Theory, by George J. Klir
© 2006 by John Wiley & Sons, Inc.

if and only if there exist nonnegative real numbers α_x^A for all nonempty sets $A \subseteq X$ and for all $x \in A$ such that

$$p_x = \sum_{x|x \in A} \alpha_x^A = m(A).$$

Using the results of Lemma C.1 and Lemma C.2, we can now address the principal issue of this appendix, the correctness of Algorithm 6.1. This issue is the subject of the following theorem.

Theorem C.1. Algorithm 6.1 stops after a finite number of steps and the output is the correct value of function $AU(Bel)$ since $\langle p_x | x \in X \rangle$ maximizes the Shannon entropy within the constraints induced by Bel.

Proof. The frame of discernment X is a finite set and the set A chosen in Step 1 of the algorithm is nonempty (note also that the set A is determined uniquely). Therefore, the "new" X has a smaller number of elements than the "old" X and so the algorithm terminates after finitely many passes through the loop of Steps 1–5.

To prove the correctness of the algorithm, we proceed in two stages. In the first stage, we show that any distribution $\langle p_x | x \in X \rangle$ that maximizes the Shannon entropy within the constraints induced by Bel has to satisfy the equality $p_x = Bel(A)/|A|$ for all $x \in A$. In the second stage we show that, for any partial distribution $\langle p_x | x \in X - A \rangle$ that satisfies the constraints induced by the "new" generalized belief function defined in Step 3 of the algorithm, the complete distribution $\langle q_x | x \in X \rangle$, defined by

$$q_x = Bel(A)/|A| \quad \text{for } x \in A \quad \text{and} \quad q_x = p_x \quad \text{for} \quad x \in X - A, \qquad \text{(C.1)}$$

satisfies the constraints induced by the original (generalized) belief function, and vice versa. That is, for any distribution $\langle q_x | x \in X \rangle$ satisfying the constraints induced by Bel and such that $q_x = Bel(A)/|A|$ for all $x \in A$, it holds that $\langle q_x | x \in X - A \rangle$ satisfies the constraints induced by the "new" generalized belief function defined in Step 3 of the algorithm. The theorem then follows by induction.

First, assume that $\langle p_x | x \in X \rangle$ is such that

$$AU(Bel) = -\sum_{x \in X} p_x \log_2 p_x,$$

$$\sum_{x \in X} p_x = Bel(X),$$

and

$$Bel(B) \le \sum_{x \in B} p_x.$$

for all $B \subset X$, and there is $y \in A$ such that $p_y \neq Bel(A)/|A|$, where A is the set chosen in Step 1 of the algorithm. That is, $Bel(A)/|A| \geq Bel(B)/|B|$ for all $B \subseteq X$, and if $Bel(A)/|A| = Bel(B)/|B|$, then $|A| > |B|$. Furthermore, we can assume that $p_y > Bel(A)/|A|$. This is justified by the following argument: If $p_y < Bel(A)/|A|$, then due to $Bel(A) \leq \Sigma_{x \in A} p_x$ there exists $y' \in A$ such that $p_{y'} > Bel(A)/|A|$, and we can take y' instead of y.

For a finite sequence $\{x_i\}_{i=0}^m$, where $x_i \in X$ and m is a positive integer, let $\mathcal{F}$ denote the set of all focal elements associated with Bel and let

$$\Phi(x_i) = \cup \{C \subseteq X \,|\, C \in \mathcal{F}, x_{i-1} \in C, \text{ and } \alpha_{x_{i-1}}^c > 0\}$$

for $i = 1, \ldots, m$, where α_x^c are the nonnegative real numbers whose existence is guaranteed by Lemma C.2 (we fix one such set of those numbers). Let

$$D = \left\{ x \in X \,\middle|\, \exists \quad m \text{ non-negative integers and } \{x_i\}_{i=0}^m \text{ such that} \right.$$
$$x_0 = y, \quad x_m = x, \quad \text{for all } i = 1, 2, \ldots, m, \quad x_i \in \Phi(x_i) \quad \text{and for all}$$
$$\left. i = 2, 3, \ldots, m, \quad \forall z \in \Phi \quad (x_{i-1}) p_z \geq p_{x_{i-2}} \right\}.$$

There are now two possibilities. Either there is $z \in D$ such that in the sequence $\{x_i\}_{i=0}^m$ from the definition of D, where $x_m = z$, it holds that $p_z < p_{m-1}$. This, however, leads to a contradiction with the maximality of $\langle p_x | x \in X \rangle$ since

$$-\sum\nolimits_{x \in X} p_x \log_2 p_x < -\sum\nolimits_{x \in X} q_x \log_2 q_x$$

by Lemma C.1, where $q_x = p_x$ for $x \in (X - \{z, x_{m-1}\})$, $q_z = p_z + \varepsilon$, and $q_{x_{m-1}} = p_{x_{m-1}} - \varepsilon$; here

$$\varepsilon \in (0, \min\{\alpha_{x_{m-1}}^c, (p_{x_{m-1}} + p_z)/2\}),$$

where C is the focal element of Bel from the definition of D containing both z and x_{m-1} and such that $\alpha_{x_{m-1}}^C > 0$. The distribution $\langle q_x | x \in X \rangle$ satisfies the constraints induced by Bel due to the Lemma C.2. Or, the second possibility is that for all $x \in D$ and all focal elements $C \subseteq X$ of Bel such that $x \in C$, whenever $\alpha_x^C > 0$, it holds that $p_z \geq p_x$ for all $z \in (C - \{x\})$. It follows from Lemma C.2 that $Bel(D) = \Sigma_{x \in D} p_x$. However, this fact contradicts the choice of A, since

$$\frac{Bel(D)}{|D|} = \frac{\sum_{x \in D} P_x}{|D|} \geq p_y > \frac{Bel(A)}{|A|}.$$

We have shown that any $\langle p_x | x \in X \rangle$ maximizing the Shannon entropy within the constraints induced by Bel, has to satisfy $p_x = Bel(A)/|A|$ for all $x \in A$.

Let Bel' denote the generalized belief function on $X - A$ defined in Step **3** of the algorithm; that is, $Bel'(B) = Bel(B \cup A) - Bel(A)$. It is really a generalized belief function. The reader can verify that its corresponding generalized basic probability assignment can be expressed by

$$m'(B) = \sum_{C \subseteq X | (C \cap (X-A)) = B} m(C)$$

for all nonempty sets $B \subseteq X - A$, and $m'(\emptyset) = 0$. Assume, then, that $\langle p_x | x \in X - A \rangle$ is such that $p_x \in [0, 1]$,

$$\sum_{x \in X-A} p_x = Bel'(X - A),$$

and

$$\sum_{x \in B} p_x \geq Bel'(B)$$

for all $B \subset X - A$. Let $\langle q_x | x \in X \rangle$ denote the complete distribution defined by Eq. (C.1). Clearly, $q_x \in [0, 1]$. Since

$$Bel'(X - A) = Bel(X) - Bel(A),$$

we have

$$Bel(X) = \sum_{x \in X-A} p_x + Bel(A)$$

$$= \sum_{x \in X-A} p_x + \sum_{x \in A} \frac{Bel(A)}{|A|}$$

$$= \sum_{x \in X} q_x.$$

From $Bel(A)/|A| \geq Bel(B)/|B|$ for all $B \subseteq X$, it follows that $\sum_{x \in B} q_x \geq Bel(C)$ for all $C \subseteq A$. Assume that $B \subseteq X$ and $B \cap (X - A) \neq \emptyset$. We know that

$$\sum_{x \in B \cap (X-A)} p_x \geq Bel'(B \cap (X - A)) = Bel(A \cup B) - Bel(A).$$

From Eq. (5.40), we get

$$Bel(A \cup B) - Bel(A) \geq Bel(B) - Bel(A \cap B).$$

Since $Bel(A)/|A| \geq Bel(A \cap B)/|A \cap B|$, we get

$$\sum_{x \in B} q_x = \sum_{x \in B \cap (X-A)} p_x + \sum_{x \in A \cap B} \frac{Bel(A)}{|A|} \geq Bel(B).$$

Conversely, assume $\langle q_x | x \in X \rangle$ is such that $q_x \in [0, 1]$, $q_x = Bel(A)/|A|$ for all $x \in A$, $\Sigma_{x \in X} q_x = Bel(X)$, and $\Sigma_{x \in B} q_x \geq Bel(B)$ for all $B \subset X$. Clearly, we have

$$Bel'(X - A) = Bel(X) - Bel(A) = \sum_{x \in X - A} q_x.$$

Let $C \subseteq X - A$. We know that $\Sigma_{x \in A \cup C} q_x \geq Bel(A \cup C)$, but it follows from this fact that

$$\sum_{x \in C} q_x \geq Bel(A \cup C) - Bel(A) = Bel'(C).$$

This means that we reduced the size of our problem, and therefore the theorem follows by induction.   ■

Several remarks regarding Algorithm 6.1 can be made:

(a) Note that we have also proved that the distribution maximizing the entropy within the constraints induced by a given belief function is unique. It is possible to prove this fact directly by using the concavity of the Shannon entropy.

(b) The condition that A has the maximal cardinality among sets B with equal value of $Bel(B)/|B|$ in Step 1 is not necessary for the correctness of the algorithm, but it speeds it up. The same is true for the condition $Bel(X) > 0$ in Step 5. Moreover, we could exclude the elements of X outside the union of all focal elements of Bel (usually called *core* within evidence theory) altogether.

(c) If $A \subset B$ and $Bel(A) = Bel(B)$, then $Bel(A)/|A| > Bel(B)/|B|$. It means that it is not necessary to work with the whole power set $\mathcal{P}(X)$; it is enough to consider $C = \{A \subseteq X \ni \{F_1, \ldots, F_l\} \subseteq \mathcal{P}(X)$ such that $m(F_i) > 0$ and $A = \bigcup_{i=1}^{l} F_i\}$.

APPENDIX D

PROPER RANGE OF GENERALIZED SHANNON ENTROPY

The purpose of this Appendix is to prove that $\bar{S}(Bel) - GH(Bel) \geq 0$ for any given belief function (or any of its equivalent representations). To facilitate the proof, which is adopted from Smith [2000], some relevant features of Algorithm 6.1 are examined first.

Applying Algorithm 6.1 to a given belief function Bel defined on $\mathcal{P}(X)$ results in the following sequence of sets $A_i \subseteq X$ and associated probabilities p_i, each obtained in the ith pass through Steps 1–5 (or the ith iteration) of the algorithm ($i \in \mathbb{N}_n$, assuming that the algorithm terminates after n passes):

- A_1 is equal to $A \subseteq X$ such that

$$\frac{Bel(A)}{|A|}$$

is maximal. If there are more such sets than one, then the one with the largest cardinality is chosen. Each element of set A_1 is assigned the probability

$$p_1 = \frac{Bel(A_1)}{|A_1|}.$$

- A_2 is equal to $A \subseteq X$ such that

$$\frac{Bel(A \cup A_1) - Bel(A_1)}{|A - A_1|}$$

Uncertainty and Information: Foundations of Generalized Information Theory, by George J. Klir
© 2006 by John Wiley & Sons, Inc.

is maximal. If there are more such sets than one, then the one with the largest cardinality $|A - A_1|$ is chosen. Each element of set $A_2 - A_1$ is assigned the probability

$$p_2 = \frac{Bel(A_2 \cup A_1) - Bel(A_1)}{|A_2 - A_1|}.$$

- A_3 is equal to $A \subseteq X$ such that

$$\frac{Bel(A \cup A_2 \cup A_1) - Bel(A_2 \cup A_1)}{|A - (A_2 \cup A_1)|}$$

is maximal. If there are more such sets than one, then the one with the largest cardinality $|A - (A_2 \cup A_1)|$ is chosen. Each element of set $A_3 - (A_1 \cup A_2)$ is assigned the probability

$$p_3 = \frac{Bel(A_3 \cup A_2 \cup A_1) - Bel(A_2 \cup A_1)}{|A_3 - (A_2 \cup A_1)|}.$$

- In general, for any $i \in \mathbb{N}_n$, A_i is equal to $A \subseteq A_i$ such that

$$\frac{Bel\left(A \cup \bigcup_{j=1}^{i-1} A_j\right) - Bel\left(\bigcup_{j=1}^{i-1} A_j\right)}{\left|A - \left(\bigcup_{j=1}^{i-1} A_j\right)\right|} \tag{D.1}$$

is maximal. If there are more such sets than one, then the one with the largest cardinality $|A - (\bigcup_{j=1}^{i-1} A_j)|$ is chosen. Each element of set $A_i - (\bigcup_{j=1}^{i-1} A_j)$ is assigned the probability

$$p_i = \frac{Bel\left(\bigcup_{j=1}^{i} A_j\right) - Bel\left(\bigcup_{j=1}^{i-1} A_j\right)}{\left|A_i - \left(\bigcup_{j=1}^{i-1} A_j\right)\right|}. \tag{D.2}$$

In the following, let m denote the basic probability assignment function associated with the considered belief function Bel. Observe that in each iteration of the algorithm, values of m for some focal elements are converted to probabilities. In pass 1, these are focal elements B such that $B \subseteq A_1$; in pass 2, they are focal elements B such that $B \subseteq A_1 \cup A_2$, but $B \not\subseteq A$. In general, in the ith iteration they are focal elements in the family

$$\mathcal{F}_i = \left\{B \subseteq X \,\middle|\, B \subseteq \bigcup_{j=1}^{i} A_j \text{ and } B \subseteq \bigcup_{j=1}^{i-1} A_j\right\}. \tag{D.3}$$

The regularities of Algorithm 6.1, expressed by Eqs. (D.1)–(D.3), make it possible to decompose $\bar{S}$ and GH into n components, $\bar{S}_i$ and GH_i, one for each iteration of the algorithm, such that

$$\bar{S}(Bel) = \sum_{i=1}^{n} \bar{S}_i(Bel), \tag{D.4}$$

$$GH(Bel) = \sum_{i=1}^{n} GH_i(Bel). \tag{D.5}$$

The proof that $\bar{S}(Bel) - GH(Bel) \geq 0$ then can be accomplished by proving that $\bar{S}_i(Bel) \geq GH_i(Bel)$ for all $i \in \mathbb{N}_n$. To pursue the proof, it is convenient to prove the following Lemma first.

Lemma D.1. In each iteration of Algorithm 6.1,

$$p_i \leq \frac{1}{\left| \bigcup_{j=1}^{i} A_j \right|}. \tag{D.6}$$

Proof. Assume that inequality (D.6) does not hold for some $i \in \mathbb{N}_n$. Then,

$$p_i > \frac{1}{\left| \bigcup_{j=1}^{i} A_j \right|}. \tag{D.7}$$

Due to the maximization of the assigned probabilities in each iteration of the algorithm, $p_j > p_i$ for all $j < i$. Hence,

$$p_j > \frac{1}{\left| \bigcup_{j=1}^{i} A_j \right|} \tag{D.8}$$

for all $j \in \mathbb{N}_i$ under the assumption (D.7).

According to Eq. (D.2), $|A_i - \bigcup_{j=1}^{i-1} A_j|$ elements of X are assigned the probability p_i in the ith iteration of the algorithm. Hence, the inequality

$$\sum_{j=1}^{i} \left| A_j - \bigcup_{k=1}^{j-1} A_k \right| p_j \leq 1 \tag{D.9}$$

must hold since all the probabilities generated by the algorithm must add to 1. However,

$$\sum_{j=1}^{i}\left|A_j - \bigcup_{k=1}^{j-1} A_k\right| p_j = |A_1|p_1 + |A_2 - A_1|\, p_2 + \cdots + \left|A_i - \bigcup_{j=1}^{i-1} A_j\right| p_i$$

$$> \frac{1}{\left|\bigcup_{j=1}^{i} A_j\right|}\left(|A_1| + |A_2 - A_1| + \cdots \right.$$
$$\left. + \left|A_i - \bigcup_{j=1}^{i-1} A_j\right|\right) \quad \text{(by Eq. (D.8))}$$

$$= \frac{1}{\left|\bigcup_{j=1}^{i} A_j\right|}\left(|A_1| + \left(|A_2 \cup A_1| - |A_1|\right) + \cdots \right.$$
$$\left. + \left(\left|\bigcup_{j=1}^{i} A_j\right| - \left|\bigcup_{j=1}^{i-1} A_j\right|\right)\right)$$

$$= \frac{\left|\bigcup_{j=1}^{i} A_j\right|}{\left|\bigcup_{j=1}^{i} A_j\right|} = 1,$$

which contradicts inequality (D.9). ∎

Theorem D.1. For any belief function *Bel* and the associated basic probability assignment function m, $\bar{S}(m) - GH(m) \geq 0$.

Proof. Assume that Algorithm 6.1 terminates after n iterations. Let

$$GH_i = \sum_{B \in \mathcal{F}_i} m(B)\log_2|B|, \tag{D.10}$$

where $\mathcal{F}_i$ is defined by Eq. (D.3). Then,

$$\sum_{i=1}^{n} GH_i = GH(m) \tag{D.11}$$

and, since $B \not\subseteq \bigcup_{j=1}^{i-1} A_j$ for all $B \in \mathcal{F}_i$,

$$\sum_{B \in \mathcal{F}_i} m(B)\log_2|B| \leq \sum m(B)\log_2\left|\bigcup_{j=1}^{i} A_j\right|. \tag{D.12}$$

Now define

$$\bar{S}_i = -\left|A_i - \bigcup_{j=1}^{i-1} A_j\right| p_i \log_2 p_i, \tag{D.13}$$

which represents the contribution to $\bar{S}(m)$ by probabilities generated in iteration i of the algorithm. That is,

$$\sum_{i=1}^{n} \bar{S}_i = \bar{S}(m). \tag{D.14}$$

Now observe that

$$Bel\left(\bigcup_{j=1}^{i} A_j\right) - Bel\left(\bigcup_{j=1}^{i-1} A_j\right) = \sum_{B \in \mathcal{F}_i} m(B). \tag{D.15}$$

Applying this equation and Eq. (D.2) to Eq. (D.13), we obtain

$$\bar{S}_i = -\left|A_i - \left(\bigcup_{j=1}^{i-1} A_j\right)\right| \frac{\sum\limits_{B \in \mathcal{F}_i} m(B)}{\left|A_i - \left(\bigcup_{j=1}^{i-1} A_j\right)\right|} \log_2 p_i$$

$$= -\sum_{B \in \mathcal{F}_i} m(B) \log_2 p_i. \tag{D.16}$$

The proof is now completed by showing that $\bar{S}_i \geq GH_i$ for all $i \in \mathbb{N}_n$:

$$\bar{S}_i = -\sum_{B \in \mathcal{F}_i} m(B) \log_2 p_i$$

$$\geq -\sum_{B \in \mathcal{F}_i} m(B) \log_2 \left(\frac{1}{\left|\bigcup_{j=1}^{i} A_j\right|}\right) \quad \text{(by Lemma 9.1)}$$

$$= \sum_{B \in \mathcal{F}_i} m(B) \log_2 \left|\bigcup_{j=1}^{i} A_j\right|$$

$$\geq \sum_{B \in \mathcal{F}_i} m(B) \log_2 |B| \quad \text{(by Eq. (D.12))}$$

$$= GH_i. \qquad\blacksquare$$

APPENDIX E

MAXIMUM OF GS_a IN SECTION 6.9

It is stated in Section 6.9 that the maximum, $GS_a^*(n)$, of the functional

$$GS_a(n) = -\sum_{i=1}^{n} m_i \log_2(m_i i), \tag{E.1}$$

where $m_i \in [0, 1]$ for all $i \in \mathbb{N}_n$ and

$$\sum_{i=1}^{n} m_i = 1, \tag{E.2}$$

is obtained when

$$m_i = (1/i)2^{-(a+1/\ln 2)} \tag{E.3}$$

for all $i \in \mathbb{N}_n$; the value of α is determined by solving the equation

$$2^{-(\alpha+1/\ln 2)} \sum_{i=1}^{n} (1/i) = 1. \tag{E.4}$$

In order to derive this maximum, it is convenient to use the method of Lagrange multipliers. Using the constraint Eq. (E.2) of values of m_i, we form the Lagrange function

$$L = -\sum_{i=1}^{n} m_i \log_2(m_i i) - \alpha \left(\sum_{i=1}^{n} m_i - 1 \right),$$

Uncertainty and Information: Foundations of Generalized Information Theory, by George J. Klir
© 2006 by John Wiley & Sons, Inc.

where α is a Lagrange multiplier. Now, setting the partial derivatives of L with respect to the individual variables m_i to zero, we obtain

$$\frac{\partial L}{\partial m_i} = -\log_2(m_i i) - (1/\ln 2) - \alpha = 0$$

for all $i \in \mathbb{N}_n$. Solving these equations for m_i, we obtain

$$m_i = (1/i)2^{-(\alpha+1/\ln 2)} \qquad (\text{E.5})$$

for all $m_i \in \mathbb{N}_n$. Adding all these equations results in

$$1 = \sum_{i=1}^{n} (1/i)2^{-(\alpha+1/\ln 2)},$$

due to Eq. (E.2). This equation can be rewritten as

$$1 = 2^{-(\alpha+1/\ln 2)} \sum_{i=1}^{n} (1/i). \qquad (\text{E.6})$$

Introducing, for convenience

$$S_n = \sum_{i=1}^{n} (1/i),$$

and solving Eq. (E.6) for α, we readily obtain

$$\alpha = -\log_2(1/s_n) - (1/\ln 2). \qquad (\text{E.7})$$

Substituting this expression for α into Eq. (E.5), we obtain

$$m_i = (1/i)2^{\log_2(1/s_n)}$$
$$= 1/s_n i.$$

Finally, substituting this expression for each m_i into Eq. (E.1), we obtain

$$GA_a^*(n) = \sum_{i=1}^{n} (1/i)(1/s_n)\log_2 s_n$$
$$= [(1/s_n)\log_2 s_n]\sum_{i=1}^{n} (1/i)$$
$$= \log_2 s_n.$$

APPENDIX F

GLOSSARY OF KEY CONCEPTS

Additive (classical) measure. A set function $\mu: C \to \mathbb{R}^+$ defined on a nonempty family C of subsets of a given set X for which $\mu(\emptyset) = 0$ and $\mu(A \cup B) = \mu(A) + \mu(B)$ for all $A, B, A \cup B \in C$ such that $A \cap B = \emptyset$.

Aggregate uncertainty. For a given convex set $\mathcal{D}$ of probability distributions, aggregate uncertainty (subsuming nonspecificity and conflict) is defined by the maximal value of the Shannon entropy within $\mathcal{D}$.

Alternating Choquet capacity of order k. A subadditive measure μ on $\langle X, C \rangle$ that satisfies the inequality

$$\mu\left(\bigcap_{j=1}^{k} A_j\right) \leq \sum_{\substack{K \subseteq \mathbb{N}_k \\ K \neq \emptyset}} (-1)^{|K|+1} \mu\left(\bigcup_{j \in K} A_j\right)$$

for all families of k sets in C.

Basic possibility function. A possibility measure restricted to singletons.

Basic probability assignment. For any given nonempty and finite set X, a function $m: \mathcal{P}(X) \to [0, 1]$ such that $m(\emptyset) = 0$, $\sum_{A \subseteq X} m(A) = 1$, and $m(A) \geq 0$ for all $A \subseteq X$.

Belief measure. Choquet capacity of infinite order.

Bit. A unit of uncertainty: one bit of uncertainty is equivalent to uncertainty regarding the truth or falsity of one elementary proposition.

Characteristic function. Function $\chi_A: X \to \{0, 1\}$ by which a subset A of a given universal set X is defined. For each $x \in X$,

$$\chi_A = \begin{cases} 1 & \text{if } x \text{ is a member of } A \\ 0 & \text{if } x \text{ is not a member of } A. \end{cases}$$

Uncertainty and Information: Foundations of Generalized Information Theory, by George J. Klir
© 2006 by John Wiley & Sons, Inc.

Choquet capacity of order k. A superadditive measure μ on $\langle X, C \rangle$ that satisfies the inequalities

$$\mu\left(\bigcup_{j=1}^{k} A_j\right) \geq \sum_{\substack{K \subseteq \mathbb{N}_k \\ K \neq \varnothing}} (-1)^{|K|+1} \mu\left(\bigcap_{j \in K} A_j\right)$$

for all families of k sets in C.

Choquet integral. Given a nonnegative, finite, and measurable function f on set X and a monotone measure on a family C of subsets of X that contains $\varnothing$ and X, the Choquet integral of f with respect to μ on a given set $A \in C$, $(C)\int_A f d\mu$, is defined by the formula

$$(C)\int_A f \, d\mu = \int_A \mu(A \cap {}^{\alpha}F) \, d\alpha,$$

where ${}^{\alpha}F = \{x \,|\, f(x) \geq \alpha\}$, $\alpha \in [0, \infty)$.

Compatibility relation. Binary relation on X^2 that is reflexive and symmetric.

Conflict. The type of information-based uncertainty that is measured by the Shannon entropy and its various generalizations.

Convex fuzzy subset A of $\mathcal{R}^n$. Fuzzy set whose α-cuts are convex subset of $\mathcal{R}^n$ in the classical sense for all $\alpha \in (0, 1]$.

Convex subset A of $\mathcal{R}^n$. For any pair of points $\mathbf{r} = \langle r_i | i \in \mathbb{N}_n \rangle$ and $\mathbf{s} = \langle s_i | i \in \mathbb{N}_n \rangle$ in A and every real number $\lambda \in [0, 1]$, the point $t = \langle \lambda r_i + (1 - \lambda) s_i | i \in \mathbb{N}_n \rangle$ is also in A.

Cutworthy property. Any property of classical sets that is fuzzified via the α-cut representation by requiring that it holds in the classical sense in all α-cuts of the fuzzy sets involved.

Defuzzification. A replacement of a given fuzzy interval by a single real number that, in the context of a given application, best represents the fuzzy interval.

Dempster–Shafer theory. Theory of uncertainty based on belief measures and plausibility measures.

Disaggregated total uncertainty. For a given set $\mathcal{D}$ of probability distributions, either the pair

$$TU = \langle GH, \bar{S} - GH \rangle$$

or, alternatively, the pair

$${}^{a}TU = \langle \bar{S} - \underline{S}, \underline{S} \rangle$$

where GH, $\bar{S}$ and $\underline{S}$ denote, respectively, the generalized Hartley measure, the aggregate uncertainty, and the minimal value of the Shannon entropy within $\mathcal{D}$.

Equivalence relation. Binary relation on X^2 that is reflexive, symmetric, and transitive.

Fuzzification. A process of imparting fuzzy structure to a definition (concept), a theorem, or even a whole theory.

Fuzziness. The type of uncertainty that is not based on information deficiency, but rather on the linguistic imprecision (vagueness) of natural language.

Fuzzy partition of set X. A finite family $\{A_i | A_i \in \mathcal{F}(X), A_i \neq \emptyset, i \in \mathbb{N}_n, n \geq 1\}$ of fuzzy subsets A_i of X such that for each $x \in X$

$$\sum_{i \in \mathbb{N}_n} A_i(x) = 1.$$

Fuzzy complement. A function $c: [0, 1] \to [0, 1]$ that is monotonic decreasing and satisfies $c(0) = 1$ and $c(1) = 0$; also it is usually continuous and such that $c(c(a)) = a$ for any $a \in [0, 1]$.

Fuzzy implication. Function J of the form $[0, 1]^2 \to [0, 1]$ that for any truth values a, b of given fuzzy propositions p, q, respectively, defines the truth value $J(a, b)$, of the proposition "if p, then q."

Fuzzy number. Normal fuzzy sets on $\mathbb{R}$ whose support is bounded and whose α-cuts are closed intervals of real numbers for all $\alpha \in (0, 1]$.

Fuzzy relation. Fuzzy subset of a Cartesian product of several crisp sets.

Fuzzy system. A system whose variables range over states that are fuzzy numbers or fuzzy intervals (or some other relevant fuzzy sets).

Generalized Hartley measure. A functional GH defined by the formula

$$GH(\mathcal{D}) = \sum_{A \subseteq X} m_{\mathcal{D}}(A) \log_2 |A|,$$

where $\mathcal{D}$ is a given convex set of probability distributions on a finite set X and $m_{\mathcal{D}}$ is the Möbius representation associated with $\mathcal{D}$.

Generalized Shannon entropy. For a given convex set $\mathcal{D}$ of probability distributions, the difference between the aggregate uncertainty and the generalized Hartley measure or, alternatively, the minimal value of the Shannon entropy within $\mathcal{D}$.

Hartley-like measure of uncertainty. The functional defined by Eq. (2.38) by which the uncertainty associated with any bounded and convex subset of $\mathbb{R}^n$ is measured.

Hartley measure of uncertainty. The functional $H(E) = \log_2 |E|$, where E is a finite set of possible alternatives and uncertainty is measured in bits.

Information transmission. In every theory of uncertainty, the difference between the sum of marginal uncertainties and the joint uncertainty.

Interval-valued probability distribution. For a given finite set X, a tuple $\langle [\underline{p}(x), \bar{p}(x)] | x \in X \rangle$ such that $\sum_{x \in X} \underline{p}(x) \leq 1$ and $\sum_{x \in X} \bar{p}(x) \geq 1$.

k-Monotone measures. An alternative name for Choquet capacities of order k.

Linguistic variable. A variable whose states are fuzzy intervals assigned to relevant linguistic terms.

Lower probability function. For any given set $\mathcal{D}$ of probability distribution functions p on a finite set X, the lower probability function $^{\mathcal{D}}\underline{\mu}$ is defined for all sets $A \subseteq X$ by the formula

$$^{\mathcal{D}}\underline{\mu}(A) = \inf_{p \in \mathcal{D}} \sum_{x \in A} p(x).$$

Measure of fuzziness of a fuzzy set. A functional that usually measures the lack of distinction between a given fuzzy set and its complement.

Möbius representation. Given a monotone measure μ on $\langle X, \mathcal{P}(X) \rangle$, where X is finite, its Möbius representation is a set function m defined for all $A \in \mathcal{P}(X)$ by the formula

$$m(A) = \sum_{B|B \subseteq A} (-1)^{|A-B|} \mu(B).$$

Monotone measure. A set function $\mu: C \to \mathbb{R}^+$ defined on a nonempty family C of a given set X for which $\mu(\varnothing) = 0$ and $\mu(A) \le \mu(B)$ for all $A, B \in C$ such that $A \subseteq B$.

Necessity measure. A superadditive measure μ for which $\mu(A \cap B) = \min\{\mu(A), \mu(B)\}$.

Nested family of crisp sets. Family of sets $\{A_1, A_2, \ldots, A_n\}$ such that $A_i \subseteq A_{i+1}$ for all $i = 1, 2, \ldots, n-1$.

Nonspecificity. The type of information-based uncertainty that is measured by the Hartley measure and its various generalizations, or, alternatively, by the difference between the maximum and minimum values of the Shannon entropy within a given convex set of probability distributions.

Partial ordering. A binary relation on X^2 that is reflexive, antisymmetric, and transitive.

Partition of X. A disjoint family $\{A_1, A_2, \ldots A_n\}$ of nonempty subsets of X such that $\bigcup_{i=1}^{n} A_i = X$.

Plausibility measure. Alternating Choquet capacity of infinite order.

Possibility measure. A subadditive measure μ for which $\mu(A \cup B) = \max\{\mu(A), \mu(B)\}$.

Possibility profile. A tuple consisting of all values of a basic possibility function in decreasing order.

Relation. A subset (crisp or fuzzy) of a Cartesian product.

Shannon cross-entropy. The functional

$$S[p(x), q(c) | x \in X] = \sum_{x \in X} p(x) \log_2 \frac{p(x)}{q(x)},$$

where p and q are probability distribution functions and X is a finite set. When $X = \mathbb{R}$, p and q are probability density functions, and

$$S[p(x), q(c) | x \in \mathbb{R}] = \int_{\mathbb{R}} p(x) \log_2 \frac{p(x)}{q(x)} dx.$$

Shannon entropy. The functional $S[p(x) | x \in X] = -\sum_{x \in X} p(x) \log_2 p(x)$, where p is a probability distribution function on a finite set X. This functional measures the amount of uncertainty associated with function p.

Standard complement of fuzzy set A. Fuzzy set whose membership function is defined by $1 - A(x)$ for all $x \in X$.

Standard intersection of fuzzy sets A and B. Fuzzy set whose membership function is defined by $\min\{A(x), B(x)\}$ for all $x \in X$.

Standard union of fuzzy sets A and B. Fuzzy set whose membership function is defined by $\max\{A(x), B(x)\}$ for all $x \in X$.

Strong α-cut of fuzzy set A. Crisp set $^{\alpha+}A = \{x | A(x) \geq \alpha\}$.

α-cut of fuzzy set A. Crisp set $^{\alpha}A = \{x | A(x) \geq \alpha\}$.

Subadditive measure. A monotone measure μ on $\langle X, C \rangle$ for which $\mu(A \cup B) \leq \mu(A) + \mu(B)$ whenever $A, B, A \cup B \in C$.

Sugeno λ-measure. A monotone measure $^{\lambda}\mu$ on $\langle X, \mathcal{P}(X) \rangle$ such that

$$^{\lambda}\mu(A \cup B) = {}^{\lambda}\mu(A) + {}^{\lambda}\mu(B) + \lambda \, {}^{\lambda}\mu(A)\mu(B)$$

for any disjoint subsets A, B of X, where $\lambda \in (-1, \infty)$.

Superadditive function. A monotone measure on $\langle X, C \rangle$ for which $\mu(A \cup B) \geq \mu(A) + \mu(B)$ whenever $A, B, A \cup B \in C$.

Triangular conorm (t-conorm). A function $u : [0, 1]^2 \rightarrow [0, 1]$ that is commutative, associative, monotone nondecreasing, and such that $u(a, 0) = a$ for all $a \in [0, 1]$. Triangular conorms qualify as union operations of fuzzy sets.

Triangular norm (t-norm). A function $i : [0, 1]^2 \rightarrow [0, 1]$ that is commutative, associative, monotone nondecreasing, and such that $i(a, 1) = a$ for all $a \in [0, 1]$. Triangular norms qualify as intersection operations of fuzzy sets.

Uncertainty-based information. The difference between a priori uncertainty and a reduced a posteriori uncertainty.

Universal set. The collection of all objects that are of interest in a given application.

Upper probability function. For any given set $\mathcal{D}$ of probability distribution functions p on a finite set X, upper probability function $^{\mathcal{D}}\overline{\mu}$ is defined for all $A \subseteq X$ by the formula

$$^{\mathcal{D}}\overline{\mu}(A) = \sup_{p \in \mathcal{D}} \sum_{x \in A} p(x).$$

APPENDIX G

GLOSSARY OF SYMBOLS

GENERAL SYMBOLS

$\{x, y, \ldots\}$	Set of elements $x, y, \ldots$
$\{x \mid p(x)\}$	Set determined by property p
$\langle x_1, x_2, \ldots, x_n \rangle$	n tuple
$[x_{ij}]$	Matrix
$[x_1, x_2, \ldots, x_n]$	Vector
$[a, b]$	Closed interval of real numbers between a and b
$[a, b), (b, a]$	Interval of real numbers closed in a and open in b
(a, b)	Open interval of real numbers
$[a, \infty]$	Set of real numbers greater than or equal to a
$A, B, C, \ldots$	Arbitrary sets (crisp or fuzzy)
X	Universal set (universe of discourse)
$\emptyset$	Empty set
$x \in A$	Element x belongs to crisp set A
χ_A	Characteristic function of crisp set A
$A(x)$	Membership grade of x in fuzzy set A
$A = B$	Equality of sets (crisp or fuzzy)
$A \neq B$	Inequality of sets (crisp or fuzzy)
$A - B$	Difference of crisp sets ($A - B = \{x \mid x \in A$ and $x \notin A\}$)
$A \subseteq B$	Set inclusion of crisp or fuzzy sets
$A \subset B$	Proper set inclusion ($A \subseteq B$ and $A \neq B$) of crisp or fuzzy sets
$SUB(A, B)$	Degree of subsethood of A in B
$\mathcal{P}(X)$	Set of all crisp subsets of X (power set)
$\mathcal{F}(X)$	Set of all fuzzy subsets of X (fuzzy power set)
$|A|$	Cardinality of crisp or fuzzy set A (sigma count)
h_A	Height of fuzzy set A

Uncertainty and Information: Foundations of Generalized Information Theory, by George J. Klir
© 2006 by John Wiley & Sons, Inc.

$\bar{A}$	Complement of crisp set A
$A \cap B$	Intersection of crisp sets
$A \cup B$	Union of crisp sets
$A \times B$	Cartesian product of crisp sets A and B
A^2	Cartesian product $A \times A$
$[a, b]^2$	Cartesian product $[a, b] \times [a, b]$
$X \rightarrow Y$	Function from X to Y
$R(X, Y)$	Relation on $X \times Y$ (crisp or fuzzy)
$R \circ Q$	Standard composition of relations R and Q (crisp or fuzzy)
$R * Q$	Standard join of relations R and Q (crisp or fuzzy)
R^{-1}	Inverse of a binary relation (crisp or fuzzy)
$<$	Less than
$\leq$	Less than or equal to (also used for a partial ordering)
$x \wedge y$	Meet (greatest lower bound) of x and y in a lattice or logic conjunction
$x \vee y$	Join (least upper bound) of x and y in a lattice or logic disjunction
$x\|y$	x given y
$x \Rightarrow y$	x implies y
$x \Leftrightarrow y$	x if and only if y
$\forall$	For all (universal quantifier)
$\exists$	There exists (existential quantifier)
Σ	Summation
Π	Product
$\max\{a_1, a_2, \ldots, a_n\}$	Maximum of $\{a_1, a_2, \ldots, a_n\}$
$\min\{a_1, a_2, \ldots, a_n\}$	Minimum of $\{a_1, a_2, \ldots, a_n\}$
i, j, k	Arbitrary identifiers (indices)
I, J, K	General sets of identifiers
$\mathbb{N}$	Set of positive integers (natural numbers)
$\mathbb{N}_n$	Set $\{1, 2, \ldots, n\}$
$\mathbb{R}$	Set of all real numbers
$\mathbb{R}^+$	Set of nonnegative real numbers
$\mathbb{Z}$	Set of all integers
$\pi(A)$	Partition of crisp set A

SPECIAL SYMBOLS

AU	Aggregate uncertainty in Dempster–Shafer theory
$^{\alpha}A$	α-Cut of fuzzy set A for some value $\alpha \in [0, 1]$
$^{\alpha+}A$	Strong α-cut of fuzzy set A for some value $\alpha \in [0, 1]$
Bel	Belief measure
c	Fuzzy complement

$(C)\int_A f\,d\mu$	Choquet integral of function f with respect to monotone measure μ
$d(A)$	Defuzzified value of fuzzy set A
$\mathcal{D}$	Convex set of probability distributions (credal set)
$\mathcal{F}$	Set of focal elements in evidence theory
$\langle \mathcal{F}, m \rangle$	Body of evidence
i	Fuzzy intersection or t-norm
m	Möbius representation of a lower probability function
Nec	Necessity measure
Nec_F	Necessity measure associated with a fuzzy proposition "V is F"
Pl	Plausibility measure
Pos	Possibility measure
Pos_F	Possibility measure associated with a fuzzy proposition "V is F"
$\mathbf{p}_X, \mathbf{p}_Y$	Marginal probability distributions
$\mathbf{r}$	Possibility profile
u	Fuzzy union or t-conorm
$f(A)$	Measure of fuzziness of fuzzy set A
GH	Generalized Hartley measure
GS	Generalized Shannon entropy
h	Averaging operation
H	Hartley measure
HL	Hartley-like measure (for convex subsets of $\mathbb{R}^n$)
S	Shannon entropy
$\underline{S}$	Minimal value of Shannon entropy within a given convex set of probability distributions (a generalized Shannon entropy)
$\bar{S}$	Aggregate uncertainty for convex sets of probability distributions
T	Information transmission
TU	Disaggregated total uncertainty $TU = \langle GH, \bar{S} - GH \rangle$
$^a TU$	Alternative disaggregated total uncertainty $^a TU = \langle \bar{S} - \underline{S}, \underline{S} \rangle$
U	U-Uncertainty (measure of nonspecificity in the theory of graded possibilities)
μ	Monotone measure
$^\lambda \mu$	Sugeno λ-measure
$\underline{\mu}$	Lower probability function
$\overline{\mu}$	Upper probability function

BIBLIOGRAPHY

Abellán, J., and Moral, S. [1999], "Completing a total uncertainty measure in the Dempster–Shafer theory." *International Journal of General Systems*, **28**(4–5), pp. 299–314.

Abellán, J., and Moral, S. [2000], "A non-specificity measure for convex sets of probability distributions." *International Journal of Uncertainty, Fuzziness, and Knowledge-Based Systems*, **8**(3), pp. 357–367.

Abellán, J., and Moral, S. [2003a], "Maximum of entropy for credal sets." *International Journal of Uncertainty, Fuzziness, and Knowledge-Based Systems*, **11**(5), pp. 587–597.

Abellán, J., and Moral, S. [2003b], "Building classification trees using the total uncertainty criterion." *International Journal of Intelligent Systems*, **18**(12), pp. 1215–1225.

Abellán, J., and Moral, S. [2004], "Range of entropy for credal sets." In M. Lopez-Díaz et al. (eds.), *Soft Methodology and Random Information Systems.* Springer, Berlin and Heidelberg, pp. 157–164.

Abellán, J., and Moral, S. [2005], "Difference of entropies as a non-specificity function on credal sets." *International Journal of General Systems*, **34**(3), pp. 203–217.

Aczél, J. [1966], *Lectures on Functional Equations and Their Applications.* Academic Press, New York.

Aczél, J. [1984], "Measuring information beyond communication theory." *Information Processing & Management*, **20**(3), pp. 383–395.

Aczél, J., and Daróczy, Z. [1975], *On Measures of Information and Their Characterizations.* Academic Press, New York.

Aczél, J.; Forte, B.; and Ng, C. T. [1974], "Why the Shannon and Hartley entropies are 'natural'." *Advances in Applied Probability*, **6**, pp. 131–146.

Alefeld, G., and Herzberger, J. [1983], *Introduction to Interval Computation.* Academic Press, New York.

Applebaum, D. [1996], *Probability and Information: An Integrated Approach.* Cambridge University Press, Cambridge and New York.

Arbib, M. A., and Manes, E. G. [1975], "A category-theoretic approach to systems in a fuzzy world." *Synthese*, **30**, pp. 381–406.

Uncertainty and Information: Foundations of Generalized Information Theory, by George J. Klir
© 2006 by John Wiley & Sons, Inc.

Ash, R. B. [1965], *Information Theory.* Interscience, New York (reprinted by Dover, New York, 1990).

Ashby, W. R. [1964], "Constraint analysis of many-dimensional relations." *General Systems Yearbook,* **9**, pp. 99–105.

Ashby, W. R. [1965], "Measuring the internal informational exchange in a system." *Cybernetica,* **8**(1), pp. 5–22.

Ashby, W. R. [1969], "Two tables of identities governing information flows within large systems." *ASC Communications,* **1**(2), pp. 3–8.

Ashby, W. R. [1970], "Information flows within coordinated systems." In J. Rose (ed.), *Progress in Cybernetics,* Vol. 1. Gordon and Breach, London, pp. 57–64.

Ashby, W. R. [1972], "Systems and their informational measures." In G. J. Klir (ed.), *Trends in General Systems Theory.* Wiley-Interscience, New York, pp. 78–97.

Atanassov, K. T. [1986], "Intuitionistic fuzzy sets." *Fuzzy Sets and Systems,* **20**(1), pp. 87–96.

Atanassov, K. T. [2000], *Intuitionistic Fuzzy Sets.* Springer-Verlag, New York.

Attneave, F. [1959], *Applications of Information Theory to Psychology.* Holt, Rinehart & Winston, New York.

Aubin, J. P., and Frankowska, H. [1990], *Set-Valued Analysis.* Birkhäuser, Boston.

Auman, R. J., and Shapley, L. S. [1974], *Values of Non-Atomic Games.* Princeton University Press, Princeton, NJ.

Avgers, T. G. [1983], "Axiomatic derivation of the mutual information principle as a method of inductive inference." *Kybernetes,* **12**(2), pp. 107–113.

Babuška, R. [1998], *Fuzzy Modeling for Control.* Kluwer, Boston.

Ban, A. I., and Gal, S. G. [2002] *Defects of Properties in Mathematics.* World Scientific, Singapore.

Bandler, W., and Kohout, L. J. [1980a], "Fuzzy power set and fuzzy implication operators." *Fuzzy Sets and Systems,* **4**(1), pp. 13–30.

Bandler, W., and Kohout, L. J. [1980b], "Semantics of implication operators and fuzzy relational products." *International Journal of Man-Machine Studies,* **12**(1), pp. 89–116.

Bandler, W., and Kohout, L. J. [1988], "Special properties, closures and interiors of crisp and fuzzy relations." *Fuzzy Sets and Systems,* **26**(3), pp. 317–331.

Bandler, W., and Kohout, L. J. [1993], "Cuts commute with closures." In: R. Lowen and M. Roubens (eds.), *Fuzzy Logic: State of the Art.* Kluwer, Dordrecht and Boston, pp. 161–167.

Banon, G. [1981], "Distinction between several subsets of fuzzy measures." *Fuzzy Sets and Systems,* **5**(3), pp. 291–305.

Bárdossy, G., and Fodor, J. [2004], *Evaluation of Uncertainties and Risks in Geology.* Springer, Berlin and Heidelberg.

Batten, D. F. [1983], *Spatial Analysis of Interacting Economics.* Kluwer-Nighoff, Boston.

Bell, D. A. [1953], *Information Theory and Its Engineering Applications.* Pitman, New York.

Bell, D. A.; Guan, J. W.; and Shapcott, C. M. [1998], "Using the Dempster-Shafer orthodonal sum for reasoning which involves space." *Kybernetes,* **27**(5), pp. 511–526.

Bellman, R. [1961], *Adaptive Control Processes: A Guided Tour.* Princeton University Press, Princeton, NJ.

Bělohlávek, R. [2002], *Fuzzy Relational Systems*. Kluwer/Plenum, New York.

Bělohlávek, R. [2003], "Cutlike semantics for fuzzy logic and its applications." *International Journal of General Systems*, **32**(4), pp. 305–319.

Ben-Haim, Y. [2001], *Information-Gap Decision Theory*. Academic Press, San Diego.

Benvenuti, P., and Mesiar, R. [2000], "Integrals with respect to a general fuzzy measure." In M. Grabish et al. (eds.), *Fuzzy Measures and Integrals*. Springer-Verlag, New York, pp. 203–232.

Bezdek, J. C.; Dubois, D.; and Prade, H. (eds) [1999], *Fuzzy Sets in Approximate Reasoning and Information Systems*. Kluwer, Boston.

Bharathi-Devi, B., and Sarma, V. V. S. [1985], "Estimation of fuzzy memberships from histograms." *Information Sciences*, **35**(1), pp. 43–59.

Bhattacharya, P. [2000], "On the Dempster-Shafer evidence theory and non-hierarchical aggregation of belief structures." *IEEE Transactions on Systems, Man, and Cybernetics*, **30**(5), pp. 526–536.

Billingsley, P. [1965], *Ergodic Theory and Information*. John Wiley, New York.

Billingsley, P. [1986], *Probability and Measure*. John Wiley, New York.

Billot, A. [1992], *Economic Theory of Fuzzy Equilibria: An Axiomatic Analysis*. Springer-Verlag, New York.

Black, M. [1937], "Vagueness: an exercise in logical analysis." *Philosophy of Science*, **4**, pp. 427–455 (reprinted in *International Journal of General Systems*, **17**(2–3), 1990, pp. 107–128).

Black, P. K. [1997], "Geometric structure of lower probabilities." In J. Goutsias, R. P. S. Mahler, and H. T. Nguyen (eds.), *Random Sets*. Springer, New York, pp. 361–383.

Blahut, R. E. [1987], *Principles and Practice of Information Theory*. Addison-Wesley, Reading, MA.

Boekee, D. E., and Van Der Lubbe, J. C. A. [1980], "The R-norm information measure." *Information and Control*, **45**(2), pp. 136–155.

Bolc, L., and Borowic, P. [1992], *Many-Valued Logics: Theoretical Foundations*. Springer-Verlag, Berlin and Heidelberg.

Bordley, R. F. [1983], "A central principle of science: optimization." *Behavioral Science*, **28**(1), pp. 53–64.

Borgelt, C., and Kruse, R. [2002], *Graphical Models: Methods for Data Analysis and Mining*. John Wiley, New York.

Brillouin, L. [1956], *Science and Information Theory*. Academic Press, New York.

Brillouin, L. [1964], *Scientific Uncertainty and Information*. Academic Press, New York.

Broekstra, G. [1976–77], "Constraint analysis and structure identification." *Annals of Systems Research I*, **5**, pp. 67–80; *II*, **6**, 1–20.

Broekstra, G. [1980], "On the foundation of GIT (General Information Theory)." *Cybernetics and Systems*, **11**, pp. 143–165.

Buck, B., and Macaulay, V. A. [1991], *Maximum Entropy in Action*. Oxford University Press, New York.

Buckley, J. J. [2003], *Fuzzy Probabilities*. Physica-Verlag/Springer-Verlag, Heidelberg and New York.

Cai, K. Y. [1996], *Introduction to Fuzzy Reliability*. Kluwer, Boston.

Cano, A., and Moral, S. [2000], "Algorithms for imprecise probabilities." In J. Kohlas and S. Moral (eds.), *Algorithms for Uncertainty and Defeasible Reasoning.* Kluwer, Dordrecht and Boston, pp. 369–420.

Caratheodory, C. [1963], *Algebraic Theory of Measure and Integration.* Chelsea, New York.

Carlsson, C.; Fedrizzi, M.; and Fuller, R. [2004], *Fuzzy Logic in Management.* Kluwer, Boston.

Cavallo, R. E., and Klir, G. J. [1979], "Reconstructability analysis of multi-dimensional relations." *International Journal of General Systems,* **5**(3), pp. 143–171.

Cavallo, R. E., and Klir, G. J. [1981], "Reconstructability analysis: evaluation of reconstruction hypotheses." *International Journal of General Systems,* **7**(1), pp. 7–32.

Cavallo, R. E., and Klir, G. J. [1982a], "Decision making in reconstructability analysis." *International Journal of General Systems,* **8**(4), pp. 243–255.

Cavallo, R. E., and Klir, G. J. [1982b], "Reconstruction of possibilistic behavior systems." *Fuzzy Sets and Systems,* **8**(2), pp. 175–197.

Chaitin, G. J. [1987], *Information, Randomness, and Incompleteness: Papers on Algorithmic Information Theory.* World Scientific, Singapore.

Chameau, J. L., and Santamarina, J. C. [1987], "Membership functions I, II." *International Journal of Approximate Reasoning,* **1**(3), pp. 287–301, 303–317.

Chateauneuf, A., and Jaffray, J. Y. [1989], "Some characterizations of lower probabilities and other monotone capacities through the use of Möbius inversion." *Mathematical Social Sciences,* **17**, pp. 263–283.

Chau, C. W. R.; Lingras, P.; and Wong, S. K. M. [1993], "Upper and lower entropies of belief functions using compatible probability functions." In J. Komorovski and Z. W. Ras (eds.), *Methodologies for Intelligent Systems.* Springer-Verlag, New York, pp. 303–315.

Chellas, B. F. [1980], *Modal Logic: An Introduction.* Cambridge University Press, Cambridge and New York.

Cherry, C. [1957], *On Human Communication.* MIT Press, Cambridge, MA.

Chokr, B. A., and Kreinovich V. Y. [1991], "How far are we from the complete knowledge? Complexity of knowledge acquisition in the Dempster-Shafer approach." *Proc. of the Fourth Univ. of New Brunswick Artificial Intelligence Workshop,* Fredericton, N.B. Canada, pp. 551–561.

Chokr, B. A., and Kreinovich, V. [1994], "How far are we from the complete knowledge? Complexity of knowledge acquisition in the Dempster-Shafer approach." In R. R. Yager, M. Federizzi, and J. Kacprzyk (eds.), *Advances in the Dempster-Shafer Theory of Evidence.* John Wiley, New York, pp. 555–576.

Choquet, G. [1953–54], "Theory of capacities." *Annales de L'Institut Fourier,* **5**, pp. 131–295.

Christensen, R. [1980–81], *Entropy Minimax Sourcebook.* Entropy Limited, Lincoln, MA.

Christensen, R. [1985], "Entropy minimax multivariate statistical modeling—I: Theory." *International Journal of General Systems,* **11**(3), pp. 231–277.

Christensen, R. [1986], "Entropy minimax multivariate statistical modeling—II: Applications." *International Journal of General Systems,* **12**(3), pp. 227–305.

Clarke, M.; Kruse, R.; and Moral, S. (eds.) [1993], *Symbolic and Quantitative Approaches to Reasoning and Uncertainty*. Springer-Verlag, Berlin and New York.

Coletti, G.; Dubois, D.; and Scozzafava, R. (eds.) [1995], *Mathematical Models for Handling Partial Knowledge in Artificial Intelligence*. Plenum Press, New York and London.

Coletti, G., and Scozzafava, R. [2002], *Probabilistic Logic in a Coherent Setting*. Kluwer, Boston.

Colyvan, M. [2004], "The philosophical significance of Cox's theorem." *International Journal of Approximate Reasoning*, **37**(1), pp. 71–85.

Conant, R. C. [1976], "Laws of information which govern systems." *IEEE Transactions on Systems, Man, and Cybernetics*, **6**(4), pp. 240–255.

Conant, R. C. [1981], "Efficient proofs of identities in N-dimensional information theory." *Cybernetica*, **24**(3), pp. 191–197.

Conant, R. C. [1988], "Extended dependency analysis of large systems—Part I: Dynamic analysis; Part II: Static analysis." *International Journal of General Systems*, **14**(2), pp. 97–141.

Cordón, O.; Herrera, F.; Hoffmann, F.; and Magdalena, L. [2001], *Genetic Fuzzy Systems: Evolutionary Tuning and Learning of Fuzzy Knowledge Bases*. World Scientific, Singapore.

Cover, T. M., and Thomas, J. A. [1991], *Elements of Information Theory*. John Wiley, New York.

Cox, R. T. [1946], "Probability, frequency, and reasonable expectation." *American Journal of Physics*, **14**(1), pp. 1–13.

Cox, R. T. [1961], *The Algebra of Probable Inference*. Johns Hopkins Press, Baltimore.

Csiszár, I., and Körner, J. [1981], *Information Theory: Coding Theorems for Discrete Memoryless Systems*. Academic Press, New York.

Daróczy, Z. [1970], "Generalized information functions." *Information and Control*, **16**(1), pp. 36–51.

De Baets, B., and Kerre, E. E. [1994], "A primer on solving fuzzy relational equations on the unit interval." *International Journal of Uncertainty, Fuzziness and Knowledge-Based Systems*, **2**(2), pp. 205–225.

De Bono, E. [1991], *I Am Right, You Are Wrong: From Rock Logic to Water Logic*. Viking Penguin, New York.

De Campos, L. M., and Bolaños, M. J. [1989], "Representation of fuzzy measures through probabilities." *Fuzzy Sets and Systems*, **31**(1), pp. 23–36.

De Campos, L. M., and Bolaños, M. J. [1992], "Characterization and comparison of Sugeno and Choquet integrals." *Fuzzy Sets and Systems*, **52**(1), pp. 61–67.

De Campos, L. M., and Huete, J. F. [1993], "Independence concepts in upper and lower probabilities." In B. Bouchon-Meunier, L. Velverde, and R. R. Yager (eds.), *Uncertainty in Intelligent Systems*. North-Holland, Amsterdam, pp. 85–96.

De Campos, L. M., and Huete, J. F. [1999], "Independence concepts in possibility theory." *Fuzzy Sets and Systems*, **103**(1&3), pp. 127–152 & 487–505.

De Campos, L. M.; Huete, J. F.; and Moral, S. [1994], "Probability intervals: a tool for uncertain reasoning." *International Journal of Uncertainty, Fuzziness and Knowledge-Based Systems*, **2**(2), pp. 167–196.

De Campos, L. M.; Lamata, M. T.; and Moral, S. [1990], "The concept of conditional fuzzy measure." *International Journal of Intelligent Systems*, **5**(3), pp. 237–246.

De Cooman, G. [1997], "Possibility theory—I, II, III." *International Journal of General Systems*, **25**(4), pp. 291–371.

De Cooman, G.; Ruan, D.; and Kerre, E. E. (eds.) [1995], *Foundations and Applications of Possibility Theory*. World Scientific, Singapore.

De Finetti, B. [1974], *Theory of Probability* (Vol. 1). John Wiley, New York and London.

De Finetti, B. [1975], *Theory of Probability*, Vol. 2. John Wiley, New York and London.

De Luca, A., and Termini, S. [1972], "A definition of a nonprobabilistic entropy in the setting of fuzzy sets theory." *Information and Control*, **20**(4), pp. 301–312.

De Luca, A., and Termini, S. [1974], "Entropy of L-fuzzy sets." *Information and Control*, **24**(1), pp. 55–73.

Delgado, M., and Moral, S. [1989], "Upper and lower fuzzy measures." *Fuzzy Sets and Systems*, **33**(2), pp. 191–200.

Delmotte, F. [2001], "Comparison of the performances of decision aimed algorithms with Bayesian and belief bases." *International Journal of Intelligent Systems*, **16**(8), pp. 963–981.

Dembo, A.; Cover, T. C.; and Thomas, J. A. [1991], "Information theoretic inequalities." *IEEE Transactions on Information Theory*, **37**(6), pp. 1501–1518.

Demicco, R. V., and Klir, G. J. [2003], *Fuzzy Logic in Geology*. Academic Press, San Diego.

Dempster, A. P. [1967a], "Upper and lower probabilities induced by a multivalued mapping." *Annals of Mathematical Statistics*, **38**(2), pp. 325–339.

Dempster, A. P. [1967b], "Upper and lower probability inferences based on a sample from a finite univariate population." *Biometrika*, **54**, pp. 515–528.

Dempster, A. P. [1968a], "A generalization of Bayesian inference." *Journal of the Royal Statistical Society*, Ser. B, **30**, pp. 205–247.

Dempster, A. P. [1968b], "Upper and lower probabilities generated by a random closed interval." *Annals of Mathematical Statistics*, **39**, pp. 957–966.

Denneberg, D. [1994], *Non-Additive Measure and Integral*. Kluwer, Boston.

Devlin, K. [1991], *Logic and Information*. Cambridge University Press, Cambridge and New York.

Di Nola, A.; Sessa, S.; Pedrycz, W.; and Sanches, E. [1989], *Fuzzy Relation Equations and Their Applications to Knowledge Engineering*. Kluwer, Dordrecht.

Dockx, S., and Bernays, P. (eds), [1965], *Information and Prediction in Science*. Academic Press, New York.

Dretske, F. I. [1981], *Knowledge and the Flow of Information*. MIT Press, Cambridge, MA.

Dretske, F. I. [1983], "Precis of knowledge and the flow of information." *Behavioral and Brain Sciences*, **6**, pp. 55–90.

Dubois, D.; Lang, J.; and Prade, H. [1994], "Possibilistic logic." In D. M. Gabbay, et al. (eds.), *Handbook of Logic in Artificial Intelligence and Logic Programming*, Vol. 3. Clarendon Press, Oxford, pp. 439–513.

Dubois, D.; Moral, S.; and Prade, H. [1997], "A semantics for possibility theory based on likelihoods." *Journal of Mathematical Analysis and Applications*, **205**, pp. 359–380.

Dubois, D.; Nguyen, H. T.; and Prade, H. [2000] "Possibility theory, probability theory and fuzzy sets: misunderstandings, bridges and gaps." In D. Dubois and H. Prade (ed.), *Fundamentals of Fuzzy Sets*. Kluwer, Boston, pp. 343–438.

Dubois, D., and Prade, H. [1978], "Operations on fuzzy numbers." *International Journal of Systems Science*, **9**(6), pp. 613–626.

Dubois, D., and Prade, H. [1979], "Fuzzy real algebra: some results." *Fuzzy Sets and Systems*, **2**(4), pp. 327–348.

Dubois, D., and Prade, H. [1980], *Fuzzy Sets and Systems: Theory and Applications*. Academic Press, New York.

Dubois, D., and Prade, H. [1982a], "Towards fuzzy differential calculus." *Fuzzy Sets and Systems*, **8**(1), pp. 1–17; **8**(2), 105–116; **8**(3), 225–233.

Dubois, D., and Prade, H. [1982b], "A class of fuzzy measures based on triangular norms." *International Journal of General Systems*, **8**(1), pp. 43–61.

Dubois, D., and Prade, H. [1985a], "Evidence measures based on fuzzy information." *Automatica*, **21**, pp. 547–562.

Dubois, D., and Prade, H. [1985b], "A review of fuzzy set aggregation connectives." *Information Sciences*, **36**(1–2), pp. 85–121.

Dubois, D., and Prade H. [1985c], "A note on measures of specificity for fuzzy sets." *International Journal of General Systems*, **10**(4), pp. 279–283.

Dubois, D., and Prade, H. [1986a], "A set-theoretic view of belief functions." *International Journal of General Systems*, **12**(3), pp. 193–226.

Dubois, D., and Prade, H. [1986b], "On the unicity of Dempster rule of combination." *International Journal of Intelligent Systems*, **1**(2), pp. 133–142.

Dubois, D., and Prade, H. [1987a], "Fuzzy numbers: an overview." In J. C. Bezdek (ed.), *Analysis of Fuzzy Information—Vol. 1: Mathematics and Logic*. CRC Press, Boca Raton, FL, pp. 3–39.

Dubois, D., and Prade, H [1987b], "The principle of minimum specificity as a basis for evidential reasoning." In B. Bouchon and R. R. Yager (eds.), *Uncertainty in Knowledge-Based Systems*. Springer-Verlag, Berlin, pp. 75–84.

Dubois, D., and Prade, H. [1987c], "Two-fold fuzzy sets and rough sets: some issues in knowledge representation." *Fuzzy Sets and Systems*, **23**(1), pp. 3–18.

Dubois, D., and Prade, H. [1987d], "Properties of measures of information in evidence and possibility theories." *Fuzzy Sets and Systems*, **24**(2), pp. 161–182.

Dubois, D., and Prade, H. [1988a], *Possibility Theory*. Plenum Press, New York.

Dubois, D., and Prade, H. [1988b], "Modelling uncertainty and inductive inference: a survey of recent non-additive probability systems." *Acta Psychologica*, **68**, pp. 53–78.

Dubois, D., and Prade, H. [1990a], "Consonant approximations of belief functions." *International Journal of Approximate Reasoning*, **4**(5–6), pp. 419–449.

Dubois, D., and Prade, H. [1990b], "Rough fuzzy sets and fuzzy rough sets." *International Journal of General Systems*, **17**(2–3), pp. 191–209.

Dubois, D., and Prade, H. [1991], "Fuzzy sets in approximate reasoning." *Fuzzy Sets and Systems*, **40**(1), pp. 143–244.

Dubois, D., and Prade, H. [1992a], "On the combination of evidence in various mathematical frameworks." In J. Flamm and T. Luisi (eds.), *Reliability Data Collection and Analysis*. Kluwer, Dordrecht and Boston, pp. 213–241.

Dubois, D., and Prade, H. [1992b], "Putting rough sets and fuzzy sets together." In R. Slowinski (ed.), *Intelligent Decision Support*. Kluwer, Boston, pp. 203–232.

Dubois, D., and Prade, H. [1992c], "Evidence, knowledge, and belief functions." *International Journal of Approximate Reasoning*, 6(3), pp. 295–319.

Dubois, D., and Prade, H. [1994], "A survey of belief revision and updating rules in various uncertainty models." *International Journal of Intelligent Systems*, 9(1), pp. 61–100.

Dubois, D., and Prade, H. [1998], "Possibility Theory: Qualitative and Quantitative Aspects." In D. M. Gabbay and P. Smets [1998–], *Handbook of Defeasible Reasoning and Uncertainty Management Systems*, Kluwer, Dordrecht, Boston, and London, Vol. 1, pp. 169–226.

Dubois, D., and Prade, H. (eds.) [1998–], *The Handbooks of Fuzzy Sets Series*. Kluwer, Boston.

Dubois, D., and Prade, H. (eds.) [2000], *Fundamentals of Fuzzy Sets*. Kluwer, Boston.

Dvořák, A. [1999], "On Linguistic approximation in the frame of fuzzy logic deduction." *Soft Computing*, 3(2), pp. 111–115.

Ebanks, B.; Sahoo, P.; and Sanders, W. [1997], *Characterizations of Information Measures*. World Scientific, Singapore.

Eckschlager, K. [1979], *Information Theory as Applied to Chemical Analysis*. John Wiley, New York.

Elsasser, W. M. [1937], "On quantum measurements and the role of the uncertainty relations in statistical mechanics." *Physical Review*, 52, pp. 987–999.

Fagin, J., and Halpern, J. Y. [1991], "A new approach to updating beliefs." In P. P. Bonissone et al. (eds.), *Uncertainty in Artificial Intelligence 6*. North-Holland, Amsterdam and New York, pp. 347–374.

Fast, J. D. [1962], *Entropy*. Gordon and Breach, New York.

Feinstein, A. [1958], *Foundations of Information Theory*. McGraw-Hill, New York.

Feller, W. [1950], *An Introduction to Probability Theory and Its Applications*, Vol. I. John Wiley, New York.

Feller, W. [1966], *An Introduction to Probability Theory and Its Applications*, Vol. II. John Wiley, New York.

Fellin, W. et al. [2005], *Analyzing Uncertainty in Civil Engineering*. Springer, Berlin and Heidelberg.

Ferdinand, A. E. [1974], "A theory of system complexity." *International Journal of General Systems*, 1(1), pp. 19–35.

Ferson, S. [2002], *RAMAS Risk Calc 4.0 Software: Risk Assessment with Uncertainty Numbers*. Lewis Publishers, Boca Raton, FL.

Ferson, S.; Ginzburg, L.; Kreinovich, V.; Myers, D.; and Sentz, K. [2003], *Constructing Probability Boxes and Dempster–Shafer Structures*. Sandia National Laboratory, SAND 2002–4015, Albuquerque, NM.

Ferson, S., and Hajagos, J. G. [2004], "Arithmetic with uncertainty numbers: rigorous and (often) best possible answers." *Reliability and Systems Safety*, 85 (1–3), pp. 135–152.

Fine, T. L. [1973], *Theories of Probability: An Examination of Foundations*. Academic Press, New York.

Fisher, R. A. [1950], *Contributions to Mathematical Statistics*. John Wiley, New York.

Forte, B. [1975], "Why the Shannon entropy?" *Symposia Mathematica*, **15**, pp. 137–152.

Frieden, B. R. [1998], *Physics from Fisher Information: A Unification*. Cambridge University Press, Cambridge and New York.

Gabbay, D. M., and Smets, P. (eds.) [1998-], *Handbook of Defeasible Reasoning and Uncertainty Management Systems*. Kluwer, Dordrecht, Boston, and London.

Gaines, B. R. [1978], "Fuzzy and Probability Uncertainty Logics." *Information and Control*, **38**, pp. 154–169.

Garner, W. R. [1962], *Uncertainty and Structure as Psychological Concepts*. John Wiley, New York.

Gatlin, L. L. [1972], *Information Theory and the Living System*. Columbia University Press, New York.

Geer, J. F., and Klir, G. J. [1991], "Discord in possibility theory." *International Journal of General Systems*, **19**(2), pp. 119–132.

Geer, J. F., and Klir, G. J. [1992], "A mathematical analysis of information preserving transformations between probabilistic and possibilistic formulations of uncertainty." *International Journal of General Systems*, **20**(2), pp. 143–176.

Georgescu-Roegen, N. [1971], *The Entropy Law and the Economic Process*. Harvard University Press, Cambridge, MA.

Gerla, G. [2001], *Fuzzy Logic: Mathematical Tools for Approximate Reasoning*. Kluwer, Boston.

Giachetti, R. E., and Young, R. E. [1997], "A parametric representation of fuzzy numbers and their arithmetic operators." *Fuzzy Sets and Systems*, **91**(2), pp. 185–202.

Gibbs, J. W. [1902], *Elementary Principles in Statistical Mechanics*. Yale University Press, New Haven (reprinted by Ox Bow Press, Woodbridge, CT, 1981).

Gil, M. A. (ed.) [2001], "Special Issue on Fuzzy Random Variables." *Information Sciences*, **133**(1–2), pp. 1–100.

Glasersfeld, E. von [1995], *Radical Constructivism: A Way of Knowing and Learning*. The Farmer Press, London.

Gnedenko, B. V. [1962], *Theory of Probability*. Chelsea, New York.

Godo, L., and Sandri, S. [2004], "Special Issue on Possibilistic Logic and Related Issues." *Fuzzy Sets and Systems*, **144**(1), pp. 1–249.

Goguen, J. A. [1967], "*L*-fuzzy sets." *Journal of Mathematical Analysis and Applications*, **18**, pp. 145–174.

Goguen, J. A. [1968–69], "The logic of inexact concepts." *Synthese*, **19**, pp. 325–373.

Goguen, J. A. [1974], "Concept representation in natural and artificial languages: axioms, extensions and applications for fuzzy sets." *International Journal of Man-Machine Studies*, **6**(5), pp. 513–561.

Goldman, S. [1953], *Information Theory*. Prentice Hall, Englewood Cliffs, NJ. (Reprinted by Dover, New York, 1968.)

Good, I. J. [1950], *Probability and the Weighting of Evidence*. Hafner, New York, Charles Griffin, London.

Good, I. J. [1962] "Subjective probability as the measure of non-measurable set." In E. Nagel (ed.), *Logic, Methodology, and Philosophy of Science*. Stanford University Press, Stanford, CA.

Good, I. J. [1983], *Good Thinking*. Minnesota University Press, Minneapolis.

Goodman, I. R., and Nguyen, H. T. [1985], *Uncertainty Models for Knowledge-Based Systems*. North-Holland, New York.

Gottwald, S. [1979], "Set theory for fuzzy sets of higher level." *Fuzzy Sets and Systems*, **2**(2), pp. 125–151.

Gottwald, S. [1993], *Fuzzy Sets and Fuzzy Logic*. Verlag Vieweg, Wiesbaden.

Gottwald, S. [2001], *A Treatise on Many-Valued Logics*. Research Studies Press, Baldock, UK.

Goutsias, J.; Mahler, R. P. S.; and Nguyen, H. T. (eds.), [1997], *Random Sets: Theory and Applications.* Springer-Verlag, New York.

Grabisch, M. [1997a], "k-order additive discrete fuzzy measures and their representation." *Fuzzy Sets and Systems*, **92**(2), pp. 167–189.

Grabisch, M. [1997b], "Alternative representations of discrete fuzzy measures for decision making." *International Journal of Uncertainty, Fuzziness and Knowledge-Based Systems*, **5**(5), pp. 587–607.

Grabisch, M. [1997c], "Fuzzy measures and integrals: a survey of applications and recent issues." In D. D. Dubois, H. Prade, and R. R. Yager (eds.), *Fuzzy Information Engineering*. John Wiley, New York, pp. 507–529.

Grabisch, M. [2000], "The interaction and Möbius representations of fuzzy measures on finite spaces, k-additive measures: a survey." In M. Grabisch et al. (eds.), *Fuzzy Measures and Integrals: Theory and Applications.* Springer-Verlag, New York, pp. 70–93.

Grabisch, M.; Murofushi, T.; and Sugeno, M. (eds.) [2000], *Fuzzy Measures and Integrals: Theory and Applications*. Springer-Verlag, New York.

Grabisch, M.; Nguyen H. T.; and Walker, E. A. [1995], *Fundamentals of Uncertainty Calculi With Applications to Fuzzy Inference*. Kluwer, Dordrecht and Boston.

Gray, R. M. [1990], *Entropy and Information Theory*. Springer-Verlag, New York.

Greenberg, H. J. [2001], "Special Issue on Representations of Uncertainty." *Annals of Mathematics and Artifical Intelligent*, **32**(1–4), pp. 1–431.

Guan, J. W., and Bell, D. A. [1991–92], *Evidence Theory and Its Applications:* Vol. 1 (1991), Vol. 2 (1992). North-Holland, New York.

Guiasu, S. [1977], *Information Theory with Applications*. McGraw-Hill, New York.

Hacking, I. [1975], *The Emergence of Probability*. Cambridge University Press, Cambridge and New York.

Hájek, P. [1993], "Deriving Dempster's rule." In L. Valverde and R. R. Yager (eds.), *Uncertainty in Intelligent Systems.* North-Holland, Amsterdam, pp. 75–83.

Hájek, P. [1998], *Metamathematics of Fuzzy Logic*. Kluwer, Boston.

Halmos, P. R. [1950], *Measure Theory*. Van Nostrand, Princeton, NJ.

Halpern, J. Y. [1999], "A counterexample of theorems of Cox and Fine." *Journal of Artificial Intelligence Research*, **10**, pp. 67–85.

Halpern, J. Y. [2003], *Reasoning about Uncertainty*. MIT Press, Cambridge, MA.

Hamming, R. W. [1980], *Coding and Information Theory*. Prentice Hall, Engelwood Cliffs, NJ.

Hansen, E. R. [1992], *Global Optimization Using Interval Analysis*. Marcel Dekker, New York.

Harmanec, D. [1995], "Toward a characterization of uncertainty measure for the Dempster-Shafer theory." In *Proceedings of the Eleventh International Conference on Uncertainty in Artificial Intelligence*, Montreal, Canada, pp. 255–261.

Harmanec, D. [1996], *Uncertainty in Dempster-Shafer Theory*. Ph.D. Dissertation in Systems Science, T.J. Watson School, Binghamton University—SUNY, Binghamton, NY.

Harmanec, D. [1997], "A note on uncertainty, Dempster rule of combination, and conflict." *International Journal of General Systems*, 26(1–2), pp. 63–72.

Harmanec, D., and Klir, G. J. [1994], "Measuring total uncertainty in Dempster-Shafer theory: a novel approach." *International Journal of General Systems*, 22(4), pp. 405–419.

Harmanec, D., and Klir, G. J. [1997], "On information-preserving transformations." *International Journal of General Systems*, 26(3), pp. 265–290.

Harmanec, D.; Klir, G. J.; and Resconi, G. [1994], "On a modal logic interpretation of Dempster-Shafer theory of evidence." *International Journal of Intelligent Systems*, 9(10), pp. 941–951.

Harmanec, D.; Klir, G. J.; and Wang, Z. [1996], "Modal logic interpretation of Dempster-Shafer theory: An infinite case." *International Journal of Approximate Reasoning*, 14(2–3), pp. 81–93.

Harmanec, D.; Resconi, G.; Klir, G. J.; and Pan, Y. [1996], "On the computation of uncertainty measure in Dempster-Shafer theory." *International Journal of General Systems*, 25(2), pp. 153–163.

Harmuth, H. F. [1992], *Information Theory Applied to Space-Time Physics*. World Scientific, River Edge, NJ.

Hartley, R. V. L. [1928], "Transmission of information." *The Bell System Technical Journal*, 7(3), pp. 535–563.

Hawkins, T. [1975], *Lebesgue's Theory of Integration: Its Origins and Development*. Chelsea, New York.

Helton, J. C., and Oberkampf [2004], "Special Issue on Alternative Representations of Epistemic Uncertainty." *Reliability Engineering & System Safety*, 85(1–3), pp. 1–369.

Hernandez, E., and Recasens, J. [2004], "Indistinguishability relation in Dempster-Shafer theory of evidence." *International Journal of Approximate Reasoning*, 37(3), pp. 145–187.

Higashi, M., and Klir, G. J. [1982], "On measures of fuzziness and fuzzy complements." *International Journal of General Systems*, 8(3), pp. 169–180.

Higashi, M., and Klir, G. J. [1983a], "Measures of uncertainty and information based on possibility distributions." *International Journal of General Systems*, 9(1), pp. 43–58.

Higashi, M., and Klir, G. J. [1983b], "On the notion of distance representing information closeness: possibility and probability distributions." *International Journal of General Systems*, 9(2), pp. 103–115.

Higashi, M.; Klir, G. J.; and Pittarelli, M. A. [1984], "Reconstruction families of possibilistic structure systems." *Fuzzy Sets and Systems*, 12(1), pp. 37–60.

Hirshleifer, J., and Riley, J. G. [1992], *The Analytics of Uncertainty and Information*. Cambridge University Press, Cambridge, MA.

Hisdal, E. [1978], "Conditional possibilities, independence and noninteraction." *Fuzzy Sets and Systems*, 1(4), pp. 283–297.

Höhle, U. [1982], "Entropy with respect to plausibility measures." In *Proceedings of the 12th IEEE International Symposium on Multiple-Valued Logic*, Paris, pp. 167–169.

Höhle, U., and Klement, E. P. (eds.) [1995], *Non-Classical Logics and Their Applications to Fuzzy Subsets*. Kluwer, Boston.

Höhle, U., and Rodabaugh, S. E. (eds.) [1999], *Mathematics of Fuzzy Sets: Logic, Topology, and Measure Theory*. Kluwer, Boston.

Huber, P. J. [1981], *Robust Statistics*. John Wiley, New York.

Hughes, G. E., and Cresswell, M. J. [1996], *A New Introduction to Modal Logic*. Routledge, London and New York.

Hyvärinen, L. P. [1968], *Information Theory for Systems Engineers*. Springer-Verlag, New York.

Ihara, S. [1993], *Information Theory for Continuous Systems*. World Scientific, Singapore.

Jaffray, J. Y. [1997], "On the maximum of conditional entropy for upper/lower probabilities generated by random sets." In J. Goutsias, R. P. S. Mahler, and H. T. Nguyen (eds.), *Random Sets: Theory and Applications*. Springer, New York, pp. 107–128.

Jaynes, E. T. [1968], "Prior probabilities." *IEEE Transactions on Systems Science and Cybernetics*, **4**(3), pp. 227–241.

Jaynes, E. T. [1979], "Where do we stand on maximum entropy?" In R. L. Levine and M. Tribus (eds.), *The Maximum Entropy Formalism*. MIT Press, Cambridge, Mass., pp. 15–118.

Jaynes, E. T. [2003], *Probability Theory: The Logic of Science*. Cambridge University Press, Cambridge, MA.

Jeffreys, H. [1939], *Theory of Probability*. Oxford University Press, Oxford.

Jelinek, F. [1968], *Probabilistic Information Theory: Discrete and Memoryless Models*. McGraw-Hill, New York.

Jiroušek, R.; Kleitner, G. D.; and Vejnarová, J. [2003], "Special Issue on Uncertainty Processing." Soft Computing, **7**(5), pp. 279–368.

Jiroušek, R., and Vejnarová, J. [2003], "General framework for multidimensional models." *International Journal of Intelligent Systems*, **18**(1), pp. 107–127.

John, R. [1998], "Type 2 fuzzy sets: an appraisal of theory and applications." *International Journal of Uncertainty, Fuzziness and Knowledge-Based Systems*, **6**(6), pp. 563–576.

Jones, B. [1982], "Determination of reconstruction families." *International Journal of General Systems*, **8**(4), pp. 225–228.

Jones, B. [1985], "Reconstructability considerations with arbitrary data." *International Journal of General Systems*, **11**(2), pp. 143–151.

Jones, B. [1986], "K-systems versus classical multivariate systems." *International Journal of General Systems*, **12**(1), pp. 1–6.

Jones, D. S. [1979], *Elementary Information Theory*. Clarendon Press, Oxford.

Josang, A., and Mc Anally, D. [2005], "Multiplication and comultiplication of beliefs." *International Journal of Approximate Reasoning*, **38**(1), pp. 19–51.

Joslyn, C. [1994], *Possibilistic Processes for Complex Systems Modeling*. Ph.D. Dissertation in Systems Science, T.J. Watson School, SUNY-Binghamton, Binghamton, NY.

Joslyn, C. [1997], "Measurement of possibilistic histograms from interval data." *International Journal of General Systems*, **26**(1–2), pp. 9–33.

Joslyn, C., and Klir, G. J. [1992], "Minimal information loss in possibilistic approximations of random sets." *Proceedings of the IEEE International Conference on Fuzzy Systems*, San Diego, pp. 1081–1088.

Jumarie, G. [1986], *Subjectivity, Information Systems: Introduction to a Theory of Relativistic Cybernetics.* Gordon and Breech, New York.

Jumarie, G. [1990], *Relative Information: Theories and Applications.* Springer-Verlag, New York.

Kåhre, J. [2002], *The Mathematical Theory of Information.* Kluwer, Boston.

Kaleva, O. [1987], "Fuzzy differential equations." *Fuzzy Sets and Systems*, **24**(3), pp. 301–317.

Kandel, A. [1986], *Fuzzy Mathematical Techniques with Applications.* Addison-Wesley, Reading, MA.

Kapur, J. N. [1983], "Twenty-five years of maximum entropy principle." *Journal of Mathematical Physical Sciences*, **17**, pp. 103–156.

Kapur, J. N. [1989], *Maximum Entropy Models in Science and Engineering.* John Wiley, New York.

Kapur, J. N. [1994/1996], "Insight into Entropy Optimization Principles." *Mathematical Science Trust Society*, New Delhi, India: Vol. I, 1994; Vol. II, 1996. (Some parts are also published in *Bulletin of the Mathematical Association of India*, **21**, 1989, pp. 1–38; **22**, 1990, pp. 1–42.)

Kapur, J. N. [1994], *Measures of Information and Their Applications.* John Wiley, New York.

Kapur, J. N. [1997], *Measures of Fuzzy Information.* Mathematical Sciences Trust Society, New Delhi.

Kapur, J. N.; Baciu, G.; and Kesavan, H. K. [1995], "The MinMax information measure." *International Journal of Systems Science*, **26**(1), pp. 1–12.

Kapur, J. N., and Kesavan, H. K. [1987], *The Generalized Maximum Entropy Principle.* Sadford Educational Press, Waterloo.

Kapur, J. N., and Kesavan, H. K. [1992], *Entropy Optimization Principles with Applications.* Academic Press, San Diego.

Karmeshu (ed.) [2003], *Entropy Measures, Maximum Entropy Principle and Emerging Applications.* Springer-Verlag, Heidelberg and New York.

Kaufmann, A. [1975], *Introduction to the Theory of Fuzzy Subsets.* Academic Press, New York.

Kaufmann, A., and Gupta, M. M. [1985], *Introduction to Fuzzy Arithmetic: Theory and Applications.* Van Nostrand Reinhold, New York.

Kendall, D. G. [1973] *Foundations of a Theory of Random Sets in Stochastic Geometry.* John Wiley, New York.

Kendall, D. G. [1974], "Foundations of a theory of random sets." In E. F. Harding and D. G. Kendall (eds.), *Stochastic Geometry.* John Wiley, New York, pp. 322–376.

Kern-Isberner, G. [1998], "Characterizing the principle of minimum cross-entropy within conditional-logical framework." *Artificial Intelligence*, **98**(1–2), pp. 169–208.

Kerre, E. E., and De Cock, M. [1999], "Linguistic modifiers: an overview." In G. Chen, M. Ying, and K. Y. Cai (eds.), *Fuzzy Logic and Soft Computing*. Kluwer, Boston.

Khinchin, A. I. [1957], *Mathematical Foundations of Information Theory*. Dover, New York.

Kingman, J. F. C., and Taylor, S. J. [1966], *Introduction to Measure and Probability*. Cambridge University Press, New York.

Kirschenmann, P. P. [1970], *Information and Reflections*. Humanities Press, New York.

Klawonn, F., and Schwecke, E. [1992], "On the axiomatic justification of Dempster's rule of combination." *International Journal of Intelligent Systems*, **7**(5), pp. 469–478.

Klement, E. P.; Mesiar, R.; and Pap, E. [2000], *Triangular Norms*. Kluwer, Dordrecht, Boston, and London.

Klement, E. P.; Mesiar, R.; and Pap, E. [2004], "Triangular norms. Position paper I. Basic analytical and algebraic properties." *Fuzzy Sets and Systems*, **143**(1), pp. 5–26.

Klir, G. J. [1976], "Identification of generative structures in empirical data." *International Journal of General Systems*, **3**(2), pp. 89–104.

Klir G. J. [1985], *Architecture of Systems Problem Solving*. Plenum Press, New York.

Klir, G. J. [1986a], "Reconstructability analysis: an offspring of Ashby's constraint theory." *Systems Research*, **3**(4), pp. 267–271.

Klir, G. J. [1986b], "The role of reconstructability analysis in social science research." *Mathematical Social Sciences*, **12**, pp. 205–225.

Klir, G. J. (ed.) [1987], "Special Issue on Measures of Uncertainty." *Fuzzy Sets and Systems*, **24**(2), pp. 139–254.

Klir, G. J. [1989a], "Is there more to uncertainty than some probability theorists might have us believe?" *International Journal of General Systems*, **15**(4), pp. 347–378.

Klir, G. J. [1989b], "Probability-possibility conversion." *Proceedings of the Third IFSA Congress*, Seattle, WA, pp. 408–411.

Klir, G. J. [1990a], "A principle of uncertainty and information invariance." *International Journal of General Systems*, **17**(2–3), pp. 249–275.

Klir, G. J. [1990b], "Dynamic aspects in reconstructability analysis: the role of minimum uncertainty principles." *Revue Internationale de Systemique*, **4**(1), pp. 33–43.

Klir, G. J. [1991], "Generalized information theory." *Fuzzy Sets and Systems*, **40**(1), pp. 127–142.

Klir, G. J. [1994], "Multivalued logics versus modal logics: alternative frameworks for uncertainty modeling." In P. P. Wang (ed.), *Advances in Fuzzy Theory and Technology*, Vol. II. *Duke Univ.*, Durham, NC, pp. 3–47.

Klir, G. J. [1995], "Principles of uncertainty: What are they? Why do we need them?" *Fuzzy Sets and Systems*, **74**(1), pp. 15–31.

Klir, G. J. [1997a], "Fuzzy arithmetic with requisite constraints." *Fuzzy Sets and Systems*, **91**(2), pp. 165–175.

Klir, G. J. [1997b], "The role of constrained fuzzy arithmetic in engineering." In B. M. Ayyub (ed.), *Uncertainty Analysis in Engineering and the Sciences*. Kluwer, Boston, pp. 1–19.

Klir, G. J. [1999], "On fuzzy-set interpretation of possibility theory." *Fuzzy Sets and Systems*, **108**(3), pp. 263–273.

Klir, G. J. [2000], *Fuzzy Sets: Fundamentals, Applications, and Personal Views*. Beijing University Press, Beijing.

Klir, G. J. [2001], "Foundations of fuzzy set theory and fuzzy logic: a historical overview." *International Journal of General Systems*, **30**(2), pp. 91–132.

Klir, G. J. [2002a], "Uncertainty in economics: the heritage of G.LS. Shackle." *Fuzzy Economic Review*, **VII**(2), pp. 3–21.

Klir, G. J. [2002b], "Basic issues of computing with granual probabilities." In T. Y. Lin, Y. Y. Yao, and L. A. Zadeh (eds.), *Data Mining, Rough Sets and Granular Computing*. Physica-Verlag/Springer-Verlag, Heidelberg and New York, pp. 339–349.

Klir, G. J. [2003], "An update on generalized information theory." *Proceedings of ISIPTA '03*, Carleton Scientific, Lugano, Switzerland, pp. 321–334.

Klir, G. J. [2005], "Measuring uncertainty associated with convex sets of probability distributions: a new approach." *Proceedings of NAFIPS '05*, Ann Arbor, MI (on CD).

Klir, G. J., and Folger, T. A. [1988], *Fuzzy Sets, Uncertainty and Information*. Prentice Hall, Englewood Cliffs, NJ.

Klir, G. J., and Harmanec, D. [1994], "On modal logic interpretation of possibility theory." *International Journal of Uncertainty, Fuzziness, and Knowledge-Based Systems*, **2**(2), pp. 237–245.

Klir, G. J., and Harmanec, D. [1995], "On some bridges to possibility theory." In G. De Cooman et al. (eds.) *Foundations and Applications of Possibility Theory*. World Scientific, Singapore, pp. 3–19.

Klir, G. J., and Mariano, M. [1987], "On the uniqueness of possibilistic measure of uncertainty and information." *Fuzzy Sets and Systems*, **24**(2), pp. 197–219.

Klir, G. J., and Pan, Y. [1998], "Constrained fuzzy arithmetic: basic questions and some answers." *Soft Computing*, **2**(2), pp. 100–108.

Klir, G. J., and Parviz, B. [1992], "Probability-possibility transformations: a comparison." *International Journal of General Systems*, **21**(3), pp. 291–310.

Klir, G. J.; Parviz, B.; and Higashi, M. [1986], "Relationship between true and estimated possibilistic systems and their reconstruction." *International Journal of General Systems*, **12**(4), pp. 319–331.

Klir, G. J., and Ramer, A. [1990], "Uncertainty in the Dempster-Shafer theory: a critical re-examination." *International Journal of General Systems*, **18**(2), pp. 155–166.

Klir, G. J., and Sentz, K. [2005], "On the issue of linguistic approximation." In A. Mehler and R. Kohler (eds.), *Aspects of Automatic Text Analysis*. Springer, Berlin and New York.

Klir, G. J., and Smith, R. M. [2001], "On measuring uncertainty and uncertainty-based information: recent developments." *Annals of Mathematics and Artificial Intelligence*, **32**(1–4), pp. 5–33.

Klir, G. J.; Wang Z.; and Harmanec, D. [1997], "Constructing fuzzy measures in expert systems." *Fuzzy Sets and Systems*, **92**(2), pp. 251–264.

Klir, G. J., and Way, E. C. [1985], "Reconstructability analysis: aims, results, open problems." *Systems Research*, **2**(2), pp. 141–163.

Klir, G. J., and Wierman, M. J. [1998], *Uncertainty-Based Information: Elements of Generalized Information Theory*. Physica-Verlag/Springer-Verlag, Heidelberg and New York (2nd ed., 1999).

Klir, G. J., and Yuan, B. [1993], "On measures of conflict among set-valued statements." *Proceedings of the 1993 World Congress on Neural Networks*, Vol. II, Portland, OR, pp. 627–630.

Klir, G. J., and Yuan, B. [1995a], *Fuzzy Sets and Fuzzy Logic: Theory and Applications.* Prentice Hall, Upper Saddle River, NJ.

Klir, G. J., and Yuan, B. [1995b], "On nonspecificity of fuzzy sets with continuous membership functions." *Proceedings 1995 International Conference on Systems, Man, and Cybernetics*, Vancouver, pp. 25–29.

Klir, G. J., and Yuan B. (eds.) [1996], *Fuzzy Sets, Fuzzy Logic, and Fuzzy Systems: Selected Papers by Lotfi A. Zadeh.* World Scientific, Singapore.

Knopfmacher, J. [1975], "On measures of fuzziness." *Journal of Mathematical Analysis and Applications*, **49**, pp. 529–534.

Kogan, I. M. [1988], *Applied Information Theory.* Gordon and Breach, New York and London.

Kohlas, J. [1991], "The reliability of reasoning with unreliable arguments." *Annals of Operations Reserach*, **32**, pp. 67–113.

Kohlas, J., and Monney, P. A. [1994], "Theory of evidence: a survey of its mathematical foundations, applications and computational aspects." *ZOR—Mathematical Methods of Operations Research*, **39**, pp. 35–68.

Kohlas, J., and Monney, P. A. [1995], *A Mathematical Theory of Hints: An Approach to the Dempster-Shafer Theory of Evidence.* Springer, Berlin.

Kohlas, J., and Moral, S. (eds.) [2000], *Algorithms for Uncertainty and Defeasible Reasoning.* Kluwer, Dordrecht and Boston.

Kolmogorov, A. N. [1950], *Foundations of the Theory of Probability.* Chelsea, New York. (First published in German in 1933.)

Kolmogorov, A. N. [1965], "Three approaches to the quantitative definition of information." *Problems of Information Transmission*, **1**, pp. 1–7.

Kong, A. [1986], *Multivariate Belief Functions and Graphical Models.* Research Report S-107, Harvard University, Cambridge, MA.

Kornwachs, K., and Jacoby, K. (eds.) [1996], *Information: New Questions to a Multidisciplinary Concept.* Academie Verlag, Berlin.

Kosko, B. [1987], "Fuzzy entropy and conditioning." *Information Science*, **40**(1), pp. 1–10.

Kosko, B. [1993a], *Fuzzy Thinking: The New Science of Fuzzy Logic.* Hyperion, New York.

Kosko, B. [1993b], "Addition as fuzzy mutual entropy." *Information Sciences*, **73**(3), pp. 273–284.

Kosko, B. [1999], *The Fuzzy Future.* Harmony Books, New York.

Kosko, B. [2004], "Probable equivalence, superpower sets, and superconditionals." *International Journal of Intelligent Systems*, **19**(12), pp. 1151–1171.

Kramosil, I. [2001], *Probabilistic Analysis of Belief Functions.* Kluwer Academic/ Plenum Publishers, New York.

Krätschmer, V. [2001], "A unified approach to fuzzy random variables." *Fuzzy Sets and Systems*, **123**(1), pp. 1–9.

Kreinovich, V.; Nguyen, H. T.; and Yam, Y. [2000], "Fuzzy systems are universal approximators for a smooth function and its derivatives." *International Journal of Intelligent Systems*, **15**(6), pp. 565–574.

Krippendorff, K. [1986], *Information Theory: Structural Models for Qualitative Data.* Sage, Beverly Hills, CA.

Kříž, O. [2003], "Envelops of a simplex of discrete probabilities." *Computing*, **7**(5), pp. 336–343.

Kruse, R. [1982a], "On the construction of fuzzy measures." *Fuzzy Sets and Systems*, **8**(3), pp. 323–327.

Kruse, R. [1982b], "A note on λ-additive fuzzy measures." *Fuzzy Sets and Systems*, **8**(2), pp. 219–222.

Kruse, R.; Gebhardt, J.; and Klawonn [1994], *Foundations of Fuzzy Systems.* John Wiley, Chichester, UK.

Kruse, R., and Meyer, K. D. [1987], *Statistics with Vague Data.* D. Reidel, Dordrecht and Boston.

Kruse, R.; Schwecke, E.; and Heinsohn, J. [1991], *Uncertainty and Vagueness in Knowledge Based Systems: Numerical Methods.* Springer-Verlag, New York.

Kruse, R., and Siegel, P. (eds.) [1991], *Symbolic and Quantitative Approaches to Uncertainty.* Springer-Verlag, New York.

Kuhn, T. S. [1962], *The Structure of Scientific Revolutions.* University of Chicago Press, Chicago.

Kullback, S. [1959], *Information Theory and Statistics.* John Wiley, New York.

Kyburg, H. E. [1961], *Probability and the Logic of Rational Belief.* Wesleyan University Press, Middletown, CT.

Kyburg, H. E. [1987], "Bayesian and non-Bayesian evidential updating." *Artificial Intelligence*, **31**, pp. 271–293.

Kyburg, H. E., and Teng, C. M. [2001], *Uncertain Inference.* Cambridge University Press, Cambridge and New York.

Kyburg, H. E., and Pittarelli, M. [1996], "Set-based Bayesianism." *IEEE Transactions on Systems, Man and Cybernetics,* Part A, **26**(3), pp. 324–339.

Lamata, M. T., and Moral, S. [1988], "Measures of entropy in the theory of evidence." *International Journal of General Systems*, **14**(4), pp. 297–305.

Lamata, M. T., and Moral, S. [1989], "Classification of fuzzy measures." *Fuzzy Sets and Systems*, **33**(2), pp. 243–253.

Lao Tsu [1972], *Tao Te Ching.* Vintage Books, New York.

Lees, E. S., and Shu, Q. [1995], *Fuzzy and Evidential Reasoning.* Physica-Verlag, Heidelberg.

Levine, R. D., and Tribus, M. (eds.) [1979], *The Maximum Entropy Formalism.* MIT Press, Cambridge, MA.

Lewis, H. W. [1997], *The Foundations of Fuzzy Control.* Plenum Press, New York.

Lewis, P. M. [1959], "Approximating probability distributions to reduce storage requirements." *Information and Control*, **2**(3), pp. 214–225.

Levi, I. [1967], *Gambling with Truth.* Knopf, New York.

Levi, I. [1980], *The Enterprise of Knowledge.* MIT Press, London.

Levi, I. [1984], *Decisions and Revisions*. Cambridge University Press, New York.

Levi, I. [1986], *Hard Choices*. Cambridge University Press, New York.

Levi, I. [1991], *The Fixation of Belief and Its Undoing*. Cambridge University Press, New York.

Levi, I. [1996], *For the Sake of Argument*. Cambridge University Press, New York.

Levi, I. [1997], *The Covenant of Reason*. Cambridge University Press, New York.

Li, M., and Vitányi, P. [1993], *An Introduction to Kolmogorov Complexity and Its Applications*. Springer-Verlag, New York.

Lin, C. T., and Lee, C. S. G. [1996], *Neural Fuzzy Systems: A Neuro Fuzzy Synergism to Intelligent Systems*. Prentice Hall, Upper Saddle River, NJ.

Lingras, P., and Wong, S. K. M. [1990], "Two perspectives of the Dempster-Shafer theory of belief functions." *International Journal of Man-Machine Studies*, **33**(4), pp. 467–488.

Lipschutz, S. [1964], *Set Theory and Related Topics*. Shawn, New York.

Loo, S. G. [1977], "Measures of fuzziness." *Cybernetica*, **20**(3), pp. 201–210.

Mackay, D. M. [1969], *Information, Mechanism and Meaning*. MIT Press, Cambridge, MA.

Madden, R. F., and Ashby, W. R. [1972], "On the identification of many-dimensional relations." *International Journal of Systems Science*, **3**, pp. 343–356.

Maeda, Y., and Ichihashi, H. [1993], "An uncertainty measure with monotonicity under the random set inclusion." *International Journal of General Systems*, **21**(4), pp. 379–392.

Maeda, Y.; Nguyen, H. T.; and Ichihashi, H. [1993], "Maximum entropy algorithms for uncertainty measures." *International Journal of Uncertainty, Fuzziness and Knowledge-Based Systems*, **1**(1), pp. 69–93.

Malinowski, G. [1993], *Many-Valued Logics*. Oxford University Press, Oxford.

Manton, K. G.; Woodbury, M. A.; and Tolley, H. D. [1994], *Statistical Applications Using Fuzzy Sets*. John Wiley, New York.

Mansuripur, M. [1987], *Introduction to Information Theory*. Prentice Hall, Englewood Cliffs, N.J.

Mareš, M. [1994], *Computation Over Fuzzy Quantities*. CRC Press, Boca Raton, FL.

Mariano, M. [1985], "The problem of resolving inconsistency in reconstructability analysis." *IEEE Workshop on Languages for Automation*, Palma de Mallorca, Spain.

Mariano, M. [1997], *Aspects of Inconsistency in Reconstructability Analysis*. Ph.D. Dissertation in System Science, Binghamton University—SUNY, Binghamton, NY.

Marichal, J., and Roubens, M. [2000], "Entropy of discrete fuzzy measures." *International Journal of Uncertainty, Fuzziness, and Knowledge-Based Systems*, **8**(6), pp. 625–640.

Martin, N. F. G., and England, J. W. [1981], *Mathematical Theory of Entropy*. Addison-Wesley, Reading, MA.

Mathai, A. M., and Rathie, P. N. [1975], *Basic Concepts of Information Theory and Statistics*. John Wiley, New York.

Matheron, G. [1975], *Random Sets and Integral Geometry*. John Wiley, New York.

Maung, I. [1995], "Two characterizations of a minimum-information principle for possibilistic reasoning." *International Journal of Approximate Reasoning*, **12**(2), pp. 133–156.

McLaughlin, D. W. [1984], *Inverse Problems*. American Mathematical Society, Providence, RI.

McNeill, D., and Freiberger, P. [1993], *Fuzzy Logic: The Discovery of a Revolutionary Computer Technology—and How It Is Changing Our World*. Simon & Schuster, New York.

Mendel, J. M. [2001], *Uncertain Rule-Based Fuzzy Logic Systems*. Prentice Hall PTR, Upper Saddle River, NJ.

Menger, K. [1942], "Statistical metrics." *Proceedings of the National Academy of Science*, **28**, pp. 535–537.

Meyerowitz, A.; Richman, F.; and Walker, E. A. [1994], "Calculating maximum-entropy probability densities for belief functions." *International Journal of Uncertainty, Fuzziness, and Knowledge-Based Systems*, **2**(4), pp. 377–389.

Miranda, E.; Cousco, I.; and Gil, P. [2003], "Extreme points of credal sets generated by 2 alternating capacities." *International Journal of Approximate Reasoning*, **33**(1), pp. 95–115.

Molchanov, I. [2005], *Theory of Random Sets*. Springer, New York.

Moles, A. [1966], *Information Theory and Esthetic Perception*. University of Illinois Press, Urbana, IL.

Möller, B., and Beer, M. [2004], *Fuzzy Randomness*. Springer, Berlin, Heidelberg, and New York.

Moore, R. E. [1966], *Interval Analysis*. Prentice Hall, Englewood Cliffs, NJ.

Moore, R. E. [1979], *Methods and Applications of Interval Analysis*. SIAM, Philadelphia.

Morderson, J. N., and Malik, D. S. [2002], *Fuzzy Automata and Languages: Theory and Applications*. Chapman & Hall/CRC, Boca Raton, Fl.

Mordeson, J. N., and Nair, P. S. [1998], *Fuzzy Mathematics: An Introduction for Engineers and Scientists*. Physica-Verlag/Springer-Verlag, Heidelberg and New York.

Murofushi, T., and Sugeno, M. [1989], "An interpretation of fuzzy measures and the Choquet integral as an integral with respect to a fuzzy measure." *Fuzzy Sets and Systems*, **29**(2), pp. 201–227.

Murofushi, T., and Sugeno, M. [1993], "Some quantities represented by the Choquet integral." *Fuzzy Sets and Systems*, **56**(2), pp. 229–235.

Murofushi, T., and Sugeno, M. [2000], "Fuzzy measures and fuzzy integrals." In M. Grabish et al. (eds.), *Fuzzy Measures and Integrals: Theory and Applications*. Springer-Verlag, New York, pp. 3–41.

Murofushi, T.; Sugeno, M.; and Machida, M. [1994], "Non-monotonic fuzzy measures and the Choquet integral." *Fuzzy Sets and Systems*, **64**(1), pp. 73–86.

Natke, H. G., and Ben-Haim, Y. (eds.) [1997], *Uncertainty: Models and Measures*. Academie Verlag, Berlin.

Nauck, D.; Klawonn, F.; and Kruse, R. [1997], *Foundations of Neuro-Fuzzy Systems*. John Wiley, New York.

Negoita, C. V., and Ralescu, D. A. [1975], *Applications of Fuzzy Sets to Systems Analysis*. Birkhäuser, Basel and Stuttgart.

Negoita, C. V., and Ralescu, D. A. [1987], *Simulation, Knowledge-Based Computing, and Fuzzy Statistics*. Van Nostrand Reinhold, New York.

Neumaier, A. [1990], *Interval Methods for Systems of Equations*. Cambridge University Press, Cambridge and New York.

Nguyen, H. T. [1978a], "On conditional possibility distributions." *Fuzzy Sets and Systems*, **1**(4), pp. 299–309.

Nguyen, H. T. [1978b], "On random sets and belief functions." *Journal of Mathematical Analysis and Applications*, **65**, pp. 531–542.

Nguyen, H. T.; Kreinovich, V.; and Shekhter, V. [1998], "On the possibility of using complex values in fuzzy logic for representing inconsistencies." *International Journal of Intelligent Systems*, **13**(8), pp. 683–714.

Nguyen, H. T., and Walker, E. A. [1997], *A First Course in Fuzzy Logic*. CRC Press, Boca Raton, FL.

Norton, J. [1988], "Limit theorems for Dempster's rule of combination." *Theory and Decision*, **25**, pp. 287–313.

Novák, V.; Perfilieva, I.; and Močkoř, J. [1999], *Mathematical Principles of Fuzzy Logic*. Kluwer, Boston.

Padet, C. [1996], "On applying information principles to fuzzy control." *Kybernetes*, **25**(1), pp. 61–64.

Pal, N. R., and Bezdek, J. C. [1994], "Measuring fuzzy uncertainty." *IEEE Transactions on Fuzzy Systems*, **2**(2), pp. 107–118.

Pan, Y. [1997a], *Calculus of Fuzzy Probabilities and Its Applications*. Ph.D. Dissertation in Systems Science, T.J. Watson School, Binghamton University—SUNY, Binghamton, NY.

Pan, Y. [1997b], "Revised hierarchical analysis method based on crisp and fuzzy entries." *International Journal of General Systems*, **26**(1–2), pp. 115–131.

Pan, Y., and Klir, G. J. [1997], "Bayesian inference based on interval probabilities." *Journal of Intelligent and Fuzzy Systems*, **5**(3), pp. 193–203.

Pan, Y., and Yuan, B. [1997], "Bayesian inference of fuzzy probabilities." *International Journal of General Systems*, **26**(1–2), pp. 73–90.

Pap, E. [1995], *Null-Additive Set Functions*. Kluwer, Boston.

Pap, E. [1997], "Decomposable measures and nonlinear equations." *Fuzzy Sets and Systems*, **92**(2), pp. 205–221.

Pap, E. (ed.) [2002], *Handbook of Measure Theory*. Elsevier, Amsterdam.

Paris, J. B. [1994], *The Uncertain Reasoner's Companion: A Mathematical Perspective*. Cambridge University Press, Cambridge, UK.

Paris, J. B., and Vencovská, A. [1989], "On the applicability of maximum entropy to inexact reasoning." *International Journal of Approximate Reasoning*, **3**(1), pp. 1–34.

Paris, J. B., and Vencovská, A. [1990], "A note on the inevitability of maximum entropy." *International Journal of Approximate Reasoning*, **4**(3), pp. 183–223.

Pavelka, J. [1979], "On fuzzy logic I, II, III." *Zeitschrift für Math. Logik und Grundlagen der Mathematik*, **25**(1), pp. 45–52; **25**(2), pp. 119–134; **25**(5), pp. 447–464.

Pawlak, Z. [1982], "Rough sets." *International Journal of Computer and Information Sciences*, **11**, pp. 341-356.

Pawlak, Z. [1991], *Rough Sets*. Kluwer, Boston.

Pedrycz, W., and Gomide, F. [1998], *An Introduction to Fuzzy Sets: Analysis and Design*. MIT Press, Cambridge, MA.

Peeva, K., and Kyosev, Y. [2004], *Fuzzy Relational Calculus: Theory, Applications and Software*. World Scientific, Singapore.

Piegat, A. [2001], *Fuzzy Modeling and Control*. Physica-Verlag/Springer-Verlag, Heidelberg and New York.

Pinsker, M. S. [1964], *Information and Information Stability of Random Variables and Processes*. Holden-Day, San Francisco.

Pittarelli, M. [1990], "Reconstructability analysis: an overview." *Revue Internationale de Systemique*, **4**(1), pp. 5–32.

Pollack, H. N. [2003], *Uncertain Science ... Uncertain World*. Cambridge University Press, Cambridge and New York.

Prade, H., and Yager, R. R. [1994], "Estimations of expectedness and potential surprize in possibility theory." *International Journal of Uncertainty, Fuzziness and Knowledge-Based Systems*, **2**(4), pp. 417–428.

Press, S. J. [2003], *Subjective and Objective Bayesian Statistics*. Wiley-Interscience, Hoboken, NJ.

Qiao, Z. [1990], "On fuzzy measure and fuzzy integral on fuzzy sets." *Fuzzy Sets and Systems*, **37**(1), pp. 77–92.

Quastler, H. [1955], *Information Theory in Psychology*. The Free Press, Glencoe, Ill.

Ragin, C. C. [2000], *Fuzzy-Set Social Science*. University of Chicago Press, Chicago.

Ramer, A. [1986], *Informational and Combinatorial Aspects of Reconstructability Analysis: A Mathematical Inquiry*. Ph.D. Dissertation in Systems Science, T.J. Watson School, Binghamton University—SUNY, Binghamton, NY.

Ramer, A. [1987], "Uniqueness of information measure in the theory of evidence." *Fuzzy Sets and Systems*, **24**(2), pp. 183–196.

Ramer, A. [1989], "Conditional possibility measures." *Cybernetics and Systems*, **20**(3), pp. 233–247.

Ramer, A. [1990a], "Axioms of uncertainty measures: dependence and independence." *Fuzzy Sets and Systems*, **35**(2), pp. 185–196.

Ramer, A. [1990b], "Information measures for continuous possibility distributions." *International Journal of General Systems*, **17**(2–3), pp. 241–248.

Ramer, A., and Klir, G. J. [1993], "Measures of discord in the Dempster-Shafer theory." *Information Sciences*, **67**(1–2), pp. 35–50.

Ramer, A., and Lander, L. [1987], "Classification of possibilistic uncertainty and information functions." *Fuzzy Sets and Systems*, **24**(2), pp. 221–230.

Ramer, A., and Padet, C. [2001], "Nonpecificity in $\mathcal{R}^n$." *International Journal of General Systems*, **30**(6), pp. 661–680.

Reche, F., and Salmerón, A. [2000], "Operational approach to general fuzzy measures." *International Journal of Uncertainty, Fuzziness, and Knowledge-Based Systems*, **8**(3), pp. 369–382.

Regan, H. M.; Ferson, S.; and Berleant, D. [2004], "Equivalence of methods for uncertainty propagation of real-valued random variables." *International Journal of Approximate Reasoning*, **36**(1), pp. 1–30.

Reichenbach, H. [1949], *The Theory of Probability*. University of California Press, Berkeley and Los Angeles.

Rényi, A. [1970a], *Foundations of Probability*. Holden-Day, San Francisco.

Rényi, A. [1970b], *Probability Theory*. North-Holland, Amsterdam (Chapter IX, "Introduction to information theory," pp. 540–616).

Rényi, A. [1987], *A Diary on Information Theory*. John Wiley, New York.

Rescher, N. [1969], *Many-Valued Logic*. McGraw-Hill, New York.

Rescher, N. [1976], *Plausible Reasoning*. Van Gorcum, Amsterdam.

Resconi, G.; Klir, G. J.; and St. Clair, U. [1992], "Hierarchical uncertainty metatheory based upon modal logic." *International Journal of General Systems*, **21**(1), pp. 23–50.

Resconi, G.; Klir, G. J.; St. Clair, U.; and Harmanec, D. [1993], "On the integration of uncertainty theories." *International Journal of Uncertainty, Fuzziness, and Knowledge-Based Systems*, **1**(1), pp. 1–18.

Resnikoff, H. L. [1989], *The Illusion of Reality*. Springer-Verlag, New York.

Reza, F. M. [1961], *Introduction to Information Theory*. McGraw-Hill, New York. (Reprinted by Dover, New York, 1994.).

Rissanen, J. [1989], *Stochastic Complexity in Statistical Inquiry*. World Scientific, Teaneck, NJ.

Rodabaugh, S. E., and Klement, E. P. (eds.) [2003], *Topological and Algebraic Structures in Fuzzy Sets*. Kluwer, Boston.

Rodabaugh, S. E.; Klement, E. P.; and Höhle, U. (eds.) [1992], *Applications of Category Theory to Fuzzy Subsets*. Kluwer, Boston.

Rosenkrantz, R. D. (ed.) [1983], *Jaynes, E. T.: Papers on Probability, Statistics and Statistical Physics*. Reidel, Boston.

Ross, T. J.; Booker, J. M.; and Parkinson, W. J. (eds.) [2002], *Fuzzy Logic and Probability Applications: Bridging the Gap*. ASA-SIAM, Philadelphia.

Rouvray, D. H. (ed.) [1997], *Fuzzy Logic in Chemistry*. Academic Press, San Diego.

Rubinstein, R. Y., and Kroese, D. P. [2004], *The Cross-Entropy Method: A Unified Approach to Combinatorial Optimization, Monte-Carlo Simulation, and Machine Learning*. Springer, New York.

Ruspini, E. H.; Bonissone, P. P.; and Pedrycz, W. (eds.) [1998], *Handbook of Fuzzy Computation*. Institute of Physics Publication, Bristol, UK, and Philadelphia.

Russell, B. [1950], *Unpopular Essays*. Simon and Schuster, New York.

Rutkowska, D. [2002], *Neuro-Fuzzy Architectures and Hybrid Learning*. Physica-Verlag/Springer-Verlag, Heidelberg and New York.

Rutkowski, L. [2004], *Flexible Neuro-Fuzzy Systems*. Kluwer, Boston.

Sanchez, E. [1976], "Resolution in composite fuzzy relation equations." *Information and Control*, **30**(1), pp. 38–48.

Sancho-Royo, A., and Verdegay, J. L. [1999], "Methods for the construction of membership functions." *International Journal of Intelligent Systems*, **14**(12), pp. 1213–1230.

Savage, L. J. [1972], *The Foundations of Statistics*. Dover, New York.

Schubert, J. [1994], *Cluster-Based Specification Techniques in Dempster-Shafer Theory for an Evidential Intelligence Analysis of Multiple Target Tracks*. Royal Institute of Technology, Stockholm.

Schweizer, B., and Sklar, A. [1983], *Probabilistic Metric Spaces*. North-Holland, New York.

Sgarro, A. [1997], "Bodies of evidence versus simple interval probabilities." *International Journal of Uncertainty, Fuzziness and Knowledge-Based Systems*, **5**(2), pp. 199–209.

Shackle, G. L. S. [1949], *Expectation in Economics*. Cambridge University Press, Cambridge.

Shackle, G. L. S. [1955], *Uncertainty in Economics and Other Reflections*. Cambridge University Press, Cambridge.

Shackle, G. L. S. [1961], *Decision, Order and Time in Human Affairs*. Cambridge University Press, New York and Cambridge.

Shackle, G. L. S. [1979], *Imagination and the Nature of Choice*. Edinburgh University Press, Edinburgh.

Shafer, G. [1976a], *A Mathematical Theory of Evidence*. Princeton University Press, Princeton, NJ.

Shafer, G. [1976b], "A theory of statistical evidence." In W. L. Harper and C. A. Hooker (eds.), *Foundations of Probability Theory, Statistical Inference, and Statistical Theories of Science*. D. Reidel, Dordrecht, pp. 365–436.

Shafer, G. [1978], "Non-additive probabilities in the work of Bernoulli and Lambert." *Archive for History of Exact Sciences*, **19**, pp. 309–370.

Shafer, G. [1979], "The allocation of probability." *Annals of Probability*, **7**(5), pp. 827–839.

Shafer, G. [1981], "Constructive probability." *Synthese*, **48**, pp. 1–60.

Shafer, G. [1982], "Belief functions and parametric models." *Journal of the Royal Statistical Society*, **B-44**, pp. 322–352.

Shafer, G. [1985], "Belief functions and possibility measures." In J. C. Bezdek (ed.), *Analysis of Fuzzy Information*. CRC Press, Boca Raton, FL.

Shafer, G. [1986], "The Combination of Evidence." *International Journal of Intelligent Systems*, **1**(3), pp. 155–179.

Shafer, G. [1990], "Perspectives on the theory and practice of belief functions." *International Journal of Approximate Reasoning*, **4**(5–6), pp. 323–362.

Shannon, C. E. [1948], "The mathematical theory of communication." *The Bell System Technical Journal*, **27**(3&4), pp. 379–423, 623–656.

Shannon, C. E., and Weaver, W. [1949], *The Mathematical Theory of Communication*. University of Illinois Press, Urbana, IL.

Shapley, L. S. [1971], "Core of convex games." *International Journal of Game Theory*, **1**(1), pp. 11–26.

Shore, J. E., and Johnson, R. W. [1980], "Axiomatic derivation of the principle of maximum entropy and the principle of minimum cross-entropy." *IEEE Transactions on Information Theory*, **26**(1), pp. 26–37.

Shore, J. E., and Johnson, R. W. [1981], "Properties of cross-entropy minimization." *IEEE Transactions on Information Theory*, **27**(4), pp. 472–482.

Sims, J. R., and Wang, Z. [1990], "Fuzzy measures and fuzzy integrals: an overview." *International Journal of General Systems*, **17**(2–3), pp. 157–189.

Slepian, D. (ed.) [1974], *Key Papers in the Development of Information Theory*. IEEE Press, New York.

Sloane, N. J., and Wyner, A. D. (eds.) [1993], *Claude Elwood Shannon: Collected Papers*. IEEE Press, Piscataway, NJ.

Smets, P. [1981], "The degree of belief in a fuzzy event." *Information Sciences*, **25**(1), pp. 1–19.

Smets, P. [1983], "Information content of an evidence." *International Journal of Man-Machine Studies*, **19**(1), pp. 33–43.

Smets, P. [1988], "Belief functions." In P. Smets et al. (eds.), *Non-standard Logics for Automated Reasoning*. Academic Press, San Diego, pp. 253–286.

Smets, P. [1990], "The combination of evidence in the transferable belief model." *IEEE Transactions on Pattern Analysis and Machine Intelligence*, **12**(5), pp. 447–458.

Smets, P. [1992a], "The transferable belief model and random sets." *International Journal of Intelligent Systems*, **7**(1), pp. 37–46.

Smets, P. [1992b], "Resolving misunderstandings about belief functions." *International Journal of Approximate Reasoning*, **6**(3), pp. 321–344.

Smets, P. [1998], "The transferable belief model for quantified belief representations." In D. M. Gabbay and P. Smets (eds.), *Handbook of Defeasible Reasoning and Uncertainty Management Systems*, Vol. 1, Kluwer, Boston, pp. 267–301.

Smets, P., and Kennes, R. [1994], "The transferable belief model." *Artificial Intelligence*, **66**, pp. 191–234.

Smith, R. M. [2000], *Generalized Information Theory: Resolving Some Old Questions and Opening Some New Ones*. Ph.D. Dissertation in Systems Science, T.J. Watson School, Binghamton University—SUNY, Binghamton, NY.

Smith, S. A. [1974], "A derivation of entropy and the maximum entropy criterion in the context of decision problems." *IEEE Transactions on Systems, Man, and Cybernetics*, **4**(2), pp. 157–163.

Smithson, M. [1989], *Ignorance and Uncertainty: Emerging Paradigms*. Springer-Verlag, New York.

Smuts, J. C. [1926], *Holism and Evolution*. Macmillan, London. (Reprinted by Greenwood Press, Westport, CT, 1973.)

Stonier, T. [1990], *Information and the Internal Structure of Universe*. Springer-Verlag, New York.

Sugeno, M. [1974], *Theory of Fuzzy Integrals and its Applications*. Ph. D. Thesis, Tokyo Institute of Technology.

Sugeno, M. [1977], "Fuzzy measures and fuzzy integrals: a survey." In M. M. Gupta, G. N. Saridis, and B. R. Gaines (eds.), *Fuzzy Automata and Decision Processes*. North-Holland, Amsterdam and New York, pp. 89–102.

Tanaka, H.; Sugihara, K.; and Maeda, Y. [2004], "Non-additive measures by interval probability functions." *Information Sciences*, **164**, pp. 209–227.

Tarantola, A. [1987], *Inverse Problem Theory*. Elsevier, New York.

Temple, G. [1971], *The Structure of Lebesgue Integration Theory*. Oxford University Press, London.

Theil, H. [1967], *Economics and Information Theory.* North-Holland, Amsterdam, and Rand McNally & Co., Chicago.

Theil, H., and Fiebig, D. G. [1984], *Exploiting Continuity: Maximum Entropy Estimation of Continuous Distributions.* Ballinger, Cambridge, MA.

Thomson, W. [1891], *Popular Lectures and Addresses.* MacMillan, London.

Tribus, M. [1969], *Rational Descriptions, Decisions and Designs.* Pergamon Press, Oxford.

Tsiporkova, E.; Boeva, V.; and De Baets, B. [1999], "Evidence measures induced by Kripke's accessibility relations." *International Journal of Uncertainty, Fuzziness and Knowledge-Based Systems,* **7**(6), pp. 589–613.

Tyler, S. [1978], *The Said and the Unsaid.* Academic Press, New York.

Vajda, I. [1989], *Theory of Statistical Inference and Information.* Kluwer, Boston.

Van der Lubbe, J. C. A. [1984], "A generalized class of certainty and information." *Information Sciences,* **32**(3), pp. 187–215.

Van Leekwijck, W., and Kerre, E. E. [1999], "Defuzzification: criteria and classification." *Fuzzy Sets and Systems,* **108**(2), pp. 159–178.

Vejnarová, J. [1991], "A few remarks on measures of uncertainty in Dempster-Shafer theory." In *Proceedings of the Workshop on Uncertainty in Expert Systems,* Alšovice, Czech Republic.

Vejnarová, J. [1998], "A note on the interval-valued marginal problem and its maximum entropy solution." *Kybernetika,* **34**(1), pp. 17–26.

Vejnarová, J. [2000], "Conditional independence relations in possibility theory." *International Journal of Uncertainty, Fuzziness, and Knowledge-Based Systems,* **8**(3), pp. 253–269.

Vejnarová, J., and Klir, G. J. [1993], "Measure of strife in Dempster-Shafer theory." *International Journal of General Systems,* **22**(1), pp. 25–42.

Verdú, S., and McLaughlin, S. W. (eds.) [2000], *Information Theory: 50 Years of Discovery.* IEEE Press, Piscataway, NJ.

Vicig, P. [2000], "Epistemic independence for imprecise probabilities." *International Journal of Approximate Reasoning,* **24**(2–3), pp. 235–250.

Viertl, R. [1996], *Statistical Methods for Non-Precise Data.* CRC Press, Boca Raton, FL.

Walker, C. L. [2003], "Categories of fuzzy sets." *Soft Computing,* **8**(4), pp. 299–304.

Walley, P. [1991], *Statistical Reasoning With Imprecise Probabilities.* Chapman & Hall, London.

Walley, P. [1996], "Measures of uncertainty in expert systems." *Artificial Intelligence,* **83**, pp. 1–58.

Walley, P. [1997], "Statistical inferences based on a second-order possibility distribution." *International Journal of General Systems,* **26**(4), pp. 337–383.

Walley, P. [2000], "Towards a unified theory of imprecise probability." *International Journal of Approximate Reasoning,* **24**(2–3), pp. 125–148.

Walley, P., and De Cooman, G. [1999], "Coherence of rules for defining conditional possibility." *International Journal of Approximate Reasoning,* **21**(1), pp. 63–107.

Walley, P., and De Cooman, G. [2001], "A behavioral model for linguistic uncertainty." *Information Sciences,* **134**(1), pp. 1–37.

Walley, P., and Fine, T. L. [1979], "Varieties of model (classificatory) and comparative probability." *Synthese*, **41**(3), pp. 321–374.

Walley, P., and Fine, T. L. [1982], "Toward a frequentist theory of upper and lower probability." *The Annals of Statistics*, **10**(3), pp. 741–761.

Wang, J., and Wang, Z. [1997], "Using neural networks to determine Sugeno measures by statistics." *Neural Networks*, **10**(1), pp. 183–195.

Wang, W.; Wang, Z.; and Klir, G. J. [1998], "Genetic algorithms for determining fuzzy measures from data." *Journal of Intelligent and Fuzzy Systems*, **6**(2), pp. 171–183.

Wang, Z., and Klir, G. J. [1992], *Fuzzy Measure Theory*. Plenum Press, New York.

Wang, Z.; Klir, G. J.; and Wang, W. [1996], "Monotone set functions defined by Choquet integral." *Fuzzy Sets and Systems*, **81**(2), pp. 241–250.

Watanabe, S. [1969], *Knowing and Guessing*. John Wiley, New York.

Watanabe, S. [1981], "Pattern recognition as a quest for minimum entropy." *Pattern Recognition*, **13**(5), pp. 381–387.

Watanabe, S. [1985], *Pattern Recognition: Human and Mechanical*. John Wiley, New York.

Weaver, W. [1948], "Science and complexity." *American Scientist*, **36**, pp. 536–544.

Webber, M. J. [1979], *Information Theory and Urban Spatial Structure*. Croom Helm, London.

Weber, S. [1984], "Decomposable measures and integrals for Archimedean *t*-conorms." *Journal of Mathematical Analysis and Applications*, **101**(1), pp. 114–138.

Weichselberger, K. [2000], "The theory of interval-probability as a unifying concept for uncertainty." *International Journal of Approximate Reasoning*, **24**(2–3), pp. 149–170.

Weichselberger, K., and Pöhlmann, S. [1990], *A Methodology for Uncertainty in Knowledge-Based Systems*. Springer-Verlag, New York.

Weir, A. J. [1973], *Lebesgue Integration and Measure*. Cambridge University Press, New York.

Weltner, K. [1973], *The Measurement of Verbal Information in Psychology and Education*. Springer-Verlag, New York.

Whittemore, B. J., and Yovits, M. C. [1973], "A generalized conceptual development for the analysis and flow of information." *Journal of the American Society for Information Sciences*, **24**(3), pp. 221–231.

Whittemore, B. J., and Yovits, M. C. [1974], "The quantification and analysis of information used in decision processes." *Information Sciences*, **7**(2), pp. 171–184.

Wierman, M. [1994], *Possibilistic Image Processing*. Ph.D. Dissertation in Systems Science, T.J.Watson School, SUNY-Binghamton, Binghamton, NY.

Wierzchoń, S. T. [1982], "On fuzzy measure and fuzzy integral." In M. M. Gupta and E. Sanchez (eds.), *Fuzzy Information and Decision Processes*. North-Holland, New York, pp. 79–86.

Wierzchoń, S. T. [1983], "An algorithm for identification of fuzzy measure." *Fuzzy Sets and Systems*, **9**(1), pp. 69–78.

Williams, P. M. [1980], "Bayesian conditionalisation and the principle of minimum information." *British Journal for the Philosophy of Science*, **31**, pp. 131–144.

Williamson, R. C., and Downs, T. [1990], "Probabilistic arithmetic. I. Numerical methods for calculation convolutions and dependency bounds." *International Journal of Approximate Reasoning*, **4**(2), pp. 89–158.

Wilson, A. G. [1970], *Entropy in Urban and Regional Modelling*. Pion, London.

Wilson, N. [2000], "Algorithms for Dempster-Shafer theory." In J. Kohlas and S. Moral (eds.), *Algorithms for Uncertainty and Defeasible Reasoning*. Kluwer, Dordrecht and Boston, pp. 421–475.

Wolf, R. G. [1977], "A survey of many-valued logic (1966–1974)." In J. M. Dunn and G. Epstein (eds.), *Modern Uses of Multiple-Valued Logic*. D. Reidel, Boston, pp. 167–323.

Wolkenhauer, O. [1998], *Possibility Theory with Applications to Data Analysis*. Research Studies Press, Tauton, UK.

Wong, S. K. M.; Wang, L. S.; and Yao, Y. Y. [1995], "On modeling uncertainty with interval structures." *Computational Intelligence*, **11**(2), pp. 406–426.

Wonneberger, S. [1994], "Generalization of an invertible mapping between probability and possibility." *Fuzzy Sets and Systems*, **64**(2), pp. 229–240.

Wyner, A. D. [1981], "Fundamental limits in information theory." *Proceedings of the IEEE*, **69**, pp. 239–251.

Yager, R. R. [1979], "On the measure of fuzziness and negation. Part I: membership in the unit interval." *International Journal of General Systems*, **5**(4), pp. 221–229.

Yager, R. R. [1980a], "On a general class of fuzzy connectives." *Fuzzy Sets and Systems*, **4**(3), pp. 235–242.

Yager, R. R. [1980b], "On the measure of fuzziness and negation. Part II: lattices." *Information and Control*, **44**(3), pp. 236–260.

Yager, R. R. (ed.) [1982a], *Fuzzy Set and Possibility Theory*. Pergamon Press, Oxford.

Yager, R. R. [1982b], "Generalized probabilities of fuzzy events from fuzzy belief structures." *Information Sciences*, **28**(1), pp. 45–62.

Yager, R. R. [1982c], "Measuring tranquility and anxiety in decision making: an application of fuzzy sets." *International Journal of General Systems*, **8**(3), pp. 139–146.

Yager, R. R. [1983], "Entropy and specificity in a mathematical theory of evidence." *International Journal of General Systems*, **9**(4), pp. 249–260.

Yager, R. R. [1984], "Probabilities from fuzzy observations." *Information Sciences*, **32**(1), pp. 1–31.

Yager, R. R. [1986], "Toward general theory of reasoning with uncertainty: nonspecificity and fuzziness." *International Journal of Intelligent Systems*, **1**(1), pp. 45–67.

Yager, R. R. [1987a], "Set based representations of conjunctive and disjunctive knowledge." *Information Sciences*, **41** (1), pp. 1–22.

Yager, R. R. [1987b], "On the Dempster-Shafer framework and new combination rules." *Information Sciences*, **41**, pp. 93–137.

Yager, R. R. [1990], "Ordinal measures of specificity." *International Journal of General Systems*, **17**(1), pp. 57–72.

Yager, R. R. [1991], "Similarity based specificity measures." *International Journal of General Systems*, **19**(2), pp. 91–105.

Yager, R. R. [2000], "On the entropy of fuzzy measures." *IEEE Transactions on Fuzzy Systems*, **8**(4), pp. 453–461.

Yager, R. R. [2004], "On the retranslation process in Zadeh's paradigm of computing with words." *IEEE Transactions on Systems, Man and Cybernetics (Part B)*, **34**(2), pp. 1184–1195.

Yager, R. R.; Fedrizzi, M.; and Kacprzyk, J. (eds.) [1994], *Advances in the Dempster-Shafer Theory of Evidence*. John Wiley, New York.

Yager, R. R., and Filev, D. P. [1994], *Essentials of Fuzzy Modeling and Control*. John Wiley, New York.

Yager, R. R.; Ovchinnikov, S.; Tong, R. M.; and Nguyen, H. T. (eds.) [1987], *Fuzzy Sets and Applications: Selected Papers by L. A. Zadeh*. John Wiley, New York.

Yaglom, A. M., and Yaglom, I. M. [1983], *Probability and Information*. Reidel, Boston.

Yang, M.; Chen, T.; and Wu, K. [2003], "Generalized belief function, plausibility function, and Dempster's combination rule to fuzzy sets." *International Journal of Intelligent Systems*, **18**(8), pp. 925–937.

Yen, J. [1990], "Generalizing the Dempster-Shafer theory to fuzzy sets." *IEEE Transactions on Systems, Man, and Cybernetics*, **20**(3), pp. 559–570.

Yeung, R. W. [2002], *A First Course in Information Theory*. Kluwer, Boston.

Yovits, M. C.; Foulk, C. R.; and Rose, L. L. [1981], "Information flow and analysis: theory, simulation, and experiments." *Journal of the American Society for Information Science.*, **32**, pp. 187–210, 243–248.

Yu, F. T. S. [1976], *Optics and Information Theory*. John Wiley, New York.

Zadeh, L. A. [1965], "Fuzzy Sets." *Information and Control*, **8**(3), pp. 338–353.

Zadeh, L. A. [1968], "Probability measures of fuzzy events." *Journal of Mathematical Analysis and Applications*, **23**, pp. 421–427.

Zadeh, L. A. [1971], "Similarity relations and fuzzy orderings." *Information Science*, **3**(2), pp. 177–200.

Zadeh, L. A. [1975–76], "The concept of a linguistic variable and its application to approximate reasoning." *Information Sciences*, **8**, pp. 199–249, 301–357; **9**, pp. 43–80.

Zadeh, L. A. [1978a], "Fuzzy sets as a basis for a theory of possibility." *Fuzzy Sets and Systems*, **1**(1), pp. 3–28.

Zadeh, L. A. [1978b], "PRUF—a meaning representation language for natural languages." *International Journal of Man-Machine Studies*, **10**(4), pp. 395–460.

Zadeh, L. A. [1981], "Possibility theory and soft data analysis." In L. Cobb and R. M. Thrall (eds.), *Mathematical Frontiers of the Social and Policy Sciences.* Westview Press, Boulder, CO, pp. 69–129.

Zadeh, L. A. [1986], "A simple view of the Dempster-Shafer theory of evidence and its implication for the rule of combination." *AI Magazine*, **7**(2), pp. 85–90.

Zadeh, L. A. [1996], "Fuzzy logic = computing with words." *IEEE Transactions on Fuzzy Systems*, **4**(2), pp. 103–111.

Zadeh, L. A. [1997], "Toward a theory of fuzzy information granulation and its centrality in human reasoning and fuzzy logic." *Fuzzy Sets and Systems*, **90**(2), pp. 111–127.

Zadeh, L. A. [1999], "From computing with numbers to computing with words—from manipulation of measurements to manipulation of perceptions." *IEEE Transactions on Circuits and Systems (I. Fundamental Theory and Applications)*, **45**(1), pp. 105–119.

Zadeh, L. A. [2002], "Toward a perception-based theory of probabilistic reasoning with imprecise probabilities." *Journal of Statistical Planning and Inference*, **105**, pp. 233–264.

Zadeh, L. A. [2005], "Toward a generalized theory of uncertainty (GTU)—An outline." *Information Sciences*, **172**(1–2), pp. 1–40.

Zimmermann, H. J. [1996], *Fuzzy Set Theory—And Its Applications*. Kluwer, Boston.

Zimmermann, H. J. [2000], "An application-oriented view of modeling uncertainty." *European Journal of Operations Research*, **122**, pp. 190–198.

Zwick, M. [2004], "An overview of reconstructability analysis." *Kybernetes*, **33**(5–6), pp. 877–905.

Zwick, R., and Wallsten, T. S. [1989], "Combining stochastic uncertainty and linguistic inexactness." *International Journal of Man-Machine Studies*, **30**(1), pp. 69–111.

SUBJECT INDEX

Uncertainty and Information: Foundations of Generalized Information Theory, by George J. Klir
© 2006 by John Wiley & Sons, Inc.

NAME INDEX

Abellán, J., 254, 255, 458
Aczél, J., 95, 96, 458
Alefeld, G., 306, 458
Apostol, T.M., 23
Applebaum, D. 458
Arbib, M. A., 350, 458
Aristotle, 101
Ash, R. B., 95, 97, 459
Ashby, W. R., 97, 411, 459, 475
Atanassov, K. T., 308, 459
Attneave, F., 95, 459
Aubin, J.P., 139, 459
Auman, R.J., 139, 459
Avgers, T.G., 409, 459

Babuška, R., 307, 459
Baciu, G., 470
Ban, A.I., 139, 459
Bandler, W., 307, 350, 459
Banon, G., 187, 459
Bárdossy, G., 308, 459
Barrett, J.D., 422
Batten, D.F., 95, 408, 459
Beer, M., 351, 476
Bell, D. A., 95, 187, 459, 467
Bellman, R., 5, 459
Bělohlávek, R., 307, 350, 460
Ben-Haim, Y., 418, 421, 460, 476
Benvenuti, P., 139, 460
Berleant, D., 479
Bernays, P., 463
Bernoulli, J., 61, 138
Bezdek, J. C., 307, 351, 460, 477

Bharathi-Devi, B., 308, 460
Bhattacharya, P., 460
Billingsley, P., 95, 137, 460
Billot, A., 308, 460
Black, M., 305, 460
Black, P.K., 460
Blahut, R. E., 95, 460
Boekee, D. E., 96, 460
Boeva, V., 482
Bolaños, M.J., 138, 139, 462
Bolc, L., 460
Bonissone, P.P., 479
Booker, J.M., 473, 479
Bordley, R.F., 410, 460
Borel, É., 137
Borgelt, C., 186, 460
Borowic, P., 307, 460
Bouchon-Meunier, B., 462
Brillouin, L., 95, 460
Broekstra, G., 411, 460
Buck, B., 408, 460
Buckley, J.J., 351, 460

Cai, K.Y., 351, 460
Cano, A., 138, 461
Cantor, G., 137
Caratheodory, C., 137, 461
Carlsson, C., 461
Cauchy, A., 137
Cavallo, R.E., 408, 411, 461
Chaitin, G.J., 23, 461
Chameau, J.- L., 308, 461
Chateauneuf, A., 138, 461

Uncertainty and Information: Foundations of Generalized Information Theory, by George J. Klir
© 2006 by John Wiley & Sons, Inc.

||||| ||||| ||||| ||||| ||||| |||||

Printed and bound by CPI Group (UK) Ltd, Croydon, CR0 4YY

27/10/2024

14580330-0004